Aquatic Chemistry

An Introduction Emphasizing
Chemical Equilibria in Natural Waters

Aquatic Chemistry
An Introduction Emphasizing Chemical Equilibria in Natural Waters

WERNER STUMM

Gordon McKay Professor of Applied Chemistry
Harvard University

JAMES J. MORGAN

Professor of Environmental Engineering Science
California Institute of Technology

WILEY-INTERSCIENCE
A Division of John Wiley & Sons, Inc.

NEW YORK · LONDON · SYDNEY · TORONTO

Copyright © 1970 by John Wiley & Sons, Inc.

Library of Congress Catalog Card Number: 79-109392

SBN 471-83495-5

Printed in the United States of America

10 9 8 7 6 5 4 3 2

Preface

This book concerns itself with chemical processes in the natural water environment. Its aim is to present a quantitative treatment of the variables which determine the composition of natural waters. We have chosen the title *Aquatic Chemistry* to emphasize the fact that we are concerned with the chemical behavior of natural aquatic systems of all kinds, including lakes, ocean waters, rivers, estuaries, groundwaters, and water treatment systems. Aquatic chemistry draws upon the basic principles of physical chemistry in order to define the pertinent variables governing the composition of natural waters and to arrive at a quantitative understanding of the relationship between these variables.

Aquatic Chemistry is addressed to those scientists and engineers (limnologists, oceanographers, geochemists, ecologists, environmental and sanitary engineers) who share an interest in, and a sense of responsibility for, the aquatic environment. It is a teaching book; rather than providing descriptive data, the methodologies of applying chemical principles to problems of natural water environments are stressed. Graphical representations and numerical examples are used extensively.

The book has its origins in the graduate teaching program of the senior author at Harvard University. The material has been developed in part for courses dealing with the chemistry of natural waters, chemical limnology, and the chemistry of water quality management. According to the authors' experience the level of treatment is suitable for advanced undergraduate or graduate students who have had a course in elementary chemistry. *Aquatic Chemistry* should also be of interest and use to those actively working in fields concerned with water quality.

The sequence of topics is from the simple to the more complex. Thus considerable attention is devoted to the systematic development of the subjects of chemical equilibrium between acids and bases before the treatment in detail of the subject of dissolved carbon dioxide whose distribution in solution and between the aqueous phase and the atmosphere is of central importance in the chemistry of natural waters. In keeping with the didactic purpose of the book, the following subjects are dealt with extensively: Interactions between solid phases and aqueous solutions; coordination chemistry, especially in

v

relation to the properties of metal ions in aqueous solutions; stability relations of pertinent oxidation and reduction components and microbial mediation in redox processes.

The theory of thermodynamic equilibrium is utilized extensively throughout the book in formulating models for aquatic systems, for handling stability relationships and computing equilibrium properties. The general basis for applying thermodynamics to the study of reactions and processes in natural water systems is set forth at the beginning of the book (Chapter Two). Here, kinetic and thermodynamic approaches to the equilibrium state are compared and the chemical thermodynamic criteria for treating spontaneous change and equilibrium are summarized. There may be a few readers who initially will find the formalism of thermodynamics somewhat abstract; they should not hesitate to proceed to the subsequent chapters; after having acquired a better appreciation of the relevance that thermodynamics assumes in aquatic chemistry, they will wish to return to Chapter Two for a thorough introduction to chemical thermodynamics of natural waters.

The regulation of the chemical composition of natural waters is considered in Chapter Eight. The principles and methods developed in the preceding chapters of the book, together with a consideration of the interaction between organisms and their abiotic environment, are applied to obtain an understanding of how natural waters achieve their observed compositions. Many of the chemical processes of interest in natural waters occur at interfaces. Hence the nature of the solid-solution interface is taken up in some detail in Chapter Nine, treating the main quantitative features of interfacial chemistry in the context of aquatic systems. Properties of the electric double layer are presented and the subject of adsorption from solution is developed. Of particular interest in relation to natural water systems is the surface chemistry of oxides, hydroxides, and oxide minerals, and the role of coagulation. In the concluding chapter we illustrate how information derived from laboratory experience can be employed to gain some understanding of the complex behavior of a few elements in natural water environments. Equilibrium and kinetic data are drawn upon in presenting case studies for phosphorus, iron, and manganese in natural waters.

Acknowledgments

The hospitality extended by the Department of Physical, Inorganic and Analytic Chemistry of the University of Berne, Switzerland, to Werner Stumm during the 1967–1968 academic year is acknowledged and appreciated. We are indebted to Dr. Paul Schindler of the University of Berne, Dr. Perry McCarty of Stanford University, and Drs. Lloyd A. Spielman and Elisabeth

Stumm-Zollinger of Harvard University for review of individual chapters. We are grateful to Joel Gordon, Y.-S. Chen, C. P. Huang, and Hermann Hahn of Harvard University and Francois Morel of Caltech for their valuable assistance with a number of tasks involved in the preparation of the manuscript.

Cambridge, Massachusetts   WERNER STUMM

Pasadena, California   JAMES J. MORGAN

April, 1970

Contents

I   Introduction: Scope of Aquatic Chemistry

Aquatic chemistry is concerned with the chemical processes affecting the distribution and circulation of chemical compounds in natural waters; its aims include the formulation of an adequate theoretical basis for the chemical behavior of ocean waters, estuaries, rivers, lakes, groundwaters and of soil water systems as well as the description of the processes involved in water treatment. Obviously, aquatic chemistry draws primarily on the fundamentals of chemistry but it is also influenced by other sciences, especially geology and biology.

This book does not cover all aspects of aquatic chemistry; a large part is devoted to the demonstration that elementary principles of physical chemistry can be used to identify some of the pertinent variables determining the composition of natural water systems. The student of chemistry is not always fully aware that the well-known laws of physical chemistry not only apply in the laboratory but regulate the course of the reactions taking place in nature. During the hydrological cycle water interacts continuously with matter. Thus a progressive differentiation of matter is achieved by processes of weathering, soil erosion and soil and sediment formation. These separation processes accomplished by nature on a large scale have been likened [1] to the sequence of separations carried out during the course of a chemical analysis. The processes — dissolution and precipitation, oxidation, and reduction, acid–base and coordinative interactions — are the same in nature as in the laboratory; although this book contains treatment of topics similar to those found in an analytical chemistry text, it endeavors to consider the spatial and temporal scales of the reactions in nature as entirely different from those in the laboratory; for example, in analytical chemistry precipitates (of frequently metastable and active compounds) are formed from strongly oversaturated solutions whereas in natural water systems the solid phase is usually formed under conditions of slight supersaturation; often crystal growth and aging continues for geological time-spans. Interfacial phenomena are particularly important because most chemical processes of significance in nature occur at phase discontinuities.

[1] K. Rankama and Th. G. Sahama, *Geochemistry*, Univ. Chicago Press, Chicago, 1950.

The actual natural water systems usually consist of numerous mineral assemblages and often of a gas phase in addition to the aqueous phase; they nearly always include a portion of the biosphere. Hence natural aquatic habitats are characterized by a complexity seldom encountered in the laboratory. In order to understand the pertinent variables out of a bewildering number of possible ones, it is advantageous to compare the real systems with their idealized counterparts.

MODELS TO OBVIATE NATURE'S COMPLEXITY. Simplified and manageable models may be used to illustrate the principal regulatory factors that control the chemical composition of natural waters and in turn the composition of the atmosphere. To be useful a model need not be realistic as long as it produces fruitful generalizations and valuable insight into the nature of aquatic chemical processes and improves our ability to describe and to measure natural water systems.

The theory of *thermodynamic equilibrium* appears to be the most expedient concept to facilitate identification of many variables relevant in determining the mineral relations and in establishing chemical boundaries of aquatic environments. Free energy concepts describe the thermodynamically stable state and characterize the direction and extent of processes that are approaching equilibrium. Discrepancies between predicted equilibrium calculations and the available data of real systems give valuable insight into those cases in which chemical reactions are not understood sufficiently, in which nonequilibrium conditions prevail, or in which the analytical data are not sufficiently accurate or specific. Such discrepancies thus provide an incentive for future research and the development of more refined models.

By comparing the actual composition of seawater (sediments + sea + air) with a model in which the pertinent components (minerals, volatiles) with which water has come into contact are allowed to reach true equilibrium, Sillén [2] in 1959 epitomized the application of equilibrium models for portraying the prominent features of the chemical composition of this system. His analysis, for example, has indicated that contrary to the traditional view, the pH of the ocean is not buffered primarily by the carbonate system; his results suggest that heterogeneous equilibria of silicate minerals comprise the principal pH buffer systems in oceanic waters. This approach and its expansion have provided a more quantitative basis for Forchhammer's suggestion of 100 years ago that the quantity of the different elements in seawater is not proportional to the quantity of elements which river water pours into the sea but is inversely proportional to the facility with which the elements in seawater are made insoluble by general chemical actions in the sea. Although inland

[2] L. G. Sillén in *Oceanography*, M. Sears, Ed., American Association for Advancement of Science, Washington, D.C., 1961.

waters represent more transitory systems than the sea, equilibrium models are also useful here for interpreting observed facts. We can obtain some limits on the variational trends of chemical composition even in highly dynamic systems, and we can speculate on the type of dissolved species and solid phases one may expect.

Natural waters indeed are open and dynamic systems with variable inputs and outputs of mass and energy for which the state of equilibrium is a construct. *Steady-state models* reflecting the time-invariant condition of a reaction system may frequently serve as an idealized counterpart of an open natural water system. The concept of free energy is not less important in dynamic systems than in equilibrium systems. The flow of energy from a higher to a lower potential or energy "drives" the hydrological and the geochemical cycles [3, 4]. The ultimate source of the energy flow is the sun's radiation.

ECOLOGY. In natural waters organisms and their abiotic environments are interrelated and interact on each other. Because of the continuous input of solar energy (photosynthesis) necessary to maintain life, ecological systems are never in equilibrium. The ecological system, or ecosystem, may be considered as a unit of the environment that contains a biological organization made up of all the organisms interacting reciprocally with the chemical and physical environment. In an ecosystem the flow of energy and of negative entropy is reflected by the characteristic trophic structure and leads to material cycles within the system (cf. E. P. Odum [5]). In a balanced ecological system a steady state of production and destruction of organic material as well as of production and consumption of O_2 is maintained.

Our understanding of nature is seriously hampered by a lack of kinetic information on reactions typically encountered in natural waters. It is partially for this reason that this book cannot avoid putting a heavy emphasis on the concept of equilibrium as compared to that of rates.

WATER AS A RESOURCE. Aquatic chemistry is of great practical importance because water is a necessary resource for man. We are not concerned with the quantity — water as a substance is abundant — but with the quality of water and its distribution. We are altering the natural pattern of our environment at an accelerating rate. Figure 1-1 shows a forecast [6] of the hydrological cycle of water use in the Los Angeles area in the year 2050. By manipulating the land, by disposing our wastes into the atmosphere and hydrosphere and

[3] B. Mason, *Principles of Geochemistry*, Wiley, New York, 1966.
[4] H. J. Morowitz, *Energy Flow in Biology*, Academic Press, New York, 1968.
[5] E. P. Odum, *Science*, **164**, 262 (1969).
[6] N. H. Brooks, in *The Next Ninety Years*, California Institute of Technology, Pasadena, Calif., 1967.

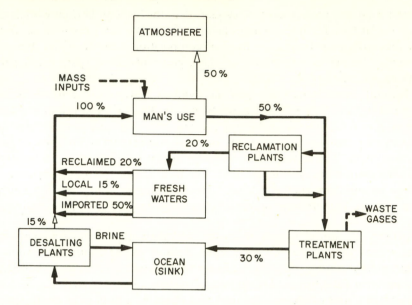

Figure 1-1 Cycle of water use in the Los Angeles area estimated for the year 2050. From Brooks [6]. Local water from ground water and mountain streams is estimated to be about 15%; imported water might be 50%; reclaimed water will be perhaps 20%; and the desalted water might be 15%. At the present time, the supply is about 61% imported, 38% local and 1% reclaimed water. Reproduced with permission from N. H. Brooks.

by accelerated combustion of fossil fuels we can hardly circumvent influencing nature and with it the hydrological cycle.

Conservation of aquatic resources cannot be achieved by avoiding man's interference with the aquatic habitat. Water pollution control cannot solely consist in waste treatment. To what extent can the ocean be used as a sink of wastes? How can we improve the fertility of the ocean and exploit its production of harvestable food for the progressively increasing world population? How can we restore the ecological balance between photosynthetic and respiratory activity in a nutritionally enriched (polluted) lake? We have to develop the ability to modify and to manipulate our aquatic environment in order to improve its quality. Despite significant developments these problems cannot be solved by technology alone as long as we lack the necessary scientific understanding. It is hoped that the following discussion enhances the learning process toward a better understanding of the aquatic environment. "Man masters nature not by force but by understanding" [7].

[7] J. Bronowski, *Science and Human Values*, Harper and Row, New York, 1965.

2   Chemical Thermodynamics of Natural Waters

2-1   Introduction

The chemical composition observed in a particular natural water is the result of a variety of chemical reactions and physico-chemical processes acting in concert. In a broad classification of natural water chemistry we can recognize acid–base reactions, gas–solution processes, precipitation and dissolution of solid phases, coordination reactions of metal ions and ligands, oxidation–reduction reactions and adsorption–desorption processes at interfaces as potentially significant. In succeeding chapters these classes of chemical reactions and processes will be elaborated and examined individually in considerable detail.

    In this chapter we set forth the general basis for applying chemical thermodynamics to the study of all reactions and processes in natural water systems. The discipline of chemical thermodynamics is concerned with variables such as chemical composition, temperature, pressure and volume, and with the relationships between them. Specifically, chemical thermodynamics provides two very important kinds of information about physico-chemical transformations and chemical reactions. First, it provides a criterion for determining whether a particular process is possible, that is, whether it can proceed spontaneously under specified conditions; in other words, we can establish the direction of natural change using thermodynamic principles. Second, chemical thermodynamics provides the mathematical framework for computing the composition of a chemically-reacting system at equilibrium. The unifying concept on which these two aspects of the thermodynamic description of chemical reactions is based is the idea of entropy production, developed by de Donder on the basis of the earlier treatments of thermodynamic equilibrium by J. Willard Gibbs and van't Hoff. In any irreversible process, for example, a chemical reaction occurring spontaneously, entropy is produced. In a state of thermodynamic equilibrium the rate of entropy production is zero. We will develop the implications of this concept further and apply it to problems of interest in aquatic chemistry. Before doing so it will prove useful to consider some general aspects of the chemistry of natural water

5

systems and their possible interpretations in terms of the kinds of systems treated in chemical thermodynamics.

2-2   Natural Water Systems and Thermodynamic Systems

The general term natural water system refers to an actual system of some complexity, consisting of an aqueous solution phase, one or more solid phases, and most often, a gas phase. The real system may be inhomogeneous over-all, although it may be sufficiently well-characterized as consisting of identifiable homogeneous regions; for example, the oceans have relatively discrete regions of rather constant temperature and density and include relatively well-isolated deep basins, often of distinct chemical compositions. Deep lakes and reservoirs often undergo a seasonal stratification during which the bottom, or hypolimnetic, waters are cold and dense and the surface, or epilimnetic, waters are warm and less dense. Distinct differences in certain chemical properties can develop in the two zones and may persist over long periods. Ground water formations may show pronounced spatial differences in principal mineral assemblages. The point of these examples is that in the real natural water systems it may be possible to identify subsystems of distinct characteristics. Obviously, adequate characterization of the real system requires some care if the pertinent mechanical, thermal and chemical variables are to be adequately reflected in experimental or conceptual models. Two properties of the real systems of particular interest in chemical thermodynamic analysis are the temperature and the pressure. The existence of significant temperature gradients in many bodies of waters prevents an exact application of chemical thermodynamic relationships for the system as a whole. However, when the temperature is sufficiently uniform over a portion of the system, local thermal equilibrium can be assumed for that part of the system. In deep bodies of water the hydrostatic pressure may become large enough to require quantitative consideration of its effects on the chemical properties. In contrast to temperature gradients, pressure gradients in natural water systems do not represent a fundamental obstacle to exact application of equilibrium relationships. As a practical point, however, spatial gradients of both kinds may represent only minor influences in comparison to temporal variations in properties of a particular system.

Thermodynamic systems are parts of the real world isolated for thermodynamic study. The parts of the real world isolated for our purposes are either entire natural water systems or specified regions within these systems, depending upon the physical and chemical features of the actual situation (chemically inhomogeneous regions, temperature profiles, etc.). In thermodynamics distinction is made between *isolated* systems, *closed* systems and *open* systems.

Isolated systems cannot exchange heat, work or matter with their environment. Closed systems can exchange energy but not matter with their environment. Changes of chemical composition within closed systems are not precluded in their definition. Open systems can exchange both energy and matter with their environment. Beyond the scope of chemical thermodynamics *per se* are *continuous* or *flow* systems. Such systems can continuously exchange matter with their surroundings. They are characterized by fluxes of both matter and energy at their boundaries. Organisms are examples of continuous or flow systems. Some reflection on the nature of certain natural water systems will suggest that they, too, are *continuous* systems. We return later to the question of the suitability of a thermodynamic representation for natural water systems viewed in this light.

To illustrate the representation of the pertinent features of natural water systems by their counterpart thermodynamic systems, let us consider a generalized model as shown in Figure 2-1. The main features of a thermodynamic model comprise a description of the composition of the aqueous solution phase, the gas phase and one or more solid phases. For the aqueous solution the actual composition is given by a set of *concentrations*. We have used the symbol $[i]$ to represent the molar concentration of constituent i. The symbols γ_i and $\{i\}$ refer to the *activity coefficient* and *activity*, respectively, of

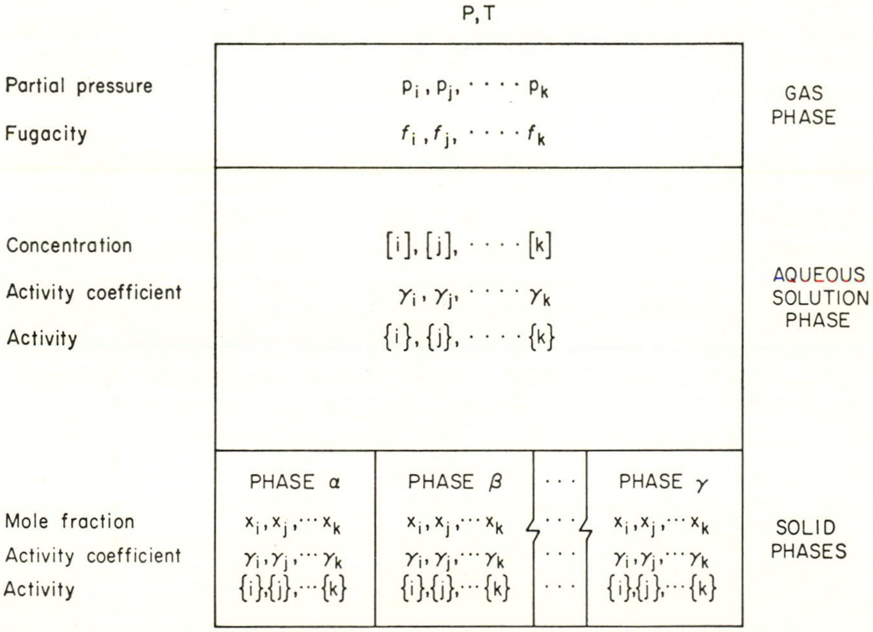

$$P, T$$

Partial pressure	$p_i, p_j, \cdots p_k$	GAS
Fugacity	$f_i, f_j, \cdots f_k$	PHASE

Concentration	$[i], [j], \cdots [k]$	
Activity coefficient	$\gamma_i, \gamma_j, \cdots \gamma_k$	AQUEOUS SOLUTION PHASE
Activity	$\{i\}, \{j\}, \cdots \{k\}$	

	PHASE α	PHASE β	$\cdots$	PHASE γ	
Mole fraction	$x_i, x_j, \cdots x_k$	$x_i, x_j, \cdots x_k$	$\cdots$	$x_i, x_j, \cdots x_k$	SOLID
Activity coefficient	$\gamma_i, \gamma_j, \cdots \gamma_k$	$\gamma_i, \gamma_j, \cdots \gamma_k$	$\cdots$	$\gamma_i, \gamma_j, \cdots \gamma_k$	PHASES
Activity	$\{i\}, \{j\}, \cdots \{k\}$	$\{i\}, \{j\}, \cdots \{k\}$	$\cdots$	$\{i\}, \{j\}, \cdots \{k\}$	

Figure 2-1 Generalized model for thermodynamic description of natural water systems.

constituent i. The need for these quantities arises from the generally *nonideal* behavior of actual aqueous solutions. For the gas phase the actual composition is given by the partial pressures of the constituents. For gases the idealized pressure is known as the fugacity. The solids are represented generally as made up of several components. For pure solids, the mole fraction, X_i, is unity and the activity coefficient and activity are constants for fixed conditions of temperature and pressure. For solids of variable composition (solid solutions), activity coefficients and activities will be required as in the case of aqueous solutions. (The matter of activity coefficients, activities and fugacities is not essential for the purpose at hand, but it is convenient to call attention to the problem of nonideality at this point. We return to it in more detail later.)

If we impose sufficient specific conditions on the model represented in Figure 2-1, we will have defined a thermodynamic system amenable to quantitative description; for example, we may specify the temperature, the total pressure for all phases and the instantaneous composition variables (pressures, concentrations, mole fractions) of the system. We can then examine the state of the entire system with respect to spontaneous change and equilibrium for each conceivable chemical reaction and phase transfer process. Alternatively, the appropriate thermodynamic conditions of chemical equilibrium for the closed system can be imposed and the equilibrium composition of the several phases computed. In what has been said thus far concerning the generalized model of Figure 2-1 it has been more or less implied that all of the features of some prototype system are represented by the thermodynamic model. Such need not be the case; indeed, such seldom is the case. The emphasis in chemistry is usually on those reactions or phase transformations which are specifically of interest. Most systems are not truly at equilibrium in the total sense. Accordingly, the constituents in our generalized model should be understood to be those abstracted from the actual system for specific consideration. Other constituents of the system may react so slowly that they can be neglected in considering chemical equilibrium, or they may be completely "uncoupled" with respect to the chemical reactions of the constituents of interest. Omissions from the composition of the model may simply reflect ignorance of species or reactions in the real system.

In addition to neglecting chemical components which do not interest us, we may also construct our thermodynamic system by isolating only certain parts of the actual natural water system; for example, as mentioned previously, we might do this because of inhomogeneities in the prototype. Figure 2-2 shows schematically some thermodynamic systems which might be isolated from the more general system described in Figure 2-1. The dashed lines represent thermodynamic boundaries between the system and its surroundings, while the solid lines represent phase boundaries. System (*a*) of

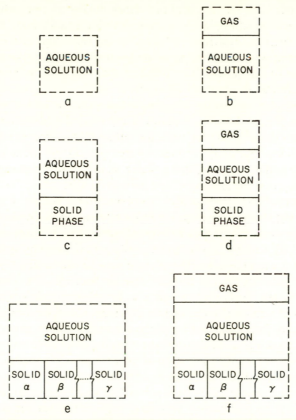

Figure 2-2 Thermodynamic systems which might be isolated from the generalized natural water system model of Figure 2-1.

Figure 2-2 represents essentially the simplest kind of thermodynamic model that one can choose for a real system: an aqueous solution which is not subject to physico-chemical reactions with any solid or gaseous phases. Such a system is an idealization, of course, in that water itself and all solutes have some finite escaping tendency, however small. Experimentally, systems of this kind are quite important in the measurement of thermodynamic properties of solutions. An aqueous solution of electrolyte in a laboratory vessel maintained at constant temperature and exposed to atmospheric pressure represents a common arrangement for studying chemical equilibria. Thermodynamic data (heats of reaction, entropies, free energies) obtained from such experimental systems can be used to interpret the behavior of real or model systems of a more complex character. System (*a*) is a closed thermodynamic system in the sense that the total mass of its contents is fixed and it does not exchange matter with its surroundings either during an irreversible process (e.g., a chemical reaction in the solution) or at equilibrium. Such closed systems are sometimes described as "open" in the sense that their contents can be changed arbitrarily by addition or removal of constituents, thus defining different states of the system. This usage of the term open system is obvious in its application to solutions of all kinds.

A closed system consisting of an aqueous solution and a gas phase is represented in Figure 2-2b. Each phase of the total closed system is an *open* phase: it can exchange matter with the other phase. For example, in the system containing the two components water and carbon dioxide, consisting of a gas phase and an aqueous phase, water molecules are exchanged between the liquid phase and the gas phase, and carbon dioxide molecules are likewise exchanged between the two phases. Experimental systems of type (b) are important in the study of aqueous solutions of electrolytes (water vapor pressure lowering) and in the description of gas solution rates or solubility equilibria. Systems of type (b) are sometimes approximated by systems of type (a) as when, for example, a solute molecule may be considered "nonvolatile" because of a slow liquid–gas transfer rate or because of a small escaping tendency. A system such as (b) might also be an appropriate representation for a portion of a natural water system in which a local equilibrium exists at the gas–solution interface, that is, where the interphase transport process is quite fast in comparison to other transport processes or to chemical reactions.

A closed system comprising a solid phase and an aqueous solution is represented by (c) of Figure 2-2. An experimental system corresponding to (c) is that in which a simple solid phase partially dissolves in water, the solid not reacting to yield an appreciably volatile component under the conditions of the system. Examples of this kind of system are afforded by two-component systems of water and such solids as the metal-ion sulfates, phosphates and silicates. A laboratory system of pure calcium carbonate solid and water at equilibrium under a nitrogen atmosphere *approximates* the definition in (c) because of the rather low partial pressure of the volatile aqueous carbon dioxide formed from the solution of the carbonate. System (c) can frequently come into play in the description of local reactions or local equilibrium at the sediment–water interface, that is, where an individual solid phase approaches equilibrium with the water mass in its vicinity. The equilibrium under consideration may also be partial: other solids might not be at equilibrium with respect to independent reactions between themselves and certain constituents of the water. The validity of local and partial equilibrium models needs to be justified through knowledge of the relative rates of various pertinent reactions and transport processes for the real system.

In actual natural water systems it will be found that the solubility behavior of the sparingly-soluble carbonates and sulfides cannot be adequately characterized by a system of type (c): the gas phase will need to be considered together with the solid and the aqueous phase. The simplest system which can account for the simultaneous influence of a solid and a gas phase upon the chemical composition of an aqueous solution is represented by (d) in Figure 2-2. Specific experimental examples are provided by the systems CO_2–H_2O–$CaCO_3$ and H_2S–H_2O–MnS. Many more examples can be cited. In (d) the over-all system is closed and each of the phases is an open system with respect to the others. Reaction of solid calcium carbonate with water leads to transfer of CO_2 to the gas phase. Reaction of calcium ions and bicarbonate ions in the aqueous phase can transfer CO_2 to the gas phase and form $CaCO_3$ solid:

$$Ca^{2+}(aq) + 2HCO_3^-(aq) = CO_2(g) + CaCO_3(s) + H_2O(l)$$

In discussing the dissolution and precipitation behavior of carbonates and sulfides in natural waters, type (c) systems are frequently invoked. In general they represent partial reaction systems, and it will be found that their application is limited.

The restriction of considering only one solid phase is removed in system (e) in which solid phases $\alpha, \beta, \ldots, \gamma$ are present in the closed system. The solid phases are open with respect both to one another and to the aqueous phase. Simultaneous chemical reactions between the solid phases and the aqueous phase influence the composition of the solution, and vice versa. No gas phase is assumed present in system (e). The model is restricted to solids of low escaping tendency or to systems in which reactions of interest take place in essential isolation with respect to the atmosphere (local reactions). An important area of application of systems of type (e) is the study of reactions between solid aluminosilicates and aqueous solutions. As examples we have the reactions between the minerals muscovite and kaolonite

$$KAl_3Si_3O_{10}(OH)_2(s) + H^+(aq) + \tfrac{3}{2}H_2O(l) = \tfrac{3}{2}Al_2Si_2O_5(OH)_4(s) + K^+(aq)$$

and between kaolinite and gibbsite

$$Al_2Si_2O_5(OH)_4(s) + 5H_2O(l) = 2H_4SiO_4(aq) + Al_2O_3 \cdot 3H_2O(s)$$

System (e) also lends itself to the description of chemical reactions between several pure solids only, since the composition of the solution phase is constant during such reactions under conditions of constant temperature and pressure; for example, in an oxygen-free system the reaction

$$3MnO_2(s) + 4Fe_3O_4(s) = Mn_3O_4(s) + 6Fe_2O_3(s)$$

Having considered a number of restricted thermodynamic systems applicable in varying degrees to the modeling of different natural water systems, we can remove all of the particular limitations of systems (a), (b), (c), (d) and (e) in Figure 2-2 and obtain the generalized system, (f), which is identical with the model of Figure 2-1. System (f) reduces to each of the special cases we have examined when the appropriate boundaries or limitations are specified. We now turn to the quantitative description of these chemical systems so that the conditions of stability and the directions of spontaneous change can be computed.

Chemical equilibrium and chemical change can be described in terms of two different approaches, kinetic or thermodynamic. From a kinetic point of view, the equilibrium state is one of dynamic balance between opposing reactions. From a thermodynamic point of view, the equilibrium state is a state of maximum stability — one of zero entropy production — toward which a *closed* system proceeds by irreversible processes. A closed chemical system proceeds to a state of chemical equilibrium through spontaneous chemical reactions and phase transformations. At equilibrium further changes in the chemical composition of the system cannot occur. The thermodynamic approach is powerful for arriving at the quantitative laws describing the effects of composition, temperature, pressure and volume upon chemical equilibrium. The kinetic approach is useful for gaining a detailed understanding of chemical reactions in homogeneous systems such as solutions and gases and the mechanisms of their approach to equilibrium. Obviously, to understand the chemical behavior of actual natural water systems we require some insight into the kinetics of the processes involved. For the equilibrium state both the kinetic and thermodynamic approaches of course yield identical descriptions of the chemical composition. We introduce our discussion of chemical equilibrium by illustrating the kinetic approach to chemical equilibrium in aqueous solution. Then we present the thermodynamic approach on which most of the analysis in this book is based.

2-3   Kinetic Approach to Chemical Equilibrium

The classical test for chemical equilibrium with respect to a particular reaction is to allow that reaction to proceed in a closed system in the "forward" direction (left to right as the reaction is written) until no net change in the composition of the system is observable and then to allow the same reaction to proceed in the "backward" direction until again no net change is observed. The composition of the reaction system should approach the same limit from both directions if chemical equilibrium is attained. It is well known that true equilibrium is difficult to achieve for certain systems. Many reactions of organic compounds occur very slowly and fail to approach an equilibrium state in short time spans. Mechanisms are frequently complicated. For reactions which do not approach equilibrium sufficiently to apply equilibrium concepts only kinetic and mechanistic descriptions of systems far from equilibrium will suffice. For a large number of reactions in aqueous solution, however, there is sufficient kinetic and equilibrium evidence to demonstrate that compositions approaching the equilibrium composition can be expected in real systems.

Consider the case of a simple, or elementary, reversible reaction at constant temperature and pressure

$$\nu_A A + \nu_B B + \cdots = \nu_C C + \nu_D D + \cdots \tag{1}$$

where the various ν's are the stoichiometric coefficients of the reactants A, B, ... and the products C, D, ... The rate of the forward reaction, v_f, is given by

$$v_f = k_f [A]^{\nu_A} [B]^{\nu_B} \cdots \tag{2}$$

and that of the backward reaction, v_b, by

$$v_b = k_b [C]^{\nu_C} [D]^{\nu_D} \cdots \tag{3}$$

where k_f and k_b are rate coefficients in the forward and backward directions, respectively, and [A], [B], etc. are molar concentrations. At equilibrium, defined as the condition of dynamic balance between the forward and backward reaction rates, $v_f = v_b$ or

$$k_f [A]^{\nu_A} [B]^{\nu_B} \cdots = k_b [C]^{\nu_C} [D]^{\nu_D} \cdots \tag{4}$$

from which

$$\left(\frac{[C]^{\nu_C} [D]^{\nu_D} \cdots}{[A]^{\nu_A} [B]^{\nu_B} \cdots} \right)_{equil} = \frac{k_f}{k_b} = K \tag{5}$$

where K is defined as the equilibrium constant for the elementary reaction and is a function of the temperature and pressure.

It is of course not generally the case that the experimentally observed rate expressions for a net chemical reaction will contain each of the reactant and product concentrations raised to a power which is identical to its stoichiometric coefficient in the net reaction. The experimental rate laws found are frequently complex, reflecting the rate-limiting steps of various mechanistic pathways when there is more than one. Even when there is but one rate-limiting step the net reaction need not be an elementary one; the rate expressions for the forward and backward reactions then cannot be deduced from the stoichiometry of the net reaction. The exponents of the concentration terms must be obtained from experiment. However, when the over-all reaction has only one rate-limiting step the equilibrium constant for the over-all reaction *is* given by the ratio of the specific rate constants for the forward and backward reaction, that is, by (5); for example, for the following over-all reaction in aqueous solution

$$H_3AsO_4 + 3I^- + 2H^+ = H_3AsO_3 + H_2O + I_3^- \qquad (6)$$

the rate law observed for I_3^- formation is

$$\frac{d[I_3^-]}{dt} = v_f = k_f[H_3AsO_4][I^-][H^+] \qquad (7)$$

and the rate law for the reverse reaction is

$$\frac{-d[I_3^-]}{dt} = v_b = k_b \frac{[H_3AsO_3][I_3^-]}{[I^-]^2[H^+]} \qquad (8)$$

The exponent of the concentrations in the rate laws is not identical with the stoichiometric coefficient of each reactant in the over-all reaction. At equilibrium $v_f = v_b$ and

$$k_f[H_3AsO_4][I^-][H^+] = k_b \frac{[H_3AsO_3][I_3^-]}{[I^-]^2[H^+]} \qquad (9)$$

Therefore

$$\frac{k_f}{k_b} = \frac{[H_3AsO_3][I_3^-]}{[H_3AsO_4][I^-]^3[H^+]^2} = K \qquad (10)$$

Independent calculations of the equilibrium constant from studies of the kinetics and from the measurement of the equilibrium composition of (6) show good agreement [1].

For reactions which have more than one rate-limiting step, (5) can still be applied if the "forward rate constant" is interpreted as the *product* of the specific rate constants for each of the consecutive steps in any path leading from reactants to products, and if the "backward rate constant" is similarly interpreted [2]. Detailed kinetic information on complicated mechanisms with many paths is very limited for reactions of interest in natural waters. For the remainder of this discussion of kinetics and equilibrium we limit ourselves to the case of reactions with one rate-limiting step.

At equilibrium we have seen that $v_f = v_b$ and $K = k_f/k_b$ [(4) and (5)]. At any composition the ratio v_f/v_b is of interest. From (2) and (3) we find

$$\frac{v_f}{v_b} = \frac{k_f}{k_b}\left(\frac{[A]^{v_A}[B]^{v_B}\cdots}{[C]^{v_C}[D]^{v_D}\cdots}\right) \qquad (11)$$

[1] J. O. Edwards, *Inorganic Reaction Mechanisms*, Benjamin, New York, 1964.
[2] T. S. Lee, in *Treatise on Analytical Chemistry*, I. M. Kolthoff and P. J. Elving, Eds., Part I, Vol. 1, Interscience, New York, 1959.

The quotient

$$\left(\frac{[C]^{\nu_C}[D]^{\nu_D}\cdots}{[A]^{\nu_A}[B]^{\nu_B}\cdots}\right)$$

at any composition of the reaction system is called the reaction quotient Q and has the same form as K, the equilibrium constant, but is not limited by the condition of equilibrium. In general, therefore,

$$\frac{v_f}{v_b} = \frac{k_f}{k_b}\frac{1}{Q} = \frac{K}{Q} \tag{12}$$

When $K = Q$, $v_f = v_b$, and equilibrium exists; when $K > Q$, $v_f > v_b$, and the direction of net reaction is from left to right; when $K < Q$, $v_f < v_b$, and the direction of net reaction is from right to left. The ratio K/Q will be encountered again in the thermodynamic treatment of chemical reactions, where K/Q will be found to give the "driving force" for spontaneous reaction.

If the forward and backward rate constants for a reaction are known it is possible to compute the net rate of approach to equilibrium for any system composition. There are wide ranges in the magnitudes of the experimentally determined rate constants for various reactions in aqueous solutions.

The forward rate constant for the reaction

$$OH^- + HCO_3^- = CO_3^{2-} + H_2O$$

is approximately 6×10^9 l mole^{-1} sec^{-1} (20°C). Equilibrium is approached extremely rapidly in this reaction. The water exchange reaction for the aquo Cr(III) ion

$$Cr(OH_2)_6^{3+} + H_2O^* = (H_2O^*)Cr(OH_2)_5 + H_2O$$

has a rate constant of about 3×10^{-6} sec^{-1} (27°C). Equilibrium in this reaction is approached rather slowly (half-times of exchange reaction of about 2 days). The aquation reaction

$$Co(NH_3)_5Cl^{2+} + H_2O = Co(NH_3)_5H_2O^{3+} + Cl^-$$

has a forward velocity rate constant of about 10^{-4} min^{-1} (25°C), corresponding to a half-life of about 110 hr. The decarboxylation of the amino acid alanine

$$CH_3CHNH_2COOH = CH_3CH_2NH_2 + CO_2$$

has a rate constant of approximately 10^{-11} sec^{-1} at ordinary temperatures, corresponding to a half-time of ca. 4000 years. These few examples may serve to indicate the chemical kinetic variations which are encountered in aquatic systems.

Example 2-1.   Describe the rates of reaction and approach to equilibrium for the reaction

$$CO_2(aq) + H_2O \underset{k_b}{\overset{k_f}{\rightleftharpoons}} HCO_3^- + H^+ \dagger \qquad (13)$$

in an aqueous solution containing initially $CO_2(aq)$ at a concentration of 1×10^{-5} mole liter^{-1} (M). For purposes of the example $CO_2(aq)$ is treated as a nonvolatile constituent and other chemical reactions are neglected.

At 25°C we choose $k_f = 3.0 \times 10^{-2}$ sec^{-1} and $k_b = 7.0 \times 10^4$ M^{-1} sec^{-1} (e.g., Kern [3]). The solution is dilute and the water concentration remains essentially unchanged during the reaction. The reversible reaction is therefore of the type

$$A \underset{k_b}{\overset{k_f}{\rightleftharpoons}} B + C \qquad (14)$$

A general rate expression has been presented by Benson [4] and is of the form

$$\frac{dx}{dt} = \alpha + \beta x - \gamma x^2 \qquad (15)$$

where x is the change in concentration of A, B and C at time t ($x = A_0 - A = B - B_0 = C - C_0$); α, β and γ are simple functions of the rate constants and the initial concentrations of reactants and products:

$$\alpha = k_f A_0 - k_b B_0 C_0; \qquad \beta = -(k_f + k_b C_0 + k_b B_0); \qquad \gamma = -k_b$$

The solution to the rate equation is of the form

$$\ln \frac{x + (\beta - q^{1/2})/2\gamma}{x + (\beta + q^{1/2})/2\gamma} = tq^{1/2} + \ln \frac{\beta - q^{1/2}}{\beta + q^{1/2}} \qquad (16)$$

where $q = \beta^2 - 4\alpha\gamma$.

For $A_0 \equiv [CO_2(aq)]_0 = 1 \times 10^{-5}$ M, $B_0 \equiv [HCO_3^-]_0 = 0$ and $C_0 \equiv [H^+]_0 = 1 \times 10^{-7}$ M, application of (16) yields the result shown in Figure 2-3a, where concentrations of $CO_2(aq)$ and HCO_3^- are plotted as a function of reaction time. The equilibrium composition corresponds to $[HCO_3^-] = 1.8 \times 10^{-6}$ M (1.8 μM) and $[CO_2(aq)] = 8.2$ μM. The equilibrium hydrogen ion concentration $[H^+]$ is approximately equal to $[HCO_3^-]$. The equilibrium composition is approached closely (to within less than 1%) in a reaction time of 20 sec. This reaction approaches equilibrium more slowly than many other acid–base reactions in solution, for example, the ionization of acetic acid or the hydrolysis of ammonia (times on the order of micro- to milliseconds).

---

† The symbol H^+ denotes the aqueous proton. The chemistry of the proton in aqueous solution is discussed in Chapter 3.

[3] D. M. Kern, *J. Chem. Educ.*, **37**, 14 (1960).
[4] S. W. Benson, *The Foundations of Chemical Kinetics*, McGraw-Hill, New York, 1960.

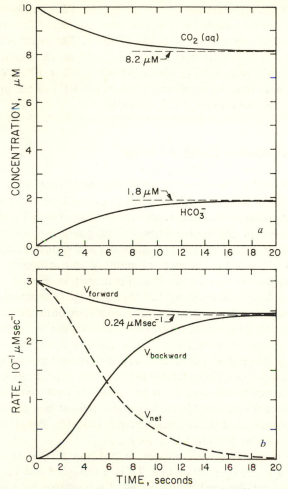

Figure 2-3 *a*: Computed concentrations of $CO_2(aq)$ and HCO_3^- as a function of time for the reversible reaction $CO_2(aq) + H_2O = HCO_3^- + H^+$ at $25°C$ in a closed aqueous system. The total concentration, $C_T = [CO_2(aq)] + [HCO_3^-]$, is 1×10^{-5} M. $CO_2(aq)$ is assumed to be nonvolatile. *b*: Computed velocities, $v_{forward}$, $v_{backward}$ and v_{net} in the reaction mixture as a function of time for $CO_2(aq) + H_2O = HCO_3^- + H^+$. At equilibrium, the velocity in both directions is 0.24 μM sec^{-1}.

The forward velocity of the reaction v_f is given by

$$v_f = k_f[CO_2(aq)] \tag{17}$$

and the backward velocity v_b by

$$v_b = k_b[HCO_3^-][H^+] \tag{18}$$

The computed velocities are shown in Figure 2-3b. The equilibrium condition $v_f = v_b$ is marked by a *net* velocity v_{net} of zero, and equal forward and backward velocities of approximately $0.24 \ \mu M \ sec^{-1}$. The equilibrium constant for (13), k_f/k_b, is approximately 4.3×10^{-7}. At a reaction time of 4 sec, application of (12) gives v_f/v_b' of about 3.5 ($Q \simeq 1.2 \times 10^{-7}$; $Q < K$). It is interesting to note that the equilibrium velocity is 80% of the initial forward velocity under the conditions of this example.

2-4   Thermodynamics of Chemical Change

From the thermodynamic point of view the equilibrium state is a state of maximum *stability* toward which a closed physico-chemical system proceeds by irreversible processes. Let us consider the notions of stability and instability as they are used in thermodynamics before going into detail about the quantitative treatment of chemical equilibrium and spontaneous change. It is instructive to examine the concepts of stability and instability for a simple mechanical system, for example, that illustrated in Figure 2-4a. Similar examples have been discussed by La Mer and by Guggenheim [5]. Three different "equilibrium" positions of a box on a horizontal surface are shown. In positions *1* and *3* the center of gravity is lower than in any infinitely near position of the box on the surface. If the position is slightly disturbed, the box returns to its original position. The gravitational potential energy is a minimum and the equilibrium is stable in each case. In position *2*, the potential energy is a maximum with respect to any infinitely near position of the box; if the position of the box in *2* is slightly disturbed the box moves away from its original position, that is, the equilibrium is unstable. Position *2* is unstable with respect to stable positions such as *1* and *3*.

Position *1* is unstable with respect to position *3*, that is, the potential energy in *3* is absolutely less than in *1*. Thus, although *1* and *3* are both stable with respect to infinitesimal disturbances, *1* is unstable with respect to the finite change which would move it to position *3*. We may describe *3* as absolutely stable (with reference to the surface) and *1* as *metastable*.

An analogy between the mechanical system and a thermodynamic system is drawn in Figure 2-4b. A hypothetical and generalized energy *or* entropy profile is shown as a function of the state of the system (determined by specifying a sufficient number of state variables, such as temperature, pressure and composition). State *b* represents an unstable equilibrium: an infinitesimal change can cause the system to move away from *b* toward states *a* or *c*, that is, toward a state of lower energy or greater entropy depending upon the defini-

[5] E. A. Guggenheim, *Thermodynamics*, 4th ed., North-Holland, Amsterdam, 1959.

A

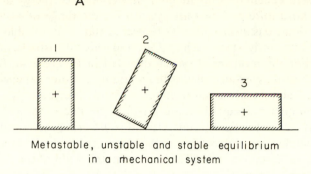

Metastable, unstable and stable equilibrium
in a mechanical system

B

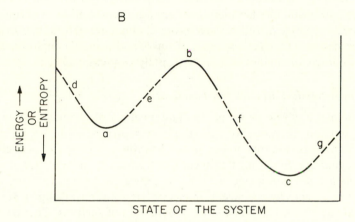

STATE OF THE SYSTEM

Metastability, instability and stability for
different energetic states of a thermodynamic system

Figure 2-4 *a*: Illustration of mechanical metastability, instability and stability. *b*: An analogy between mechanical and thermodynamic stability concepts. ("Energy" and entropy have opposite senses on the ordinate.) The state of the system may depend on pressure, temperature and chemical composition.

tion of the system. If the system under consideration is an isolated system, natural or spontaneous changes are associated with an increase in the total entropy of the isolated system. If the system is a closed system at constant temperature and pressure, the direction of possible change is that accompanied by a decrease of a particular energy function, the Gibbs free energy.

States *a* and *c* are stable with respect to infinitesimally near states of the system. However, state *a* is unstable with respect to a finite change of state to *c*. We may say that state *c* is *absolutely stable* (absolute maximum entropy of an isolated system or absolute minimum Gibbs free energy of a closed, constant temperature and pressure system) for the range of system states considered. State *a* is *metastable* with respect to state *c*; it is stable with respect to *d* and *e*. Obviously, states such as *d*, *e*, *f* and *g* are all unstable, non-equilibrium states. For the mechanical system depicted in Figure 2-4*a*, the ordinate of Figure 2-4*b* may be identified with gravitational potential energy: then positions *1*, *2* and *3* correspond to states *a*, *b* and *c*, respectively.

Thermodynamic equilibria are found to be stable (state *c*) or metastable (state *a*), but not unstable (state *b*); that is, natural processes proceed toward equilibrium states, not away from them. It is necessary to emphasize however that thermodynamics in itself will tell us nothing about the rate of approach to the stable or metastable state. For that, kinetic information is required. *Stability* and unstability are thermodynamic concepts and have nothing to do with rates of change directly. *Spontaneous* changes are natural processes: not rapid processes. The kinetic analogs of stable and unstable might be taken to be *inert* (slowly changing) and *labile* (rapidly changing) systems.

Thermodynamic Variables and Functions

A systematic development of the fundamentals of chemical thermodynamics is not our purpose. We will simply set down in summary fashion the basic definitions and the major generalizations of classical and chemical thermodynamics. Since our treatment is introductory in character, a number of details are not considered here. Some topics are treated in detail in later chapters (e.g., activity coefficients in electrolyte solutions, Chapter 3; electrochemical equilibria, Chapter 7; thermodynamics of surfaces, Chapter 9). For

Table 2-1   Some Basic Thermodynamic Definitions and Relationships

Fundamental Thermodynamic Variables of a System

T	absolute temperature, an intensive state property
S	entropy, an extensive state property
P	pressure, an intensive state property
V	volume, an extensive state property

Transfer of Energy between a System and Its Environment

q	heat absorbed by a system from its environment
w	work done on a system by its environment

Characteristic Thermodynamic Functions of the State of a System

E	internal energy, an extensive state property of a system
H	enthalpy, an extensive state property of a system
F	Helmholtz free energy, an extensive state property of a system
G	Gibbs free energy, an extensive state property of a system

The First and Second Principles of Thermodynamics

The first principle of thermodynamics:

$$dE = dq + dw \tag{1}$$

The second principle of thermodynamics: There exists a function S of the state of a system called entropy. The entropy of a system is the sum of the entropies of its parts.

The total entropy change of a system is the sum of changes inside the system, d_iS, and the flow of entropy between the system and its environment, d_eS

$$dS = d_iS + d_eS \tag{2}$$

$$d_eS = \frac{dq}{T} \tag{3}$$

For a reversible or equilibrium process in a system

$$d_iS = 0 \tag{4}$$

For an irreversible or natural process in a system

$$d_iS > 0 \tag{5}$$

Stated another way, since according to (4) and (5), $d_iS \geq 0$

$$dS \geq \frac{dq}{T} \tag{6}$$

Definitions of Thermodynamic State Functions

Enthalpy	$H = E + PV$	(7)
Helmholtz free energy	$F = E - TS$	(8)
Gibbs free energy	$G = H - TS = F + PV$	(9)

Fundamental Relationships between Thermodynamic State Functions for a Closed System of Fixed Composition in which only Pressure–Volume Work is Involved

$$dE = TdS - PdV \tag{10}$$

$$dH = TdS + VdP \tag{11}$$

$$dF = -SdT - PdV$$

$$dG = -SdT + VdP \tag{12}$$

Criteria for Reversible and Irreversible Processes in Isothermal Systems Involving Only Pressure–Volume Work

A constant temperature and constant volume system

For a reversible process	$dF = 0$	(13)
For an irreversible or natural process	$dF < 0$	(14)
For an impossible process	$dF > 0$	(15)

A constant temperature and constant pressure system

For a reversible process	$dG = 0$	(16)
For an irreversible or natural process	$dG < 0$	(17)
For an impossible process	$dG > 0$	(18)

Additional Useful Relationships Among Thermodynamic Quantities

For an isobaric process in which only pressure–volume work is involved

$$H_2 - H_1 \equiv \Delta H = q \tag{19}$$

The heat capacity, C_P

$$C_P = \left(\frac{\partial H}{\partial T}\right)_P \tag{20}$$

Variation of free energy with pressure

$$\left(\frac{\partial G}{\partial P}\right)_T = V \tag{21}$$

Variation of free energy with temperature

$$\left(\frac{\partial G}{\partial T}\right)_P = -S \tag{22}$$

$$\left(\frac{\partial (G/T)}{\partial T}\right)_P = \frac{H}{T^2} \tag{23}$$

Relationship between Gibbs Free Energy Change and non-Pressure–Volume Work

If a system does not work $-w'$ (other than reversible pressure–volume work) on its surroundings at constant pressure and temperature,

For a reversible process $dw' = dG$ $\hspace{2cm}$ (24)

For a natural or irreversible process $-dw' < -dG$ $\hspace{1cm}$ (25)

---

extensive developments of chemical thermodynamics the reader is referred to such works as those of Guggenheim, Prigogine and Defay, Lewis and Randall, Klotz, Denbigh, and Harned and Owen. A list of selected references is given at the end of this chapter.

The notions of thermodynamic systems and of phases (homogeneous parts of a system) have already been introduced (Section 2-2). Specification of the state of a thermodynamic system requires a certain number of thermodynamic state variables or functions. Table 2-1 lists the defined fundamental thermodynamic variables and functions and the process variables, heat and work (for the present time it is assumed that only pressure–volume work is involved). The entropy, volume, internal energy, enthalpy, Helmholtz free energy and Gibbs free energy are *extensive* state properties, that is, they depend upon the quantity of material (number of moles) in a phase or in a system of several phases. Pressure and temperature (as well as density) are *intensive* state properties (independent of the quantity of material in the system).

The relationship between the internal energy of a system, heat flow to a system and work done on the system constitutes the first principle of thermodynamics [(1) or Table 2-1]. The definitions of the extensive thermodynamic state functions H, F and G are given in (7), (8) and (9), respectively.

Spontaneous Change and Equilibrium

The second principle of thermodynamics is stated in Table 2-1 in terms of the internal entropy production d_iS and the flow of entropy to a system from its surroundings d_eS.

The combined statement of (4) and (5) of Table 2-1

$$d_iS \geq 0 \qquad (19)$$

constitutes the most general statement of the second law in chemical thermo-dynamics. An alternative statement is provided by (6). For an isolated system $dq = 0$. Then (6) reduces to

$$dS \geq 0 \qquad (20)$$

for an adiabatic process. For a natural adiabatic process, $dS > 0$; for an equilibrium (reversible) adiabatic process, $dS = 0$. Comparison with Figure 2-4b should now illustrate a specific interpretation of the criteria for stability and instability in terms of the entropy of an isolated system. For an isolated system, the condition of equilibrium is that S has reached its maximum values consistent with the fixed values of E and V for the system.

Other criteria of equilibrium for specific systems in terms of extensive state properties can be formulated by combining (1), (2), (3), (6) and the appropriate equation from (7), (8) and (9) of Table 2-1. The following alternative condi-tions for equilibrium are obtained:

$$\text{For given } E \text{ and } V, \text{ that } S \text{ is a } maximum \qquad (21a)$$

$$\text{For given } H \text{ and } P, \text{ that } S \text{ is a } maximum \qquad (21b)$$

$$\text{For given } S \text{ and } V, \text{ that } E \text{ is a } minimum \qquad (21c)$$

$$\text{For given } S \text{ and } P, \text{ that } H \text{ is a } minimum \qquad (21d)$$

$$\text{For given } T \text{ and } V, \text{ that } F \text{ is a } minimum \qquad (21e)$$

$$\text{For given } T \text{ and } P, \text{ that } G \text{ is a } minimum \qquad (21f)$$

The last two equilibrium conditions are of special interest to chemists. In particular, constant temperature and pressure conditions are found in many chemical reactions in nature and in the laboratory. Equations 13–15 of Table 2-1 summarize the criteria for reversible and natural processes under constant T and V conditions. The Helmholtz free energy F is then the characteristic function. The criteria for equilibrium and natural processes under isothermal and isobaric conditions are summarized in (16)–(18) of Table 2-1. The Gibbs free energy G is the characteristic function. The derivation of the criteria in terms of G is as follows. From (2) of Table 2-1

$$d_iS = dS - d_eS \tag{22}$$

combining with (3) of Table 2-1

$$d_iS = dS - \frac{dq}{T} \tag{23}$$

The first law is

$$dq = dE - dw = dE + PdV \tag{24}$$

Then, combining (23) and (24)

$$d_iS = dS - (dE + PdV)/T$$
$$= \frac{TdS - dE - PdV}{T} \tag{25}$$

From (7) and (8) of Table 2-1

$$G = E + PV - TS \tag{26}$$

Under constant T and P conditions

$$dG = dE + PdV - TdS \tag{27}$$

Therefore, comparing (25) and (27)

$$d_iS = -\frac{dG}{T} \tag{28}$$

For an equilibrium or reversible process under constant T and P conditions $d_iS = 0$ and $dG = 0$ which is (16) of Table 2-1.

For a spontaneous or natural process at constant T and P $d_iS > 0$ and $dG < 0$ which is (17) of Table 2-1. $d_iS < 0$ or $dG > 0$ corresponds to an impossible process.

The relationships summarized in (1)–(18) of Table 2-1 are generally applicable to physico-chemical changes of all kinds: phase changes, transport processes, chemical reactions and surface phenomena. We will wish to state the appropriate form of the thermodynamic relationships for systems of variable compositions, for example, heterogeneous systems and chemically-reacting systems.

Systems of Variable Composition

Extension of the definitions of the characteristic state functions E, H, F and G to open phases or phases of variable composition is accomplished by adding a term to account for changes in the energy of the system resulting from a change in its chemical content. The term is $\sum_i \mu_i dn_i$, where μ_i is the chemical potential of component i introduced by Gibbs, and dn_i is the infinitesimal change in the number of moles of i. The chemical potential is an intensive property. Equations 1–4 of Table 2-2 define the characteristic extensive state

Table 2-2   Thermodynamic Relationships for Closed Systems Comprising Open Phases of Variable Composition

Fundamental Equations

$$dE = TdS - PdV + \sum_i \mu_i dn_i \qquad (1)$$

$$dH = TdS + VdP + \sum_i \mu_i dn_i \qquad (2)$$

$$dF = -SdT - PdV + \sum_i \mu_i dn_i \qquad (3)$$

$$dG = -SdT + VdP + \sum_i \mu_i dn_i \qquad (4)$$

The Chemical Potential, μ_i

$$\mu_i = \left(\frac{\partial E}{\partial n_i}\right)_{S,V,n_j} = \left(\frac{\partial H}{\partial n_i}\right)_{P,S,n_j} = \left(\frac{\partial F}{\partial n_i}\right)_{T,V,n_j} = \left(\frac{\partial G}{\partial n_i}\right)_{T,P,n_j} \qquad (5)$$

Partial Molar Quantities

$$\left(\frac{\partial G}{\partial n_i}\right)_{T,P,n_j} = \bar{G}_i = \mu_i \qquad (6)$$

$$\left(\frac{\partial V}{\partial n_i}\right)_{T,P,n_j} = \bar{V}_i \qquad (7)$$

$$\left(\frac{\partial S}{\partial n_i}\right)_{T,P,n_j} = \bar{S}_i \qquad (8)$$

$$\left(\frac{\partial H}{\partial n_i}\right)_{T,P,n_j} = \bar{H}_i \qquad (9)$$

Gibbs Free Energy of a Closed Multicomponent System at Constant T and P

   Most generally

$$G = \sum_\alpha \sum_i \mu_i{}^\alpha n_i{}^\alpha \qquad (10)$$

where the summation is carried out for all species and all phases of the system.

Conditions of Isothermal, Isobaric Equilibrium or Spontaneous Change in Closed Multicomponent Systems

Equilibrium

$$dG = -SdT + VdP + \sum_i \mu_i dn_i = 0$$

or
$$\sum_i \mu_i dn_i = 0 \tag{11}$$

or
$$G \text{ is a minimum} \tag{12}$$

Spontaneous change

$$dG = -SdT + VdP + \sum_i \mu_i dn_i < 0$$

or
$$\sum_i \mu_i dn_i < 0 \tag{13}$$

Gibbs-Duhem Relationship for Phases of Variable Composition

$$-SdT + VdP - \sum_i n_i d\mu_i = 0 \tag{14}$$

---

functions for open phases. Equation 5 explicitly defines the chemical potential. Integration of the basic equation for dG for a single phase at constant T and P [(4), Table 2-2] yields

$$G = \sum_i n_i \mu_i \tag{29}$$

which shows explicitly the nature of the Gibbs free energy as an extensive system property. The chemical potential is a partial *molar* quantity, as seen from (5) and (6) of Table 2-2. The symbol $\bar{G}_i$ is sometimes used to denote the partial molar Gibbs free energy of a component. Other pertinent partial molar quantities are defined by (7), (8) and (9) of Table 2-2. For a one-component phase, a pure substance, the chemical potential is equal to the molar Gibbs free energy (6), that is, the partial molar Gibbs energy of a pure substance is identical to the Gibbs energy per mole [6]. For mixtures, the physical meaning of μ_i is the increase in the Gibbs free energy of a phase at constant temperature and pressure on addition of one mole of component i to an infinitely large sample of the phase, the composition thus remaining unchanged. Analogous meanings of the chemical potential are evident for systems with other constraints (constant S and V, constant T and V, etc.).

[6] Experimentally derived values of partial molar Gibbs free energies have been tabulated for many substances and for components in different phases. An example is the compilation of W. M. Latimer, *Oxidation Potentials*, 2nd ed., Prentice-Hall, Englewood Cliffs, N.J., 1952.

For a multicomponent system consisting of one to several phases, for example, system f of Figure 2-2, the change in the Gibbs free energy is obtained by summing (4) of Table 2-2 over all components and all phases:

$$dG = - \sum_{\alpha} S^{\alpha} dT^{\alpha} + \sum_{\alpha} V^{\alpha} dP^{\alpha} + \sum_{\alpha} \sum_{i} \mu_i^{\alpha} dn_i^{\alpha} \qquad (30)$$

Integration for constant temperature and pressure yields the total Gibbs free energy of the system [(10), Table 2-2].

There is a necessary relationship between simultaneous changes of temperature, pressure and the chemical potentials. The relationship is obtained from the total differential of G, (29).

$$dG = \sum_{i} n_i d\mu_i + \sum_{i} \mu_i dn_i \qquad (31)$$

Comparison with (4) of Table 2-2 yields

$$- SdT + VdP - \sum_{i} n_i d\mu_i = 0$$

(14) of Table 2-2, which is known as the Gibbs-Duhem equation. If there are k substances in a particular phase, out of the $k + 2$ intensive variables only $k + 1$ can vary independently.

From the general criteria of equilibrium and spontaneous change stated above in terms of E, H, F and G, it is evident from (1)–(4) of Table 2-2 that the general condition of equilibrium in a closed system of variable composition is

$$\sum_{i} \mu_i dn_i = 0$$

which is (11) of Table 2-2. In writing (11) of Table 2-2 it is assumed that the summation is complete (over all phases as well as over all components). We summarize the conditions for chemical change in constant temperature, constant pressure systems:

At equilibrium: $\quad \sum_{i} \mu_i dn_i = 0, \quad G$ is a minimum

Spontaneous change: $\sum_{i} \mu_i dn_i < 0$

In general: $\quad \sum_{i} \mu dn_i \le 0$

$$(32)$$

Compare these conditions with the generalized energy profile in Figure 2-4*b*.

Phase Equilibria and Transfer

In a system of several phases in thermal equilibrium at constant temperature each phase is an open system: it can exchange chemical components with neighboring phases. For the transfer of a component i from phase β to phase α we have $dn_i^\alpha = -dn_i^\beta$, and $dG \leq 0$ for constant temperature and pressure. Applying (11) and (13) of Table 2-2 we obtain

$$(\mu_i^\alpha - \mu_i^\beta)dn_i^\alpha \leq 0 \tag{33}$$

If $\mu_i^\beta > \mu_i^\alpha$, the direction of spontaneous transfer is from phase β to phase α: dn_i^α and $\mu_i^\beta - \mu_i^\alpha$ have the same sign. Each chemical species i tends to move from a phase in which its chemical potential is higher to one in which it is lower. For $(\mu_i^\alpha - \mu_i^\beta)dn_i^\alpha = 0$ we have the equilibrium state, or $\mu_i^\alpha = \mu_i^\beta$ (equilibrium). The free energy of a system of several phases will be a minimum when the mobile components are distributed so that their chemical potentials are equal in all phases.

Example 2-2.   Consider the two phases of pure calcium carbonate, calcite and aragonite at a pressure of 1 atm and 25°C. The molar Gibbs free energies of the pure phases are

$$\bar{G}_{\text{calcite}} \quad = \mu_{\text{calcite}} \quad = -269.78 \text{ kcal mole}^{-1}$$
$$\bar{G}_{\text{aragonite}} = \mu_{\text{aragonite}} = -269.53 \text{ kcal mole}^{-1}$$

The chemical potentials favor $CaCO_3$ transfer from aragonite to calcite, that is, dn_{calcite} will be positive. One might think that the process

$$CaCO_3(\text{aragonite}) = CaCO_3(\text{calcite})$$

should be treated as a chemical reaction; formally it makes no difference. It should be noted that there is no equilibrium condition for those two pure substances: $\mu_i^\alpha \neq \mu_i^\beta$. The transfer or reaction is spontaneous as long as aragonite remains under the specified temperature and pressure conditions.†

Before treating other and perhaps more interesting cases of phase equilibria we will need to know how to compute the chemical potential of a component in solutions and in gas mixtures. The treatment of chemical reactions in general also requires this information. Therefore we summarize the various definitions of chemical potentials in different kinds of phases before proceeding further with the consideration of equilibrium and irreversible chemical change.

Chemical Potentials in Various Phases

The chemical potential μ_i, of component i in an aqueous solution, solid solution or gas mixture, may generally be written as

$$\mu_i = \mu_i^\circ + RT \ln \{i\} \tag{34}$$

† At other pressures and temperatures it is possible to find $\mu_{\text{calcite}} = \mu_{\text{aragonite}}$ [(21) and (22), Table 2-1].

where μ_i^o is a constant under specified conditions of temperature and pressure and $\{i\}$ is the *activity*. Equation 34, due to G. N. Lewis (1907), is in fact a definition of the activity which may be considered an idealized concentration. The chemical potential may also be expressed by

$$\mu_i = \mu_i^o + RT \ln X_i \gamma_i \tag{35}$$

in which X_i is the mole fraction of component i and γ_i is an activity coefficient for component i. Expressions of the same form as (35) are written with the mole fraction replaced by a partial pressure, a molal concentration or a molar concentration. In each instance the μ_i^o, the γ_i and the concentration scale (partial pressure, mole fraction, molal or molar) are defined consistently with respect to one another. The magnitudes of μ_i^o and $RT \ln X_i \gamma_i$ are arbitrary in the sense that only their sum μ_i is fixed. The meaning of $\{i\}$ in (34) is defined by the designation of a *standard* state and a *reference* state so that

in the *standard* state, $\{i\} = 1$ and $\mu_i = \mu_i^o$

in the *reference* state, $\gamma_i = 1$ and $\{i\} = X_i$, etc.

The choice of standard states and reference states is a matter of convention and convenience for all choices within the framework of (34) and (35) are equally valid from a thermodynamic viewpoint. For extensive treatment of the various problems and conventions associated with activities, activity coefficients and standard states the reader may consult the discussions of Lewis and Randall, Harned and Owen, Robinson and Stokes, and Sillén.

Table 2-3   Definitions of Chemical Potentials in Different Phases

GAS PHASE

An Ideal Gas

$$\mu_i = \mu_i^o + RT \ln P/P^o \tag{1a}$$

if $P^o = 1$

$$\mu_i = \mu_i^o + RT \ln P \tag{1b}$$

μ_i^o is a function of T

A Nonideal Gas

$$\left. \begin{array}{l} \mu_i = \mu_i^o + RT \ln f \\ f/P \to 1 \quad \text{as } P \to 0 \end{array} \right\} \tag{2}$$

f is an idealized pressure called the fugacity

An Ideal Gas Mixture

$$\mu_i = \mu_i^o + RT \ln p_i \tag{3}$$

where p_i is the partial pressure of component i

A Nonideal Gas Mixture

$$\left.\begin{array}{l}\mu_i = \mu_i^\circ + RT \ln f_i \\ f_i/p_i \rightarrow 1 \quad \text{as } P \rightarrow 0\end{array}\right\} \tag{4}$$

f_i is the fugacity of component i

PURE SOLID PHASE

$$\mu_i = \mu_i^\circ \tag{5}$$

where μ_i° is a function of T and P

PURE LIQUID PHASE

$$\mu_i = \mu_i^\circ \tag{6}$$

where μ_i° is a function of T and P

LIQUID OR SOLID SOLUTIONS OF NONELECTROLYTES

An Ideal Solution

$$\mu_i = \mu_i^\circ + RT \ln X_i \tag{7}$$

X_i is the mole fraction of i, μ_i° is a function of T and P

A Nonideal Solution

$$\mu_i = \mu_i^\circ + RT \ln \gamma_i X_i \tag{8}$$

γ_i is an activity coefficient. A convention is required to specify the conditions under which γ_i becomes unity and the solution approaches ideal behavior.

For the component or components whose mole fraction can approach unity in the solution:

$$\left.\begin{array}{l}\mu_i = \mu_i^\circ + RT \ln \gamma_i X_i \\ \gamma_i \rightarrow 1 \quad \text{as } X_i \rightarrow 1\end{array}\right\} \tag{8a}$$

For the solutes:

$$\left.\begin{array}{l}\mu_i = \mu_i^\circ + RT \ln \gamma_i X_i \\ \gamma_i \rightarrow 1 \quad \text{as } X_i \rightarrow 0\end{array}\right\} \tag{8b}$$

The molal and molar concentration scales are commonly used for aqueous solutions. In terms of molal concentration, m_i, for solutes:

$$\left.\begin{array}{l}\mu_i = \mu_i^\circ + RT \ln \gamma_i' m_i \\ \gamma_i' \rightarrow 1 \text{ as } m_i \rightarrow 0\end{array}\right\} \tag{9}$$

In terms of molar concentrations, $[i]$, for solutes:

$$\left.\begin{array}{l}\mu_i = \mu_i^* + RT \ln f_i [i] \\ f_i \rightarrow 1 \quad \text{as } [i] \rightarrow 0\end{array}\right\} \tag{10}$$

In general for solutes $\gamma_i \neq \gamma_i' \neq f_i$ and $\mu_i^\circ \neq \mu_i^\circ \neq \mu_i^\circ$. We usually will not distinguish between the different standard potentials by particular symbols, and will use the general notation μ_i°. The numerical values of the different standard potential scales, concentration scales and the activity coefficient scales are specifically coupled; the scales must be specified in each situation.

SOLUTIONS OF ELECTROLYTES

A Solution of Neutral Electrolyte $M_{\nu_+} A_{\nu_-}$

$$\mu = \nu_+ \mu_+ + \nu_- \mu_- \tag{11}$$

where μ is the chemical potential of the electrolyte, defined by

$$\mu \equiv \left(\frac{\partial G}{\partial m} \right)_{T,P,n_0}$$

where m is the molality of the electrolyte and n_0 the moles of solvent
   The chemical potentials of the ions are defined by

$$\left.\begin{array}{l} \mu_+ = \mu_+^\circ + RT \ln \gamma_+ m_+ \\ \mu_- = \mu_-^\circ + RT \ln \gamma_- m_- \\ \gamma_+, \gamma_- \to 1 \quad \text{as } m \to 0 \end{array}\right\} \tag{12}$$

where the m's and γ's are defined for the individual ions. In terms of the *defined* properties for the individual ions

$$\mu = \nu_+ \mu_+^\circ + \nu_- \mu_-^\circ + RT \ln \gamma_+^{\nu_+} \gamma_-^{\nu_-} m_+^{\nu_+} m_-^{\nu_-} \tag{13}$$

   The mean ion activity coefficient, $\gamma_\pm$, is defined by

$$\gamma_\pm^{\nu} = \gamma_+^{\nu_+} \gamma_-^{\nu_-} \tag{14}$$

where $\nu \equiv \nu_+ + \nu_-$
   The *mean ionic molality*, $m_\pm$, is defined by

$$m_\pm^{\nu} = m_+^{\nu_+} m_-^{\nu_-} \tag{15}$$

   In (13), (14) and (15) the *physically significant* (measurable) quantities are $\nu_+ \mu_+^\circ + \nu_- \mu_-^\circ$, $\gamma_\pm$ and $m_\pm$.

$$\gamma_\pm \to 1 \quad \text{as } m_\pm \to 0 \tag{16}$$

   Equation (13) may also be written

$$\mu = \mu_\pm^\circ + RT \ln \gamma_\pm^{\nu} m_\pm^{\nu} \tag{17}$$

where $\mu_\pm^\circ \equiv \nu_+ \mu_+^\circ + \nu_- \mu_-^\circ$

An Electrolyte in a Mixed Electrolyte Solution

For an electrolyte $M_{\nu_+} A_{\nu_-}$ in solution containing different electrolytes $(NX)_1$, $(NX)_2$, ..., $(NX)_i$

$$\mu = \mu_\pm^\circ + RT \ln \gamma_\pm^{\nu} m_\pm^{\nu}{}_{,MA} \tag{18}$$

Two different conventions can be defined for $\gamma_\pm$ and $\mu_\pm^\circ$. The *infinite dilution convention*:

$$\gamma_\pm > 1 \quad \text{as } m_{\pm,MA} + \sum_i m_{\pm,NX_i} \to 0 \tag{18a}$$

The *constant ionic medium convention*

$$\gamma_\pm \to 1 \quad \text{as } m_{\pm,MA} \to 0 \tag{18b}$$

Individual Ions in Mixed Electrolyte Solutions

$$\mu_i = \mu_i^\circ + RT \ln \gamma_i m_i \tag{19}$$

Infinite Dilution Convention

$$\gamma_i \to 1 \quad \text{as } I \to 0 \tag{19a}$$

where $I = \frac{1}{2} \sum Z_i^2 C_i \simeq \frac{1}{2} \sum Z_i^2 m_i$, C_i being the molar concentration and Z_i the charge number of each ion i, the summation being taken over all ions in the solution.

Constant Ionic Medium Convention

$$\gamma_i \to 1 \quad \text{as } m_i \to 0 \tag{19b}$$

Table 2-3 is a summary of some of the more widely used definitions of chemical potentials for gases, solids, liquids, solutions of nonelectrolytes and solutions of electrolytes. The contents of the table are for the most part self-explanatory. In summarizing the definitions and conventions for electrolyte solutions we have usually referred to the molal scale. Similar statements apply for the molar scale. For pure solids, liquids and solutions we have indicated μ_i° as a function of T and P. This is again a matter of convention and will not affect the use of the conventional standard state Gibbs free energy data, which are for a standard pressure of 1 atm. Lewis and Randall [7] write

$$\mu_i = \mu_i^\circ + RT \ln \Gamma_i + RT \ln X_i \gamma_i \tag{36}$$

and incorporate the effect of pressure in Γ_i which they call the activity of the reference state. The use of Γ_i allows a distinction to be made between effects of pressure on γ_i and on the "standard" term chemical potential.

Example 2-3.   Consider a two-phase system such as that shown in b of Figure 2-2. The partial pressure of CO_2 in the gas phase is 0.001 atm; the concentration of molecular CO_2 in the aqueous phase is 1.0×10^{-5} M. If the temperature is 25°C and the total pressure 1 atm, is the system at equilibrium with respect to CO_2 transfer between the two phases?

The standard chemical potential of CO_2 in the gas phase is -94.26 kcal mole^{-1}. The standard value in aqueous solution is -92.31 kcal mole^{-1}. We can compute $\mu_{CO_2}^{gas}$ from (3) of Table 2-3 and $\mu_{CO_2}^{aq}$ from (10) of the same table. Assuming ideal behavior, we have

$$\mu_{CO_2}^{gas} = \mu_{CO_2}^{gas,0} + RT \ln p_{CO_2}$$
$$= -94.26 + 1.364 \log (0.001)$$
$$= -98.35 \text{ kcal mole}^{-1}$$

and

$$\mu_{CO_2}^{aq} = \mu_{CO_2}^{aq,0} + RT \ln [CO_2(aq)]$$
$$= -92.31 + 1.364 \log (1.0 \times 10^{-5})$$
$$= -99.13 \text{ kcal mole}^{-1}$$

---

[7] G. N. Lewis and M. Randall, *Thermodynamics*, revised by K. S. Pitzer and L. Brewer, McGraw-Hill, New York, 1961.

The chemical potential of CO_2 is greater in the gas phase than in the aqueous phase:

$$\mu_{CO_2}^{g} - \mu_{CO_2}^{aq} = -98.35 - (-99.13) = 0.78 \text{ kcal mole}^{-1}$$

The direction of spontaneous CO_2 transfer is from the gas to the aqueous phase.

Chemical Reactions and Equilibria

We have already stated the general condition for chemical equilibrium in an isothermal, isobaric system:

$$\sum_i \mu_i dn_i = 0$$

which was obtained from the basic condition for equilibrium in terms of the internal entropy production d_iS:

$$d_iS = 0$$

We also have the result that, for a spontaneous or natural process,

$$d_iS > 0$$

or

$$\sum_i \mu_i dn_i < 0$$

Now we apply these results to the particular case of chemical reactions.

A single chemical reaction can be represented as

$$\nu_A A + \nu_B B + \cdots = \nu_C C + \nu_D D + \cdots \tag{1}$$

where the ν's are the stoichiometric coefficients (relative molar numbers) of the reactants and products of the reaction and A, B, ..., C, D, ... stand for the chemical symbols (molecular weights) of the reactants and products. More concisely, the reaction can be written as

$$\sum_i \nu_i M_i = 0 \tag{37}$$

where M_i is the molecular weight of component i and ν_i is the stoichiometric coefficient of i. The ν_i are positive for products and negative for reactants. The change in the Gibbs free energy resulting from changes in the number of moles of each component dn_i as a result of chemical reaction is

$$dG = \sum_i \mu_i dn_i \tag{38}$$

where the summation is taken over all components and all phases.

We introduce a variable called the extent of the reaction, ξ, which is defined by

$$dn_i = v_i d\xi \tag{39}$$

and

$$n_i = n_i^\circ + v_i \xi \tag{40}$$

where n_i is the number of moles of i at any extent of reaction and n_i° is the number of moles of i under initial conditions ($\xi = 0$). The units of ξ are *moles*; the v_i are dimensionless numbers. Note that ξ and v_i are arbitrary; the product of the two is physically meaningful for a reaction in a defined system (specified chemical content in terms of moles).

The total free energy of the system in which the chemical reaction occurs is

$$G = \sum_i n_i \mu_i = \sum_i (n_i^\circ + v_i \xi) \mu_i \tag{41}$$

From (38) the change in the Gibbs energy is, in terms of the extent of reaction

$$dG = \sum_i v_i \mu_i d\xi \tag{42}$$

The internal entropy production is then given by

$$d_i S = -\frac{dG}{T} = -\frac{1}{T} \sum_i v_i \mu_i d\xi \tag{43}$$

The quantity $\sum_i v_i \mu_i$ is called the *free energy change of the reaction* ΔG by Lewis and Randall. The quantity $-\sum_i v_i \mu_i$ is called the *affinity* A by de Donder. Thus

$$\Delta G = -A = \sum_i v_i \mu_i \tag{44}$$

From (42)

$$\frac{dG}{d\xi} = \Delta G = -A \tag{45}$$

The free energy of the reaction is the rate of change of the Gibbs free energy of the system with respect to the extent of the reaction. ΔG and A are intensive state variables of a chemical reaction system.

From (43) and (44)

$$d_i S = -\frac{\Delta G}{T} d\xi = \frac{A}{T} d\xi \tag{46}$$

Therefore at equilibrium for the reaction $d_i S = 0$, $\Delta G = 0$ or $A = 0$; the rate of change of the Gibbs energy is zero, and the system free energy is a minimum at equilibrium. For a spontaneous reaction $d_i S > 0$ and $\Delta G < 0$ or $A > 0$.

Summarizing, for a chemical reaction $\sum_i v_i M_i = 0$

equilibrium:
$$d_i S = 0;\ \Delta G = -A = \sum_i v_i \mu_i = <0$$

$$\tag{47}$$

spontaneous reaction: $d_i S > 0;\ \Delta G = \sum_i v_i \mu_i < 0;\ A = -\sum_i v_i \mu_i > 0$

It is useful to distinguish immediately between chemical reactions involving pure substances only (solids and liquids) and chemical reactions involving mixtures (gases, liquid solutions, solid solutions). For the former type of

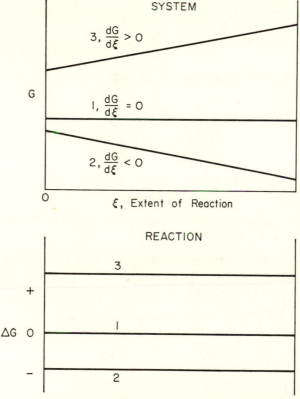

Figure 2-5 Variation of the Gibbs free energy function G and the Gibbs free energy change of the reaction ΔG with extent of reaction for reactions between pure substances, for example, solids or solids and liquids.

reaction $\mu_i = \mu_i^o(P,T)$, and at given temperature and pressure application of (47) leads to a particularly simple result: $\Delta G = \sum \nu_i \mu_i^o$ is constant for any extent of reaction at fixed T and P. The reaction is either at equilibrium ($\Delta G = 0$) or spontaneous in one direction. Figure 2-5 illustrates the variation of the Gibbs energy G and the free energy of reaction ΔG with the extent of reaction for a chemical reaction between pure substances. For case 1, $\Delta G = 0$ and equilibrium exists for any relative amounts of products and reactants. For case 2, $\Delta G < 0$ and the reactants form products spontaneously. For case 3, $\Delta G > 0$ and conversion of reactants to products is impossible; the reverse reaction is spontaneous.

Example 2-4.   What is the stability relationship for the reaction involving the oxides pyrolusite, magnetite, manganite and hematite

$$2MnO_2(s) + 2Fe_3O_4(s) = Mn_2O_3(s) + 3Fe_2O_3(s)$$

at 25°C and 1 atm?

$$\Delta G = \sum_i \nu_i \mu_i = \sum_i \nu_i \mu_i^o = \sum_i \nu_i \overline{G}_{f,i}^o$$

The following free energy data are available [6]:

	$\overline{G}_f^o$, kcal mole^{-1}
$MnO_2(s)$	−111.1
$Mn_2O_3(s)$	−212.3
$Fe_3O_4(s)$	−242.4
$Fe_2O_3(s)$	−177.1

$$\Delta G = -2(-111.1) - 2(-242.4) + 1(-212.3) + 3(-177.1)$$
$$\Delta G = -36.6 \text{ kcal mole}^{-1}$$

Pyrolusite and magnetite are unstable with respect to manganite and hematite. The reaction is type 2, Figure 2-5.

Example 2-5.   Which is the stable solid phase at 25°C and 1 atm, $Mn_2O_3(s)$ or $Mn(OH)_3(s)$?

A reaction relating the solids is

$$Mn_2O_3(s) + 3H_2O = 2Mn(OH)_3(s)$$

	$\overline{G}_f^o$, kcal mole^{-1}
$Mn_2O_3(s)$	−212.3
$Mn(OH)_3(s)$	−181.0
H_2O	−56.690

$$\Delta G = -1(-212.3) - 3(-56.690) + 2(-181.0)$$
$$\Delta G = +20.4 \text{ kcal mole}^{-1}$$

Mn_2O_3 is the stable phase. This system corresponds to type 3 of Figure 2-5.

THE EQUILIBRIUM CONSTANT. The relationship between the Gibbs free energy of a reaction and the composition of the system is obtained by substituting the expression for the chemical potential in terms of activity

$$\mu_i = \mu_i^\circ + RT \ln \{i\} \tag{34}$$

into the expression for ΔG

$$\Delta G = \sum_i \nu_i \mu_i \tag{44}$$

The direct result is

$$\Delta G = \sum_i \nu_i \mu_i^\circ + RT \sum_i \nu_i \ln \{i\} \tag{48}$$

or

$$\Delta G = \Delta G^\circ + RT \ln \prod_i \{i\}^{\nu_i} \tag{49}$$

where

$$\Delta G^\circ = \sum_i \nu_i \mu_i^\circ \tag{50}$$

is the standard state Gibbs free energy of the reaction. In terms of (1), (49) becomes

$$\Delta G = \Delta G^\circ + RT \ln \frac{\{C\}^{\nu_C}\{D\}^{\nu_D} \cdots}{\{A\}^{\nu_A}\{B\}^{\nu_B} \cdots} \tag{51}$$

The reaction quotient is defined as

$$Q = \prod_i \{i\}^{\nu_i} = \frac{\{C\}^{\nu_C}\{D\}^{\nu_D} \cdots}{\{A\}^{\nu_A}\{B\}^{\nu_B} \cdots} \tag{52}$$

At equilibrium $\Delta G = 0$ and the numerical value of Q is K, the equilibrium constant. Then from (51)

$$\Delta G^\circ = -RT \ln K \tag{53}$$

Under any conditions

$$\Delta G = RT \ln \frac{Q}{K} \tag{54}$$

and the affinity is given by

$$A = RT \ln \frac{K}{Q} \tag{55}$$

Note the analogous character of (55) and (12), which was developed from the point of view of chemical kinetics. According to (55), when $K/Q > 1$ there is an energetic driving force for an irreversible process approaching the equilibrium state, which is characterized by $K/Q = 1$ and $A = 0$. According to (12), when $K/Q > 1$, $v_f/v_b > 1$ and $v_{\text{net}} > 0$, and the system is proceeding to the equilibrium state. Note also that neither (55) nor (12) gives the *rate* of the reaction. The specific kinetic information is contained in the forward and backward rate constants themselves, but not in their ratio k_f/k_b, the equilibrium constant.

To conclude our discussion of the equilibrium constant and homogeneous phases of variable composition we mention the variation of the Gibbs free energy of reacting mixtures with the extent of the reaction. Substitution of (34) into (41) yields

$$G = \sum_i (n_i^\circ + v_i\xi)(\mu_i^\circ + RT \ln \{i\}) \qquad (56)$$

The Gibbs free energy versus ξ for reactions involving solutions and gas mixtures is characterized by a minimum value at the equilibrium composition. We have already seen that the minimum value of G corresponds to $\Delta G = 0$, ΔG being the derivative of G with respect to ξ. The $(n_i^\circ + v_i\xi)RT \ln \{i\}$ terms in (56) account for the free energy of mixing. For reactions between pure

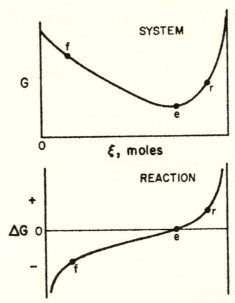

Figure 2-6 Variation of the Gibbs free energy function G and the Gibbs free energy change of the reaction ΔG for a single reaction in a system of variable composition, for example, heterogeneous reactions or reactions in solution.

substances (Figure 2-5) these terms are absent and there is no minimum in the G function. Figure 2-6 depicts generalized relationships between the G function, the Gibbs free energy of reaction and the extent of reaction for a single chemical reaction mixture (compare with Figures 2-4b and 2-5). Point e corresponds to the equilibrium state for the reaction. Point f represents a nonequilibrium state from which the system may proceed to equilibrium spontaneously, the reaction proceeding from left to right as written. Point r represents a state in which the reaction is spontaneous in the reverse direction (right to left as written).

Example 2-6.   Analyze the system previously considered in Example 2-1 in terms of the Gibbs energy function, the Gibbs free energy of the reaction and the related enthalpy and entropy terms.

The reaction (25°C, 1 atm) is

$$CO_2(aq) + H_2O = HCO_3^- + H^+$$

The initial ($\xi = 0$) composition per liter of solution is (neglecting the ionization of water)

$$n_{CO_2(aq)} = 1 \times 10^{-5} \text{ mole}$$
$$n_{H_2O} = 55.4 \text{ moles}$$
$$n_{HCO_3^-} = 0 \text{ mole}$$
$$n_{H^+} = 0 \text{ mole}$$

As in Example 2-1, we neglect computation of the hydration of $CO_2(aq)$ and we treat CO_2 as nonvolatile for purposes of the example (system a, Figure 2-2). The stoichiometric coefficients are: $CO_2(aq)$, -1; H_2O, -1; HCO_3^-, $+1$; and H^+, $+1$. The standard partial molar free energies of formation, standard partial molar enthalpies of formation and standard partial molar entropies for the reactants and products are as follows (from Latimer [6]):

Species	$\bar{G}_f^\circ$, kcal mole^{-1}	$\bar{H}_f^\circ$, kcal mole^{-1}	$\bar{S}^\circ$, cal mole^{-1} deg^{-1}
$CO_2(aq)$	-92.31	-98.69	29.0
$H_2O(l)$	-56.690	-68.317	16.716
$HCO_3^-(aq)$	-140.31	-165.18	22.7
$H^+(aq)$	0.0	0.0	0.0

The extent of reaction ξ for this system is numerically equal to the number of moles of HCO_3^- in solution. When $\xi = 1 \times 10^{-5}$ moles the reactants are completely converted to products. The Gibbs free energy of the solution may be computed by application of (56), which gives

$$G = n_{CO_2}(\mu_{CO_2}^\circ + RT \ln \{CO_2\}) + n_{H_2O}(\mu_{H_2O}^\circ + RT \ln \{H_2O\})$$
$$+ n_{HCO_3^-}(\mu_{HCO_3^-}^\circ + RT \ln \{HCO_3^-\}) + n_{H^+}(\mu_{H^+}^\circ + RT \ln \{H^+\})$$

The computation of G for a reaction mixture must be in accord with the Gibbs-Duhem relationship (Table 2-2, equation 14). In general, the mole fraction scale is the logical choice for the expression of activities in order to satisfy the Gibbs-Duhem relationship straightforwardly. When $\sum \nu_i$ for a chemical reaction in a mixture is not zero, the total number of moles in the mixture is a function of the extent of reaction. If *concentrations* are used to express activities, the computation of G requires a correction term whenever $\sum \nu_i$ is not zero. It can be shown that for *dilute* solutions or constant ionic media

$$G \cong \text{constant} + \sum [i]\mu_i^* + RT \sum [i] \ln [i] - RT \sum [i]$$

Alternatively, if activities are expressed in mole fractions and the corresponding standard state potentials for a mole fraction scale, μ_i°, are used, then

$$G = \sum [i]\mu_i^\circ + RT \sum [i] \ln X_i$$

In our example, the mole fractions can be equated with the activities since the solution is quite dilute. We have then, for $CO_2(aq)$

$$\{CO_2\} = X_{CO_2} = \frac{n_{CO_2}}{n_{H_2O} + n_{CO_2} + n_{HCO_3^-} + n_{H^+}}$$

Similar expressions obtain for the mole fractions of H_2O, HCO_3^- and H^+. The standard state chemical potentials ($\bar{G}_f^\circ$) for $CO_2(aq)$, HCO_3^- and H^+ are given for a molal (approximately the same as molar) scale. The corresponding values on a mole fraction scale are obtained by adding the quantity $RT \ln 55.4$ to each of the tabulated molal-scale values.

We have $n_{CO_2} = 1 \times 10^{-5} - \xi$, $n_{H_2O} = 55.4 - \xi$, $n_{HCO_3} = \xi$ and $n_{H^+} = \xi$. Substitution of various values for ξ leads to the computed Gibbs function for the reaction system shown in Figure 2-7a, where G is plotted with reference to the initial free energy, G_0. Analogous computations have been made for the system enthalpy, $H - H_0$, and the product of the system entropy and the absolute temperature, $T(S - S_0)$. These functions also are shown in Figure 2-7a. Recalling that for a system of constant temperature and pressure the equilibrium condition in terms of the system free energy is $dG/d\xi = 0$, we find that the equilibrium extent of reaction is 1.8 μmoles per liter, that is, $1.8 \, \mu M \, HCO_3^-$. The $CO_2(aq)$ concentration at equilibrium is $1 \times 10^{-5} - \xi_{equil}$, or $8.2 \, \mu M$.

In proceeding from the initial state to the equilibrium state ($dG/d\xi < 0$ for $0 < \xi < \xi_{equil}$) the closed constant-temperature system absorbs 3.3 millicalories of heat per liter of solution from its surroundings (e.g., a heat reservoir). The entropy of the closed system increases by 0.019 mcal deg^{-1} and the Gibbs free energy of the closed system decreases by 2.4 mcal $liter^{-1}$. In terms of the *total isolated system* consisting of the *solution plus surroundings*, the equilibrium state represents a state of maximum entropy increase from the

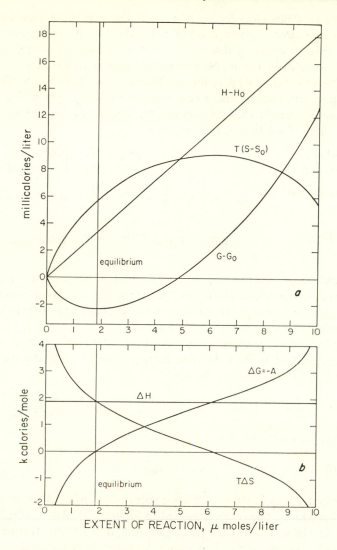

Figure 2-7   *a*: Gibbs function, system enthalpy and system entropy variations for the reaction $CO_2(aq) + H_2O = HCO_3^- + H$ at 25°C in a closed aqueous system at various extents of reaction. The total concentration, $C_T = [CO_2(aq)] + [HCO_3^-]$, is 1×10^{-5} M. $CO_2(aq)$ is assumed to be nonvolatile. The extent of reaction is numerically equal to the concentration of HCO_3^- in solution. Equilibrium is at the minimum value of G and corresponds to 1.8 μM HCO_3^-. *b*: Gibbs free energy change ΔG, enthalpy change ΔH and entropy change expressed as $T\Delta S$ for the reaction $CO_2(aq) + H_2O = HCO_3^- + H^+$. At equilibrium $\Delta G = 0$. ΔH is independent of composition for the dilute solution involved.

initial state of 0.008 mcal deg^{-1}. Note that the maximum increase in the entropy of the closed system itself ($\xi \simeq 6 \mu M$) does not represent equilibrium: the entropy of the system itself continues to increase for compositions beyond the equilibrium state. Note also that there is an increase in enthalpy in going from the initial state to the equilibrium state. The driving force is an increase in system entropy and a consequent decrease in system free energy. We may summarize the changes between the initial state and the equilibrium state for 1 liter of solution as follows:

heat absorbed, $\qquad\qquad\qquad\qquad q = H_{eq} - H_0 = 3.3$ mcal

entropy change, $\qquad\qquad\qquad\qquad\quad S_{eq} - S_0 = 0.019$ mcal deg^{-1}

environmental entropy change, $\qquad (S_{eq} - S_0)_e = \dfrac{q}{T} = \dfrac{H_{eq} - H_0}{T} = 0.011$ mcal deg^{-1}

internal entropy change, $\qquad\qquad (S_{eq} - S_0)_i = \dfrac{-(G - G_0)}{T} = 0.008$ mcal deg^{-1}

If we start with pure products — HCO_3^- and H^+ in aqueous solution — the approach to equilibrium is favored both by a decrease in the enthalpy and an increase in the entropy: the decrease in free energy in going from products to the equilibrium state is 15.1 mcal liter^{-1}.

Having looked at the system in terms of the total Gibbs energy, enthalpy and entropy we turn to the Gibbs free energy of the reaction and the affinity. Applying (51) to this example, we have

$$\Delta G = \mu^{\circ}_{HCO_3^-} + \mu^{\circ}_{H^+} - \mu^{\circ}_{CO_2} - \mu^{\circ}_{H_2O} + RT \ln \frac{\{HCO_3^-\}\{H^+\}}{\{CO_2\}\{H_2O\}}$$

or for a liter of solution

$$\Delta G = \mu^{\circ}_{HCO_3^-} + \mu^{\circ}_{H^+} - \mu^{\circ}_{CO_2} - \mu^{\circ}_{H_2O} + RT \ln \frac{(\xi)(\xi)}{(1 \times 10^{-5} - \xi)}$$

ΔG is computed as a function of ξ upon substitution of the $\bar{G}_f^{\circ}$'s for the reactants and products. The result is plotted in Figure 2-7b. Also plotted are ΔH and $T\Delta S$ for the reaction. The equilibrium state is at ΔG (or A) equal to zero [(47), above]. As $\Delta G = dG/d\xi$, we note the expected correspondence of the minimum in the $G - G_0$ function and the zero point of ΔG. Recalling the definition of G, we note that $\Delta H = T\Delta S$ at equilibrium ($1.8 \mu M$ HCO_3^-). The standard free energy change of the reaction, ΔG°, is

$$\Delta G^{\circ} = \sum_i \nu_i \mu_i^{\circ} = \sum_i \nu_i \bar{G}_{f,i}^{\circ}$$
$$= 1(-140.31) + 1(0.0) - 1(-92.31) - 1(56.690)$$
$$= 8.69 \text{ kcal mole}^{-1}$$

The equilibrium constant is computed from (53)

$$\ln K = \frac{-\Delta G°}{RT}$$

$$\log K = \frac{-8.69}{1.364} = -6.36$$

The equilibrium composition (ξ) is readily computed from

$$K = \frac{\{HCO_3^-\}\{H^+\}}{\{CO_2\}\{H_2O\}} \simeq \frac{[HCO_3^-][H^+]}{[CO_2]}$$

or

$$K \simeq \frac{(\xi)(\xi)}{1 \times 10^{-5} - \xi}$$

The result is of course

$$\xi = 1.8 \times 10^{-6} \, M$$

Simultaneous Reactions

It is rather seldom that we are interested in only one chemical reaction or transfer process in a natural water system. We usually deal with simultaneous reactions. The arguments for a single chemical reaction are easily extended to the case of several simultaneous reactions in the same system. If there are r chemical reactions or other physico-chemical processes in a constant-T, constant-P system, the complete equilibrium state is defined by

$$\Delta G_1 = 0, \Delta G_2 = 0, \ldots, \Delta G_r = 0 \tag{57a}$$

or by

$$A_1 = 0, A_2 = 0, \ldots, A_r = 0 \tag{57b}$$

In terms of the Gibbs function of the system, G, the complete equilibrium condition is

$$\frac{\partial G}{\partial \xi_1} = 0, \frac{\partial G}{\partial \xi_2} = 0, \ldots, \frac{\partial G}{\partial \xi_r} = 0 \tag{58}$$

Conditions (57) and (58) are equivalent of course.

For a system of simultaneous reactions not at complete equilibrium the total internal entropy production $d_i S$ is positive, or $\sum_i \Delta G < 0$. Two distinct cases may be considered. First, there is the straightforward case where none of the reactions in the system have any common reactants or products.

Obviously, all reactions in such a system can be treated completely independently since there is no interlocking or coupling. The total system is a superposition of the individual reactions. The second case is where there are common reactants and products among one or more reactions. Then the direction of the actual change for a particular reaction may not be dictated by its own ΔG value alone. For the independent reactions 1 and 2

$$\sum_i \nu_{i,1} M_i = 0$$

and

$$\sum_i \nu_{i,2} M_i = 0$$

where at least one reactant, i, is common to 1 and 2, both reactions can proceed spontaneously ($v_1 > 0$ and $v_2 > 0$) when $\Delta G_1 > 0$ and $\Delta G_2 < 0$, if $\Delta G_1 + \Delta G_2 < 0$. This simply means that *over-all* reactions must be considered in deciding upon the direction of spontaneous change for each independent reaction when there are interlocking constituents. This is the typical situation in natural water systems. Hydrogen ion and hydroxide ion, oxygen, electrons, bicarbonate ion and several metal ions are among the major species involved in quite a large number of independent reactions in aquatic systems. Hydrogen ions and electrons in particular are important "master variables" in the description of aquatic equilibria.

Example 2-7a.   A constant ionic medium solution contains $10^{-4.0}$ M free CO_3^{2-} ion, $10^{-2.3}$ M free Ca^{2+} ion, and $10^{-5.3}$ M of the Ca^{2+} and CO_3^{2-} ion pair, $CaCO_3(aq)$. What is the state of the system with respect to the following two reactions?

$$CaCO_3(s) = Ca^{2+} + CO_3^{2-} \tag{i}$$
$$Ca^{2+} + CO_3^{2-} = CaCO_3(aq) \tag{ii}$$

The equilibrium constant (25°C) for (i) is $10^{-7.0}$ and for (ii) $10^{2.0}$.
  For (i)

$$\Delta G = RT \cdot \ln \frac{Q}{K} = 1.364 \log \frac{10^{-6.3}}{10^{-7.0}} = 0.95 \text{ kcal mole}^{-1}$$

  For (ii)

$$\Delta G = 1.364 \log \frac{10^{1.0}}{10^{2.0}} = -1.36 \text{ kcal mole}^{-1}$$

  For the sum of (i) and (ii)

$$CaCO_3(s) = CaCO_3(aq)$$
$$\Delta G = 0.95 - 1.36 = -0.41 \text{ kcal mole}^{-1}$$

Reactions i and ii would both proceed to the right under the initial solution conditions.

Example 2-7b.   At $25°C$, $10 \mu M$ concentrations of $CO_2(aq)$, $NaHCO_3$, HCl, $NH_3(aq)$ and NH_4Cl are mixed instantaneously in aqueous solution. In what directions do the reactions

$$CO_2(aq) + H_2O = HCO_3^- + H^+ \qquad \text{(i)}$$

and

$$NH_3(aq) + H^+ = NH_4^+ \qquad \text{(ii)}$$

proceed initially? The applicable equilibrium constants are $10^{-6.3}$ for (i) and $10^{9.3}$ for (ii).

For (i) we have

$$\Delta G = 1.364 \log \frac{Q}{K} = 1.364 \log \frac{(10^{-5})(10^{-5})/(10^{-5})}{10^{-6.3}} = 1.77 \text{ kcal mole}^{-1}$$

For (ii) we have

$$\Delta G = 1.364 \log \frac{(10^{-5})/(10^{-5})(10^{-5})}{10^{9.3}} = -5.86 \text{ kcal mole}^{-1}$$

The sum of (i) and (ii)

$$CO_2(aq) + NH_3(aq) + H_2O = HCO_3^- + NH_4^+$$

has a reaction free energy of -4.09 kcal mole^{-1}. Reactions i and ii both proceed to the right under the initial conditions.

Driving Force for Chemical Change

In an individual reaction or in a system of interlocking reactions the driving force for chemical change is a negative value of the Gibbs free energy reaction ΔG. The free energy change is composed of an enthalpy and an entropy contribution

$$\Delta G = \Delta H - T\Delta S \qquad (59)$$

or for standard state conditions

$$\Delta G° = \Delta H° - T\Delta S° \qquad (60)$$

The driving force for a reaction (stability of the products with respect to reactants) can be the result of a negative ΔH, a positive ΔS or both. The condition for equilibrium is that $\Delta G = \Delta H - T\Delta S = 0$ or

$$\Delta H = T\Delta S \qquad (61)$$

A particular example of this condition was noted in Example 2-6, Figure 2-7*b*. The magnitude and direction of the driving force for a reaction obviously depend upon the magnitude and sign of the enthalpy change, the magnitude and sign of the entropy change, and the temperature. For $\Delta H < 0$ and $\Delta S < 0$ as well as for $\Delta H > 0$ and $\Delta S > 0$ the possibility of spontaneous reaction depends upon T.

The driving force for the ionization of aqueous CO_2 (Example 2-6) was found to be the result of an increase in entropy: $T\Delta S > \Delta H$ at 25°C. Under standard state conditions at 25°C the process

$$CO_2(g) = CO_2(aq)$$

is favored by an enthalpy decrease: $\Delta H° = -4.64$ kcal mole^{-1}. However, the decrease in entropy results in $T\Delta S° = -6.66$ and the standard free energy change is $+1.94$ kcal mole^{-1}. The small stability difference in favor of calcite with respect to aragonite

$$CaCO_3(\text{aragonite}) = CaCO_3(\text{calcite}), \quad \Delta G° = -0.25 \text{ kcal mole}^{-1}$$

is associated with an entropy increase of 1 cal mole^{-1} deg^{-1}.

The precipitation of barium sulfate

$$Ba^{2+} + SO_4^{2-} = BaSO_4(s)$$

is characterized by the following thermodynamic properties (kcal mole^{-1}, 25°C):

$$\Delta G° = -12.1, \quad \Delta H° = -4.6, \quad T\Delta S° = +7.5$$

For the precipitation of ferric phosphate

$$Fe^{3+} + PO_4^{3-} = FePO_4(s)$$
$$\Delta G° = -24.4, \quad \Delta H° = +18.7, \quad T\Delta S° = 43.1$$

The reaction is endothermic. The precipitation of ferric hydroxide is favored both by $T\Delta S° = 30.1$ kcal mole^{-1} and by $\Delta H° = -20.6$ kcal mole^{-1}, with a resulting $\Delta G°$ of -50.7 kcal mole^{-1} (25°C). Large standard entropy changes in the precipitation of a metal ion are associated with increase in the randomness of water (decreased aquation of metal ions).

Ion association reactions and chelation reactions of aqueous metal ions are generally characterized by significant entropy increases (decreased orientation of solvent molecules and configurational entropy). For example the ion-pair reaction [8]

$$Co(NH_3)_5H_2O^{3+} + SO_4^{2-} = Co(NH_3)_5(H_2O)SO_4^{1+}$$

[8] G. Anderegg, *Chimia*, **22**, 477 (1968).

has $\Delta H° = 0$ kcal mole^{-1} and $\Delta S° = +16.4$ cal mole^{-1} deg^{-1} at 25°C. The association reaction

$$Cu^{2+} + P_3O_{10}^{5-} = CuP_3O_{10}^{3-}$$

has $\Delta H = +4.9$ kcal mole^{-1} and $\Delta S = +59$ cal mole^{-1} deg^{-1} with a resulting ΔG of -12.48 kcal mole^{-1} (20°C, 0.1 M constant ionic medium).

The chelation reactions of Ca^{2+} and Mg^{2+} with ethylenediamine tetraacetate (Y^{4-}) have the following thermodynamic properties (0.1 M constant ionic medium, 25°C):

$$Ca^{2+} + Y^{4-} = CaY^{2-}$$
$$\Delta H = -8, \quad \Delta S = 22, \quad \Delta G = -14.2, \quad \log K = 10.4$$

$$Mg^{2+} + Y^{4-} = MgY^{2-}$$
$$\Delta H = 2, \quad \Delta S = 47, \quad \Delta G = -11.8, \quad \log K = 8.6$$

The binding of aqueous protons to the anions of weak acids is generally favored by sizeable entropy increases. For example the reaction

$$SO_4^{2-} + H^+ = HSO_4^-$$

has $\Delta S° = 26.3$ cal deg^{-1} mole^{-1} and $\Delta H° = 5.2$ kcal mole^{-1}.

It is clear that the enthalpy and entropy changes of reactions in aqueous systems vary greatly. The enthalpy change does not serve as a criterion of spontaneous reaction. Free energy changes provide the only general description of the driving force of reactions.

2-5 Thermodynamic Data

The basic data needed for equilibrium calculations and for ascertaining the direction of spontaneous reaction are the partial molar or molar free energies of formation of substances under well-defined conditions. The partial molar free energy of formation $\bar{G}_f°$ is the free energy change accompanying the formation of a substance from the elements in their standard states. For ions in aqueous solutions the free energies of formation are based on the arbitrary assignment of zero values to $\bar{G}_f°$, $\bar{H}_f°$ and $\bar{S}°$ for the aqueous hydrogen ion. The partial molar free energy of formation at a given temperature is equal to the standard chemical potential at the same temperature and 1 atm pressure:

$$\mu_i°(T, 1\ atm) = \bar{G}_{f,i}°(T, 1\ atm)$$

The effects of temperature and pressure on the chemical potential are described by (22) and (21) of Table 2-1.

A rather complete thermodynamic description of each substance involved in a chemical reaction, enabling its chemical behavior to be described over a wide range of conditions, would include the following: $\bar{G}_f^\circ$, $\bar{H}_f^\circ$, $\bar{S}^\circ$, C_p° (the heat capacity) and $\bar{V}^\circ$ (the partial molar volume). These thermodynamic data permit calculation or estimation of equilibrium constants for reactions under a variety of conditions:

$$\Delta G^\circ = \sum_i \nu_i \mu_i^\circ \qquad (50)$$

$$RT \ln K = -\Delta G^\circ \qquad (53)$$

Extensive, critical compilations of thermodynamic data for substances are available. Two widely-used sources are Latimer's text [6] and the National Bureau of Standards Circular 500 [9]. Thermodynamic data have been obtained in a variety of ways: calorimetry, chemical measurements on reactions at equilibrium, electromotive force measurements of galvanic cells, calculations of the entropy based on the third law of thermodynamics and estimations of ion entropies. Thus, through application of the basic relationship

$$\Delta G^\circ = \Delta H^\circ - T\Delta S^\circ \qquad (60)$$

standard free energies, hence equilibrium constants, can be calculated for many chemical reactions which have not been studied directly. On the other hand, many enthalpy and entropy values have been obtained from chemical equilibrium measurements rather than from thermal data. The reader should consult the extensive treatments on the subject of thermodynamic data and their interrelationships, for example Latimer [6], Lewis and Randall [7] and Klotz [10].

Stability constants of metal–ion complexes, dissociation constants of acids, solubility products of solids and equilibrium constants or potentials for oxidation–reduction reactions have been compiled. The most extensive collection of equilibrium constant data presently available is *Stability Constants of Metal–Ion Complexes*, compiled by Sillén and Martell [11]. It contains separate compilations of data on inorganic and organic ligands. Generally speaking, the equilibrium constants found in the chemical literature are defined in terms of either the infinite dilution reference state or the constant ionic medium reference state (see Table 2-3). Both kinds of equilibrium constant are equally valid from a thermodynamic point of view. In terms of the Gibbs free energies of reaction

[9] F. D. Rossini, D. D. Wagman, W. H. Evans, S. Levine and I. Jaffe, *Selected Values of Chemical Thermodynamic Properties*, Natl. Bur. Std. Circ. 500, 1952. Partial revisions in Natl. Bur. Std. Technical Note 270-3, 1968. U.S. Dept. Commerce, Washington, D.C.
[10] I. Klotz, *Chemical Thermodynamics*, Benjamin, New York, 1964.
[11] L. G. Sillén and A. E. Martell, *Stability Constants of Metal–Ion Complexes*, The Chemical Society, London, 1964.

$\Delta G°$ (infinite dilution scale) = $\Delta G'$ (constant ionic medium scale) + constant

$$(62)$$

where the constant reflects the differences in the activity coefficients of the reactants and products as defined on each scale.

Example 2-8. From available enthalpy and entropy data [6] compute the equilibrium constant for the solution of gaseous ammonia in water at 25°C. The standard molar enthalpies of formation and the standard molar entropies are

	$\bar{H}_f°$, kcal mole^{-1}	$\bar{S}°$, cal deg^{-1} mole^{-1}
$NH_3(g)$	-11.04	46.01
$NH_3(aq)$	-19.32	26.3

For the process

$$NH_3(g) = NH_3(aq)$$
$$\Delta H° = -19.32 - (-11.04) = -8.28 \text{ kcal mole}^{-1}$$
$$\Delta S° = 26.3 - 46.01 = -19.7 \text{ cal mole}^{-1} \text{ deg}^{-1}$$
$$T\Delta S° = 298.16(-19.7) = -5.88 \text{ kcal mole}^{-1}$$

The standard free energy change is

$$\Delta G° = \Delta H° - T\Delta S° = -8.28 - (-5.88)$$
$$= -2.40 \text{ kcal mole}^{-1}$$

The equilibrium constant is computed from

$$-RT \ln K = \Delta G°$$

or

$$\log K = \frac{-\Delta G°}{1.364} = \frac{2.40}{1.364} = 1.75$$

From this result we may compute the equilibrium partial pressure of $NH_3(g)$ over a solution containing 1×10^{-3} M $NH_3(aq)$. Assuming ideal behavior

$$\frac{[NH_3(aq)]}{p_{NH_3}} = K = 56$$

$$p_{NH_3} = \frac{1 \times 10^{-3}}{56} = 1.78 \times 10^{-5} \text{ atm}$$

Example 2-9. The equilibrium constant for the solubility of precipitated $MnCO_3(s)$ in water has been determined [12] to be 3.92×10^{-11} (25°C,

[12] J. J. Morgan, in *Principles and Applications of Water Chemistry*, S. D. Faust and J. V. Hunter, Eds., Wiley, New York, 1967.

1 atm). From available data on the standard free energies of formation of $Mn^{2+}(aq)$ and $CO_3^{2-}(aq)$ compute the standard free energy of formation of $MnCO_3(s)$.

The standard free energy of the reaction

$$MnCO_3(s) = Mn^{2+}(aq) + CO_3^{2-}(aq)$$

is

$$\Delta G^\circ = -RT \ln K = -1.364 \log (3.92 \times 10^{-11})$$
$$= +14.20 \text{ kcal mole}^{-1}$$

The standard free energies of $Mn^{2+}(aq)$ and $CO_3^{2-}(aq)$ are -54.4 and -126.22 kcal mole^{-1}, respectively [6]. The desired free energy is obtained from

$$G^\circ_{f,MnCO_3(s)} = \bar{G}^\circ_{f,Mn^{2+}} + \bar{G}^\circ_{f,CO_3^{2-}} - \Delta G^\circ$$
$$= -54.4 - 126.22 - 14.20$$
$$= -194.8 \text{ kcal mole}^{-1}$$

Example 2-10.   Compute the equilibrium constant and standard free energy of reaction for the conversion of calcite to hydroxylapatite. Equilibrium constants for the following two reactions are known at 25°C:

(i) $Ca_5OH(PO_4)_3(s) = 5Ca^{2+} + 3PO_4^{3-} + OH^-$ $\log K = -55.6$
(ii) $CaCO_3(s) = Ca^{2+} + CO_3^{2-}$ $\log K = -8.4$

The standard free energy change for (i) is $\Delta G^\circ_{(i)} = -1.364(-55.6) = +75.8$ kcal mole^{-1}; for (ii) $\Delta G^\circ_{(ii)} = -1.364(-8.4) = +11.5$ kcal mole^{-1}. The desired reaction is

(iii) $5CaCO_3(s) + 3PO_4^{3-} + OH^- = Ca_5OH(PO_4)_3(s) + 5CO_3^{2-}$

The standard free energy change of (iii), recalling that $\Delta G^\circ = \sum_i \nu_i \mu_i^\circ$, is

$$\Delta G^\circ_{(iii)} = 5\Delta G^\circ_{(ii)} - \Delta G^\circ_{(i)}$$

Therefore,

$$\Delta G^\circ_{(iii)} = 5(11.5) - 75.8 = -18.3 \text{ kcal mole}^{-1}$$

and the log of the equilibrium constant for (iii) is

$$\log K_{(iii)} = -\frac{(-18.3)}{1.364} = 13.4$$

The form of the equilibrium constant is

$$K_{(iii)} = \frac{\{CO_3^{2-}\}^5}{\{PO_4^{3-}\}^3\{OH^-\}}$$

2-6   Influences of Temperature and Pressure on Chemical Equilibria

The chemical potential of a pure substance, a substance in a gas mixture or a dissolved substance is a function of both temperature and pressure. The basic relationships for temperature are [Table 2-1, (22)]

$$\left(\frac{\partial G}{\partial T}\right)_P = -S$$

and for pressure [Table 2-1, (21)]

$$\left(\frac{\partial G}{\partial P}\right)_T = V$$

In terms of the Gibbs free energy change for reactions involving single-component phases the total differential is

$$d\Delta G = \left(\frac{\partial \Delta G}{\partial T}\right)_P dT + \left(\frac{\partial \Delta G}{\partial P}\right)_T dP \tag{63}$$

or

$$d\Delta G = -\Delta S dT + \Delta V dP \tag{64}$$

For equilibrium with respect to temperature and pressure changes (64) leads to the well-known Clapeyron equation

$$\frac{dP}{dT} = \frac{\Delta S}{\Delta V} = \frac{\Delta H}{T\Delta V} \tag{65}$$

and to the approximate Clausius-Clapeyron relationship

$$\frac{d\ln P}{dT} \simeq \frac{\Delta \bar{H}_{\text{vap}}}{RT^2} \tag{66}$$

For systems of variable composition the total differential of the equilibrium constant for any reaction or process is given by

$$Rd\ln K = -\sum \nu_i \left(\frac{\partial \mu_i^\circ/T}{\partial T} dT + \frac{\partial \mu_i^\circ/T}{\partial P} dP\right) \tag{67}$$

or

$$Rd\ln K = \frac{\Delta H^\circ}{T^2} dT - \frac{\Delta V^\circ}{T} dP \tag{68}$$

The range of temperatures encountered in most aquatic systems is perhaps from about 0°C to close to 100°C. Most natural waters will fall between 0 and 40–50°C. The range of pressure is from about 1 atm at the earth's surface to about 1100 atm in the ocean depths. Differences in the temperature, and to a lesser degree the pressure, may have some significant influences on the equilibrium properties and spontaneous changes in aquatic systems. We will treat the influences of temperature and pressure individually and rather concisely. In many cases adequate thermodynamic data are not yet available for a complete examination of these effects, particularly the effects of pressure. In general, however, temperature exerts the greater influence.

Influence of Temperature

Garrels and Christ [13] have provided a thorough discussion of the effects of temperature on mineral stability relationships in the light of available thermodynamic data. In general, for the temperature range from 0 to 25°C for reactions involving *pure solids* or *pure liquids* there are not large changes in the values of $\Delta H°$ and $\Delta S°$ with temperature, that is, $\Delta C_P°$ is generally small (apart from phase changes). The free energy of reaction at temperature T is then approximated by

$$\Delta G_T° \simeq \Delta H_{298°}° - T\Delta S_{298°}° \tag{69}$$

A plot of $\Delta G°$ versus T yields a straight line; the slope of the line is usually small; for example, for the reaction

$$CaCO_3(\text{calcite, 1 atm}) = CaCO_3(\text{aragonite, 1 atm})$$

the values of $\Delta H°$ and $\Delta S°$ at 25°C are -0.04 kcal mole^{-1} and -1.0 cal mole^{-1} deg^{-1}, respectively. Then

$$\Delta G_{298}° = -0.04 - 298(-1.0) = +0.25 \text{ kcal mole}^{-1}$$

and

$$\Delta G_{373}° = -0.04 - 373(-1.0) = +0.33 \text{ kcal mole}^{-1}$$

The small entropy difference for the calcite–aragonite reaction results in a rather small difference in free energies over a 100° range of low temperatures. Larger entropy differences are involved for reactions involving aluminosilicate minerals and related oxides; for example, the following standard entropies are available [14]

	$\bar{S}°$, cal deg^{-1} mole^{-1}
Kaolinite, $Al_2Si_2O_5(OH)_4$	45.53
Gibbsite, $Al_2O_3 \cdot 3H_2O$	16.75
Quartz, $SiO_2(s)$	9.88
Albite, $Na(AlSi_3O_8)$	3.86

[13] R. M. Garrels and C. L. Christ, *Solutions, Minerals, and Equilibria*, Harper and Row, New York, 1965.
[14] R. A. Robie and D. R. Waldbaum, *Thermodynamic Properties of Minerals and Related Substances*, Geological Survey Bull. 1259, Washington, D.C., 1968.

It might be expected that greater effects of temperature on stabilities can be involved for some reactions involving these and related phases. Of particular interest is the influence of temperature on the weathering reaction of the clay minerals and the formation of new phases in the ocean. Unfortunately, entropy and enthalpy data on the dissolved constituents involved in these reactions are greatly lacking at present.

When heat capacity data are available as a function of temperature exact calculations are possible:

$$H_2 - H_1 = \int_{T_1}^{T_2} C_P dT \tag{70}$$

and

$$S_2 - S_1 = \int_{T_1}^{T_2} C_P d \ln T \tag{71}$$

For constant ΔC_P for a reaction

$$\Delta H_2 - \Delta H_1 = \Delta C_P (T_2 - T_1) \tag{72}$$

and

$$S_2 - S_1 = C_P \ln \frac{T_2}{T_1} \tag{73}$$

THE EQUILIBRIUM CONSTANT. Table 2-4 summarizes the pertinent relationships for describing the influence of temperature on the equilibrium constant of a chemical reaction or phase equilibrium. The thermodynamic information of interest includes $\Delta H°$ for the reaction, $\Delta C_P°$ for the reaction and the variation of $\Delta C_P°$ with temperature. For a number of reactions of interest in aquatic systems heat capacity data are limited or unavailable. However, sufficient enthalpy data are available for most reactions of interest to provide at least a limited assessment of temperature influences. Enthalpy data are available in Latimer's text [6], in the compilation of Robie and Waldbaum, [14], in NBS Circular 500 [9], and to a limited extent in *Stability Constants of Metal–Ion Complexes* [11] (several other sources of heat data are referenced by Garrels and Christ [13]). Nancollas [15] has summarized and discussed enthalpy data for ion pairs and complexes. Enthalpy data are obtained by direct calorimetry and by measurements of equilibrium properties of chemical reactions, electro-chemical cells and transfer processes.

Equations 4, 7 and 10 of Table 2-4 suggest that a plot of the logarithm of the equilibrium constant (or a representative equilibrium activity) of a reaction versus the reciprocal of absolute temperature can yield information

[15] G. H. Nancollas, *Interactions in Electrolyte Solutions*, Elsevier, New York, 1966.

Table 2-4   Influence of Temperature on the Equilibrium Constant

The basic relationships are

$$\frac{d \ln K}{dT} = \frac{\Delta H^\circ}{RT^2} \tag{1}$$

$$\ln \frac{K_2}{K_1} = \int_{T_1}^{T_2} \frac{\Delta H^\circ}{RT^2} \, dT \tag{2}$$

When ΔH° is independent of temperature

$$\ln \frac{K_2}{K_1} = \frac{\Delta H^\circ}{R} \left(\frac{1}{T_1} - \frac{1}{T_2} \right) \tag{3}$$

or

$$\ln K = -\frac{\Delta H^\circ}{RT} + \text{constant} \tag{4}$$

When the heat capacity of the reaction, ΔC_p°, is independent of temperature

$$\Delta H_2^\circ = \Delta H_1^\circ + \Delta C_p^\circ (T_2 - T_1) \tag{5}$$

Integration of (1) then yields

$$\ln \frac{K_2}{K_1} = \frac{\Delta H_1^\circ}{R} \left(\frac{1}{T_1} - \frac{1}{T_2} \right) + \frac{\Delta C_p^\circ}{R} \left(\frac{T_1}{T_2} - 1 - \ln \frac{T_1}{T_2} \right) \tag{6}$$

or

$$\ln K = B - \frac{\Delta H_0}{RT} + \frac{\Delta C_p^\circ}{R} \ln T \tag{7}$$

where ΔH_0 and B are constants

When ΔC_p° is a function of temperature

If the heat capacity of each reactant and product is given by an expression of the form

$$C_p^\circ = a_i + b_i T + c_i T^2 \tag{8}$$

then the heat capacity of the reaction is given by

$$\frac{d \Delta H^\circ}{dT} = \Delta C_p^\circ = \Delta a + \Delta b T + \Delta c T^2 \tag{9}$$

Integration of (9) and (1) yields

$$\ln K = B - \frac{\Delta H_0}{RT} + \frac{\Delta a}{R} \ln T + \frac{\Delta b}{2R} T + \frac{\Delta c}{6R} T^2 \tag{10}$$

where ΔH_0 and B are constants, and $\Delta a = \sum_i \nu_i a_i$, etc.

concerning $\Delta H°$. For many reactions $\Delta C_P°$ is close to zero and $\Delta H°$ is essentially independent of temperature [(4), Table 2-4], and a linear plot of log K versus $1/T$ is obtained over an appreciable temperature range. The equilibrium constant can then be computed readily by the simple relationship of (3) in Table 2-4. When $\Delta C_P°$ is constant over a range of temperature, (6) of Table 2-4 can be used to compute the equilibrium constant temperature coefficient. The

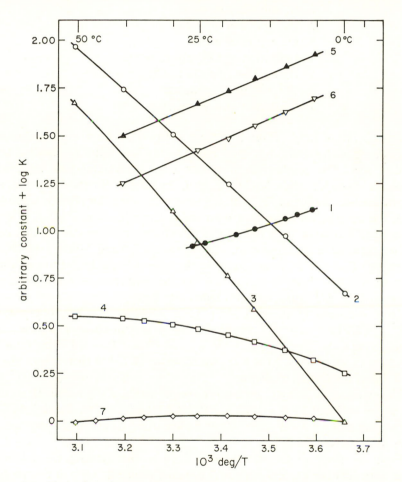

Figure 2-8   Plots of the log of the equilibrium constant versus reciprocal of absolute temperature for seven aqueous physico-chemical processes. The processes and the values of the arbitrary constant, C, are: (1) $O_2(g) = O_2(aq)$, C = 2.8; (2) $H_2O(l) = H_2O(g)$, C = 0; (3) $H_2O(l) = H^+ + OH^-$, C = +14.93; (4) $CO_2(aq) + H_2O = HCO_3^- + H^+$, C = +6.29; (5) $CaCO_3(s) = Ca^{+2} + CO_3^{=}$, C = +9.01; (6) $CO_2(g) = CO_2(aq)$, C = +2.89; (7) $CH_3COOH(aq) = CH_3COO^- + H^+$, C = +4.7807.

most general case, in which ΔC_P° is a function of temperature, is described by (10) of Table 2-4.

To illustrate the influence of temperature on the equilibrium of reactions and processes in aquatic systems we have selected representative examples. Figure 2-8 contains plots of log K versus the reciprocal of absolute temperature for seven different aqueous equilibria.

1. O_2 solubility in water
2. Vaporization of water
3. Ionization of water
4. Ionization of CO_2(aq) in water
5. Dissolution of solid $CaCO_3$ in water
6. CO_2 solubility in water
7. Ionization of acetic acid in water

The temperature range covered is from 0 to 50°C. An arbitrary constant (specific for each system) has been added to log K in order to obtain a convenient plot. The equilibrium constants refer to the standard states and reference states for solids, liquids, gases and solutions. Solution activities are on the infinite dilution scale. With the exception of the data for O_2 solubility [16], the equilibrium data have been taken from the compilations already referred to. Processes 2, 5 and 6 have essentially constant ΔH° values over the temperature range considered, while processes 3, 4 and 7 show a significant variation of ΔH° with temperature. In the case of 7, the ionization of acetic acid, there is a reversal in sign of ΔH° as the temperature varies. Process 1 shows a slight ΔH° variation with temperature.

From the slope of the plot for oxygen solubility at 25°C, one obtains $\Delta H^\circ \simeq -3.9$ kcal mole^{-1}. From the enthalpy data in Latimer [6], the computed value is $\Delta H^\circ = \overline{H}_{f,O_2(aq)}^\circ - \overline{H}_{f,O_2(g)}^\circ = -3.8 - 0 = -3.8$ kcal mole^{-1}. The heat capacity of O_2(g) is a known function of T, but there appear to be no $\overline{C}_P^\circ$ data for O_2(aq) available.

The equation describing the ion product of water, K_W (process 3), is [17]: log $K_W = (4470.99/T) + 6.0875 - 0.01706T$ [compare (10), Table 2-4]. The curvilinear plot for 3 in Figure 2-8 can be linearized by use of the appropriate constants in (10), Table 2-4.

From the slope of the straight line fitted to the data for the solubility constant of calcite (process 5) we may obtain a value for ΔH° for the solution

[16] J. C. Morris, W. Stumm and H. Galal, *J. Sanit. Eng. Div., Am. Soc. Civil Engrs.*, *87*, p. 81 (Jan. 1961).
[17] H. S. Harned and B. B. Owen, *The Physical Chemistry of Electrolytic Solutions*, 3rd ed., Reinhold, New York, 1958.

reaction. The slope is $+1050$ deg. From (4), Table 2-4, the slope is equal to $-\Delta H^\circ/2.303R$. Therefore we have

$$-\frac{\Delta H^\circ}{2.303 \times 1.987 \text{ cal mole}^{-1} \text{ deg}^{-1}} = 1050 \text{ deg}$$

and

$$\Delta H^\circ = -4.8 \text{ kcal mole}^{-1}$$

We may note that this result is not in exact agreement with ΔH° computed from tabulated standard enthalpies of formation, for example, in NBS Circular 500 [9]:

	$\bar{H}^\circ_{f,298.15}$, kcal mole^{-1}
Calcite	-288.45
Ca^{2+}(aq)	-129.77
$CO_3{}^{2-}$(aq)	-161.63

These data yield

$$\Delta H^\circ = -129.77 - 161.63 - (-288.45) = -2.95 \text{ kcal mole}^{-1}$$

On the other hand, Sillén [18] selected $\log K = -8.20$ at 5°C and $\log K = -8.32$ at 25°C as the infinite dilution scale equilibrium constants for $CaCO_3(s) = Ca^{2+} + CO_3$. These K's correspond to a ΔH° for the reaction of -2.2 kcal mole^{-1}. Apparent discrepancies in ΔH values of the order of 1 to 2 kcal mole^{-1} may seem quite troublesome. However, they can be sorted out with more measurements, and these discrepancies are almost certainly less significant than our frequent lack of detailed information with regard to the chemical identity of actual solid phases and dissolved species in the natural systems. Under the actual conditions of salt concentration (but not considering the pressure conditions) encountered in seawater, $\log K$ for the same solubility equilibrium has been estimated as -5.67 and -6.19 for 5 and 25°C, respectively. The greater effect of temperature on the relative solubilities in the ionic medium of seawater reflects the influence of temperature upon activity coefficients on the infinite dilution scale:

$$\frac{d \ln \gamma_i}{dT} = \frac{\bar{H}^\circ_i - \bar{H}_i}{RT^2} \tag{74}$$

where $\bar{H}_i$ is the partial molar enthalpy in the *actual* solution and $\bar{H}^\circ_i$ is the standard state value [compare (74) and (1) of Table 2-4]. From a practical point of view it appears desirable to establish the temperature dependences

[18] L. G. Sillén, in *Oceanography*, M. Sears, Ed., American Association for the Advancement of Science, Washington, D.C., 1961.

of solubility equilibria, acid–base equilibria and complex formation equilibria in seawater at a constant ionic medium.

In summarizing our brief discussion we want to emphasize that aqueous equilibria may be shifted to the right or to the left by an increase in temperature. The direction and magnitude of the shift depends upon the sign and magnitude of the heat of the reaction as a function of temperature, and the magnitude is affected by the salt content of the water, ΔH being a function of the ionic medium. From (1) of Table 2-4 the *approximate* magnitude of the change in log K per degree is $0.0029\Delta H$ at $0°C$, 0.0025 at $25°C$ and 0.0021 at $50°C$. In the examples given in Figure 2-8 we have illustrated a range of possibilities for the effect of temperature on equilibrium in the range from 0 to $50°C$.

Effect of Pressure

The basic relationships for describing the effects of pressure on physico-chemical equilibrium are set down in Table 2-5. A comprehensive treatment of the physico-chemical effects of pressure has been provided by Hamann [19]. Calculations of the effect of pressure on ionic equilibria in aqueous solutions were detailed by Owen and Brinkley [20]. Distèche and Distèche [21] and Pytkowicz [22] recently have made detailed investigations into the effects of pressure on chemical equilibria in seawater. In this brief discussion we merely state the basis for understanding the effects of pressure and attempt to illustrate some of the features of a few reaction systems of interest. The reader should consult the references mentioned for details of measurement techniques and for compilations of data.

Table 2-5   Influence of Pressure on Chemical Equilibrium in Aqueous Systems

Basic relationships

$$\left(\frac{\partial \mu_i}{\partial P}\right)_T = \overline{V}_i \tag{1a}$$

$$\left(\frac{\partial \mu_i^\circ}{\partial P}\right)_T = \overline{V}_i^\circ \tag{1b}$$

where $\overline{V}_i$ and $\overline{V}_i^\circ$ are the partial molar volumes of i under actual conditions and under defined standard state conditions, respectively

$$\overline{k}_i^\circ = \left(\frac{\partial \overline{V}_i^\circ}{\partial P}\right)_T \tag{2}$$

[19] S. D. Hamann, *Physico-Chemical Effects of Pressure*, Butterworths, London, 1957.
[20] B. B. Owen and S. R. Brinkley, Jr., *Chem. Revs.*, **29**, 461 (1941).
[21] A. Distèche and S. Distèche, *J. Electrochem. Soc.*, **114**, 330 (1967).
[22] C. Culberson and R. Pytkowicz, *Limnol. Oceanog.*, **13**, 403 (1968).

where $\bar{k}_i^\circ$ is the standard partial molar compressibility

$$\Delta V = \sum \nu_i \bar{V}_i \qquad (3a)$$

$$\Delta V^\circ = \sum \nu_i \bar{V}_i^\circ \qquad (3b)$$

where ΔV and ΔV° are the volume changes of reaction under actual and under defined standard state conditions, respectively

$$\left(\frac{\partial \ln K}{\partial P}\right)_T = -\frac{\Delta V^\circ}{RT} \qquad (4)$$

where K is the equilibrium constant

When ΔV° is independent of pressure

$$\ln \frac{K_P}{K_1} = -\frac{\Delta V^\circ (P - 1)}{RT} \qquad (5)$$

When Δk° is independent of pressure

$$\ln \frac{K_P}{K_1} = \frac{1}{RT}[-\Delta V^\circ(P - 1) + \tfrac{1}{2}\Delta k^\circ(P - 1)^2] \qquad (6)$$

where $\Delta k^\circ = \sum_i \nu_i \bar{k}_i^\circ$

When Δk° is a function of pressure

$$RT \ln \frac{K_P}{K_1} = -\Delta V^\circ(P - 1) + \Delta k^\circ(B + 1)(P - 1) - (B + 1)^2 \ln\left(\frac{B + P}{B + 1}\right) \qquad (6a)$$

where B is independent of pressure and depends upon the temperature [20]

For aqueous solutions specifically

$$\mu_i = \mu_i^\circ + RT \ln \gamma_i m_i \qquad (7)$$

$$\left(\frac{\partial \ln K}{\partial P}\right)_{T,m} = -\frac{\Delta V^\circ}{RT} \qquad (8)$$

$$\left(\frac{\partial \ln \gamma_i}{\partial P}\right)_{T,m} = \frac{\bar{V}_i - \bar{V}_i^\circ}{RT} \qquad (9)$$

$$\left(\frac{\partial \ln K'}{\partial P}\right)_{T,m} = \frac{\Delta V}{RT} \qquad (10)$$

where $K' = \prod_i m_i^{\nu_i} = K/\prod_i \gamma_i^{\nu_i}$

Example 2-11. Consider the effect of pressure on the chemical potential of water. What is the approximate chemical potential of water at 25°C and 1000 atm? We will obtain an approximate answer by assuming that the partial molar compressibility is independent of pressure. Applying (1a) and (2) of Table 2-5 we obtain

$$\mu_P - \mu_1 = \int_1^P \bar{V}_{H_2O} dP$$

$$\simeq \bar{V}_{H_2O}^\circ(P - 1) + \tfrac{1}{2}\bar{k}_{H_2O}^\circ(P - 1)^2$$

The molar volume of water at 1 atm is 18.02 cm^3 mole^{-1} and $\bar{k}_{H_2O}^{\circ}$ is 8.2 × 10^{-4} cm^3 mole^{-1} atm^{-1} [20]. Therefore

$$\mu_P - \mu_1 \simeq 18,200 \text{ cm}^3 \text{ atm mole}^{-1}$$

$$\simeq 0.44 \text{ kcal mole}^{-1}$$

The chemical potential, or molar Gibbs free energy, is approximately 0.44 kcal mole^{-1} more positive at 1000 atm than at 1 atm. Recalling that, for any substance we can write

$$\mu_i = \mu_i^{\circ} + RT \ln \{i\} \tag{34}$$

it is a matter of convention whether we now increase the value of the standard state chemical potential (1 atm) by 0.44 kcal mole^{-1} or define the activity of the water at 1000 atm to be

$$\{H_2O\} \simeq \ln^{-1} \left(\frac{0.44}{0.001987 \times 298.15} \right)$$

$$\simeq 2.1$$

compared to unity at 1 atm. A more accurate answer (corresponding to an activity of 2.06) is obtained by graphical integration of

$$\int_1^{1000} \bar{V}_{H_2O} dP$$

Owen and Brinkley [20] computed the effect of pressure on the equilibrium constant for the ionization of water, both for pure water and for 0.725 M sodium chloride as ionic medium. For pure water they obtained, in part,

P, bars†	$K_{W,P}/K_{W,1}$		
	5°C	25°C	45°C
1	1	1	1
200	1.24	1.202	1.16
1000	2.8	2.358	2.0

The results obtained for 0.725 M NaCl were not greatly different from these. At 25°C and 1000 atm the ratio K_P/K_1 was 2.1 instead of 2.358.

The standard partial molar volumes of the participants in the solution (ionization) of calcite at 25°C

$$CaCO_3(s) = Ca^{2+} + CO_3^{2-}$$

† 1 atm = 1.01325 bar.

are, according to Owen and Brinkley [20]:

	$\overline{V}°$, cm^3 mole^{-1}
CaCO$_3$(s)	36.9
Ca^{2+}†	-17.7
CO$_3{}^{2-}$	-3.7

Neglecting the compressibility of calcite we may estimate the change in chemical potential of the solid at a pressure of say 1000 atm. We have

$$\mu_P - \mu_1 = \int_1^P \overline{V}_{CaCO_3(s)} dP \simeq \overline{V}_{CaCO_3(s)}(P - 1)$$

$$\simeq 36.9 \times 999 \simeq 36{,}860 \text{ cm}^3 \text{ atm mole}^{-1} \simeq 0.89 \text{ kcal mole}^{-1}$$

This increase in chemical potential corresponds to an activity (relative to the 1 atm condition) of about 4.5.

The equilibrium constant for the solution of CaCO$_3$(s) at elevated pressure may be estimated by applying the appropriate equations in Table 2-5. The volume change of the reaction is

$$\Delta V° = -17.7 - 3.7 - 36.9 = -58.3 \text{ cm}^3 \text{ mole}^{-1}$$

At a pressure of 1000 atm the ratio of the calcite solubility equilibrium constant at that pressure to that at 1 atm is

$$\frac{K_P}{K_1} = 8.1$$

The volume change of reaction is negative for most ionization processes. Therefore the effect of increased pressure is generally to increase the extent of ionization; for example, Owen and Brinkley [20] report the following computed pressure effects in pure water medium for 1000 atm and 25°C

Reaction	K_P/K_1
CaSO$_4$(s) = Ca^{2+} + SO$_4{}^{2-}$	5.8
H$_2$O + CO$_2$(aq) = HCO$_3{}^-$ + H$^+$	3.2
HCO$_3{}^-$ = CO$_3{}^{2-}$ + H$^+$	2.7
CH$_3$COOH = CH$_3$COO$^-$ + H$^+$	1.4

Since the partial molar volumes of dissolved species are a function of the ionic strength of the medium, it is expected that the pressure effects upon equilibrium constants will be different in pure water and in a medium such as seawater (or other high salt solution). Some of the results of Owen and

† The partial molar volumes of ions tabulated by Owen and Brinkley are based on the convention $\overline{V}_{H^+}° = 0$.

Brinkley [20] will serve to indicate the extent of the differences for calcite solubility:

$$K_{calcite,P}/K_{calcite,1} \ (25°C)$$

P, bars	Pure Water	0.725 M NaCl
1	1	1
500	3.2	2.8
1000	8.1	6.7

The basis for these differences is seen in (6a), (9) and (10) of Table 2-5. On the basis of these results supersaturation of seawater would be expected to decrease significantly with depth. A complete analysis of supersaturation and equilibrium requires consideration of other reactions or other species in addition to the simple ions (e.g., ion pairs). More recent work [21, 23, 24] can be consulted for such considerations and for newer data on partial molar volumes in seawater.

From this brief discussion of pressure effects on equilibria we can conclude that only in the deep oceans (and to a lesser degree in aquifers under high pressures) are such effects of great significance.

2-7 Equilibrium Models and Calculation of Complex Equilibria

Now that the elements of the basic framework for the application of chemical thermodynamics to the study of natural water systems have been presented, it is appropriate to remark briefly on the integration of the basic concepts into equilibrium models. Recall Figures 2-1 and 2-2, where the main features of a generalized model for the thermodynamic description of natural water systems are outlined. The models of interest are closed systems (with some degree of completeness, as illustrated in Figure 2-2) at constant temperature and pressure. Variation of pressure with depth is not an obstacle to applying the equilibrium relationships since mechanical equilibrium can be safely assumed. Temperature variations present a more serious difficulty because of heat flows caused by temperature gradients and chemical reactions. However, local thermal equilibrium or steady state is probably not an inappropriate approximation for some situations. For large systems, such as the oceans and large lakes, an "average" temperature for different depths or for different depths and seasons (for the lakes) may be a reasonable assumption for thermodynamic calculations.

[23] I. W. Duedall and P. K. Weyl, *Limnol. Oceanog.*, **12**, 52 (1967).
[24] F. J. Millero, *Limnol. Oceanog.*, **14**, 376 (1969).

The steps for constructing and interpreting an isothermal, isobaric thermo-dynamic model for a natural water system are quite simple in principle. The components to be incorporated are identified, and the phases to be included are specified. The components and phases selected "model" the real system and must be consistent with pertinent thermodynamic restraints; that is, the identification of the maximum number of unknown activities with the number of independent relationships that describe the system (equilibrium constant for each reaction, stoichiometric conditions, electroneutrality condition in the solution phase). With the phase-composition requirements identified, and with adequate thermodynamic data (free energies, equilibrium constants) available, chemical equilibrium in the closed system is then assumed (the affinity of each reaction is taken as zero), and the composition variables (activities, partial pressures, mole fractions) of the system are computed. In some applications the model used is a series of imaginary closed systems, a few components being added in sequence and the conditions at chemical equilibrium being computed after each such addition. Alternatively, the assumed model is a single closed system at chemical equilibrium, and a set of simultaneous equations is solved to yield the over-all composition of the model. Finally, the problem may be explicitly formulated to find the com-position of the system which minimizes the total Gibbs free energy and the system. In each of these formal approaches, comparing the equilibrium composition of the model with that of the real natural water system is the important step by which one hopes to gain a clearer understanding of the chemical behavior of the real system.

We may summarize the quantitative operations involved in applying thermodynamic concepts in three distinguishable approaches, which we will epitomize as

    (I) the "equilibrium constant, or K approach";
   (II) the Gibbs free energy of reaction and reaction quotient, or "ΔG and Q approach"; and
  (III) the "total Gibbs function minimization, or G and ξ approach."

In (I) the equilibrium constant for each *independent* reaction considered is written together with the stoichiometric (mole balance and electroneutrality) equations, and a set of nonlinear and linear equations is solved to obtain the equilibrium composition of the system. In (II) the free energy for each reac-tion, ΔG, is computed from

$$\Delta G = \Delta G^\circ + RT \ln Q \tag{51}$$

or

$$\Delta G = RT \ln \frac{Q}{K} \tag{54}$$

subject to the stoichiometric constraints. At equilibrium, $\Delta G = 0$ and the composition is then the equilibrium composition. In (III) the expression for the total Gibbs energy function of the system is written as a function of system composition, and the function is then minimized, subject to the stoichiometric constraints. Although all of these approaches are equivalent at equilibrium, they are operationally different. Approach (II) is advantageous for iterative comparison of actual states with the equilibrium state of a reaction and for simulating the approach to equilibrium. Each approach lends itself to computer programming solutions. In the case of (III) system optimization programs can be applied. In this chapter we have applied each of these approaches to rather simple systems of individual reactions (e.g., Figure 2-7). For more complex systems discussed throughout this book we employ, in general, approach (I), "the equilibrium constant, or K approach." The value of graphical techniques in understanding equilibria and in obtaining solutions for systems of fair complexity will be emphasized. For a detailed discussion of the different mathematical approaches a review paper by Zeleznik and Gordon [25] may be helpful. An example of the application of approach (I) to a fairly complex system is described in Chapter 6, in which a computer solution that uses a matrix description of the chemical species and the Newton method has been employed.

2-8   Natural Water Systems as Continuous Systems: Rates and Equilibria

Real systems may or may not be closely approximated by equilibrium models. Significant discrepancies between the actual chemical composition and the composition predicted from an equilibrium model are frequently encountered. One hopes that an equilibrium model may give a useful first approximation to the real system. There are several possible reasons for differences between the real system and the equilibrium model. Briefly, the causes of the discrepancies can be categorized as follows.

Differences Between Continuous Systems and Closed Systems

Natural water systems are continuous or flowing systems. Flows of matter and energy exist in the real system. Concentration and temperature gradients exist. The time-invariant state of a continuous system with flows at the boundaries is the *steady state*. This state might be poorly approximated by the *equilibrium* state of a closed system.

[25] F. J. Zeleznik and S. Gordon, *Ind. Eng. Chem.*, **60**, 27 (1968).

Limitations of Thermodynamic Information

Pertinent chemical equilibria may have been ignored in formulating the model or important species in solid or solution phases not considered. Thermodynamic data on assumed species and phases may be incorrect or inadequate. Temperature, pressure and activity-concentration corrections may require refinement. Significant temperature gradients exist at times in nearly all bodies of water. Adequate data on temperature dependence of equilibria are often not available. Pressure correction requires knowing the partial molar volumes of the species involved.

Inadequate Chemical Characterization of the Species in the Real System

Analytical data on certain chemical elements in nature may be inadequate to distinguish between various physical and chemical forms (dissolved versus suspended, oxidized versus reduced, monomeric versus polymeric). Analytical deficiencies relate particularly to the highly reactive nonconservative elements: for example, iron, manganese, phosphorus, carbon, nitrogen, aluminum and other metal ions.

Slow Rates of Chemical Reaction

The rates of certain chemical reactions may be such that equilibrium is only slowly attained in the real system. Even when a *closed* system represents a reasonable approximation to the nature of the system, the slowness of some chemical reactions is a cause of discrepancy between the real system and an equilibrium model.

We consider briefly the matter of slow reaction rates and rates of flow to continuous systems in order to give a feeling for the possibilities of applying equilibrium models to advantage. In considering equilibria and kinetics in natural water systems it is useful to recall that different time scales need to be identified with chemical reactions in different systems. Relatively short times (days to weeks) are available for approaching equilibrium in rivers, smaller lakes, reservoirs and estuaries. Times for reaction in large lakes, seas and perhaps typical ground waters are of the order of tens to hundreds of years. In ocean waters the reaction time may range from thousands of years to long geological periods (millions of years). Reactions highly favorable from a thermodynamic viewpoint can be quite unfavorable kinetically, for example, the reaction

$$CH_3COOH(aq) = 2H_2O + 2C(graphite) \qquad K = 10^{13} \ (25°C)$$

is highly spontaneous ($\Delta G° = -17.9$) but almost infinitely slow. The phenomenon of metastability depends on a slow or negligible rate for a particular reaction of a phase or species. The same substance may, however, participate in other reactions which attain equilibrium, for example,

$$A \underset{}{\overset{\text{rapid}}{\rightleftharpoons}} B \quad (\Delta G < 0) \tag{75}$$

$$A \xrightarrow{\text{extremely slow}} C \quad (\Delta G < 0) \tag{76}$$

Thus for certain reactions the point f in Figure 2-6 may represent an observable state of the system because of unfavorable chemical kinetics.

Among reactions that are "slow" on a relevant time scale and in particular environments are certain metal–ion oxidations, oxidation of sulfides, sulfate reduction, various metal–ion polymerizations (e.g., vanadium, aluminum), aging of hydroxide and oxide precipitates, precipitation of metal–ion silicates and carbonates (e.g., dolomites), conversions among aluminosilicates (e.g., feldspar–kaolinite) and solution or precipitation of quartz. Some of these reactions can be accelerated greatly by biological catalysis (e.g., sulfate reduction, metal–ion oxidations). Some quantitative aspects of kinetics have been touched on in Section 2-3.

Equilibrium and Steady State

In view of the formal differences between closed and continuous systems, what general conclusions can be drawn about the applicability of equilibrium concepts in understanding and describing the chemical behavior of continuous systems? Since equilibrium is the time-invariant state of a *closed* system, the question is under what conditions do flow systems approximate closed systems. An example will illustrate the general relationship. Consider a reversible, first-order reaction

$$A \underset{k'}{\overset{k}{\rightleftharpoons}} B \tag{77}$$

which takes place in a completely mixed volume V, into which A and B are introduced at concentration $[A]_0$ and $[B]_0$ and from which A and B are withdrawn, the volume rate of flow being q. The material balance rate expressions for A and B can be written as

$$\frac{d[A]}{dt} = \frac{q[A]_0}{V} - \frac{q[A]}{V} - k[A] + k'[B] \tag{78}$$

and

$$\frac{d[B]}{dt} = \frac{q[B]_0}{V} - \frac{q[B]}{V} + k[A] - k'[B] \tag{79}$$

In the time-invariant condition

$$\frac{d[A]}{dt} = \frac{d[B]}{dt} = 0 \tag{80}$$

The ratio of [B] to [A] for this condition is

$$\frac{[B]}{[A]} = \frac{k(V/q)([A]_0 + [B]_0) + [B]_0}{k'(V/q)([A]_0 + [B]_0) + [A]_0} \tag{81}$$

The equilibrium constant and the first-order rate constants are related by

$$K = \frac{k}{k'} \tag{82}$$

It is clear from (81) and (82) that the steady-state ratio [B]/[A] approaches K as the flows to the system become small with respect to the rates of chemical reaction. In the simple case in which $[B]_0 = 0$ we obtain

$$\frac{[A]}{[B]} = \frac{1}{K} + \frac{1}{k(V/q)} \tag{83}$$

Thus, when the *residence* time $\tau_R = V/q$ is sufficiently large relative to the appropriate time scale of the *reaction*, the time-invariant condition of the well-mixed volume considered approaches chemical equilibrium. (The treatment can be extended to other reversible reactions and to irreversible reactions.) This requirement might be difficult to satisfy for a number of slowly reacting constituents in the total volume of many fresh water or estuarine systems where residence times may be small and mixing incomplete. It might be sufficiently well-satisfied for a number of elements in the greater part of large-volume and low-flow systems; for example, the major oceans, the Great Lakes and certain ground water formations. Equilibrium is a reasonable assumption for such substances as sodium, potassium, magnesium, calcium, carbonate, chloride, sulfate and fluoride in long-residence time systems.

For nearly all kinds of systems it is argued that there exist regions or environments in which, at times, the time-invariant condition closely approaches equilibrium, even though gradients exist throughout the system as a whole. The concept of local equilibrium is a fundamental assumption in thermodynamic theories of irreversible processes. Local equilibrium conditions might be expected to develop for certain kinetically rapid species and phases at sediment–water interfaces in fresh, estuarine and marine environments. In contrast, other local environments, such as the photosynthetically active surface regions of nearly all lakes and ocean waters and the biologically active regions of soil–water systems are clearly far removed from total system equilibrium.

READING SUGGESTIONS

Denbigh, K. G., *The Thermodynamics of the Steady State*, Wiley, New York, 1951.

Denbigh, K. G., *The Principles of Chemical Equilibrium: with Applications in Chemistry and Chemical Engineering*, 2nd ed., Cambridge Univ. Press, 1966.

Edwards, J. O., *Inorganic Reaction Mechanisms: An Introduction*, Benjamin, New York, 1964.

Garrels, R. M., and C. L. Christ, *Solutions, Minerals, and Equilibria*, Harper and Row, New York, 1965.

Guggenheim, E. A., *Thermodynamics: An Advanced Treatment for Chemists and Physicists*, 4th ed., North-Holland, Amsterdam, 1959.

Harned, H. S., and B. B. Owen, *The Physical Chemistry of Electrolytic Solutions*, 3rd ed., Reinhold, New York, 1958.

Klotz, I. M., *Chemical Thermodynamics: Basic Theory and Methods*, revised ed., Benjamin, New York, 1964.

Latimer, W. M., *The Oxidation States of the Elements and Their Potentials in Aqueous Solutions*, 2nd ed., Prentice-Hall, New York, 1952.

Lee, T. S., "Chemical Equilibrium and the Thermodynamics of Reactions," in *Treatise on Analytical Chemistry*, Part I, Vol. 1, I. M. Kolthof and P. J. Elving, Eds., Interscience, New York, 1959, pp. 185–275.

Lewis, G. N., and M. Randall, *Thermodynamics*, 2nd ed., revised by K. S. Pitzer and L. Brewer, McGraw-Hill, New York, 1961.

Prigogine, I., *Introduction to Thermodynamics of Irreversible Processes*, 2nd ed., Interscience, New York, 1961.

Prigogine, I., and R. Defay, *Chemical Thermodynamics*, translated by D. H. Everett, Longmans, London, 1954.

Robinson, R. A., and R. H. Stokes, *Electrolyte Solutions: The Measurement and Interpretation of Conductance, Chemical Potential and Diffusion in Solutions of Simple Electrolytes*, 2nd ed., Butterworths, London, 1959.

Rossini, F. D., et al., *Selected Values of Chemical Thermodynamic Properties*, Natl. Bur. Std. Circ. 500, U.S. Dept. Commerce, Washington, D.C., 1952.

Sillén, L. G., and A. E. Martell, *Stability Constants of Metal–Ion Complexes*, Special Publ., No. 17, The Chemical Society, London, 1964.

Waser, J., *Basic Chemical Thermodynamics*, Benjamin, New York, 1966.

3   Acids and Bases

3-1   Introduction

pH values for most mineral-bearing waters are known to lie generally within the narrow range 6 to 9 and to remain very nearly constant for any given water. The composition of natural waters is influenced by the interactions of acids and bases. According to Sillén [1] one might say that the ocean is the result of a gigantic acid–base titration; acids that have leaked out of the interior of the earth are titrated with bases that have been set free by the weathering of primary rock. $[H^+]$ of natural waters is of great significance in all chemical reactions associated with the formation, alteration and dissolution of minerals. The pH of the solution will determine the direction of the alteration process.

Biological activities such as photosynthesis and respiration and physical phenomena such as natural or induced turbulence with concomitant aeration influence pH regulation through their respective abilities to decrease and increase the concentration of dissolved carbon dioxide. Besides photosynthesis and respiration, other biologically mediated reactions affect the H^+ ion concentrations of natural waters. Oxygenation reactions lead to a decrease in pH, whereas processes such as denitrification and sulfate reduction tend to increase pH.

Because of the ubiquitousness of carbonate rocks and the equilibrium reactions of CO_2, bicarbonate and carbonate are present as bases in most natural waters. In addition small concentrations of the bases borate, phosphate, arsenate, ammonia and silicate may be present in the solution. Volcanoes and certain hot springs may yield strongly acid water by adding gases like HCl and SO_2. Free acids also enter natural water systems as a result of the disposal of industrial waters. Similarly, acidity is imparted through hydrolysis of multivalent metal ions. Other acidic constituents are boric acid, silicic acid and ammonium ions. From the point of view of interaction with solid and dissolved bases, the most important acidic constituent is CO_2, which forms H_2CO_3 with water.

Proton-transfer reactions are usually very fast (half-lives less than milliseconds). Equilibria characterizing H^+ ion transfer reactions are among the

[1] L. G. Sillén, AAAS Publ. 67, Washington, D.C., 1961, p. 549.

simplest types of equilibrium models. In this chapter we devote considerable space to develop the idea of chemical equilibrium using acid–base reactions as examples. This gives us an opportunity to discuss the various activity conventions in some detail. We also demonstrate numerical and graphical methods to compute the equilibrium composition of simple acid–base transfer systems. The graphical procedures using pH as a master variable permit us to survey conveniently the interrelationships of the equilibrium concentrations of individual solute species.

Hydrogen ion regulation in natural waters is provided by numerous homogeneous and heterogeneous buffer systems. It is important to distinguish in these systems between intensity factors (pH) and capacity factors (e.g., the total acid or base neutralizing capacity). The buffer intensity is found to be an implicit function of both these factors. In this chapter we discuss acid–base equilibria primarily from a general and didactic point of view. In Chapter 4 we address ourselves more specifically to the dissolved carbonate system.

3-2   The Nature of Acids and Bases

Arrhenius (1887) proposed that an acid is a substance whose water solution contains an excess of hydrogen ions and a base contains an excess of hydroxide ions in aqueous medium. Arrhenius assumed that the excess H^+ ions or OH^- ions in a water solution resulted from dissociation of the acid or base as it is introduced into the water. In the light of more recent knowledge the Arrhenius concept is limited. First, it applies to ionic solutions only. Second, the dissociation theory is no longer tenable since we now know, for example, that salts such as NaOH exist as ions even in the crystalline state. Furthermore, it is well known today that a hydrogen ion, that is, a proton, cannot exist as a bare ion in water solution. Theoretical calculations show that a proton would strongly react with a water molecule to form a hydrated proton, a hydronium or hydroxonium ion (H_3O^+). Actually the H_3O^+ ion in an aqueous solution is itself associated through hydrogen bonds with a variable number of H_2O molecules: $(H_7O_3)^+$, $(H_9O_4)^+$, etc. The formula H_3O^+ or H^+ is generally used, however, to denote a hydrated hydrogen ion. Formulas analogous to H_3O^+ for the solvated proton are used for other solvents; for example, NH_4^+ in liquid ammonia or $C_2H_5OH_2^+$ in ethanol. The hydroxide ion is also strongly hydrated in aqueous solutions. Similarly, metal ions do not occur as bare metal ions but as aquo complexes.

THE BRØNSTED CONCEPT. The fact that hydrogen ions cannot exist unhydrated in water solution is incompatible with the notion of a simple dissociation of an acid, the ionization of an acid in water may more logically be represented as a reaction of the acid with the water

$$HCl + H_2O = H_3O^+ + Cl^- \tag{1}$$

The function of a base in its reaction with water is the opposite of that of an acid

$$NH_3 + H_2O = NH_4^+ + OH^- \tag{2}$$

This general idea has led to the very broad concept of acids and bases proposed by Brønsted. (The same concept was suggested by Lowry.) Accordingly an acid is simply defined as any substance that can donate a proton to any other substance, and a base as any substance that accepts a proton from another substance, that is, an acid is a proton donor and a base is a proton acceptor. Thus a proton transfer can occur only if an acid reacts with a base

$$
\begin{aligned}
Acid_1 &= Base_1 + proton \\
proton + Base_2 &= Acid_2 \\
\hline
Acid_1 + Base_2 &= Acid_2 + Base_1
\end{aligned}
\tag{3}
$$

Further illustrations of such proton transfers are:

$$Acid_1(A_1) + Base_2(B_2) = Acid_2(A_2) + Base_1(B_1)$$
$$(solvent)$$

Perchloric acid	$HClO_4$	$+ H_2O$	$= H_3O^+$	$+ ClO_4^-$	(4a)
Carbonic acid	H_2CO_3	$+ H_2O$	$= H_3O^+$	$+ HCO_3^-$	(4b)
Bicarbonate	HCO_3^-	$+ H_2O$	$= H_3O^+$	$+ CO_3^{2-}$	(4c)
Ammonium	NH_4^+	$+ H_2O$	$= H_3O^+$	$+ NH_3$	(4d)
Ammonium	NH_4^+	$+ C_2H_5OH$	$= C_2H_5OH_2^+$	$+ NH_3$	(4e)
Acetic acid†	HAc	$+ NH_3$	$= NH_4^+$	$+ Ac^-$	(4f)
Water	H_2O	$+ H_2O$	$= H_3O^+$	$+ OH^-$	(4g)
Water	H_2O	$+ NH_3$	$= NH_4^+$	$+ OH^-$	(4h)

† HAc and Ac^- stand for acetic acid and the acetate ion, respectively.

In the same way the reaction of bases accepting protons from acids can be illustrated:

$$B_1 + A_2 = B_2 + A_1$$
$$(solvent)$$

Ammonia†	NH_3	$+ H_2O$	$= OH^-$	$+ NH_4^+$	(5a)
Cyanide	CN^-	$+ H_2O$	$= OH^-$	$+ HCN$	(5b)
Bicarbonate	HCO_3^-	$+ H_2O$	$= OH^-$	$+ H_2CO_3$	(5c)
Carbonate	CO_3^{2-}	$+ H_2O$	$= OH^-$	$+ HCO_3^-$	(5d)
Ammonia	NH_3	$+ C_2H_5OH$	$= C_2H_5O^-$	$+ NH_4^+$	(5e)
Amine	RNH_2	$+ HAc$	$= Ac^-$	$+ RNH_3^+$	(5f)
Hydroxide	OH^-	$+ NH_3$	$= NH_2^-$	$+ H_2O$	(5g)

† The ammonia molecule is represented by NH_3 rather than NH_4OH. The best evidence available indicates that no NH_4OH molecule exists. NH_3 may be loosely bound (hydrogen bonding) to a number of H_2O molecules.

Reaction (4g) illustrates the self-ionization of the solvent water, that is, water is both a proton donor and a proton acceptor, an acid and a base in the Brønsted sense. Similarly, self-ionization in liquid NH_3 is represented by

$$NH_3 + NH_3 = NH_4^+ + NH_2^-$$

and self-ionization in the solvent H_2SO_4 is represented by

$$H_2SO_4 + H_2SO_4 = H_3SO_4^+ + HSO_4^-$$

Reactions of a salt constituent, cation or anion, with water (4d, 4e, 5b, 5c, 5d) have been referred to as *hydrolysis* reactions. Within the framework of the Brønsted theory, the term hydrolysis is no longer necessary since in principle there is no difference involved in the protolysis of a molecule and that of a cation or anion to water.

Metal ions, like hydrogen ions, do exist in aqueous medium as hydrates. Many metal ions coordinate 4 or 6 molecules of H_2O per ion. H_2O can act as a weak acid. The acidity of the H_2O molecules in the hydration shell of a metal ion is much larger than that of water. This enhancement of the acidity of the coordinated water may, in a primitive model, be visualized as the result of the repulsion of the protons of H_2O molecules by the positive charge of the metal ion or as a result of the immobilization of the lone electron pair of the hydrate–H_2O molecule. Thus hydrated metal ions are acids

$$[Al(H_2O)_6]^{3+} + H_2O = H_3O^+ + [Al(OH)(H_2O)_5]^{2+} \qquad (6)$$

To a first approximation their acidity increases with the decrease of the radius and an increase of charge of the central ion. Similarly, the acidity of boric acid H_3BO_3 or $B(OH)_3$ can be represented formally as

$$H_3BO_3(H_2O)_x + H_2O = B(OH)_4^-(H_2O)_{x-1} + H_3O^+ \qquad (7)$$

because the borate ion is $B(OH)_4^-$ and not $H_2BO_3^-$. Acidic properties of some substances in aqueous solutions can be interpreted in terms of proton transfer only by assuming hydration of the substance and loss of protons from the primary hydration shell. This is demonstrated in the acidity of metal ions.

In the illustration given above for proton transfer reactions (4 and 5) Acid$_1$ and Base$_1$ or Acid$_2$ and Base$_2$ form *conjugate acid–base pairs*. Thus chloride is the conjugate base of hydrogen chloride; the latter is the conjugate acid of chloride. The conjugate acid of the base water is the hydronium ion, and the conjugate base of the acid water is the hydroxide ion.

Many acids can donate more than one proton. Examples are H_2CO_3, H_3PO_4 and $[Al(H_2O)_6]^{3+}$. These acids are referred to as *polyprotic acids*. Similarly, bases that can accept more than one proton, for example, OH^-, CO_3^{2-} and NH_2^- are polyprotic bases. Many important substances, for example, proteins or polyacrylic acids, so-called polyelectrolytic acids or bases, contain a large number of acidic or basic groups.

The *Lewis concept* of acids and bases (G. N. Lewis, 1923) interprets the combination of acids with bases in terms of the formation of a coordinate covalent bond. A Lewis acid can accept and share a lone pair of electrons donated by a Lewis base. Because protons readily attach themselves to lone electron pairs, Lewis bases are also Brønsted bases. Lewis acids, however, include a large number of substances in addition to proton donors: for example, metal ions, acidic oxides or atoms. The Lewis concept will be discussed more fully in Chapter 6.

3-3 The Strength of an Acid or Base

The strength of an acid or base is measured by its tendency to donate or accept a proton, respectively. Thus a weak acid is one that has a weak proton donating tendency; a strong base is one that has a strong tendency to accept protons. It is, however, difficult to define an "absolute" strength of an acid or base since the extent of proton transfer (protolysis) depends not only on the tendency of proton donation by $Acid_1$, but also on the tendency of proton acceptance by $Base_2$. Under these circumstances the *relative* strengths of acids are measured with respect to a standard $Base_2$ — usually the solvent. In aqueous solutions, the acid strength of a conjugate acid–base pair, $HA-A^-$, is measured relative to the conjugate acid–base system of H_2O, that is, $H_3O^+-H_2O$. In a similar way the relative base strength of a conjugate base–acid pair, $B-HB^+$, is defined in relation to the base–acid system of water, OH^--H_2O.

The rational measure of the strength of the acid HA relative to H_2O as proton acceptor is given by the equilibrium constant for the proton transfer reaction

$$HA + H_2O = H_3O^+ + A^-; \quad K_1 \qquad (8)$$

which may be represented formally by two steps:

$$HA = proton + A^-; \quad K_2 \qquad (9)$$

$$H_2O + proton = H_3O^+; \qquad K_3 \qquad (10)$$

Because the concentration (activity) or water is essentially constant in dilute aqueous solutions ($\sim 55\ M$), the hydration of the proton can be ignored in defining acid–base equilibria. Because the equilibrium activity of the proton and of H_3O^+ are not known separately, the thermodynamic convention sets the standard free energy change $\Delta G°$ for reaction 10 equal to zero, that is, $K_3 = 1$. In dealing with dilute solutions we can, because of this convention, represent the aquo hydrogen ion by $H^+(aq)$, or more conveniently by H^+; that is,

$$[H^+] \equiv [H^+(aq)] = \sum_n [H(H_2O)_n{}^+(aq)] \qquad (11)$$

and the free energy change ΔG involved in the proton transfer reaction 8 may be expressed in terms of the equilibrium constant of reaction 9, that is, the acidity constant of the acid HA, K_{HA}. Ignoring activity coefficients, we have

$$K_2 = K_1 = K_2 K_3 = K_{HA} = \frac{[H^+][A^-]}{[HA]} \tag{12}$$

which upon rearrangement gives

$$pH = pK_{HA} + \log \frac{[A^-]}{[HA]} \tag{13}$$

Recalling that $\Delta G° = -RT \ln K$, we can write

$$pH = \frac{\Delta G°}{2.3RT} + \log \frac{[A^-]}{[HA]} \tag{14}$$

or

$$\frac{\Delta G}{2.3RT} = \frac{\Delta G°}{2.3RT} + \log \frac{[A^-]}{[HA]} \tag{15}$$

(where 2.3 is approximately ln 10).

pH *as a Measure of the Proton Free Energy Level*

After *arbitrarily* assigning a value of zero for the standard free energy of formation of water, a scale of relative free energies of proton transfers can be established (14 and 15). Figure 3-1 gives such a relative scale for a few representative acid–base systems in aqueous solutions. The proton-transfer energy may be expressed in several ways: in terms of electronvolts (or proton-volts) per proton, in kcal mole^{-1} or in terms of pH. The concept of proton-free energy levels has been advanced by Gurney [2]. It can also be applied to electrons and complex-forming ligands [3]. In Figure 3-1 the scale on the left counts downward and measures the energy required for the transfer of one mole of protons from an acid to H_2O. The scale on the right (counting upward) measures the energy required to transfer a mole of protons from water to a base. For example, to raise a proton from a water molecule to CO_3^{2-} takes 0.22 eV. A corresponding energy (-0.22 eV) is gained if a proton is transferred from HCO_3^- to OH^-. When the population of occupied or unoccupied proton levels of a system is equal, that is, when $[HCO_3^-] = [CO_3^{2-}]$ the pH is given by $pK_{HCO_3^-}$, cf. (13). Thus the pH is a measure of the average proton-free energy per proton and depends on the relative occupation of levels with protons and the energy values of the levels present [2, 3].

[2] R. W. Gurney, *Ionic Processes in Solution*, McGraw-Hill, New York, 1953.
[3] C. N. Reilley, in *Electroanalytical Principles* (by R. W. Murray and C. N. Reilley), Wiley-Interscience, New York, 1963.

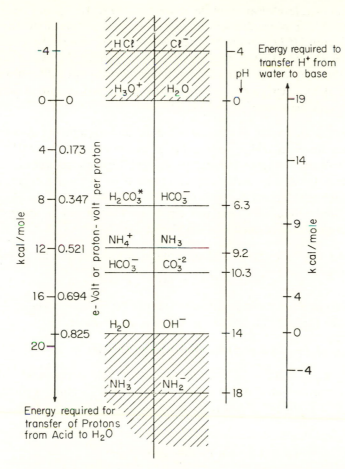

Figure 3-1   Proton-free energy levels. pH is a measure (intensity) of the average proton free energy per proton and depends on the relative proton population and energy values of the levels present.

If protons are added to the system represented in Figure 3-1, they will first tend to fill up the lowest vacant level, that is, the level of lowest free energy, which is that of hydroxide ion. After most of the OH^- ions are converted to H_2O, the next highest level, that is, CO_3^{2-}, will become occupied upon continued addition of protons. This process continues until the H_2O level becomes occupied with protons. Depopulation with protons occurs in a similar way from high to low proton-free energy levels.

The separation between the free energy levels of two systems gives the energy that must be supplied to raise a proton from a lower occupied to a higher occupied level. For example, if we dissolve CO_3^{2-} (from Na_2CO_3) in water, HCO_3^- ions are formed because some protons are lifted from H_2O to the vacant HCO_3^- level by thermal agitation. Since the energy gap between H_2O and H_2CO_3 is rather large (many times kT; $kT = 0.59$ kcal mole^{-1}) the thermal agitation at room temperature is not sufficient to maintain more than a few protons in the higher level, that is, only a very small fraction would be present as H_2CO_3.

STRONG ACIDS AND BASES. As Figure 3-1 illustrates, in aqueous systems some acids are stronger than H_3O^+ and some bases are stronger than OH^-. Water exerts a leveling influence because of its very high concentration, and pH values much lower than zero or much higher than 14 cannot be achieved in dilute aqueous solutions. In such solutions acids stronger than H_3O^+ and bases stronger than OH^- are not stable as protonated or deprotonated species, respectively (shaded areas in Figure 3-1).

SELF-IONIZATION OF WATER. In all aqueous solutions the autoprotolysis

$$H_2O + H_2O = H_3O^+ + OH^- \tag{16}$$

has to be considered. In dilute aqueous solutions ($\{H_2O\} = 1$) the equilibrium constant for (16), usually called the ion product of water, is

$$K_W = \{OH^-\}\{H_3O^+\} \equiv \{OH^-\}\{H^+\} \tag{17}$$

At 25°C, $K_W = 1.008 \times 10^{-14}$ and the pH $= 7.00$ corresponds to exact neutrality in pure water ($[H^+] = [OH^-]$). Because K_W changes with temperature (Table 3-1) the pH of neutrality also changes with temperature.

Table 3-1   Ion Product of Water†

°C	K_W	pK_W
0	0.12×10^{-14}	14.93
15	0.45×10^{-14}	14.35
20	0.68×10^{-14}	14.17
25	1.01×10^{-14}	14.00
30	1.47×10^{-14}	13.83
50	5.48×10^{-14}	13.26

† $\log K_W = 4470.99/T + 6.0875 - 0.01706T$ ($T =$ absolute temperature). From Harned and Owen, *The Physical Chemistry of Electrolytic Solutions*, Reinhold, New York, 1958.
Reproduced with permission from Reinhold Publishing Corp.

Equation 17 interrelates the acidity constant of an acid with the basicity constant of its conjugate base; for example, for the acid–base pair, HB^+–B, the basicity constant for the reaction

$$B + H_2O = OH^- + BH^+$$

is

$$K_B = \frac{\{HB^+\}\{OH^-\}}{\{B\}}$$

and the acidity constant for

$$HB^+ + H_2O = H_3O^+ + B$$

is

$$K_{HB^+} = \frac{\{H^+\}\{B\}}{\{HB^+\}}$$

Thus

$$K_{HB^+} = \frac{\{H^+\}\{B\}}{\{HB^+\}} = \frac{\{H^+\}\{OH^-\}}{K_B}$$

or

$$K_W = K_{HB^+} \cdot K_B \tag{18}$$

Thus either the acidity or basicity constant describes fully the protolysis properties of an acid–base pair. The stronger the acidity of an acid, the weaker is the basicity of its conjugate base and vice versa. For illustration purposes Table 3-2 lists a series of acids and bases in the order of their relative strength.

Composite Acidity Constants

It is not always possible to specify a protolysis reaction unambiguously in terms of the actual acid or base species. As it is possible to ignore the extent of hydration of the proton in dilute aqueous solutions, the hydration of an acid or base species can be included in a composite acidity constant; for example, it is difficult analytically to distinguish between $CO_2(aq)$ and H_2CO_3. The equilibria are

$$H_2CO_3 = CO_2(aq) + H_2O; \qquad K = \frac{\{CO_2(aq)\}}{\{H_2CO_3\}} \tag{19}$$

$$H_2CO_3 = H^+ + HCO_3^-; \qquad K_{H_2CO_3} = \frac{\{H^+\}\{HCO_3^-\}}{\{H_2CO_3\}} \tag{20}$$

Table 3-2 Acidity and Basicity Constants of Acids and Bases in Aqueous Solutions (25°C)

Acid†		$-$Log Acidity Constant, pK (approximate)	Base‡	$-$Log Basicity Constant, pK (approximate)
$HClO_4$	Perchloric acid	-7	ClO_4^-	21
HCl	Hydrogen chloride	~-3	Cl^-	17
H_2SO_4	Sulfuric acid	~-3	HSO_4^-	17
HNO_3	Nitric acid	-1	NO_3^-	15
H_3O^+	Hydronium ion	0	H_2O	14
H_3PO_4	Phosphoric acid	2.1	$H_2PO_4^-$	11.9
$[Fe(H_2O)_6]^{3+}$	Aquo ferric ion	2.2	$[Fe(H_2O)_5(OH)]^{2+}$	11.8
CH_3COOH	Acetic acid	4.7	CH_3COO^-	9.3
$[Al(H_2O)_6]^{3+}$	Aquo aluminum ion	4.9	$[Al(H_2O)_5(OH)]^{2+}$	9.1
$H_2CO_3^*$	Carbon dioxide(*)	6.3	HCO_3^-	7.7
H_2S	Hydrogen sulfide	7.1	HS^-	6.9
$H_2PO_4^-$	Dihydrogen phosphate	7.2	HPO_4^{2-}	6.8
$HOCl$	Hypochlorous acid	7.6	OCl^-	6.4
HCN	Hydrogen cyanide	9.2	CN^-	4.8
H_3BO_3	Boric acid	9.3	$B(OH)_4^-$	4.7
NH_4^+	Ammonium ion	9.3	NH_3	4.7
$Si(OH)_4$	O-Silicic acid	9.5	$SiO(OH)_3^-$	4.5
HCO_3^-	Bicarbonate	10.3	CO_3^{2-}	3.7
H_2O_2	Hydrogen peroxide	—	HO_2^-	2.3
$SiO(OH)_3^-$	Silicate	12.6	$SiO_2(OH)_2^{2-}$	1.4
HS^-	Bisulfide	14	S^{2-}	0
H_2O	Water	14	OH^-	0
NH_3	Ammonia	~23	NH_2^-	-9
OH^-	Hydroxide ion	~24	O^{2-}	~-10
CH_4	Methane	~34	CH_3^-	~-20

† In order of decreasing acid strength.
‡ In order of increasing base strength.

and a combination of (19) and (20) gives

$$\frac{\{H^+\}\{HCO_3^-\}}{(\{H_2CO_3\} + \{CO_2(aq)\})} = \frac{K_{H_2CO_3}}{1 + K} = K_{H_2CO_3^*} \tag{21}$$

where $K_{H_2CO_3^*}$ is the composite acidity constant. Under conditions in which activities can be considered equal to concentration, the sum $\{H_2CO_3\} + \{CO_2\}$ is approximately equal to the sum of the concentrations, $[H_2CO_3] + [CO_2(aq)]$. $[H_2CO_3^*]$ is defined as the analytic sum of $[CO_2(aq)]$ and $[H_2CO_3]$. The true H_2CO_3 is a much stronger acid ($pK_{H_2CO_3} = 3.8$) than the composite $H_2CO_3^*$ ($pK_{H_2CO_3^*} = 6.3$) because less than 0.3% of the CO_2 is hydrated at 25°C.

An amino acid can lose two protons by two different paths:

$$^+NH_3R\ COOH
\begin{array}{c}
\nearrow\ \ NH_3^+\ RCOO^-\ \ \searrow \\
\\
\searrow\ \ NH_2\ RCOOH\ \ \nearrow
\end{array}
NH_2\ RCOO^- \tag{22}$$

Four microscopic constants can be defined, but potentiometrically only two composite (macroscopic) acidity constants can be determined.

3-4   Activity and pH Scales

In dealing with quantitative aspects of chemical equilibrium, we inevitably are faced with the problem either of evaluating or maintaining constant the activities of the ions under consideration. G. N. Lewis (1907) defined the chemical activity of a solute A, $\{A\}$, and its relationship to chemical concentration of that solute, $[A]$, by

$$\mu_A = k_A + RT \ln \{A\} = k_A + RT \ln [A] + RT \ln f_A \tag{23}$$

where μ_A is the chemical potential of species A and k_A is a constant that identifies the concentration scale adopted (moles liter^{-1}, moles kilogram^{-1} or mole fraction) and corresponds to the value of μ at $\{A\} = 1$. It will be recognized that k_A corresponds to the more familiar standard state chemical potential μ_A° and to the standard partial molar Gibbs energy $\bar{G}_A^\circ$. (See Section 2-4 and Table 2-3.)

Any activity can be written as the product of a concentration and activity coefficient. Here we usually express concentration in terms of moles liter^{-1} of solution (molarity, M). Occasionally, concentration may be expressed in

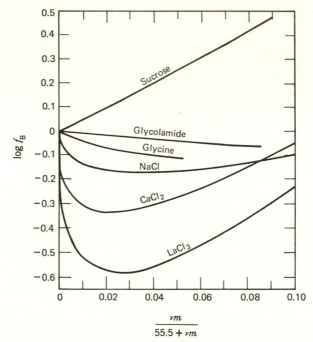

Figure 3-2   A comparison of activity coefficients of electrolytes and nonelectrolytes as a function of concentration. m = moles of solute per kg of solvent (molality); ν = number of moles of ions formed from one mole of electrolyte; 1 kg solvent contains 55.5 moles of water. From Robinson and Stokes (1955). Reproduced with permission from Butterworths, Inc., London, England.

terms of moles kilogram^{-1} of solvent (molality, m). Usually the difference between molarity and molality is small in dilute solutions, especially in comparison to the uncertainties involved in determining equilibrium constants or in estimating activity coefficients; for example, a 1.00 m solution of NaCl is 0.98 M (25°C).

As already summarized in Table 2-3, two activity scales are useful:

THE INFINITE DILUTION SCALE. This activity convention is defined in such a way that the activity coefficient $f = \{A\}/[A]$ approaches unity as the concentration of all solutes approaches zero.

THE IONIC MEDIUM SCALE. This convention can be applied to solutions that contain a "swamping" concentration of inert electrolyte in order to maintain a constant ionic medium. The activity coefficient $f = \{A\}/[A]$ becomes unity as the solution approaches the pure ionic medium, that is, when all concentrations other than the medium ions approach zero. If the concentration of the medium electrolyte is more than approximately 10 times the concentration of the species under consideration, activity coefficients remain very close to 1. Thus generally no extrapolation is necessary.

As (23) illustrates, a change in the activity scale convention merely changes k_A. In an ideal constant ionic medium (23) becomes

$$\mu_A = k'_A + RT \ln [A] \tag{24}$$

As pointed out by Sillén [4], both activity scales are thermodynamically equally well defined. In constant ionic medium, activity ($\approx$ concentration) can frequently be determined by means of emf methods. In the last few decades an increasing number of researchers in aqueous solution chemistry have been using the ionic medium activity scale.

pH CONVENTIONS. The original definition of pH (Sørensen, 1909) was

$$pH = -\log [H^+] \tag{25}$$

Within the infinite dilution concept pH may be defined in terms of hydrogen ion activity:

$$p^aH = -\log \{H^+\} = -\log [H^+] - \log f_{H^+} \tag{26}$$

Within the ionic medium convention, Sørensen's original definition (25) may be used operationally because one can usually measure $-\log [H^+] = -\log \{H^+\}$ rather accurately (Sillén [4]). An electrode system is calibrated with solutions of known concentrations of strong acid (e.g., $HClO_4$) which are adjusted with an electrolyte to the appropriate ionic strength,† the observed potentiometer (pH meter) reading being compared with $[H^+]$.

An operational definition endorsed by the International Union of Pure and Applied Chemistry (IUPAC) and based on the work of Bates [5] determines pH relative to that of a standard buffer (where pH has been estimated in terms of p^aH) from measurements on cells with liquid junction. This operational pH is not rigorously identical with p^aH defined in (26) because liquid junction potentials and single ion activities cannot be evaluated with nonthermodynamic assumptions. In dilute solutions of simple electrolytes (ionic strength, $I < 0.1$) the measured pH corresponds to within ± 0.02 to p^aH [5]. Measurement of pH by emf methods is discussed in Chapter 7.

---

† Ionic strength, I, is a measure of the interionic effect resulting primarily from electrical attraction and repulsions between the various ions; it is defined by the equation $I = \frac{1}{2} \sum_i C_i Z_i^2$. The summation is carried out for all types of ions, cations and anions, in the solution.

[4] L. G. Sillén, in *Equilibrium Concepts in Natural Water Systems*, Advances in Chemistry Series, No. 67, American Chemical Society, Washington, D.C., 1967, p. 45.

[5] R. Bates, in *Treatise on Analytical Chemistry*, I. M. Kolthoff and P. J. Elving, Eds., Part I, Volume 1, Interscience, New York, 1959, p. 361.

ACTIVITY OF WATER. If the standard state for water in aqueous solutions is taken as the pure solvent, then for all dilute solutions $\{H_2O\} = 1$. In salt solutions (e.g., seawater) the activity of water may then be defined as the ratio of the vapor pressure of the salt solution p_s to that of pure water p_{H_2O}, that is $\{H_2O\} = p_s/p_{H_2O}$, but even in seawater $\{H_2O\}$ does not fall below 0.98.

Operational Acidity Constants

According to the different activity conventions the following equilibrium expressions may be defined.

1. For the infinite dilution scale

$$K = \frac{\{H^+\}\{B\}}{\{HB\}} \tag{27}$$

In this and subsequent equations charges are omitted; B can be any base of any charge.

2. In a constant ionic medium, the concentration quotient becomes the equilibrium constant

$$^cK = \frac{[H^+][B]}{[HB]} \tag{28}$$

More exactly, on the ionic medium scale the equilibrium constant may be defined as the limiting value (as the solution composition approaches that of the pure ionic medium) for the equilibrium concentration quotient L. In usual measurements at low reactant concentrations, the deviations of L from ^cK are usually smaller than the experimental errors; hence it is preferable to set $L = K$ rather than to extrapolate (Sillén, *loc. cit.*).

3. A so-called "mixed acidity constant" is frequently used

$$K' = \frac{\{H^+\}[B]}{[HB]} \tag{29}$$

This convention is most useful when pH is measured according to the IUPAC convention (pH $\approx$ p^{a}H) but the conjugate acid–base pair is expressed in concentrations.

In a similar way to the expression of activity as a product of concentration and activity coefficient, an acidity constant can be expressed in terms of a product of an equilibrium concentration quotient $L = ([H^+][B])/[HB]$ and an activity coefficient factor. Correspondingly we can interrelate the various constants defined above in the following way:

$$K = L\frac{f_{H^+}f_B}{f_{HB}} \tag{30}$$

$$^cK = \lim L \quad \text{(pure ionic medium)} \tag{31}$$

$$K' = K\frac{f_{HB}}{f_B} \tag{32}$$

Expressions for Activity Coefficients

The theoretical expressions based on the Debye-Hückel limiting law together with more empirical expressions are given in Table 3-3. In defining the mean activity coefficient $f_\pm$ of a solute, z^2 in the equations of Table 3-3 should be replaced by $z_+ z_-$ where the charges are taken without regard of the sign.

Single ion activity coefficients are constructs; they are not measurable individually; only ratios or products of ionic activity coefficients are measurable. The use of single ion activity coefficients greatly simplifies calculations.

Table 3-4 lists some activity coefficients as calculated from the extended Debye-Hückel limiting law for various values of I. In dilute solutions such calculated values agree well with experimental data for mean activity coefficients of simple electrolytes. At higher concentrations the Davies equation usually represents the experimental data better.

Table 3-3   Individual Ion Activity Coefficients

Approximation	Equation †		Approximate Applicability [ionic strength (M)]
Debye-Hückel	$\log f = -AZ^2\sqrt{I}$	(1)	$< 10^{-2.3}$
Extended Debye-Hückel	$= -AZ^2\dfrac{\sqrt{I}}{1 + Ba\sqrt{I}}$	(2)	$< 10^{-1}$
Güntelberg	$= -AZ^2\dfrac{\sqrt{I}}{1 + \sqrt{I}}$	(3)	$< 10^{-1}$ useful in solutions of several electrolytes
Davies	$= -AZ^2\left(\dfrac{\sqrt{I}}{1 + \sqrt{I}} - 0.2I\right)$	(4)‡	< 0.5

† I (ionic strength) $= \frac{1}{2}\sum C_i Z_i^2$; $A = 1.82 \times 10^6(\epsilon T)^{\frac{2}{3}}$ (where ϵ = dielectric constant); $A \approx 0.5$ for water at 25°C; Z = charge of ion; $B = 50.3(\epsilon T)^{-\frac{1}{2}}$; $B \approx 0.33$ in water at 25°C; a = adjustable parameter (Ångström units) corresponding to the size of the ion. (See Table 3-4.)
‡ Davies has recently proposed 0.3 (instead of 0.2) as a coefficient for the last term in parenthesis.

Table 3-4 Parameter a and Individual Ion Activity Coefficients

Ion Size Parameter, a, Ångströms, $= 10^{-8}$ cm†	Ion	Activity Coefficients Calculated with (2) of Table 3-3 for Ionic Strength				
		10^{-4}	10^{-3}	10^{-2}	0.05	10^{-1}
9	H^+	0.99	0.97	0.91	0.86	0.83
	Al^{3+}, Fe^{3+}, La^{3+}, Ce^{3+}	0.90	0.74	0.44	0.24	0.18
8	Mg^{2+}, Be^{2+}	0.96	0.87	0.69	0.52	0.45
6	Ca^{2+}, Zn^{2+}, Cu^{2+}, Sn^{2+}, Mn^{2+} Fe^{2+}	0.96	0.87	0.68	0.48	0.40
5	Ba^{2+}, Sr^{2+}, Pb^{2+}, CO_3^{2-}	0.96	0.87	0.67	0.46	0.38
4	Na^+, HCO_3^-, $H_2PO_4^-$, CH_3COO^-	0.99	0.96	0.90	0.81	0.77
	SO_4^{2-}, HPO_4^{2-}	0.96	0.87	0.66	0.44	0.36
	PO_4^{3-}	0.90	0.72	0.40	0.16	0.10
3	K^+, Ag^+, NH_4^+, OH^-, Cl^- ClO_4^-, NO_3^-, I^-, HS^-	0.99	0.96	0.90	0.80	0.76

† After J. Kielland, *J. Am. Chem. Soc.*, **59**, 1675 (1937).
Reproduced with permission from American Chemical Society.

In natural water systems and under many experimental conditions, several electrolytes are present together. The limiting laws for activity coefficients can no longer be applied satisfactorily for electrolyte mixtures of unlike charge types. For estimating unknown activity coefficients, Güntelberg proposed that a in (2) of Table 3-3 be taken as 3.0 Å, resulting in a formula containing no adjustable parameter. For aqueous systems containing several electrolytes, (3) of Table 3-3 is most useful.

Activity coefficients defined within the infinite dilution activity scale cannot be formulated theoretically for the ionic medium of seawater. Since the oceans contain an ionic medium of practically constant composition, the ionic medium activity scale might be used advantageously in studying acid–base and other equilibria in seawater.

Example 3-1 *Individual Activity Coefficients.* Estimate $p^\alpha H$ of a 10^{-3} M HCl solution that is 0.05 M in NaCl. The ionic strength is 0.051. Using (3) of Table 3-3, $-\log f_{H^+} = 0.092$. Thus $p^\alpha H = -\log [H^+] + 0.092 = 3.09_2$. Approximately the same result would be obtained from the extended Debye-Hückel limiting law [(1), Table 3-3] $p^\alpha H = -\log [H^+] + 0.113 = 3.11_3$.

Salt Effects on Acidity Constants

The equilibrium condition of an acid–base reaction is influenced by the ionic strength of the solution. With the help of the formulations of the Debye-

Table 3-5   Nonideality Corrections for Mixed Acidity Constants; Numerical Values of Second Term in (33)

Charge Acid, z_{HB}	Charge Base, z_B		Ionic Strength, $I(M)$							
			0.0001	0.0005	0.001	0.002	0.005	0.01	0.05	0.1
+1 0	0 −1	+ −	0.005	0.01	0.015	0.02	0.03	0.05	0.09	0.12
+2 −1	+1 −2	+ −	0.015	0.03	0.05	0.06	0.10	0.14	0.27	0.36
+3 −2	+2 −3	+ −	0.02	0.05	0.08	0.11	0.17	0.23	0.46	0.60

Hückel theory or with empirical expressions (Table 3-3) an estimate of the magnitude of the salt effect can be obtained. For example, the mixed acidity constant K' has been related to K (i.e., the acidity constant valid at infinite dilution) by (32). Using the Güntelberg approximation for the single ion activities we can write instead of (32)

$$pK' = pK + \frac{0.5(z_{HB}^2 - z_B^2)\sqrt{I}}{1 + \sqrt{I}} \tag{33}$$

The numerical values for the correction term [i.e., the second term in (33)] are given in Table 3-5. Since we are frequently interested in the equilibrium concentrations of the various species in a solution of a given pH ($\approx p^aH$), it is convenient to convert K into K'. With the operational K', calculations can be carried out for all species in terms of concentrations with the exception of H^+. In equilibrium calculations, concentration conditions and charge balance or proton conditions are correct only if formulated in terms of concentrations.

Examples 3-2   *Effect of Ionic Strength on* pK'. Estimate the effect of ionic strength on the successive mixed acidity constants (pK' values) of a dilute ($< 10^{-3} M$), neutralized tri-basic acid (e.g., H_3PO_4) in a constant ionic medium of 0.01 M Na_2SO_4 solution. The contribution of the acid and its bases to the ionic strength can be neglected. $I = \frac{1}{2}([Na^+] + 4[SO_4^{2-}]) = 0.03 \ M$. Using (33), we have

$$pK_1' = pK_1 - 0.07$$
$$pK_2' = pK_2 - 0.21$$
$$pK_3' = pK_3 - 0.35$$

3-5   Numerical Equilibrium Calculations

The quantitative evaluation of the systematic relations that determine equilibrium concentrations (or activities) of a solution constitutes a purely mathematical problem which is amenable to exact and systematic treatment.

Any acid–base equilibrium can be described by a system of fundamental equations. The appropriate set of equations is comprised of the equilibrium constant (or mass law) relationships (which define the acidity constants and the ion product of water) and any two equations describing the constitution of the solution, for example, equations describing a concentration and an electroneutrality or proton condition. Table 3-6 gives the set of equations and their mathematical combination for pure solutions of acids, bases or ampholytes in monoprotic or diprotic systems. The problem of the precise calculation in all possible cases has been treated in a truly general way by Ricci [6].

The principle of such equilibrium calculations is best explained by a series of illustrative examples. In this and subsequent examples, a temperature of 25°C is assumed and conditions such that all activity coefficients are equal to one.

Example 3-3   pH *and Equilibrium Composition of a Monoprotic Acid.* Calculate the pH $(= -\log [H_3O^+])$ of a $5 \times 10^{-4} M$ aqueous boric acid solution $[B(OH)_3]$.

In attacking such a problem, it is convenient to proceed systematically through a number of steps.

1. Establish all the *species present* in the solution: H_3O^+, OH^-, $B(OH)_3$, $B(OH)_4^-$. For brevity we call $B(OH)_3 = HB$ and $B(OH)_4^- = B^-$. Since four chemical species in addition to H_2O are involved in solution, four independent mathematical equations are necessary to interrelate the equilibrium concentrations.

2. The *equilibrium constants* that relate the concentrations of the various species must be found. For the boric acid solution the following equilibrium constants can be used.

$$[H^+][OH^-] = K_W = 10^{-14} \tag{i}$$

$$\frac{[H^+][B^-]}{[HB]} = K = 7 \times 10^{-10} \tag{ii}$$

3. A *concentration condition* or a mass balance must be established. Since the analytical concentration C of the HB solution is known, we can write

$$C = 5 \times 10^{-4} M = [HB] + [B^-] \tag{iii}$$

The analytical concentration is the total number of moles of a pure substance that has been added to 1 liter of solution. In our example 5×10^{-4} moles of HB has been added per liter of solution. Some of the HB has protolyzed to form B^-, but the sum of $[HB] + [B^-]$ must equal the number of moles (5×10^{-4}) that was added originally.

[6] J. E. Ricci, *Hydrogen Ion Concentration*, Princeton Univ. Press, Princeton, N.J., 1952.

Table 3-6 [H$^+$] of Pure Aqueous Acids, Bases or Ampholytes

I. Monoprotic

Species†: HA   A   H$^+$ OH$^-$

Equilibrium constants‡:
$$[H^+][A]/[HA] = K \tag{1}$$
$$[H^+][OH^-] = K_W \tag{2}$$

Concentration condition:
$$[HA] + [A] = C \tag{3}$$

	Acid	Base
Proton condition§	$[H^+] = [A] + [OH^-]$ (4)	$[HA] + [H^+] = [OH^-]$ (5)
Numerical solution	$[H^+]^3 + [H^+]^2 K - [H^+](CK + K_W) - KK_W = 0$ (6)	$[H^+]^3 + [H^+]^2(C + K) - [H^+]K_W - KK_W = 0$ (7)

II. Diprotic

Species†: H$_2$X   HX   X   H$^+$ OH$^-$

Equilibrium constants‡:
$$[H^+][HX]/[H_2X] = K_1 \tag{8}$$
$$[H^+][X]/[HX] = K_2 \tag{9}$$
$$[H^+][OH^-] = K_W \tag{2}$$

Concentration condition:
$$[H_2X] + [HX] + [X] = C \tag{10}$$

Acid (H$_2$X)

Proton condition§:
$$[H^+] = [HX] + 2[X] + [OH^-] \tag{11}$$

Numerical solution:
$$[H^+]^4 + [H^+]^3 K_1 + [H^+]^2 \times (CK_1 + K_1K_2 + K_W) - [H^+]K_1(CK_2 + K_W) - K_1K_2K_W = 0 \tag{14}$$

Ampholyte (NaHX)

Proton condition§:
$$[H_2X] + [H^+] = [X] + [OH^-] \tag{12}$$

Numerical solution:
$$[H^+]^4 + [H^+]^3(C + K_1) + [H^+]^2 \times (K_1K_2 - K_W) - [H^+]K_1(CK_2 + K_W) - K_1K_2K_W = 0 \tag{15}$$

Base (Na$_2$X)

Proton condition§:
$$2[H_2X] + [HX] + [H^+] = [OH^-] \tag{13}$$

Numerical solution:
$$[H^+]^4 + [H^+]^3(2C + K_1) + [H^+]^2 \times (CK_1 + K_1K_2 - K_W) - [H^+]K_1K_2 - K_1K_2K_W = 0 \tag{16}$$

---

† Charges are omitted for acid or base species. Equations given are independent of charge type of the acid.

‡ Equilibrium constants are either ^cK or are defined in terms of the constant ionic medium activity scale.

§ Instead of the proton condition, the electroneutrality equation can be used. Independent of charge type, a combination of electro-neutrality and concentration condition gives the proton condition. Na in NaHX or Na$_2$X is used as a symbol of a nonprotolyzable cation (Li$^+$, K$^+$, ….).

So far we have established three independent equations. One additional relation has to be found in order to have as many equations as unknowns.

4. This additional equation follows from the fact that the solution must be electrically neutral. This can be expressed in a charge balance, or in an *electroneutrality equation*. The total number of positive charges per unit volume must equal the total number of negative charges, that is, the molar concentration of each species is multiplied by its charge:

$$[H^+] = [B^-] + [OH^-] \tag{iv}$$

Instead of writing the electroneutrality equation we can derive a relation called the *proton condition*. If we started making our solution from pure H_2O and HB, we can state that after equilibrium has been reached the number of excess protons must be equal to the number of proton deficiencies. Excess or deficiency of protons is counted with respect to a "zero level" representing the species that were added, that is, H_2O and HB. The number of excess protons is equal to $[H^+]$; the number of proton deficiencies must equal $[B^-] + [OH^-]$. This proton condition gives, as in (iv), $[H^+] = [B^-] + [OH^-]$.

5. Four equations (i–iv) for four unknown concentrations must be solved simultaneously. The exact solution, although straightforward, is tedious. Frequently the problem is readily solved by making suitable approximations. But in order to know what approximations can be made, we have to have a qualitative knowledge of the chemical system within the given concentration range.

We first discuss the exact numerical solution; then we illustrate what we mean by making approximations. In the next section we discuss a method of graphical representation which is very expedient in making such calculations.

EXACT NUMERICAL SOLUTION. A numerical approach may start out by eliminating $[OH^-]$ in (i) and (iv). After this substitution we have

$$[H^+] = \frac{K_w}{([H^+] - [B^-])} \tag{v}$$

We now solve (iii) for [HB] and substitute in (ii), eliminating [HB]:

$$[H^+][B^-] = K(C - [B^-]) \tag{vi}$$

Now we solve (v) for $[B^-]$ and substitute the result in (vi) to obtain a single equation in $[H^+]$:

$$[H^+]^3 + K[H^+]^2 - [H^+](CK + K_w) - KK_w = 0 \tag{vii}$$

A value for $[H^+]$ can be obtained by trial and error from this equation:

$$[H^+]^3 + [H^+]^2(7 \times 10^{-10}) - [H^+](3.6 \times 10^{-13}) - (7 \times 10^{-24}) = 0 \tag{vii$a$}$$

Most numerical methods for solving polynomial equations are based on an initial guess for the answer followed by some iterative procedure for obtaining successively better approximations. A preliminary guess may be obtained by neglecting one or more terms in (vii*a*). A very convenient way to obtain a root for this equation is to plot various values of $f([H^+])$ versus $[H^+]$ and to locate $[H^+]$ where $f([H^+])$ crosses the zero axis. (Alternatively the Newton-Raphson method [7] as well as computer methods [8] can be used to achieve a rather systematic convergence of the successive approximations to the required root.)

The value obtained by trial and error ($[H^+] = 6.1 \times 10^{-7}\ M$) is substituted into (i) to give $[OH^-]$ and into (v) to obtain $[B^-]$. Then $[HB]$ is calculated from (iii). The following results obtain

$$[H^+] = 6.10 \times 10^{-7}\ M\ (pH = 6.21), \qquad [OH^-] = 1.64 \times 10^{-8}\ M$$
$$[B^-] = 5.94 \times 10^{-7}\ M; \qquad [HB] = 4.99 \times 10^{-4}\ M$$

APPROXIMATE SOLUTION. The equations to be solved are the same as those given before. We look for terms in additive equations that are negligible. (Multiplicative terms, even if very small cannot of course be neglected.) It is always necessary to check the final result to determine whether the assumptions made were justifiable.

Our solution presumably is acid, that is, $[H^+] > [OH^-]$; we might therefore try to neglect $[OH^-]$ in (iv) as a first approximation. Then we get

$$[H^+] = [B^-] \tag{viii}$$

Considering (7) a combination of (ii) and (iii) gives

$$\frac{[H^+]^2}{C - [H^+]} = K \tag{ix}$$

which after conversion into the general quadratic equation, $ax^2 + bx + c = 0$ gives

$$[H^+]^2 + [H^+]K - KC = 0 \tag{x}$$

The solution of this quadratic equation according to

$$x = \frac{[-b \pm (b^2 - 4ac)^{1/2}]}{2a}$$

however, frequently meets with some difficulties. Sometimes $4ac \gg b^2$ or $b^2 \gg 4ac$, and the numerical result becomes imprecise because it is computed

[7] J. G. Eberhart and T. R. Sweet, *J. Chem. Educ.*, **37**, 422 (1966).
[8] A. J. Bard and D. M. King, *J. Chem. Educ.*, **42**, 127 (1965).

as the difference of two nearly equal numbers. If $b^2 \ll 4ac$, the quadratic term can be neglected and $x = [-b \pm (-4ac)^{1/2}]/2a$; if $b^2 \gg 4ac$, the linear term is negligible and $x = \pm(-c/a)^{1/2}$. In solving this example the latter approximation can be used: $[H^+] = \pm(KC)^{1/2} = \pm(7 \times 10^{-10} \times 5 \times 10^{-4})^{1/2} = 5.92 \times 10^{-7} M$. Only the positive value is physically meaningful.

Frequently an approximation can also be made in the concentration condition (iii). Either the acid is completely protolyzed ($[B^-] \gg [HB]$) or remains predominantly in the acid form ($[HB] \gg [B^-]$). Since HB is a rather weak acid the latter condition may be true:

$$C = [HB] = 5 \times 10^{-4} M \tag{xi}$$

Substituting this together with the assumption made in (viii) into (ii) gives $[H^+]^2/C = K$, and $[H^+] = 5.92 \times 10^{-7} M$. This answer is the same as that obtained using the approximation of (viii) only. With the approximations used, the following are obtained

$$[H^+] = 5.92 \times 10^{-7} M \ (\text{pH} = 6.23); \qquad [OH^-] = 1.69 \times 10^{-8} M$$
$$[B^-] = 5.92 \times 10^{-7} M; \qquad [HB] = 5.0 \times 10^{-4} M$$

Comparing the approximate results with the "exact" results given before, we see that the values for $[H^+]$ and $[OH^-]$ are off by about 3%. For most calculations of practical interest such an error is tolerable and frequently smaller than the uncertainty of the equilibrium constants used or the activity corrections made, so we may say that the approximations introduced by (viii) and (xi) were justified.

Example 3-4   pH *of a Strong Acid.*   Compute the equilibrium concentrations of all the species of a $2 \times 10^{-4} M$ HCl solution (25°C). This problem is essentially analogous to the previous one and can be approached the same way:

1. Mass laws:

$$[H^+][OH] = K_W = 10^{-14} \tag{i}$$

$$\frac{[H^+][Cl^-]}{[HCl]} = K = 10^{+3.0} \tag{ii}$$

2. Concentration condition:

$$[HCl] + [Cl^-] = C = 2 \times 10^{-4} \tag{iii}$$

3. Electroneutrality or proton condition:

$$[H^+] = [Cl^-] + [OH^-] \tag{iv}$$

As in Example 3-3, the exact numerical solution would lead first to (v)

$$[H^+]^3 + K[H^+]^2 - (K_W + CK)[H^+] - K_W K = 0 \tag{v}$$

Solving this by trial and error gives

$$[H^+] = 2.0 \times 10^{-4}\,M; \qquad [HCl] = 4 \times 10^{-11}\,M$$
$$[Cl^-] = 2.0 \times 10^{-4}\,M; \qquad [OH^-] = 5.0 \times 10^{-11}\,M$$

Since HCl has a large protolysis constant, $[H^+] \gg [OH^-]$ and correspondingly $[OH^-] \ll [Cl^-]$, the electroneutrality condition reduces to $[H^+] = [Cl^-]$. Furthermore, $[HCl] \ll [Cl^-]$; correspondingly the reaction $HCl + H_2O = H_3O^+ + Cl^-$ has gone very far to the right. As shown in this example the ratio of acid to base, that is, $[HCl]/[Cl^-]$, is extremely small. Strong acids are virtually completely protolyzed.

Example 3-5 *Weak Base.* Compute the equilibrium concentrations of a $10^{-4.5}\,M$ sodium acetate (NaAc) solution.

1. Species:

$$HAc,\ Ac^-,\ H^+,\ OH^-,\ Na^+$$

2. Constants:

$$\frac{[H^+][Ac^-]}{[HAc]} = K \qquad \text{or} \qquad \frac{[HAc][OH^-]}{[Ac^-]} = K_B \qquad \text{(i)}$$
$$pK = 4.70, \qquad pK_B = 9.30$$
$$[H^+][OH^-] = K_w$$

3. Concentration condition:

$$[HAc] + [Ac^-] = C = [Na^+] \qquad \text{(ii)}$$

4. Electroneutrality:

$$[Na^+] + [H^+] = [Ac^-] + [OH^-];\ \text{since } [Na^+] = [HAc] + [Ac^-]\ \text{(iii)}$$

this becomes

$$[HAc] + [H^+] = [OH^-] \qquad \text{(iv)}$$

and is identical with the proton condition.

Approximations in the proton condition similar to those used in the previous examples are not permissible in this case. However, an approximation might be possible in the concentration condition. Because Ac^- is a weak base, $[Ac^-] > [HAc]$ and $[Ac^-] \approx C$. If we combine this approximation with the proton condition and the two equilibrium constants we obtain

$$[H^+]^2(C + K) = KK_w \qquad \text{(v)}$$

which results in $[H^+] = 10^{-7.2}$. Subsequent calculations give $[HAc] = 10^{-7.01}$ and $[Ac^-] = 10^{-4.51}$, thus confirming that the result is consistent with the approximation made.

If one is not aware of the possibility of an approximation, one can always attempt to solve the exact equation [(7), Table 3-6].

Example 3-6 *Diprotic System; Ampholyte.* Calculate the equilibrium pH of a solution prepared by diluting $10^{-3.7}$ mole of sodium hydrogen phthalate

to 1 liter with water $\left(C_6H_4 \begin{array}{c} \text{COOH} \\ \diagup \\ \diagdown \\ \text{COONa} \end{array} = \text{NaHP} \right)$. The acidity constants at

the appropriate temperature (25°C) for H_2P and HP^- are $10^{-2.95}$ and $10^{-5.41}$, respectively.

1. Species:

$$H_2P, \ HP^-, \ P^{2-}, \ H^+, \ OH^-, \ Na^+$$

2. Equilibrium constants:

$$\frac{[H^+][HP^-]}{[H_2P]} = K_1; \qquad \frac{[H^+][P^{2-}]}{[HP^-]} = K_2 \qquad \text{(i)}$$

$$[H^+][OH^-] = K_W$$

3. Concentration condition:

$$P_T = [H_2P] + [HP^-] + [P^{2-}] \qquad \text{(ii)}$$

4. Proton condition:

$$[H_2P] + [H^+] = [P^{2-}] + [OH^-] \qquad \text{(iii)}$$

Solution of (15) (from Table 3-6) by trial and error gives pH = 4.55 and a closer analysis shows that $[H_2P] < [HP^-] > [P^{2-}]$ and that the proton condition can be approximated by $[H^+] > [H_2P]$ and $[P^{2-}] \gg [OH^-]$.

Example 3-7 *Mixture of Acid and Base.* Calculate the pH of a solution containing 10^{-3} moles of NH_4Cl and 2×10^{-4} moles of NH_3 per liter of aqueous solution.

1. Species:

$$NH_4^+, \ NH_3, \ H^+, \ OH^-, \ HCl, \ Cl^-$$

2. Equilibria, in addition to the ion product of water:

$$\frac{[NH_3][H^+]}{[NH_4^+]} = K = 10^{-9.3} \qquad \text{(i)}$$

As shown by Example 3-4, the acidity equilibrium for HCl can be ignored because $[HCl] \ll [Cl^-]$. Similarly HCl can be neglected in the subsequent concentration and proton conditions.

3. Concentration condition:

$$[NH_4^+] + [NH_3] = C_0(NH_4Cl) + C_0(NH_3) = 1.2 \times 10^{-3} \, M \quad \text{(ii)}$$

4. Electroneutrality condition:

$$[NH_4^+] = [Cl^-] + [OH^-] - [H^+] \quad \text{(iii)}$$

Because $[Cl^-] = C_0(NH_4Cl)$ and considering the concentration condition we can also write

$$[NH_3] = C_0(NH_3) - [OH^-] + [H^+] \quad \text{(iv)}$$

Combining these equations with the acidity equilibrium we obtain

$$[H^+] = K \frac{C_0(NH_4Cl) + [OH^-] - [H^+]}{C_0(NH_3) - [OH^-] + [H^+]} \quad \text{(v)}$$

Neglecting as a justified approximation $[H^+]$ and $[OH^-]$ in numerator and denominator gives

$$[H^+] = 2.5 \times 10^{-9}; \qquad pH = 8.6$$

Example 3-8 *Volatile Acid or Base.*   Estimate the pH of an aqueous electrolyte solution exposed to a partial pressure of NH_3 of 10^{-4} atm. Equilibrium constants valid at this temperature are $p^c K_{NH_4^+} = 9.5$; $\log K_H = 1.75$ (Henry's law constant $K_H = [NH_3(aq)]/p_{NH_3}$), $p^c K_w = 14.2$. The information given by the equilibrium constants can be rearranged as

$$NH_3(aq) + H^+ = NH_4^+ \qquad -\log {}^c K_{NH_4^+} = 9.5 \qquad \text{(i)}$$
$$NH_3(g) = NH_3(aq) \qquad \log K_H = 1.75 \qquad \text{(ii)}$$
$$H_2O = H^+ + OH^- \qquad \log {}^c K_w = -14.2 \qquad \text{(iii)}$$

Summing up the reaction formulas and the $\log K$ values we obtain

$$NH_3(g) + H_2O = NH_4^+ + OH^- \qquad \log K = -2.95 \qquad \text{(iv)}$$

that is, $[NH_4^+][OH^-]/p_{NH_3} = 10^{-2.95}$. At equilibrium the proton condition is $[NH_4^+] + [H^+] = [OH^-]$, or $[NH_4^+] \approx [OH^-]$. Thus $[OH^-]^2/10^{-4} = 10^{-2.95}$ which gives $[OH^-] = 10^{-3.5}$ and pH $\simeq$ 10.5.

$[H^+]$ *of Pure Aqueous Acids, Bases or Ampholytes*

As derived in the preceding examples, exact algebraic solutions for $[H^+]$ of monoprotic and diprotic acid–base systems are given in Table 3-6.

3-6 pH as a Master Variable; Equilibrium Calculations Using a Graphical Approach

Surveys of the influence of master variables, such as pH, and the rapid solution of even complicated equilibria can be accomplished with relative facility by graphic representation of equilibrium data. The concepts of graphic representation of equilibrium relationships were first introduced by Bjerrum [9] in 1914 and have more recently been developed and popularized by Sillén [10].

As we have seen, a direct numerical approach is often quite difficult because rigorous simultaneous solutions of equilibrium relationships lead to equations of third, fourth or higher order; these equations are obviously not amenable to convenient numerical resolution.

The simplest example of the application of graphic representation of equilibrium data is that for acid–base equilibria involving a monoprotic acid, such as the acid HA, for which the equilibrium expression for solution in water may be written in terms of a concentration acidity constant, that is, an acidity constant valid at the appropriate temperature and corrected for activity by, for example, the Güntelberg approximation:

$$^cK = \frac{[H^+][A^-]}{[HA]} \tag{34}$$

For the purpose of illustration it has been assumed that ^cK has the value 10^{-6} ($p^cK = 6$) and that a quantity of HA sufficient to give an exactly 10^{-3} M solution has been added to the water; the total concentration C_T of soluble A-containing species in the water at any position of equilibrium is then 10^{-3} M, or

$$C_T = 10^{-3} M = [HA] + [A^-] \tag{35}$$

The control variable in any acid–base equilibrium is pH; hence, it is desirable to represent graphically the equilibrium relationships of all species as functions of pH. For any value of pH the unknowns in the present example are, of course, [HA] and [A$^-$], each of which may now be expressed in terms of the known quantities C_T and [H$^+$] by combining (34) and (35) as follows

$$[HA] = \frac{C_T[H^+]}{^cK + [H^+]} \tag{36}$$

and

$$[A^-] = \frac{C_T{}^cK}{^cK + [H^+]} \tag{37}$$

[9] N. Bjerrum, *Sammlung Chem. u. Chem-techn. Vorträge*, **21**, 575 (1914).
[10] L. G. Sillén, in *Treatise in Analytical Chemistry*, I. M. Kolthoff and P. J. Elving, Eds., Interscience, New York, 1959.

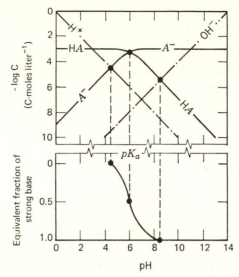

Figure 3-3 *Construction of logarithmic diagram and titration curve for a monoprotic acid.* The method of constructing logarithmic equilibrium diagrams and titration curves by graphic representation of mass law and electroneutrality relations is illustrated. For this example a monoprotic acid (HA) with a $p^c K$ value of 6 and a total concentration of 10^{-3} moles liter^{-1} was used. The three points on the curve for titration of the acid with strong base are given by:

1. $f = 0$: $[H^+] = [A^-] + [OH^-]$, pH = 4.5 (neglect $[OH^-]$).
2. $f = 0.5$: pH = pK = 6.
3. $f = 1.0$: $[HA] + [H^+] = [OH^-]$, pH = 8.5 (neglect $[H^+]$).

It is convenient in the construction of the equilibrium diagram to consider first the asymptotes of the individual curves of solute concentration against pH and in this manner to determine the slopes of the separate sections of each curve. After examining one or two examples of this, the method becomes quite obvious and it is not usually necessary to go through any computations; however, for purposes of illustration, the method is discussed one step at a time (see Figure 3-3).

For values of pH less than $p^c K$, the values of $^c K$ in the denominators of (36) and (37) is much smaller than the asymptotic value of $[H^+]$ and may be neglected. Taking logarithms, (36) then becomes

$$\log [HA] = \log C_T \tag{38}$$

Thus the slope of that portion of the curve of HA against pH in the region pH < $p^c K$ is zero.

Similarly (37), relating $[A^-]$ to pH, may be written

$$\log [A^-] = \log C_T - p^c K + pH \tag{39}$$

Differentiation of (39) with respect to pH yields $d \log [A^-]/d\text{pH} = 1$; thus the slope of that part of the curve representing the variation of $[A^-]$ with pH in the region $\text{pH} < p^c K$ is unity.

Consideration of the asymptotes of the sections of the two curves in the region $\text{pH} > p^c K$ permits us to ignore the quantity $[H^+]$ in the denominators of (36) and (37). Equation 36 then becomes

$$\log [A^-] = \log C_T \tag{41}$$

hence, $d \log [A^-]/d\text{pH} = 0$. In a similar way it can be shown that for $\text{pH} > pK$, $d \log [HA]/d\text{pH} = -1$.

The slopes of the straight-line plots for each solute species against pH have now been calculated from the asymptotic values for the curves in the regions of $\text{pH} < p^c K$ and $\text{pH} > p^c K$. None of these curve asymptotes is rigorous in the immediate region of $\text{pH} = p^c K_a$. At this point $\log [HA] = \log [A^-] = \log (C_T/2)$. Therefore, at $\text{pH} = pK_a$ the curves must intersect at an ordinate value of $(\log C_T - \log 2)$, or 0.3 unit below the ordinate value of $\log C_T$.

Computing the equilibrium composition of a 10^{-3} M HA solution, we simply have to find where on the graph the appropriate proton condition is fulfilled. The condition to be satisfied is

$$[H^+] = [A^-] + [OH^-] \tag{42}$$

Equation 42 is fulfilled at the intersection of the $[H^+]$ line with the $[A^-]$ line because obviously at this point $[OH^-] \ll [A^-]$. Equilibrium $[H^+]$ and the concentrations of all other species at this $[H^+]$ can be read directly from the logarithmic concentration diagram:

$$-\log [H^+] = -\log [A^-] = 4.5, \quad -\log [HA] = 3.0$$

If the diagram is drawn on graph paper (where one logarithmic unit corresponds to about 2 cm) the result can be read within an accuracy of better than ± 0.05 logarithmic units and the relative error $(d\Delta[X]/d[X] = 2.3 \log \Delta[X])$ is smaller than 10%. The slight loss of accuracy involved in substituting graphical for numerical procedures is usually not significant. If a very exact answer is necessary the graphical procedure will immediately show which concentrations can be neglected in the numerical calculations.

The same graph can be used to compute the equilibrium concentrations of a 10^{-3} M solution of NaA. In this case the proton condition is

$$[HA] + [H^+] = [OH^-] \tag{43}$$

This condition is fulfilled at the intersection $[HA] = [OH^-]$; since $[H^+]$ is 1000 times smaller than $[HA]$ it can be neglected in (10). This point gives $-\log [H^+] = 8.5$, $-\log [HA] = 5.5$ and $-\log [A^-] = 3.0$

The proton conditions of (42) and (43) correspond to the two equivalence points in acid–base titration systems. The half titration point is usually (not always) given by pH = pK. Thus the qualitative shape of the titration curve can be sketched readily along these three points (Figure 3-3).

Diprotic Acid–Base Systems

For more complicated equilibria the merit of the graphic method is obvious. Figure 3-4 illustrates a logarithmic pH–concentration diagram for a *diprotic acid* with acidity constants (pK_1 = 7.0 and pK_2 = 13.0) representative of hydrogen sulfide, H_2S(aq). A combination of the equilibrium expressions for the two acidity constants with the concentration condition (S_T = [H_2S] + [HS^-] + [S^{2-}]) gives the equations that define the log concentration–pH

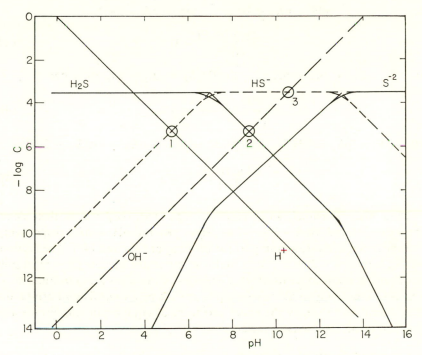

Figure 3-4   Equilibrium diagram for a diprotic acid system. Proton conditions:
1. Solution of H_2S: [H^+] = [HS^-] + 2[S^{2-}] + [OH^-].
2. Solution of NaHS: [H_2S] + [H^+] = [S^{2-}] + [OH^-].
3. Solution of Na_2S: [H_2S] + [HS^-] + [H^+] = [OH^-].
Equilibrium composition:
1. pH = pHS$^-$ = 5.3; pH_2S = 3.5; pS^{2-} = 12.3.
2. pH = 8.7; pH_2S = 5.3; pS^{2-} = 7.5; pHS$^-$ = 3.5.
3. pH = 10.5; pS^{2-} = 5.8; pH_2S = 7; pHS$^-$ = 3.5.

dependence of $[H_2S]$, $[HS^-]$ and $[S^{2-}]$

$$[H_2S] = \frac{S_T}{1 + K_1/[H^+] + K_1K_2/[H^+]^2} \tag{44}$$

$$[HS^-] = \frac{S_T}{[H^+]/K_1 + 1 + K_2/[H^+]} \tag{45}$$

$$[S^{2-}] = \frac{S_T}{[H^+]^2/K_1K_2 + H^+/K_2 + 1} \tag{46}$$

Considering (44), it is apparent that the log H_2S–pH line can be constructed as a sequence of three linear asymptotes that prevail in the three pH regions:

I: $pH < pK_1 < pK_2$; $\log [H_2S] = \log S_T$

$$\frac{d \log [H_2S]}{d\mathrm{pH}} = 0 \tag{47}$$

II: $pK_1 < pH < pK_2$; $\log [H_2S] = pK_1 + \log S_T - pH$

$$\frac{d \log [H_2S]}{d\mathrm{pH}} = -1 \tag{48}$$

III: $pK_1 < pK_2 < pH$; $\log [H_2S] = pK_1 + pK_2 + \log S_T - 2pH$

$$\frac{d \log [H_2S]}{d\mathrm{pH}} = -2 \tag{49}$$

These linear portions can be readily constructed; they change their slopes from 0 to -1 and from -1 to -2 at $pH = pK_1$ and $pH = pK_2$, respectively.

Similar considerations apply to the plotting of (45) and (46). The sections having slopes of -2 or $+2$ are usually unimportant because they occur only at extremely small concentrations. Diagrams of the types given in Figures 3-3 and 3-4 are not only useful in evaluating specific positions of equilibrium, but they permit us to survey the entire spectrum of equilibrium conditions as a function of pH as a master variable.

Example 3-9 *Strong Acid.* Estimate the equilibrium composition of a 10^{-2} M HCl and 10^{-2} M NaCl solution, respectively. (Assume an acidity constant of $K \approx 10^{+3}$; see Figure 3-5.) For the 10^{-2} M acid solution the proton condition (A) is $[H^+] = [Cl^-] + [OH^-]$, which is given by the intersection $[H^+] = [Cl^-]$. For the corresponding salt solution, the proton condition (B) is $[HCl] + [H^+] = [OH^-]$ which becomes, since HCl is negligible, $[H^+] = [OH^-]$.

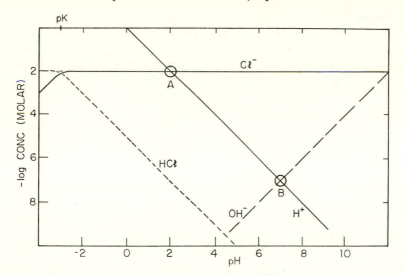

Figure 3-5 Equilibrium diagram for a strong acid (25°C) (Example 3-9). A: 10^{-2} M HCl; $[H^+] = [Cl^-] = 10^{-2}$ M, $[HCl] = 10^{-7}$ M. B: 10^{-2} M NaCl; $[H^+] = [OH^-] = 10^{-7}$ M, $[Cl^-] = 10^{-2}$ M, $[HCl] = 10^{-12}$ M.

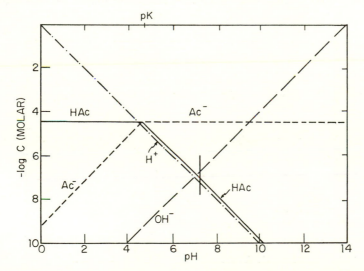

Figure 3-6 Equilibrium Composition of $10^{-4.5}$ M NaAc (Example 3-10). Proton condition: $[HAc] + [H^+] = [OH^-]$. $[H^+] = 10^{-7.2}$ M; $[HAc] = 10^{-7.0}$ M; $[HAc] = 10^{-4.5}$ M.

Example 3-10 *Weak Base.* Compute the equilibrium concentrations in a $10^{-4.5}$ M sodium acetate (NaAc) solution; pK (25°C) = 4.70. Figure 3-6 plots the expressions for the acidity constant and the ion product of water as well as the concentration condition. For a pure solution of NaAc the following proton condition is valid: $[HAc] + [H^+] = [OH^-]$.

It is obvious from the graph that no approximation in the proton condition is possible. In order to find the point where the proton condition is fulfilled, we move slightly to the right of the intersection of log $[HAc]$ with log $[OH^-]$ and find by trial and error where the proton condition is fulfilled, that is, at pH = 7.2: $10^{-7.0} + 10^{-7.2} \approx 10^{-6.8}$.

Example 3-11 *Mixture of Two Acids.* Find the pH of a 10^{-3} M NH_4Cl solution to which 4×10^{-5} M methyl orange indicator in the acid form (HIn) has been added: $pK_{NH_4^+} = 9.2$; $pK_{HIn} = 4.0$. The logarithmic diagram is constructed by superimposing the equilibrium diagrams for the two acid–base systems (Figure 3-7). The proton condition $[H^+] = [NH_3] + [In^-] + [OH^-]$ defines the composition of the solution. The first intersection of the $[H^+]$ line is with the $[In^-]$ line. The other species on the right-hand side of the proton condition can be neglected. At equilibrium we have: pH = 4.5; $pIn^- = 4.5$; pHIn = 4.9; $pNH_4^+ = 3$; $pNH_3 = 7.7$.

It is interesting to note, and should be kept in mind if one attempts to measure pH with an indicator, that the addition of an indicator acid to a poorly buffered solution may markedly affect the pH of the solution. The

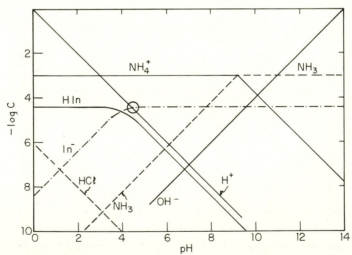

Figure 3-7 Equilibrium composition of solution containing 10^{-3} M NH_4Cl and 4×10^{-4} M methyl orange (HIn) (Example 3-11).

10^{-3} M NH$_4$Cl solution in the absence of an indicator (proton condition: [H$^+$] = [NH$_3$] + [OH$^-$]) has a pH of 6.1.

Example 3-12 *Diprotic Base.* Compute the equilibrium composition of a $10^{-1.3}$ M sodium phthalate (Na$_2$P) solution (25°C). pK_1 = 2.95; pK_2 = 5.41. Using the Güntelberg approximation, we first convert the pK values into p$^c K$ values. In order to calculate the ionic strength, we assume that [P^{2-}] > [HP$^-$]. Thus $I = \frac{1}{2}([\text{Na}^+] + 4[\text{P}^{2-}]) = 0.15$.

$$p^c K_1 = pK_1 - \frac{\sqrt{I}}{1 + \sqrt{I}} = 2.95 - 0.28 = 2.67$$

$$p^c K_2 = pK_2 - \frac{2\sqrt{I}}{1 + \sqrt{I}} = 5.41 - 0.56 = 4.85$$

$$p^c K_w = pK_w - \frac{\sqrt{I}}{1 + \sqrt{I}} = 14.00 - 0.28 = 13.72$$

The logarithmic pH concentration diagram (Figure 3-8) is now constructed using p$^c K$ values. Note that the log [OH$^-$] line intersects the log $C = 0$ at pH = 13.72. The proton condition for a solution of Na$_2$P is: 2[H$_2$P] + [HP$^-$] + [H$^+$] = [OH$^-$]. This proton condition is fulfilled at $-\log$ [H$^+$] = 8.60. The other species are present at the following concentrations

$$[\text{P}^{2-}] = 10^{-1.3} \ M; \qquad [\text{HP}^-] = 10^{-5.1} \ M; \qquad [\text{H}_2\text{P}] < 10^{-6} \ M \ (10^{-8.85} \ M)$$

$$-\log \{\text{H}^+\} = -\log [\text{H}^+] + \frac{0.5\sqrt{I}}{1 + \sqrt{I}} = 8.76$$

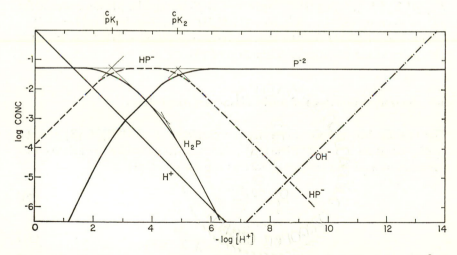

Figure 3-8 Equilibrium composition of diprotic acid (Example 3-12). A 5×10^{-2} M sodium phthalate solution (Na$_2$P) has a pH of 8.6. Proton condition: 2[H$_2$P] + [HP$^-$] + [H$^+$] = [OH$^-$]; [HP$^-$] $\approx$ [OH$^-$].

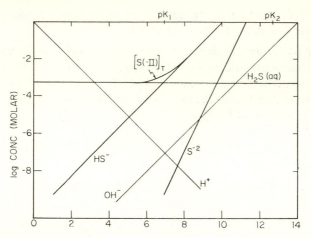

Figure 3-9 Distribution of S(-II) species of a solution in equilibrium with $p_{H_2S} = 5 \times 10^{-3}$ atm (Example 3-13).

Example 3-13 *Volatile Diprotic Acid.* Estimate the distribution of H_2S, HS^- and S^{2-} as a function of pH for an aqueous solution (20°C) in equilibrium with an atmosphere containing 0.5% (by volume) of H_2S. Equilibrium constants: $p^cK_{H_2S} = 6.9$; $p^cK_{HS^-} = 12.7$; $\log K'_H = \log ([H_2S]/p_{H_2S}) = -1.0$. Since Henry's law is fulfilled over the entire pH range, $[H_2S] = 10^{-3.3}$; $p_{H_2S} = 10^{-2.3}$ (see Figure 3-9). The line for $[HS^-]$ is defined by $[HS^-] = K_{H_2S}[H_2S]/[H^+]$; $(d \log [HS^-]/d\mathrm{pH} = +1)$. Similarly the pH dependence for S^{2-} is given by $[S^{2-}] = K_{H_2S}K_{HS^-}[H_2S]/[H^+]^2$; $(d \log [S^{2-}]/d\mathrm{pH} = +2)$.

The sum of all the $S(-II)$ species $([S(-II)_T] = [H_2S] + [HS^-] + [S^{2-}])$ is also given in the diagram as a function of pH.

3-7 Ionization Fractions of Acids, Bases and Ampholytes

The relation between the concentration condition $C = [HB] + [B]$ and the acidity constant $K = [H^+][B]/[HB]$ permits us to calculate the relative distribution of acid and conjugate base as a function of pH

$$\alpha_B = \alpha_1 = \frac{[B]}{C} = \frac{K}{K + [H^+]} \tag{50}$$

$$\alpha_{HB} = \alpha_0 = \frac{[HB]}{[C]} = \frac{[H^+]}{K + [H^+]} \tag{51}$$

where $\alpha_1 + \alpha_0 = 1$. Historically, α_1 has been called the degree of dissociation; it might be better to call it the ionization fraction or the degree of protolysis.

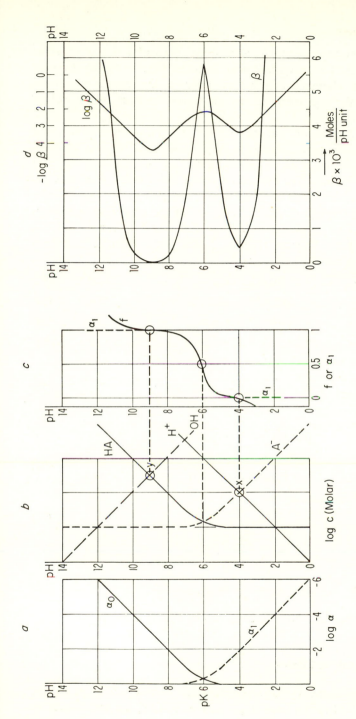

Figure 3-10 *The titration curve and the buffer intensity are related to the equilibrium species distribution.* For a monoprotic acid–base system (HB-B), *a* gives a semilogarithmic distribution diagram. α_1 and α_0 are the ionization fractions of B and HB, respectively. *b*: The log concentration–pH diagram is the super-imposition of the semilogarithmic distribution diagram (a) and the concentration condition (log C = constant). Points X and Y correspond to the equivalence points in alkalimetric or acidimetric titration curve. *c*: Plot of the titration curve. The equivalence points (X and Y) and the half-titration point pH = pK are as given in *b*. The equivalence fraction of the titrant added, *f*, shows, over a significant portion of the titration curve ($0.1 < f < 0.0$), the same dependence upon pH as α_1. *d*: The buffer intensity β, corresponding to the inverse slope of the titration curve (dc_B/dpH), can be computed from a log concentration–pH diagram (a) by multiplying by 2.3 the sum of all concentrations represented by a line of slope $+1$ or -1 at that particular pH in the diagram. (See Section 3-9.)

α_0 has been called the degree of formation of the acid and a plot of α_0 versus pH has been called the formation function, or better the *distribution diagram*. Figure 3-10 gives a schematic distribution diagram. Sometimes it is more convenient to plot the logarithms of the ionization fractions as a function of pH.

Since the ionization fractions are independent of total concentration, their tabulation or graphical representation is very convenient when calculations with the same equilibrium system have to be carried out repeatedly or with more complicated systems. The concentration of the species [HB] and [B] can then always be abbreviated by $[HB] = C\alpha_0$ and $[B] = C\alpha_1$, respectively. The logarithmic equilibrium diagram, discussed in the preceding section, is essentially the additive combination of the line $\log C = $ const. and the logarithmic distribution diagram (Figure 3-10b). In a diprotic acid–base system we define similarly

$$[H_2A] = C\alpha_0 \tag{52}$$

$$[HA^-] = C\alpha_1 \tag{53}$$

$$[A^{2-}] = C\alpha_2 \tag{54}$$

the subscript of α refers to the number of protons lost from the most protonated species. The α values are implicit functions of $[H^+]$:

$$\alpha_0 = \frac{1}{1 + K_1/[H^+] + K_1K_2/[H^+]^2} \tag{55}$$

$$\alpha_1 = \frac{1}{[H^+]/K_1 + 1 + K_2/[H^+]} \tag{56}$$

$$\alpha_2 = \frac{1}{[H^+]^2/K_1K_2 + [H^+]/K_2 + 1} \tag{57}$$

The α values are interrelated by

$$\alpha_0 + \alpha_1 + \alpha_2 = 1 \tag{58}$$

$$\alpha_0 = \frac{[H^+]}{K_1}\alpha_1 \tag{59}$$

$$\alpha_1 = \frac{[H^+]}{K_2}\alpha_2 \tag{60}$$

α values for polyprotic acid–base systems can be readily derived.

3-8   Titration of Acids and Bases

In the titration of an aqueous solution containing C moles per liter of an acid HA with a quantity of strong base (C_B), such as NaOH, the titration curve can be readily deduced because at any point in the titration the following condition of electroneutrality must be fulfilled

$$[Na^+] + [H^+] = [A^-] + [OH^-]$$

or   (61)

$$C_B = [A^-] + [OH^-] - [H^+]$$

With this equation and the logarithmic concentration–pH diagram, the titration curve relating pH to the quantity of base added can be constructed. In Figure 3-10c, pH is plotted as a function of the equivalent fraction of the titrant (strong base) added:

$$f = \frac{C_B}{C} = \frac{[Na^+]}{C} \tag{62}$$

Equation 61 can be rearranged into

$$C_B = C\alpha_1 + [OH^-] - [H^+] \tag{63}$$

$$f = \alpha_1 + \frac{[OH^-] - [H^+]}{C} \tag{64}$$

where α_1 is the degree of protolysis (ionization fraction) (see Section 3-7), $\alpha_1 = K/(K + [H^+]) = [A^-]/C$.

If we want to consider any dilution resulting from the addition of V ml of strong base to V_0 ml of solution containing a concentration C_0 of acid HA before dilution, we simply have to introduce a dilution factor and substitute for C in the above equations

$$C = C_0 \frac{V_0}{V + V_0} \tag{65}$$

For the titration of a C-molar solution of the conjugate base (i.e., the salt of a strong base with the weak acid HA, such as KA) with a strong acid (i.e., HCl), the curve for variation of pH with quantity of acid (C_A) added can be derived similarly from the electroneutrality condition

$$[K^+] + [H^+] = [A^-] + [OH^-] + [Cl^-] \tag{66}$$

$$C + [H^+] = [A^-] + [OH^-] + C_A \tag{67}$$

$$C_A = [HA] + [H^+] - [OH^-] \tag{68}$$

$$C_A = C\alpha_0 + [H^+] - [OH^-] \tag{69}$$

where $\alpha_0 = [H^+]/(K + [H^+])$ (see Section 3-7).

The equivalent fraction of the titrant (strong acid) added, $g = C_A/C$, can be given as an implicit function of H^+ by

$$g = \frac{C_A}{C} = \alpha_0 + \frac{[H^+] - [OH^-]}{C} \tag{70}$$

Comparison of (70) with (64) shows that $g = 1 - f$. Equations 63 and 69 can be generalized into

$$C_B - C_A = C\alpha_1 + [OH^-] - [H^+]$$

or   (71)

$$C_A - C_B = C\alpha_0 + [H^+] - [OH^-]$$

Either equation of (71) can be used to evaluate pH changes that result from the addition of strong acid or strong base to a monoprotic weak acid–base system, or to characterize the titration curve of a mixture of a strong acid or base with a weak acid or base. The equivalence points marked X and Y in Figure 3-10b correspond to equilibrium conditions prevailing in pure equimolar solutions of HA ($g = 1$ and $f = 0$) and NaA ($f = 1$ and $g = 0$). In other words, at the equivalence point ($f = 1$) of an alkalimetric titration of HA with NaOH, the solution cannot be distinguished from an equimolar solution of the salt NaA (proton condition: $[HA] + [H^+] = [OH^-]$). Correspondingly, in the acidimetric titration of the salt, NaA, with HCl, the proton condition at the equivalence point ($f = 0$) is identical with that of an equimolar solution of HA ($[H^+] = [A^-] + [OH^-]$).

Example 3-14 *Acidimetric Titration of Strong and Weak Base.*   Describe the acidimetric titration curve for a solution of the following composition: $[Na^+] = 3.0 \times 10^{-3} M$; $[HOCl] + [OCl^-] = 2.0 \times 10^{-3} M$; $[OH^-] = 1.0 \times 10^{-3} M$.

The titration curve can be drawn with the help of (70). $C_B = 1.0 \times 10^{-3} M$. The equivalence points at $C_A = 1.0 \times 10^{-3} M$ and $C_A = 3.0 \times 10^{-3} M$ are at pH values 9.5 and 5.2, respectively, the second equivalence point being somewhat sharper than the first one (Figure 3-11).

Because proton conditions corresponding to equivalent points $f = 0$ and $f = 1$ can be readily identified, titration curves can be sketched very expediently with the help of the log concentration diagram.

The principles outlined above can be readily extended to *multiprotic acids*. The alkalimetric titration of an acid H_2L^+ added as the salt $H_2L^+X^-$ (e.g., an amino acid, $RNH_2COOH = HL$) is given by the electroneutrality condition

$$C_B + [H_2L^+] + [H^+] = [L^-] + [OH^-] + [X^-] \tag{72}$$

which can be rearranged with the concentration condition

$$C = [H_2L^+] + [HL] + [L^-] = [X^-] \tag{73}$$

to give

$$C_B = [HL] + 2[L^-] + [OH^-] - [H^+] \tag{74}$$

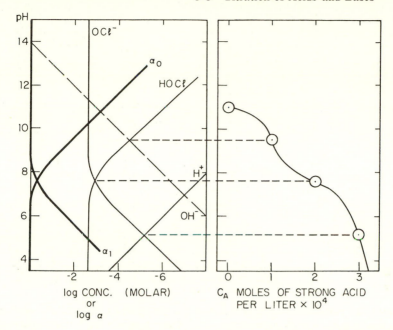

Figure 3-11 Acidimetric titration of a strong base and a weak base (Example 3-14). The solution to be titrated corresponds to "Eau de Javell" made by dissolving Cl_2 in NaOH. $[Na^+] = 3 \times 10^{-3} M$; $[HOCl] + [OCl^-] = 2 \times 10^{-3} M$, $[OH^-] = 1 \times 10^{-3} M$. The first equivalence point (equivalent to the strong base) is at pH $\simeq$ 9.5.

The relation is expressed more generally by (75) and (76)

$$C_B = C(\alpha_1 + 2\alpha_2) + [OH^-] - [H^+] \tag{75}$$

$$f = \frac{C_B}{C} = \alpha_1 + 2\alpha_2 + \frac{[OH^-] - [H^+]}{C} \tag{76}$$

where the α values are defined as $\alpha_1 = [HL]/C$ and $\alpha_2 = [L^-]/C$ and $\alpha_0 = [H_2L^+]/C$. In such a diprotic system three equivalence points may be defined. In Figure 3-12a the points x, y and z correspond to the equivalence points $f = 0$ (proton condition: $[H^+] = [HL] + 2[L^-] + [OH^-]$); $f = 1$ ($[H_2L^+] + [H^+] = [L^-] + [OH^-]$) and $f = 2$: ($2[H_2L^+] + [HL] + [H^+] = [OH^-]$), respectively.

Similarly, the equation describing the acidimetric titration curve of the base ML (where M^+ does not protolyze) can be derived from the concentration condition (73) and the electroneutrality condition:

$$[M^+] + [H_2L^+] + [H^+] = [L^-] + [OH^-] + C_A \tag{77}$$

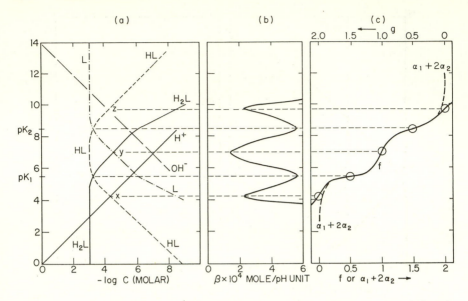

Figure 3-12 Equilibrium composition, buffer intensity and titration curve of diprotic acid–base system. a: Species distribution. b: Buffer intensity. c: Titration curve. The equivalence points, x, y and z (a) are representative of the composition of pure solutions of H_2L, NaH and Na_2L, respectively, and correspond to minima in the buffer intensity. The smaller the buffer intensity, the steeper is the titration curve.

The morphology of the titration curve is given by

$$g = 2 - f = \frac{C_A}{C} = 2\alpha_0 + \alpha_1 + \frac{[H^+] - [OH^-]}{C} \qquad (78)$$

The alkalimetric or acidimetric titration of the ampholyte HL is described simply by the appropriate portions of (76) and (78), respectively.

For a mixture of protolysis systems the variation in $[H^+]$ as a result of the addition of strong acid and/or strong base can be defined by:

$$C_B - C_A = {}^IC({}^I\alpha_1 + 2{}^I\alpha_2 + \cdots)$$
$$+ {}^{II}C({}^{II}\alpha_1 + 2{}^{II}\alpha_2 + \cdots) + \cdots + [OH^-] - [H^+] \qquad (79)$$

3-9 Buffer Intensity and Neutralizing Capacity

The slope of a titration curve (pH versus C_B) is related to the tendency of the solution at any point in the titration curve to change pH upon addition of base. The buffer intensity at any point of the titration is inversely proportional to the slope of the titration curve at that point and may be defined as

$$\beta = \frac{dC_B}{d\text{pH}} = -\frac{dC_A}{d\text{pH}} \tag{80}$$

where dC_B and dC_A are the numbers of moles liter^{-1} of strong acid or strong base required to produce a change in pH of $d\text{pH}$. β has also been called the buffer capacity or the buffer index.

Buffer Intensity

Obviously the buffer intensity can be expressed numerically by differentiating the equation defining the titration curve with respect to pH. For a monoprotic acid–base system [see (61) and (63)]:

$$\beta = \frac{dC_B}{d\text{pH}} = \frac{d[A^-]}{d\text{pH}} + \frac{d[OH^-]}{d\text{pH}} - \frac{d[H^+]}{d\text{pH}} = C\frac{d\alpha_1}{d\text{pH}} + \frac{d[OH^-]}{d\text{pH}} - \frac{d[H^+]}{d\text{pH}} \tag{81}$$

The terms on the right-hand side of (81) can be differentiated as follows

$$-\frac{d[H^+]}{d\text{pH}} = \frac{-d[H^+]}{-(1/2.3)d\ln[H^+]} = 2.3[H^+] \tag{82}$$

$$\frac{d[OH^-]}{d\text{pH}} = \frac{d[OH^-]}{(1/2.3)d\ln[OH^-]} = 2.3[OH^-] \tag{83}$$

$$C\frac{d\alpha_1}{d\text{pH}} = C\frac{d[H^+]}{d\text{pH}}\frac{d\alpha_1}{d[H^+]} = 2.3C\frac{K[H^+]}{(K+[H^+])^2} \tag{84}$$

Because $\alpha_1 = [H^+]/(K+[H^+])$ and $\alpha_0 = K/(K+[H^+])$, the right-hand side of (84) can be expressed in terms of ionization fractions or concentrations of [HA] and [A$^-$]:

$$C\frac{d\alpha_1}{d\text{pH}} = 2.3\alpha_1\alpha_0 \quad C = 2.3\frac{[HA][A^-]}{C} \tag{85}$$

Summing up the individual terms of (81) results in (86)

$$\beta = \frac{dC_B}{d\text{pH}} = 2.3([H^+] + [OH^-] + C\alpha_1\alpha_0)$$

$$= 2.3\left([H^+] + [OH^-] + \frac{[HA][A^-]}{[HA] + [A^-]}\right) \tag{86}$$

The terms on the right-hand side of (86) are in the logarithmic concentration–pH diagram, and β can readily be computed (Figure 3-10d). Maximum buffer intensity occurs (inflection point of titration curve) where $d^2\alpha_1/d(\text{pH})^2 = 0$. This occurs when $\alpha_1 = \alpha_0$, or where [HA] = [A$^-$] and pH = pK. Accordingly, *buffers* usually made by mixing an acid and its conjugate base have their

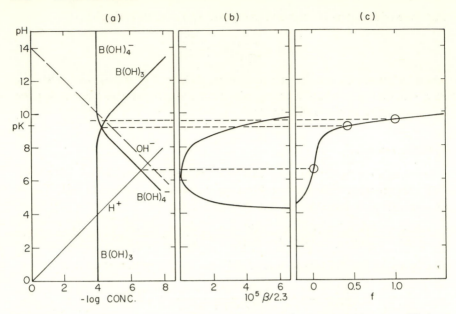

Figure 3-13 Alkalimetric titration of a weak acid (10^{-4} M boric acid). a: Equilibrium distribution. b: Buffer intensity. c: Alkalimetric titration. No pH jump occurs at the equivalence point ($f = 1$) because of buffering by OH$^-$ ions.

maximum buffer intensity at a pH where [HA] = [A$^-$]. The pH of a solution containing $C_{HA} M$ HA and $C_{NaA} M$ NaA corresponds to a point in the titration curve and can be readily computed by (87) (which can be derived with the help of the electroneutrality condition; see Example 3-7).

$$[H^+] = K \frac{C_{HA} - [H^+] + [OH^-]}{C_{NaA} + [H^+] - [OH^-]} \tag{87}$$

Aqueous solutions are well buffered at either extreme of the pH scale. If in an alkalimetric or acidimetric titration curve the pH at the equivalence point falls into a pH range where the buffer intensity caused by [H$^+$] or [OH$^-$] exceeds that of the other protolytes, obliteration of a pH jump at the equivalence point results (see Figure 3-13). The concept of pH buffers can be extended to ions other than H$^+$. Metal ion buffers will be discussed in Chapter 6.

If various acid–base pairs, HA, A; HB, B; etc., are present in the solution the buffer intensity is given by

$$\beta = 2.3\{[H^+] + [OH^-] + C_A \alpha_{HA} \alpha_A + C_B \alpha_{HB} \alpha_B + \cdots\}$$
$$= 2.3\left\{[H^+] + [OH^-] + \frac{[HA][A]}{[HA] + [A]} + \frac{[HB][B]}{[HB] + [B]} + \cdots\right\} \tag{88}$$

POLYPROTIC SYSTEMS. In the same fashion as (86) has been derived, expressions for the buffer intensity of polyprotic acid–base systems can be developed. In Table 3-7 the buffer intensity of a diprotic acid–base system is derived. A polyprotic acid can be treated the same way as a mixture of individual monoprotic acids; for example, for the dibasic acid H_2C,

$$\beta \simeq 2.3 \left\{ [H^+] + [OH^-] + \frac{[H_2C][HC^-]}{[H_2C] + [HC^-]} + \frac{[HC^-][C^{2-}]}{[HC^-] + [C^{2-}]} \right\} \quad (89)$$

This approximate equation holds with an error of less than 5% if $K_1/K_2 > 100$ (Ricci, *loc. cit.*). Note that in the last term of (86) and in the two last terms of (88) and (89), one of the concentrations in the denominator can usually be neglected: for example, at pH $< pK$, (86) becomes $\beta \approx 2.3 ([H^+] + [OH^-] + [A^-])$; at pH $> pK$, it reduces to $\beta \approx 2.3 ([H^+] + [OH^-] + [HA])$. In accord with this well-justified approximation, the buffer intensity of a solution can be computed from a logarithmic equilibrium diagram by multiplying by 2.3 the sum of all concentrations represented by a line of slope $+1$ or -1 at that particular pH in the diagram.

Table 3-7   Titration Curve and Buffer Intensity of a Two-Protic Acid (H_2C)†

I. Definitions

$$C = [H_2C] + [HC^-] + [C^{2-}] \qquad \alpha_0 = 1/[1 + K_1/[H^+] + K_1K_2/[H^+]^2]$$
$$[H_2C] = C\alpha_0 \qquad\qquad \alpha_1 = 1/[1 + [H^+]/K_1 + K_2/[H^+]]$$
$$[HC^-] = C\alpha_1 \qquad\qquad \alpha_2 = 1/[1 + [H^+]/K_2 + [H^+]^2/K_1K_2]$$
$$[C^{2-}] = C\alpha_2$$

II.

$$\frac{d\alpha_1}{dpH} = 2.3\alpha_1(\alpha_0 - \alpha_2); \qquad \frac{d\alpha_2}{dpH} = 2.3\alpha_2(\alpha_1 + 2\alpha_0); \qquad \frac{d\alpha_0}{dpH} = -2.3\alpha_0(\alpha_1 + 2\alpha_2)$$

III. Titration Curve‡

$$C_B - C_A = [\text{Alkalinity}] = [HC^-] + 2[C^{2-}] + [OH^-] - [H^+]$$
$$C_B - C_A = 2C - [\text{Acidity}] = C(\alpha_1 + 2\alpha_2) + [OH^-] - [H^+]$$

IV. Buffer Intensity

$$\beta = \frac{dC_B}{dpH} = \frac{-dC_A}{dpH} = \frac{d[\text{Alkalinity}]}{dpH} = \frac{-d[\text{Acidity}]}{dpH}$$

$$= \frac{d[OH^-]}{dpH} - \frac{d[H^+]}{dpH} + \frac{Cd(\alpha_1 + 2\alpha_2)}{dpH}$$

$$= 2.3\{[H^+] + [OH^-] + C[\alpha_1(\alpha_0 + \alpha_2) + 4\alpha_2\alpha_0]\}$$

$$\beta = 2.3 \left\{ CK_1[H^+] \frac{[H^+]^2 + 4K_2[H^+] + K_1K_2}{([H^+]^2 + K_1[H^+] + K_1K_2)^2} + [H^+] + [OH^-] \right\}$$

---

† The equations given are rigorous. In very good approximation, a polyprotic acid can be treated the same way as a mixture of individual monoprotic acids [see (89)].
‡ The terms Alkalinity and Acidity will be discussed more fully in Section 4-4.

The concept of buffer intensity considered above may be extended and defined in a generalized way for the incremental addition of a constituent to a closed system at equilibrium. Thus in addition to the buffer intensity with respect to strong acids or bases, buffer intensities with respect to weak acids and bases and for heterogeneous systems may be defined. In general

$$\beta_{C_j}^{C_i} = \frac{dC_i}{dpH} \tag{90}$$

where $\beta_{C_j}^{C_i}$ is the buffer intensity for adding C_i incrementally to a system of constant C_j; for example, $\beta_{CaCO_{3(s)}}^{C_{CO_2}}$ measures the tendency of a solution in contact and equilibrium with solid $CaCO_3$ to resist a pH change resulting from the addition or withdrawal of CO_2. In principle, the concept can be further extended to the buffering of metal ions; that is, to the stability of a water with respect to the concentration of other ions and parameters such as

$$\beta = \frac{dC_i}{dpCa}$$

can be elucidated. The buffer intensity can always be found analytically by differentiating the appropriate function of C_i for the system with respect to the pH or pMe. The buffer intensity is an intrinsic function of the pH or pMe.

Acid and Base Neutralizing Capacity

Operationally we might define as a base neutralizing capacity [BNC] the equivalent sum of all the acids that can be titrated with a strong base to an equivalence point. Similarly, the acid neutralizing capacity [ANC] can be determined from the titration with strong acid to a preselected equivalence point. At every equivalence point a particular proton condition defines a reference level of protons. Conceptually [BNC] measures the concentration of all the species containing protons in excess minus the concentration of the species containing protons in deficiency of the proton reference level, that is, it measures the net excess of protons over a reference level of protons. Similarly [ANC] measures the net deficiency of protons.

In an aqueous monoprotic acid–base system [ANC] is defined by the right-hand side of (61) or (63) in Section 3-8.

$$[ANC] = [A^-] + [OH^-] - [H^+]$$

or

$$= C\alpha_1 + [OH^-] - [H^+] \tag{91}$$

The reference level is defined by the composition of a pure solution of HA in H_2O ($f = 0$, [ANC] = 0), which is defined by the proton condition $[H^+] = [A^-] + [OH^-]$. (In this and subsequent equations, the charge type of the acid is unimportant; the equation defining the net proton excess or deficiency can always be derived from a combination of the concentration condition and the condition of electroneutrality.) Thus in a solution containing a mixture of HA and NaA, [ANC] is a conservative capacity parameter. It must be expressed in concentrations (and not activities). Addition of HA (a species defining the reference level) does not change the proton deficiency and thus does not affect [ANC].

In the same monoprotic acid–base system the base neutralizing capacity with respect to the reference level ($f = 0$) of a NaA solution (proton condition: $[HA] + [H^+] = [OH^-]$) is defined by

$$[BNC] = [HA] + [H^+] - [OH^-]$$
$$= C\alpha_0 + [H^+] - [OH^-] \tag{92}$$

Example 3-15 [ANC] *of Buffer Solutions.* Compare [ANC] of the following NH_4^+, NH_3 buffer solutions

a. $[NH_4^+] + [NH_3] = 5 \times 10^{-3} M$, pH = 9.3

b. $[NH_4^+] + [NH_3] = 10^{-2} M$, pH = 9.0

Despite the lower pH, solution b has a slightly larger base neutralizing capacity than solution a. With a pK value of 9.3, the α values are 0.5 and 0.33, and the corresponding [ANC] capacities are 2.5×10^{-3} and 3.3×10^{-3} equivalents per liter, for a and b, respectively.

In a *multiprotic acid–base system* various reference levels ($f = 0, 1, 2, \ldots$) may be defined; for example, in a sulfide-containing solution the acid neutralizing capacity with reference to the equivalence point defined by the pH of a pure H_2S solution ($f = 0$, $g = 2$) is

$$[ANC]_{f=0} = [HS^-] + 2[S^{2-}] + [OH^-] - [H^+]$$
$$= S_T(\alpha_1 + 2\alpha_2) + [OH^-] - [H^+] \tag{93}$$

where S_T is the sum of the S(-II) species ($[H_2S] + [HS^-] + [S^{2-}]$).

The base neutralizing capacity of the phosphoric acid system with reference to the equivalence point $f = 2$ (solution of Na_2HPO_4 with the proton condition: $2[H_3PO_4] + [H_2PO_4^-] + [H^+] = [PO_4^{3-}] + [OH^-]$) is given by

$$[BNC]_{f=2} = 2[H_3PO_4] + [H_2PO_4^-] + [H^+] - [PO_4^{3-}] - [OH^-]$$
$$= P_T(2\alpha_0 + \alpha_1 - \alpha_3) + [H^+] - [OH^-] \tag{94}$$

These relations can be generalized into (95) and (96)

$$[BNC]_{f=n} = C[n\alpha_0 + (n-1)\alpha_1 + (n-2)\alpha_2 + (n-3)\alpha_3 + \cdots]$$
$$+ [H^+] - [OH^-] \tag{95}$$

and

$$[ANC]_{f=n} = C[n\alpha_0 + (1 - n)\alpha_1 + (2 - n)\alpha_2 + (3 - n)\alpha_3 + \cdots] \\ - [H^+] + [OH^-] \quad (96)$$

(As mentioned before, α_x refers to the ionization fraction of the species that has lost x protons from the most protonated acid species; f defines the equivalence points, the point at the lowest pH being $f = 0$.)

The ANC and BNC concept can be extended readily to mixed acid–base systems. The proton-free energy level diagram (Figure 3-1) might be used conveniently for illustration; for example, a natural carbonate bearing water containing some NH_4^+ and borate has an $[ANC]_{f=0}$ (reference, pure CO_2 solution) of the equivalent sum of all the bases (right-hand side of diagram) that have proton-unpopulated energy levels of HCO_3^- or less, minus the equivalent sum of all the acids of energy levels higher than $H_2CO_3^*$; that is,

$$[ANC]_{f=0} = [HCO_3^-] + 2[CO_3^{2-}] + [NH_3] + [B(OH)_4^-] \\ + [OH^-] - [H^+] \quad (97)$$
$$= C_T(\alpha_1 + 2\alpha_2) + [NH_3] + [B(OH)_4^-] + [OH^-] - [H^+]$$

[ANC] and [BNC] are very useful in defining and characterizing an acid–base system. The proton-free energy level (Figure 3-1), that is, the pH of the system, is independent of the quantity of the solution (*intensity factors*). On the other hand, the number of protons (added coulometrically or with strong acids) required to attain a certain pH represents a *capacity factor* because it is proportional to the quantity of the solution. [ANC] is an integration of the buffer intensity over a pH range

$$[ANC] = \int_{f=n}^{f=x} \beta d\mathrm{pH} \quad (98)$$

and gives us a conservative parameter that is not affected by temperature and pressure.

Any acid–base system of unknown distribution can be characterized fully with the help of two parameters. For example, in a solution of phosphates (Na-salts) the equilibrium composition with regard to the six species (H_3PO_4, $H_2PO_4^-$, HPO_4^{2-}, PO_4^{3-}, H^+ and OH^-) can be resolved completely if the concentration of at least two of the species or two of certain combinations thereof are evaluated analytically; capacity factors such as [ANC] and [BNC] are especially valuable for defining acid–base systems in terms of conservative parameters. They can be determined frequently with ease and relatively good accuracy; thus, the discrepancy between conceptual and operational definition is very small.

In carbonate systems and in natural waters [ANC] is referred to as *alkalinity*, while [BNC] is called *acidity*. In the context of natural waters these terms will be discussed in the next chapter.

READING SUGGESTIONS

Butler, J. N., *Ionic Equilibrium, a Mathematical Approach*, Addison-Wesley, Reading, Mass., 1964. This text presents a unified rigorous treatment of equilibria with a large number of realistic examples and problems used throughout the text.

Butler, J. N., *Solubility and pH Calculations*, Addison-Wesley, Reading, Mass., 1964. Less extensive than the work referred to above. Covers the mathematics of the simplest ionic equilibria.

Garrels, R. M., and C. L. Christ, *Solutions, Minerals and Equilibria*, Harper and Row, New York, 1965. This very important book includes a detailed interpretation of activity coefficients in mixed electrolyte solutions.

Hägg, G., *Die Theoretischen Grundlagen der Analytischen Chemie*, Birkhäuser Verlag, Basel, 1950. Consequent application of Brønsted concept. Utilization of logarithmic equilibrium diagrams for calculating protolysis equilibria.

King, E. J., *Acid–Base Equilibria*, Macmillan, New York, 1965. Comprehensive survey of current experimental investigations and interpretations of acid–base equilibria; pays special attention to polyelectrolytic acids.

Laitinen, H. A., *Chemical Analysis; An Advanced Text and Reference*, McGraw-Hill, New York, 1960. The principles of acid–base equilibria are covered in a rigorous fashion.

Morris, J. C., "The Acid Ionization Constant of HOCl from 5 to 35°," *J. Phys. Chem.*, **70**, 3798 (1966). An example for measuring acidity constants from spectrophotometric data.

Ramette, R. W., "Equilibrium Constants from Spectrophotometric Data," *J. Chem. Educ.*, **44**, 647 (1967).

Ricci, J. E., *Hydrogen Ion Concentration*, Princeton Univ. Press, Princeton, N.J., 1952. A most systematic, uniform and rigorous presentation of the purely mathematical problems involved in the quantitative relations determining the H^+ ion concentration in aqueous solutions.

Robinson, R. A., and R. H. Stokes, *Electrolyte Solutions*, Butterworths, London, 1959. Detailed review on the nature of electrolyte solutions and theoretical treatment based on theory of Debye-Hückel and its later developments by Onsager and Fuoss.

Seel, F., *Grundlagen der Analytischen Chemie (unter besonderer Berücksichtigung der Chemie in wässerigen Systemen)*, Verlag Chemie, Weinheim, 1955. An excellent discussion of equilibria and graphical methods of calculation.

Sillén, L. G., "Graphic Presentation of Equilibrium Data," in *Treatise on Analytical Chemistry*, Part I, Volume 2, I. M. Kolthoff and P. J. Elving, Eds., Interscience, New York, 1959, Chapter 8. Logarithmic and distribution diagrams with a master variable as well as predominance area diagrams with two master variables and their use are fully discussed.

Sillén, L. G., and A. E. Martell, *Stability Constants of Metal–Ion Complexes*, Special Publ., No. 17, The Chemical Society, London, 1964. Extensive but noncritical compilation of equilibrium (including acidity) constants.

Sillén, L. G., "Master Variables and Activity Scales," in *Equilibrium Concepts in Natural Water Systems*, Advances in Chemistry Series, No. 67, American Chemical Society, Washington, D.C., 1967. Compares merits of infinite dilution activity scale and ionic medium activity scale.

Waser, J., "Acid–Base Titration and Distribution Curves," *J. Chem. Educ.*, **44**, 275 (1967).

Yingst, A., "Evaluation of Titration Analyses with Log Concentration Diagram," *J. Chem. Educ.*, **44**, 601 (1967).

PROBLEMS

3-1 Two acids, of approximately 10^{-2} M concentration, are titrated separately with a strong base and show the following pH at the end point (equivalence point, $f = 1$):

$$HA: pH = 9.5$$
$$HB: pH = 8.5$$

(i) Which one (HA or HB) is the stronger acid?
(ii) Which one of the conjugate bases (A^- or B^-) is the stronger base?
(iii) Estimate the pK values for the acids HA and HB.

3-2 Hypochlorous acid (HOCl) has an acidity constant $K = 3 \times 10^{-8}$ (25°C). The strength of a sodium hypochlorite solution can be determined by titration with a strong acid. Sketch a titration curve for a 10^{-3} M solution, indicating pH at the beginning and at the end point of the titration.

3-3 A 4×10^{-3} M solution of an acid HX has a pH of 2.4. What is the pH of an equimolar solution of the Na^+ salt of its conjugate base?

3-4 Calculate $-\log [H^+]$ and $-\log [OH^-]$ of a solution of 0.14 M ammonia:
(i) at 25°C
(ii) at 100°C
($^c K_B$ at 25°C and at 100°C = 1.8×10^{-5}).

3-5 What is the acidity produced by the addition of 1 g of pyritic agglomerate (FeS_2) to 1 liter of distilled water? Assume that Fe(II) and S_2^{2-} are oxidized to Fe(III) and SO_4^{2-}, respectively.

3-6 Report qualitatively the following titration curves:
(i) pH versus f, strong acid titrated with strong base
(ii) $[H^+]$ versus f, strong acid titrated with strong base
(iii) pH versus f, weak base titrated with strong acid
(iv) pH versus f, weak acid titrated with strong base

3-7 Arrange the following solutions in order of increasing buffer intensities:
(a) 10^{-3} M $NH_3-NH_4^+$, pH = 7
(b) 10^{-3} M $NH_3-NH_4^+$, pH = 9.2
(c) 10^{-3} M $H_2CO_3^*-HCO_3^--CO_3^{2-}$, pH = 8.2
(d) 10^{-3} M $H_2CO_3^*-HCO_3^-$, pH = 6.3

ANSWERS TO PROBLEMS

3-1 (i) HB; (ii) A^-; (iii) pK_{HA} = 7, pK_{HB} = 5

3-2 $g = 0$, pH = 9.3; $g = 0.5$, pH = 7.5; $g = 1.0$, pH = 5.3

3-3 pH = 7

3-4 $-\log [OH^-]$ (25°C) = $-\log [OH^-]$ (100°C) = 2.8
$-\log [H^+]$ (25°C) = 11.2; $-\log [H^+]$ (100°C) = 9.2
For the calculation $^c K_W = 10^{-14}$ (25°C) and $^c K_W = 10^{-12}$ (100°C) have been used.

3-5 3.3×10^{-2} equivalents liter^{-1}. [For stoichiometry of the reaction see (11)–(13) in Section 3, Chapter 10.]

3-6

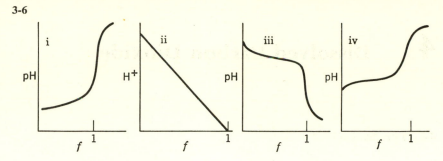

3-7 $a < c < d < b$

4   Dissolved Carbon Dioxide

4-1   Introduction

Inorganic constituents dissolved in fresh waters have their origin in minerals and the atmosphere. Carbon dioxide from the atmosphere provides an acid that reacts with the bases of the rocks. The water may also lose dissolved carbon to the sediments by precipitation reactions. Representative dissolution and precipitation reactions with $CaCO_3(s)$ and a feldspar are

$$CaCO_3(s) + CO_2 + H_2O \rightleftharpoons Ca^{2+} + 2HCO_3^- \tag{1}$$
calcite

$$NaAlSi_3O_8(s) + CO_2 + 1\tfrac{1}{2}H_2O \rightleftharpoons Na^+ + HCO_3^- + 2H_4SiO_4$$
albite
$$+ \tfrac{1}{2}Al_2Si_2O_5(OH)_4(s) \tag{2}$$
kaolinite

In these and similar reactions HCO_3^- and CO_3^{2-} (alkalinity) are imparted to or withdrawn from the water. CO_2 is also added to the atmosphere by volcanic activity and through the combustion of fossil fuels. Carbon dioxide is reduced in the course of photosynthesis and set free during respiration and oxidation of organic matter; it thus occupies a unique position in the biochemical exchange between water and biomass. Dissolved carbonate species participate in homogenous and heterogeneous acid–base and exchange reactions with the lithosphere and the atmosphere. Such reactions are significant in regulating the pH and the composition of natural waters.

Table 4-1 gives a survey on the distribution of carbon in its various forms in the atmosphere and biosphere. A discussion of the geochemical significance of carbon and its cycle is postponed to Chapter 7. Here we describe the distribution of CO_2, HCO_3^- and CO_3^{2-} in the gas and the solution phase and discuss the analytical chemistry of the dissolved carbonate system. In a water isolated from its sediments and rocks, the presence of the dissolved carbonate species and the exchange of CO_2 with the atmosphere are primarily responsible for the maintenance of near-neutral pH conditions. This buffer mechanism will be examined and capacities for acid and base neutralization defined. Idealized equilibrium models for the CO_2 dissolution and acid–base reactions are emphasized in order to account for the distribution of the

118

Table 4-1   Amount of Carbon in Sedimentary Rocks, Hydrosphere, Atmosphere and Biosphere†

	Total on Earth in 10^{18} moles C	In Units of atmospheric CO_2, A_0
Sediments		
Carbonate	1,530	28,500
Organic carbon	572	10,600
Land		
Organic carbon	0.065	1.22
Ocean		
$CO_2 + H_2CO_3$	0.018	0.3
$HCO_3{}^-$	2.6	48.7
$CO_3{}^{2-}$	0.33	6.0
Dead organic	0.23	4.4
Living organic	0.0007	0.01
Atmospheric		
$CO_2(A_0)$	0.0535	1.0

†Atmospheric $CO_2 = 0.032\%$ by volume in dry air. $p_{CO_2} = 3 \times 10^{-4}$ atm. From data given by Rubey [1], Hutchinson [2], Sverdrup et al. [3] and quoted by Revelle and Suess [4].
Reproduced with permission from Almqvist and Wiksells, Uppsala, Sweden.

carbonate species. A simplified steady-state system for the transfer of CO_2 into and out of the ocean is considered for estimating the exchange rate of CO_2 at the gas–water interface.

4-2   Dissolved Carbonate Equilibria

We first consider a system that is closed to the atmosphere; i.e., we treat $H_2CO_3^*$ as a nonvolatile acid. For simple aqueous carbonate solutions the interdependent nature of the equilibrium concentrations of the six solute components — CO_2, H_2CO_3, $HCO_3{}^-$, $CO_3{}^{2-}$, H^+ and OH^- — can be

[1] W. W. Rubey, *Bull. Geol. Soc. Amer.*, **62**, 1111 (1951).
[2] G. E. Hutchinson, in *The Earth as a Planet*, G. Kuiper, Ed., Univ. Chicago Press, Chicago, 1954.
[3] H. U. Sverdrup, M. W. Johnson and R. H. Fleming, *The Oceans*, Prentice-Hall, New York, 1942.
[4] R. Revelle and H. E. Suess, *Tellus*, **9**, 18 (1957).

described completely by a system of six equations. The appropriate set of equations is comprised of four equilibrium relationships [which define the hydration equilibrium of CO_2 ($H_2CO_3 = CO_2(aq) + H_2O$), the first and second acidity constants of H_2CO_3 and the ion product of water] and any two equations describing a concentration and an electroneutrality or proton condition. As shown earlier (Section 3-3), it is convenient to define for the aqueous carbonate system a composite acidity constant for all dissolved CO_2, hydrated or not. For the total analytical concentration of dissolved CO_2 we write $[H_2CO_3^*] = [CO_2(aq)] + [H_2CO_3]$; the number of equations necessary to describe the distribution of solutes reduces to five. These equations together with the relationships that describe the distribution of solutes and representative logarithmic distribution diagrams are given in Table 4-2 and

Table 4-2   The Equilibrium Distribution of Solutes in Aqueous Carbonate Solution (System Closed to the Atmosphere)

Species

$$CO_2(aq), \quad H_2CO_3, \quad HCO_3^-, \quad CO_3^{2-}, \quad H^+, \quad OH^-$$
$$[H_2CO_3^*] = [CO_2 \cdot aq] + [H_2CO_3]$$

Equilibrium constants†

$$[CO_2(aq)]/[H_2CO_3] = K \tag{1}$$

$$[H^+][HCO_3^-]/[H_2CO_3^*] = K_1 \tag{2}$$

$$[H^+][HCO_3^-]/[H_2CO_3] = K_{H_2CO_3} \tag{2a}$$

$$[H^+][CO_3^-]/[HCO_3^-] = K_2 \tag{3}$$

$$[H^+][OH^-] = K_W \tag{4}$$

Concentration condition

$$C_T = [H_2CO_3^*] + [HCO_3^-] + [CO_3^{2-}] \tag{5}$$

Ionization fractions‡

$$[H_2CO_3^*] = C_T\alpha_0 \quad [HCO_3^-] = C_T\alpha_1 \quad [CO_3^{2-}] = C_T\alpha_2$$

$$\alpha_0 = \left(1 + \frac{K_1}{[H^+]} + \frac{K_1 K_2}{[H^+]^2}\right)^{-1} \tag{6}$$

$$\alpha_1 = \left(\frac{[H^+]}{K_1} + 1 + \frac{K_2}{[H^+]}\right)^{-1} \tag{7}$$

$$\alpha_2 = \left(\frac{[H^+]^2}{K_1 K_2} + \frac{[H^+]}{K_2} + 1\right)^{-1} \tag{8}$$

† Equilibrium constants are defined for a constant ionic medium activity scale (Section 3-4).

‡ Equations 6 to 8 are derived as 55-57 in Section 3-7.

Table 4-2   Continued

Proton conditions of pure solutions (equivalence points)†

$$[H^+] = [HCO_3^-] \atop + 2[CO_3^{2-}] + [OH^-] \quad (9)$$

$$[H_2CO_3^*] + [H^+] \atop = [CO_3^{2-}] + [OH^-] \quad (10)$$

$$2[H_2CO_3^*] + [HCO_3^-] \atop + [H^+] = [OH^-] \quad (11)$$

$[H^+]$ at equivalence points‡

$$[H^+] = (C_T K_1 \atop + K_W)^{0.5} \quad (12)$$

$$[H^+] \simeq [K_1(K_2 \atop + K_W/C_T)]^{0.5} \quad (13)$$

$$[H^+] \simeq K_W/2C_T \atop + [K_W^2/2C_T \atop + K_2 K_W/C_T]^{0.5} \quad (14)$$

$$[H^+] \simeq (C_T K_1)^{0.5} \quad (15)$$

$$[H^+] \simeq (K_1 K_2)^{0.5} \quad (16)$$

$$[H^+] \simeq (K_2 K_W/C_T)^{0.5} \quad (17)$$

Titration:
Alkalimetric

$$f = C_B/C_T = \alpha_1 + 2\alpha_2 + ([OH^-] - [H^+])/C_T \tag{18}$$

Acidimetric

$$g = C_A/C_T = 2 - f = 2\alpha_0 + \alpha_1 - ([OH^-] - [H^+])/C_T \tag{19}$$

Acid and base neutralizing capacity:‖

Alkalinity $$[Alk] = C_T(\alpha_1 + 2\alpha_2) + [OH^-] - [H^+] \tag{20}$$

Acidity $$[Acy] = C_T(\alpha_1 + 2\alpha_0) + [H^+] - [OH^-] \tag{21}$$

Buffer intensity#

$$\beta_{C_T}^{CB} = 2.3\{[H^+] + [OH^-] + C_T[\alpha_1(\alpha_0 + \alpha_2) + 4a_2\alpha_0]\} \tag{22}$$

---

† Na^+ (in $NaHCO_3$ or Na_2CO_3) is used as a symbol of a nonprotolyzable cation (Na^+, Li^+, K^+,...). In Figure 4-1 the equivalence points corresponding to (9), (10) and (11) are marked x, y and z, respectively.

‡ The exact numerical solution is of the fourth degree in $[H^+]$ (see Table 3-6). For all practical purposes ($C_T > 10^{-6}$ M), (9)–(11) are sufficiently exact. If $C_T > 10^{-5}$ M, (15), and if $C_T \geq 10^{-3}$ M, (16) and (17) may be used.

‖ For derivation and further discussion see Sections 3-9 and 4-4.

See Sections 3-9 and 4-5.

Figure 4-1, respectively. These relationships are readily derived (see Sections 3-5 to 3-7); Table 4-2 is patterned after Table 3-6. Equilibrium constants valid for different temperatures and activity conventions are tabulated in Tables 4-7, 4-8 and 4-9. The acidity constants of true H_2CO_3, $K_{H_2CO_3}$, and the composite acidity constant of $H_2CO_3^*$, K_1, are interrelated [see (1), (2) and (2a), Table 4-2] by

$$K_1 = \frac{K_{H_2CO_3}}{1 + K} \tag{3}$$

where K is the constant describing the hydration equilibrium [(1), Table 4-2]. At 25°C K is in the order of 650. Thus (3) can be simplified to

$$K_1 \simeq \frac{K_{H_2CO_3}}{K} \tag{4}$$

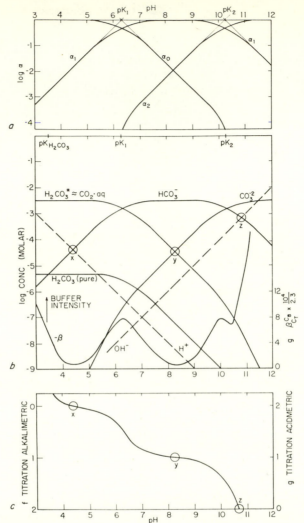

Figure 4-1 Distribution of solute species in aqueous carbonate system. This figure has been constructed under the assumption that $C_T = \text{constant} = 10^{-2.5}$ M. The following equilibrium constants (typical for a fresh water of 25°C and corrected for an ionic strength of $I \approx 5 \times 10^{-3}$ to 10^{-2} M) have been used: $pK_{\text{(hydration)}} = -2.8$, $pK_1 = 6.3$, $pK_2 = 10.25$.

a: Ionization fractions as a function of pH.

b: Logarithmic equilibrium diagram. Because $[CO_2(aq)] \gg [H_2CO_3]$, $[CO_2aq] \approx [H_2CO_3^*]$. Note that H_2CO_3 is a much stronger acid than $CO_2(aq)$ or $H_2CO_3^*$. Pure H_2CO_3 has a pK value (where $[H_2CO_3] = [HCO_3^-]$) of $pK_{H_2CO_3} = 3.5$. The equivalence points corresponding to pure solutions (C_T-molar) of $H_2CO_3^*$, $NaHCO_3$ and Na_2CO_3 are marked x, y and z, respectively. Buffer intensity is plotted as a function of pH.

c: Alkalimetric or acidimetric titration curve. Note that no pH jump occurs at the equivalence point z, because at this point the buffer intensity caused by high $[OH^-]$ is too large.

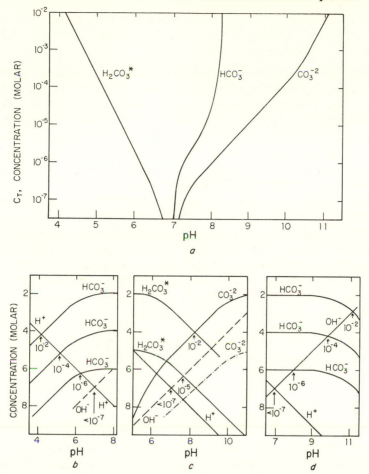

Figure 4-2   pH of pure solutions of $CO_2(H_2CO_3^*)$, $NaHCO_3$ and Na_2CO_3 at various dilutions (see Example 4-1). The curves in a can be computed by (12)–(16) or with the help of logarithmic equilibrium diagrams. Figures b, c and d give sketches for pure CO_2, $NaHCO_3$ and Na_2CO_3 solutions, respectively. For a few concentrations the intersections that characterize the appropriate proton conditions are indicated by arrows.

Correspondingly, the concentration of $CO_2(aq)$ is nearly identical with the analytical concentration of $H_2CO_3^*$. These relations are also obvious from the logarithmic equilibrium diagram (Figure 4-1). The diagram illustrates that true H_2CO_3 is quite a strong acid ($pK = 3.5$).

EQUIVALENCE POINTS.   The pH of the pure solutions of the acid, the ampholyte and the base, corresponding to the equivalence points in an alkalimetric titration, are characterized by the appropriate proton conditions [(9)–(11) of

Table 4-2]. The points marked x, y and z in Figure 4-1 correspond to equilibrium conditions prevailing in $C_T = 10^{-2.5}$ M solutions of CO_2, $NaHCO_3$ and Na_2CO_3, respectively.

Example 4-1 pH *of Pure* CO_2, $NaHCO_3$ *and* Na_2CO_3 *Solutions.*   Estimate the pH as a function of concentration for pure solutions of CO_2, $NaHCO_3$ and Na_2CO_3, respectively. Assume a closed system; that is, treat $H_2CO_3^*$ as a nonvolatile acid. The answer is given in Figure 4-2 where pH is plotted for the different solutions as a function of log C_T. The numerical calculation of these curves is based on (12)–(14) of Table 4-2, which characterize the appropriate proton conditions. The pH values at the equivalence points can also be readily estimated from logarithmic equilibrium sketches (Figure 4-2b, c and d). For example the proton condition of a pure $NaHCO_3$ solution is given by $[H_2CO_3^*] + [H^+] = [CO_3^{2-}] + [OH^-]$. As Figure 4-2c illustrates, at high concentrations ($C_T > 10^{-3}$ M) the equilibrium is characterized by the simplified condition $[H_2CO_3^*] \approx [CO_3^{2-}]$. In this concentration range, $NaHCO_3$ solutions maintain a constant pH (ignoring small changes that may result from activity variations). In a more dilute concentration range (10^{-4} M > $C_T > 10^{-7}$ M) the pH of pure $NaHCO_3$ solutions is characterized by the appropriate proton condition $[H_2CO_3^*] \approx [OH^-]$. As $C_T \rightarrow 0$, the electroneutrality condition becomes $[H^+] \approx [OH^-]$ and neutrality (pH $\simeq$ 7) exists. Pure solutions of CO_2 ($C_T > 10^{-6}$ M) and of Na_2CO_3 ($C_T > 10^{-3}$ M) have dpH$/d$ log C_T values of -0.5 and $+0.5$, respectively.

4-3   Dissolution of CO_2

If we now open the aqueous carbonate system to the atmosphere, we have to consider the equilibrium between the gas and the solution phase. Such an equilibrium can be formulated in terms of an equilibrium relationship. Because of different conventions used in expressing the concentration (activity) in the gas and the solution phase, equilibrium constants with different dimensions are commonly used to characterize the dissolution equilibrium. Table 4-3 gives the various equilibrium expressions and shows how the equilibrium constants are interrelated.

Ideally these expressions should be written in terms of activities and fugacities; then the gas solution equilibrium is independent of the salinity of the solution. For example, the activity of $CO_2(aq)$ ($= H_2CO_3^*$) is identical for fresh water and for seawater that have been equilibrated with an atmosphere containing the same p_{CO_2}.† However, since activity coefficients for uncharged

† This has been confirmed experimentally, for example, for oxygen by Mancy [5]. Amperometric membrane electrodes measure oxygen activity because the response is proportional to the activity gradient across the membrane.

Table 4-3   Solubility of Gases

Example:† $\qquad\qquad\qquad\qquad CO_2(g) \rightleftharpoons CO_2(aq)$

Assumptions: Gas behaves ideally; $[CO_2(aq)] = [H_2CO_3^*]$

I. Expressions for Solubility Equilibrium ‡
  (*1*) Distribution (mass law) constant, K_D:

$$K_D = [CO_2(aq)]/[CO_2(g)] \quad \text{(dimensionless)} \qquad (1)$$

  (*2*) Henry's law constant, K_H:
    In (1), $[CO_2(g)]$ can be expressed by Dalton's law of partial pressure:

$$[CO_2(g)] = p_{CO_2}/RT \qquad (2)$$

  Combination of (1) and (2) gives

$$[CO_2(aq)] = (K_D/RT)p_{CO_2} = K_H p_{CO_2} \qquad (3)$$

  where $K_H = K_D/RT$ (mole liter^{-1} atm^{-1})
  (*3*) Bunsen absorption coefficient, α_B:

$$[CO_2(aq)] = (\alpha_B/22.414)p_{CO_2} \qquad (4)$$

  where $22.414 = RT/p$ (liter mole^{-1}) and

$$\alpha_B = K_H \times 22.414 \text{ (atm}^{-1}) \qquad (5)$$

II. Partial Pressure and Gas Composition

$$p_{CO_2} = x_{CO_2}(P_T - w) \qquad (6)$$

where x_{CO_2} = mole fraction or volume fraction in dry gas, P_T = total pressure and
w = water vapor pressure

† Same types of expressions apply to other gases.
‡ The equilibrium constants defined by (1)–(4) are actually constants only if the equilibrium expressions are formulated in terms of activities and fugacities.

species become larger than 1 (salting-out effect), the concentration of $CO_2(aq)$ is smaller in the salt solution than in the dilute aqueous medium.

   The concentration (activity) of CO_2 (or any other gas or volatile substance) in solution can always be expressed either in terms of concentration units or in terms of the "partial pressure of the gas in solution," that is, that pressure of CO_2 in the gas phase with which the sample would be in equilibrium. $CO_2(aq)$ and p_{CO_2} are interrelated by Henry's law [(3), Table 4-3]. If the aqueous system is in equilibrium with the gas phase, the "partial pressure of the gas in solution" is equal to the partial pressure of the gas in the gas phase.

[5] K. H. Mancy, W. C. Westgarth and D. A. Okun, 17th Industrial Waste Conference, Purdue University, May 1962.

Aqueous Carbonate System Open to the Atmosphere with Constant p_{CO_2}

A very simple model showing some of the characteristics of the carbonate system of natural waters is provided by equilibrating pure water with a gas phase (e.g., the atmosphere) containing CO_2 at a constant partial pressure. One may then vary the pH by the addition of strong base or strong acid, thereby keeping the solution in equilibrium with p_{CO_2}. This simple model has its counterpart in nature when CO_2 reacts with bases of rocks (i.e., with silicates, clays).

Figure 4-3 shows the distribution of the solute species of such a model. A partial pressure of CO_2 ($p_{CO_2} = 10^{-3.5}$ atm) representative of the atmosphere and equilibrium constants valid at 25°C have been assumed. The equilibrium concentration of the individual carbonate species can be expressed as a function of p_{CO_2} and pH. By combining (6), (7) or (8) (from Table 4-2) with Henry's law

$$[H_2CO_3^*] = K_H p_{CO_2} \tag{5}$$

one obtains

$$C_T = \frac{1}{\alpha_0} K_H p_{CO_2} \tag{6}$$

$$[HCO_3^-] = \frac{\alpha_1}{\alpha_0} K_H p_{CO_2} = \frac{K_1}{[H^+]} K_H p_{CO_2} \tag{7}$$

and

$$[CO_3^{2-}] = \frac{\alpha_2}{\alpha_0} K_H p_{CO_2} = \frac{K_1 K_2}{[H^+]^2} K_H p_{CO_2} \tag{8}$$

It follows from these equations that in a logarithmic concentration–pH diagram (Figure 4-3) the lines of $H_2CO_3^*$, HCO_3^- and CO_3^{2-} have slopes of 0, +1 and +2, respectively (compare Figure 3-9).

If we equilibrate pure water with CO_2, the system is defined by two independent variables,† for example, temperature and p_{CO_2}. In other words, the equilibrium concentrations of all the solute components can be calculated with the help of Henry's law, the acidity constants and the proton condition or charge balance if, in addition to temperature, one variable, such as p_{CO_2}, $[H_2CO_3^*]$ or $[H^+]$, is given.

---

† This is in accord with the Gibbs phase rule. The system consists of two components, $C = 2$ (H_2O and CO_2) and two phases ($P = 2$). The number of independent variables, F, is given by $F = C + 2 - P = 2$. (For further application of the phase rule see Section 5-7.)

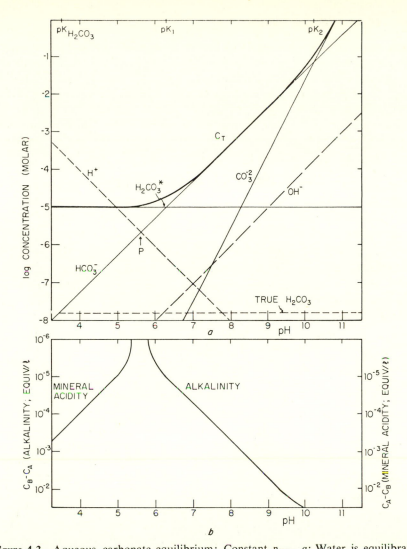

Figure 4-3 Aqueous carbonate equilibrium; Constant p_{CO_2}. *a*: Water is equilibrated with the atmosphere ($p_{CO_2} = 10^{-3.5}$ atm) and the pH is adjusted with strong base or strong acid. Equations 1–4 of Section 4-3 with the constants (25°C) $pK_H = 1.5$, $pK_1 = 6.3$, $pK_2 = 10.25$, pK (hydration of CO_2) $= -2.8$ have been used. The pure CO_2 solution is characterized by the proton condition $[H^+] = [HCO_3^-] + 2[CO_3^{2-}] + [OH^-]$ (see point *P*) and the equilibrium concentrations $-\log [H^+] = -\log [HCO_3^-] = 5.65$; $-\log [CO_2aq] = -\log [H_2CO_3^*] = 5.0$; $-\log [H_2CO_3] \approx 7.8$; $-\log [CO_3^{-2}] = 8.5$.

b: At pH values different from that of a pure CO_2 solution, the solution contains either alkalinity or mineral acidity, depending on the concentration of strong acid C_A or strong base C_B that had to be added to reach these pH values. In accordance with (11) of Section 4-3, alkalinity and acidity (logarithmic units) are plotted as a function of pH.

The pH of a pure aqueous CO_2 solution is defined by the charge balance (see Figure 4-3) $[H^+] = [HCO_3^-] + 2[CO_3^{2-}] + [OH^-]$. If we keep the system in equilibrium with a constant p_{CO_2}, we can vary the pH only by adding a base C_B or an acid C_A. The electroneutrality condition must be adjusted for such an addition.

Considering that C_B is equivalent to the concentration of a monovalent cation (e.g., $[Na^+]$ from NaOH) and that C_A is equivalent to the concentration of a monovalent anion (e.g., $[Cl^-]$ from HCl):

$$C_B + [H^+] = [HCO_3^-] + 2[CO_3^{2-}] + [OH^-] + C_A \qquad (9)$$

which can be arranged to

$$C_B - C_A = C_T(\alpha_1 + 2\alpha_2) + [OH^-] - [H^+] \qquad (10)$$

$$[Alk] = C_B - C_A = \frac{K_H p_{CO_2}}{\alpha_0}(\alpha_1 + 2\alpha_2) + [OH^-] - [H^+] \qquad (11)$$

$C_B - C_A$ is the acid neutralizing capacity of the solution with respect to the pure solution of CO_2 and is thus the *alkalinity*, [Alk]. If $C_A > C_B$, then $C_A - C_B$ represents the base neutralizing capacity (with respect to the same reference point) or the *mineral acidity*. For the conditions selected, alkalinity and mineral acidity are plotted as a function of pH in Figure 4-3b.

The model has three independent variables. Therefore, it is sufficient to give T, p_{CO_2} and alkalinity to define the system. As (10) shows, for example, the pH (implicitly contained in α values) is given by [Alk] and p_{CO_2}.

Waters in equilibrium with the atmosphere ($p_{CO_2} = 3 \times 10^{-4}$ atm) and having pH values similar to those of natural waters (6.5 to 9) have alkalinities between 10^{-2} and 10^{-5} equivalents liter^{-1}. Seawater having an alkalinity of 2.5 meq liter^{-1} brought into equilibrium with the atmosphere should have a $-\log[H^+]$ of approximately 8.2. This has been estimated by using for seawater $p^c K_1 = 6.0$. The gas composition in soils is quite different from that in the atmosphere; because of respiration by organisms, the CO_2 composition in soils is up to a few hundred times higher than that in the atmosphere. Thus, when brought to the surface, groundwaters are usually oversaturated with CO_2 and tend to lose CO_2.

Example 4-2a pH *of Solutions at Given* p_{CO_2}. Estimate the pH of the following solutions that have been equilibrated with the atmosphere ($p_{CO_2} = 10^{-3.5}$ atm)

(i) 10^{-3} M KOH
(ii) 10^{-3} M NaHCO$_3$
(iii) 5×10^{-4} M Na$_2$CO$_3$
(iv) 5×10^{-4} M MgO

All the solutions have the same alkalinity $[\text{Alk}] = 10^{-3}$ equiv liter^{-1}. They must all have the same pH at equilibrium with the same p_{CO_2}. Using the constants given in Figure 4-3, we calculate by (11) a pH of 8.3. This answer can also be found from Figure 4-3b.

Example 4-2b *Equilibration with the Gas Phase.* A liter of a 2×10^{-3} M NaHCO$_3$ solution is brought into contact with 10 ml nitrogen. How much CO$_2$ will be in the gas phase after equilibration (temp. $= 25°$C, p$K_\text{H} = 1.5$, p$K_1 = 6.3$, p$K_2 = 10.2$)?

The distribution equilibrium (i) and a mass balance (ii) need to be considered.

$$\frac{[\text{CO}_2(\text{aq})]}{[\text{CO}_2(\text{g})]} = K_\text{H}RT = 0.75 \tag{i}$$

$$\frac{[\text{CO}_2(\text{aq})]}{\alpha_0} + 0.01[\text{CO}_2(\text{g})] = 2 \times 10^{-3} \tag{ii}$$

The volume fraction $V_{\text{gas}}/V_{\text{aq}}$ is 0.01. If this fraction is small the second term in (ii) is negligible, that is, the solution composition is not changed appreciably and α_0 at equilibrium with the gas phase is equal to α_0 before equilibration with N$_2$ ($\alpha_0 \simeq 1 \times 10^{-2}$). Thus $[\text{CO}_2(\text{g})] = 2.67 \times 10^{-5}$ M, that is, the total gas phase contains 2.82×10^{-7} mole CO$_2$.

If the composition of the solution changes appreciably as a result of the transfer of CO$_2$ to the gas phase, a third equation is necessary. This, most conveniently, is an expression of alkalinity which does not change as a result of CO$_2$ loss.

4-4 Conservative Quantities: Alkalinity, Acidity and C_T

Acidimetric or alkalimetric titrations of a carbonate-bearing water to the appropriate endpoints represent operational procedures for determining alkalinity and acidity, that is, the equivalent sum of the bases that are titratable with strong acid and the equivalent sum of the acids that are titratable with strong base. Alkalinity and acidity are then capacity factors that represent, respectively, the acid and base neutralizing capacities (ANC and BNC; see Section 3-9) of an aqueous carbonate system. For solutions that contain no protolysis system other than that of aqueous carbonate, alkalinity is a measure of the quantity of strong acid per liter required to attain a pH equal to that of a C_T-molar solution of H$_2$CO$_3^*$. Alternatively, acidity is a measure of the quantity per liter of strong base required to attain a pH equal to that of a C_T-molar solution of Na$_2$CO$_3$.

The pH values at the respective equivalence points (around 4.5 and 10.3) for titrations of alkalinity and acidity represent approximate thresholds beyond which most life processes in natural waters are seriously impaired. Thus alkalinity and acidity are convenient measures for estimating the maximum capacity of a natural water to neutralize acidic and caustic wastes without permitting extreme disturbance of biologic activities in the water.

If mineral acid is added to a natural water beyond equivalence point x ($f = 0$) (compare Figures 4-1 and 4-4), the H^+ added will remain as such in solution. Such a water is said to contain *mineral acidity*. Correspondingly a water with a pH higher than point z ($f = 2$) contains *caustic alkalinity*.

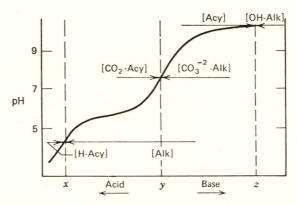

Figure 4-4 Conservative quantities: alkalinity and acidity as acid neutralizing and base neutralizing capacity. These parameters can be determined by acidimetric and alkalimetric titration to the appropriate end points. The equations given below define the various capacity factors rigorously.

Term	End point	Definition	
I Acid Neutralizing Capacity (ANC)			
Caustic alkalinity	$f=2$	$[OH^- \text{-Alk}] = [OH^-] - [HCO_3^-] - 2[H_2CO_3^*] - [H^+]$	(1)
Carbonate alkalinity	$f=1$	$[CO_3^{2-}\text{-Alk}] = [OH^-] + [CO_3^{2-}] - [H_2CO_3^*] - [H^+]$	(2)
Alkalinity	$f=0$	$[\text{Alk}] = [HCO_3^-] + 2[CO_3^{2-}] + [OH^-] - [H^+]$	(3)
II Base Neutralizing Capacity (BNC)			
Mineral acidity	$f=0$	$[\text{H-Acy}] = [H^+] - [HCO_3^-] - 2[CO_3^{2-}] - [OH^-]$	(4)
CO_2-acidity	$f=1$	$[CO_2\text{-Acy}] = [H_2CO_3^*] + [H^+] - [CO_3^{2-}] - [OH^-]$	(5)
Acidity	$f=2$	$[\text{Acy}] = 2[H_2CO_3^*] + [HCO_3^-] + [H^+] - [OH^-]$	(6)

III Combinations

$f = 2 - g$	(7)	$[\text{Alk}] + [CO_2\text{-Acy}] = C_T$	(11)
$[\text{Alk}] + [\text{H-Acy}] = 0$	(8)	$[\text{Alk}] + [\text{Acy}] = 2C_T$	(12)
$[\text{Acy}] + [OH\text{-Alk}] = 0$	(9)	$[\text{Alk}] - [CO_3\text{-Alk}] = C_T$	(13)
$[CO_3\text{-Alk}] + [CO_2\text{-Acy}] = 0$	(10)	$[CO_2\text{-Acy}] - [\text{H-Acy}] = C_T$	(14)

Titration to the intermediate equivalence point y ($f = 1$) ("phenolphthalein endpoint") is a measure of the CO_2 acidity in an alkalimetric titration and of CO_3^{2-} alkalinity in an acidimetric titration.

The equations given in Figure 4-4 are of analytical value because they represent rigorous conceptual definitions of the acid neutralizing and the base neutralizing capacities of carbonate systems. As discussed in Section 3-12, the definitions of alkalinity and acidity algebraically express the proton excess or proton deficiency of the system with respect to a reference proton level (equivalence point).† The definitions can be readily amplified to account for the presence of buffering components other than carbonates. For example, in the presence of borate and ammonia the definition for alkalinity becomes

$$[Alk] = [HCO_3^-] + 2[CO_3^{2-}] + [B(OH)_4^-] + [NH_3] + [OH^-] - [H^+] \quad (12)$$

As will be discussed later (Section 4-6) alkalimetric and acidimetric titrations to the equivalence points $f = 0$ and $f = 1$ give inherently accurate values of base and acid neutralizing capacity. As Figure 4-4 and (11)–(14) of this figure suggest, C_T ($= [H_2CO_3^*] + [HCO_3^-] + [CO_3^{2-}]$) can also be obtained from an acidimetric or alkalimetric titration (difference between the equivalence points $f = 0$ and $f = 1$).

Although individual concentrations or activities, such as $[H_2CO_3^*]$, and pH are dependent on pressure and temperature, ANC, BNC and C_T are conservative properties that are pressure and temperature independent. (Alkalinity, acidity and C_T must be expressed in concentrations, e.g., molarity or molality.)

Furthermore, these conservative quantities remain constant for selected changes in the chemical composition. The case of the addition or removal of dissolved carbon dioxide is of particular interest in natural waters because of the effect on pH of biochemical metabolism of carbon. A very simplified and generalized reaction scheme for carbon metabolism is

$$nCO_2 + nH_2O \underset{\text{respiratory activity}}{\overset{\text{photosynthetic activity}}{\rightleftharpoons}} (CH_2O)_n + nO_2$$

Respiratory activities of aquatic biota thus contribute carbon dioxide to natural water systems whereas photosynthetic activities decrease the concentrations of this weak acid. Any increase in carbon dioxide or, more rigorously, any increase in $[H_2CO_3^*]$, $dC_{H_2CO_3^*}$, increases both the acidity of the system and C_T, the total concentration of dissolved carbonic species. Unlike the case

---

† In a more exact interpretation we must be aware that the formulas for the species may include medium ions. In other words $[HCO_3^-]$ includes the concentration of "free HCO_3^-" as well as the concentration of complex bound HCO_3^-, that is, $[HCO_3^-] = [\text{true } HCO_3^-] + [NaHCO_3] + [MgHCO_3^+] \cdots$. Similarly $[CO_3^{2-}] = [\text{true } CO_3^{2-}] + [MgCO_3] + [NaCO_3^-]$.

for the addition of strong acid, however, alkalinity remains unaffected by increases or decreases in $[H_2CO_3^*]$. The fact that alkalinity is unaffected by CO_2 can be understood if we consider that alkalinity measures the proton deficiency with respect to the reference proton level $CO_2 - H_2O$. An analogous argument considers that the addition of CO_2 does not affect the charge balance (which inherently defines alkalinity) of the solution. It can be shown in a similar way that acidity remains unaffected by the addition or removal of $CaCO_3(s)$ or $Na_2CO_3(s)$. Acidity is thus a valuable capacity parameter for solutions in equilibrium with calcite. C_T, on the other hand, remains unchanged in a closed system upon addition of strong acid or strong base. Table 4-4 summarizes the effect of various chemical changes upon the capacity parameters. For each chemical change one capacity parameter that remains independent of this change can be found.

CAPACITY DIAGRAMS. Graphs using variables with conservative properties $(C_T, [Alk]$ or $[Acy])$ as coordinates can be used expediently to show contours of pH, $[H_2CO_3^*]$, $[HCO_3^-]$, etc. On such diagrams, as shown by Deffeyes [6], the addition or removal of base, acid, CO_2, HCO_3^- or CO_3^{2-} is a vector property. These graphs can be used to facilitate equilibrium calculations.

Figures 4-5 and 4-6 show two such diagrams. In Figure 4-5 [Alk] is plotted as a function of C_T. The construction of the diagram can be derived readily from the definition of [Alk]:

$$[Alk] = [HCO_3^-] + 2[CO_3^{2-}] + [OH^-] - [H^+] \qquad (13)$$

$$[Alk] = C_T(\alpha_1 + 2\alpha_2) + [OH^-] - [H^+] \qquad (14)$$

Table 4-4   Change in Capacity Parameters as a Result of Chemical Changes†

Capacity	Addition of Molar Increments of				
	C_B	C_A	$H_2CO_3^*$ (CO$_2$)	$NaHCO_3$	$CaCO_3$ or Na_2CO_3
C_T, M	0	0	+1	+1	+1
[Alk], eq liter^{-1}	+1	−1	0	+1	+2
[Acy], eq liter^{-1}	−1	+1	+2	+1	0
[CO$_2$−Acy]	−1	+1	+1	0	−1
[CO$_3^{2-}$−Alk]	+1	−1	−1	0	+1
[H−Acy]	−1	+1	0	−1	−2
[OH−Alk]	+1	−1	−2	−1	0

† Examples: $dC_B/dC_T = 0$; $d[H_2CO_3^*]/d[Acy] = +2$; $d[NaHCO_3]/d[Alk] = +1$.

[6] K. S. Deffeyes, *Limnol. Oceanog.*, **10**, 412 (1965).

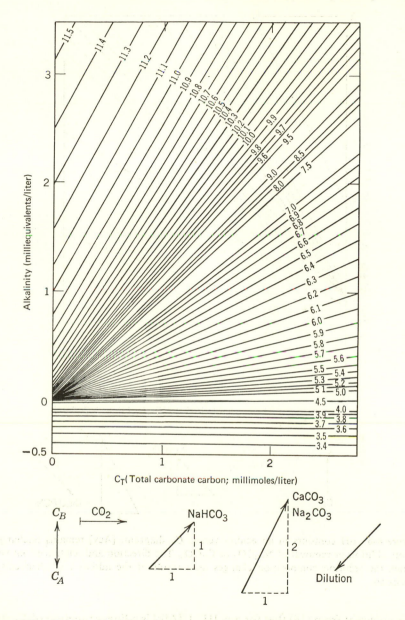

Figure 4-5   pH contours in alkalinity versus C_T diagram. The point defining the solution composition moves as a vector in the diagram as a result of the addition (or removal) of CO_2, $NaHCO_3$ and $CaCO_3$ (Na_2CO_3) or C_B and C_A. [After K. S. Deffeyes, *Limnol. Oceanog.*, **10**, 412 (1965).]

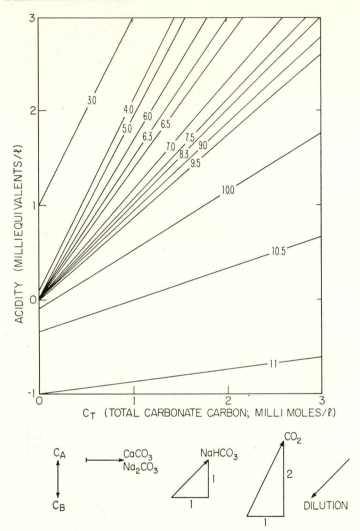

Figure 4-6 pH contours in an acidity versus C_T diagram. [Acy] remains unchanged upon addition or removal of Na_2CO_3 or $CaCO_3$. The direction and relative extent with which the solution composition changes as a result of the addition of chemicals is indicated.

It is apparent from (14) that for any $[H^+]$, [Alk] is a linear function of C_T; the system is completely defined by specifying C_T and [Alk]. For each value of C_T and [Alk], pH is fixed.

As Deffeyes points out, most of the usefulness of these diagrams stems from the fact that changes in the solution move the point representing the solution composition in definite directions on the diagram. Adding strong acid decreases [Alk] without changing C_T. Vertical lines on Figure 4-5 therefore give alkalimetric or acidimetric titration curves for each C_T value. On the other hand addition or removal of CO_2 increases or decreases C_T without changing [Alk] and the point moves to the right by the amount of CO_2 added. The change in solution composition (pH) by changes in CO_2, caused for example by respiration or photosynthesis, can be readily elucidated because the horizontal line defines a curve of titration with CO_2 for any alkalinity. Figure 4-5 gives the vector nature of the changes in C_T and [Alk] with the addition or removal of various substances.

Figure 4-6 gives a plot of [Acy] versus C_T:

$$[Acy] = 2[H_2CO_3^*] + [HCO_3^-] + [H^+] - [OH^-] \qquad (15)$$

$$[Acy] = C_T(2\alpha_0 + \alpha_1) + [H^+] - [OH^-] \qquad (16)$$

In this diagram the ordinate value ([Acy]) is unaffected by the addition or removal of crystalline carbonates such as $CaCO_3$ or Na_2CO_3. This diagram is of value in evaluating the effect of precipitation or dissolution of $CaCO_3$. Examples 4-3 and 4-4 illustrate the application of these diagrams for equilibrium calculations.

In solutions that contain bases and acids other than H^+, OH^- and the carbonate species, the contours in the diagrams will become displaced vertically. For example, if [Alk] contains other bases [see (12)] the intercept of the pH lines with the $C_T = 0$ line will be displaced vertically proportional to the concentration of the noncarbonate base.

Example 4-3   *Mixing of Waters.*   Two waters (A: pH = 6.1, [Alk] = 1.0 meq liter^{-1}; B: pH = 9, [Alk] = 2 meq liter^{-1}) are mixed in equal proportions. What is the pH of the mixture if no CO_2 was lost to the atmosphere? Using Figure 4-5 we find $C_T = 2.8$ mM and $C_T = 1.9$ mM for A and B, respectively. The mixture has [Alk] = 1.5 meq liter^{-1} and $C_T = 2.35$ mM; its pH from Figure 4-5 is pH = 6.6.

Example 4-4   *Increase of* pH *by Addition of Base or Removal of* CO_2.   The pH of a surface water having an original alkalinity of 1 meq liter^{-1} and a pH of 6.5 (25°C) is to be raised to pH 8.3. The following methods are considered:

(i) increase of pH with NaOH

(ii) increase of pH with Na_2CO_3

(iii) increase of pH by removal of CO_2 (i.e., in a cascade aerator)

For each case the chemical change and the final composition of the solution are given by

Change per Liter of Solution	pH	Final Composition		
		C_T, mM	Alk, meq liter^{-1}	Acy, meq liter^{-1}
Original solution	6.5	1.7	1.0	2.4
(i) $+0.7$ mmole NaOH	8.3	1.7	1.7	1.7
(ii) $+0.7$ mmole Na_2CO_3	8.3	2.4	2.4	2.4
(iii) -0.7 mmole CO_2	8.3	1.0	1.0	1.0

By first using Figure 4-5 we find the intersection of the pH $= 6.5$ line with the [Alk] $= 1$ (meq liter^{-1}) line and see that $C_T = 1.7 \times 10^{-3}\ M$. Assuming that no CO_2 is exchanged with the atmosphere, the quantity of NaOH (C_B) necessary to reach pH $= 8.3$ can be found directly from the graph (vertical displacement) as ~ 0.7 meq liter^{-1}.

For the pH increase with Na_2CO_3 we have to draw a line with slope 2 from the intersection (pH $= 6.5$; [Alk] $= 1$ meq liter^{-1}). About 0.7 mM liter^{-1} of Na_2CO_3 are necessary to attain a water of pH $= 8.3$.

This part of the problem could have been solved more conveniently with the help of Figure 4-6. According to this figure a water of pH $= 6.5$ and $C_T = 1.7 \times 10^{-3}$ has an [Acy] $= 2.35$ meq liter^{-1}. By drawing a horizontal line through this point ([Acy] $= 2.35$ meq liter^{-1}, pH $= 6.5$), one sees that ~ 0.7 mM liter^{-1} of Na_2CO_3 are needed to reach pH $= 8.3$.

Finally, a pH of 8.3 can also be attained by decreasing $C_T(CO_2)$ and maintaining a constant alkalinity. At the horizontal 1.0 meq liter^{-1} line (Figure 4-5) the difference between the points pH 6.5 and 8.3 corresponds to 0.7 mM liter^{-1} of CO_2.

The table summarizing the results of this example illustrates that a given pH change may be attained by different pathways, and that the final composition depends on the pathway.

Algebraic Approach

The graphs cannot always be read with sufficient precision. The uncertainty of the answer is large, especially for points in an unbuffered pH region (pH 7.5–9). The problems can of course be dealt with numerically, but the graphs may provide useful guidelines on how to attack a problem numerically.

Specifying any two independent variables at a given temperature and pressure (i.e., equilibrium constants at a given temperature and pressure) defines the composition of a carbonate solution. Recognizing how capacity parameters are affected by chemical changes (Table 4-4), a suitable pair of variables can generally be selected. Table 4-5 gives readily derivable equations that describe the titration of the carbonate system with base (acid), CO_2 or Na_2CO_3 ($CaCO_3$) while either C_T, [Alk], [Acy] or p_{CO_2} is kept constant. Corresponding titration curves are plotted in Figure 4-7.

Example 4-5 pH-*Change Resulting from Photosynthetic* CO_2 *Assimilation.*
1. As a result of photosynthesis a surface water with an alkalinity of 8.5×10^{-4} eq liter^{-1} showed within a 3-hr period a pH variation from 9.0 to 9.5. What is the rate of net CO_2 fixation? (Assume a closed aqueous system, that is, no exchange of CO_2 with atmosphere and no deposition of $CaCO_3$, $pK_1 = 6.3$, $pK_2 = 10.2$, $pK_H = 1.5$.)

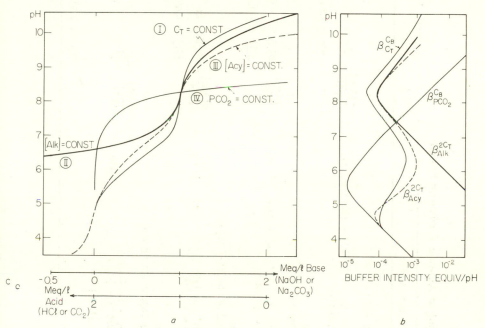

Figure 4-7 Buffering of aqueous carbonate systems of constant C_T, [Alk], [Acy] and p_{CO_2}, respectively. *a*: Titration curves. I: Titration of 10^{-3} M C_T solution with strong acid or strong base [(1) of Table 4-5]. II: Addition or removal of CO_2 to solution of [Alk] = 10^{-3} eq/liter^{-1} [(4) of Table 4-5]. III: Addition or removal of Na_2CO_3($CaCO_3$) to solution of [Acy] = 10^{-3} eq/liter^{-1} [(7) of Table 4-5]. IV: Titration of solution in equilibrium with $p_{CO_2} = 10^{-3.5}$ atm with strong acid or base [(10 of Table 4-5]. Curves constructed with the following constants: $pK_H = 1.5$; $pK_1 = 6.3$; $pK_2 = 10.2$.
b: Buffer intensities versus pH.

Table 4-5 Buffering of Aqueous Carbonate System

Titration Curve	Buffer Intensity
I. C_T = constant Addition of strong acid or base:	
(1) $C_B - C_A = C_T(\alpha_1 + 2\alpha_2) + [OH^-] - [H^+]$ $dC_B = d[Alk]$	(2) $\dfrac{d[Alk]}{dpH} = \beta_{C_T}^{C_B} = 2.3\{C_T[\alpha_1(\alpha_0 + \alpha_2) + 4\alpha_2\alpha_0] + [H^+] + [OH^-]\}$ (3) $\beta_{C_T}^{C_B} \approx 2.3\alpha_1(\alpha_0 + \alpha_2)C_T$
II. [Alk] = constant Addition or removal of CO_2:	
(4) $C_T = ([Alk] - [OH^-] + [H^+])/(\alpha_1 + 2\alpha_2)$ $dC_T = \frac{1}{2}d[Acy]$	(5) $\dfrac{d[Acy]}{dpH} = \beta_{Alk}^{2C_T} \simeq -4.6 \dfrac{\alpha_1\alpha_0 + \alpha_1\alpha_2 + 4\alpha_2\alpha_0}{(\alpha_1 + 2\alpha_2)^2}[Alk]$ (6) $\beta_{Alk}^{2C_T} \approx -4.6[(\alpha_0 + \alpha_2)/\alpha_1][Alk]$
III. [Acy] = constant Addition or removal of $NaCO_3(CaCO_3)$:	
(7) $C_T = ([Acy] - [H^+] + [OH^-])/(\alpha_1 + 2\alpha_0)$ $dC_T = \frac{1}{2}d[Alk]$	(8) $\dfrac{d[Alk]}{dpH} = \beta_{Acy}^{2C_T} \simeq 4.6 \dfrac{\alpha_1\alpha_0 + \alpha_1\alpha_2 + 4\alpha_2\alpha_0}{(\alpha_1 + 2\alpha_0)^2}[Acy]$ (9) $\beta_{Acy}^{2C_T} \approx 4.6[(\alpha_0 + \alpha_2)/\alpha_1][Acy]$
IV. p_{CO_2} = constant Addition of strong base or strong acid:	
(10) $C_B - C_A = (K_H(\alpha_1 + 2\alpha_2)/\alpha_0)p_{CO_2} + [OH^-] - [H^+]$ $dC_B = d[Alk]$	(11) $\dfrac{d[Alk]}{dpH} = \beta_{pCO_2}^{C_B} = 2.3\left\{K_H p_{CO_2}\dfrac{\alpha_1(\alpha_0 + \alpha_2) + 4\alpha_2\alpha_0 + (\alpha_1 + 2\alpha_2)^2}{\alpha_0} + [H^+] + [OH^-]\right\}$ (12) $\beta_{pCO_2}^{C_B} \approx 2.3K_H p_{CO_2}/\alpha_0 \approx 2.3C_T$

With the help of (4, Table 4-5) [with $(\alpha_1 + 2\alpha_2)_{pH=9.0} = 1.059$, $(\alpha_1 + 2\alpha_2)_{pH=9.5} = 1.165$], we compute C_T at pH = 9 and 9.5 as 7.94×10^{-4} M and 7.01×10^{-4} M, respectively. Thus the net CO_2 fixation is about 3.1×10^{-5} mole CO_2 liter^{-1} hr^{-1}.

2. In part 1 of this example it was assumed that alkalinity remained constant. In a more careful analysis it was detected that the alkalinity decreased by 3×10^{-5} eq liter^{-1} as a result of $CaCO_3$ precipitation. Considering the equations

$$[\text{Alk}]_{pH=9.5} = [\text{Alk}]_{pH=9} - 3 \times 10^{-5} \qquad \text{(i)}$$

and

$$C_{T,pH=9.5} = C_{T,pH=9} - [CO_2]_{\text{photosynthesis}} - 1.5 \times 10^{-5} \qquad \text{(ii)}$$

we calculate $\Delta[CO_2]_{\text{photosynthesis}} = 1.03 \times 10^{-4}$ M, or an hourly loss of CO_2 of 3.4×10^{-5} mole liter^{-1}.

The Effect of Other Bases

Many other dissolved buffer components occur in natural waters. Dissolved silicates, borates, ammonia, organic bases, sulfides and phosphates are among the bases that may be titrated in an acidimetric titration of alkalinity. Similarly, noncarbonic acids such as H_2S, polyvalent metal ions and organic acids may influence the base neutralizing capacity of natural waters. Usually, however, the concentrations of each of these noncarbonate components are very small in comparison to the carbonate species. Perhaps the most relevant species in seawater are the borates ($10^{-3.4}$ M) and in many fresh waters the silicates (10^{-3} to 10^{-4} M).

In the pH range of most natural waters, usually only a small fraction of the silica Si_T or boron species B_T become part of alkalinity or acidity. Because boric acid $[B(OH)_3]$ and orthosilicic acid $[Si(OH)_4]$ are rather weak acids $[pK_{B(OH)3} = 9; pK_{Si(OH)4} = 9.5]$ they have very little effect on the acidimetric titration curve (see Figure 4-8). Applying (12), we have

$$[\text{Alk}] = C_T(\alpha_1 + 2\alpha_2) + [OH^-] - [H^+] + B_T\alpha_{B^-} + Si_T\alpha_{Si^-} \qquad \text{(17)}$$

where

$$\alpha_{B^-} = \frac{[B(OH)_4^-]}{B_T}$$

$$\qquad \text{(18)}$$

$$\alpha_{Si^-} = \frac{[SiO(OH)_3^-]}{Si_T}$$

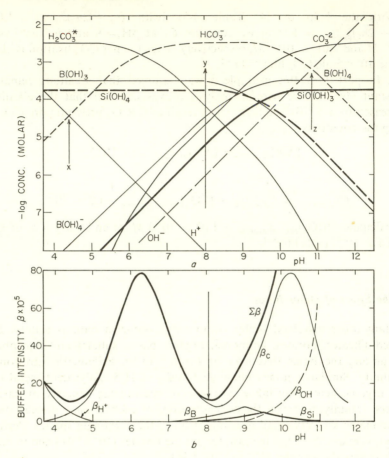

Figure 4-8 Effect of boric acid and silicic acid upon carbonate equilibria. *a*: Distribution of species as a function of pH. *b*: Buffer intensity of individual acid–base components as a function pH.

$10^{-3.5}$ *M* B_T and $10^{-3.75}$ *M* Si_T (values representative of seawater) have very little effect on buffer intensity and on the acidimetric titration curve. The equivalence point *y* ($f = 1$, end point in CO_2–acidity determination) (corresponding to the minimum in the buffer intensity) is lowered by 0.22 pH units.

The total buffer intensity is simply the sum of the contributions of the buffer intensities of the individual buffer components.

Because in waters of pH < 9, α_{B^-} and α_{Si^-} are much smaller than one, these constituents usually become only a minor portion of alkalinity. For precise calculations of carbonate equilibria a correction must be applied:

$$[\text{Alk}] - \text{X} = C_T(\alpha_1 + 2\alpha_2) + [\text{OH}^-] - [\text{H}^+] \qquad (19)$$

where X now includes the equivalent sum of the bases that can be titrated with strong acid, that is, $B(OH)_4^-$ and $SiO(OH)_3^-$. In seawater about 20% of the boron species are present as borate. Because silica exists predominantly as o-silicic acid in most natural waters, it usually does not become a part of the alkalinity.

4-5 Buffer Intensity

Buffer intensity controls the magnitude of shifts in the pH of solutions. The role of the dissolved or homogeneous carbonate system in effecting buffer action in natural waters is unquestionably significant, especially for waters that have become isolated from their environment. We shall see later, how-ever, that dissolved carbonates are not solely responsible for pH regulations in natural waters because many heterogeneous chemical, biochemical and geochemical processes contribute significantly to the buffering of these waters.

In Table 4-5 equations for the buffer intensity of the aqueous carbonate system are listed. As explained in Section 3-9, the buffer intensity is found analytically by differentiating the appropriate function of the system (equa-tions for the titration curves, Table 4-5) with respect to pH. Figure 4-7 shows a number of computed buffer intensity versus pH relationships for millimolar (C_T) or millinormal ([Alk] or [Acy]) solutions at 25°C. Besides the familiar buffer intensity with respect to strong acid or base, $\beta_{C_T}^{C_B}$, buffer intensities with respect to CO_2 as an acid, $\beta_{\text{alk}}^{2C_T}$ (5, Table 4-5) and with respect to Na_2CO_3 as a base, $\beta_{\text{acy}}^{2C_T}$ (8, Table 4-5) are given in this figure. In the lower pH range (pH < 7.5) the buffer intensity of a water with respect to CO_2 is much larger than the buffer intensity with respect to a strong acid. Equations 10 and 11 (Table 4-5) and Figures 4-3 and 4-7 illustrate that for a water which remains in equilibrium with CO_2 of the atmosphere, the huge reservoir of CO_2 imparts a significant buffering action upon waters of high pH (pH > 7.5).

In the pH range of most natural waters, noncarbonate buffer components exert little influence upon pH regulation. As Figure 4-8 illustrates, the total buffer intensity β^{C_B} is simply the sum of the contributions of the buffer intensities of the individual buffer components. β^{C_B} caused by boron and silica species is insignificant at pH values below 8.5. The presence of borate or silicate may shift the $f = 1$ equivalence point to slightly more acidic pH values. The equivalence point is now defined by the proton condition

$$[H_2CO_3^*] + [H^+] = [CO_3^{2-}] + [B(OH)_4^-] + [SiO(OH)_3^-] + [OH^-] \quad (20)$$

In Figure 4-8 the point y is shifted from pH = 8.27 (absence of B and Si) to pH 8.05 (presence of $10^{-3.5} M$ B_T and $10^{-3.75} M$ Si_T).

4-6   Analytic Considerations

Complete resolution of simple carbonate systems in terms of the equilibrium concentrations of each of the five individual components (H^+, OH^-, H_2CO_3, HCO_3^- and CO_3^{2-}), is possible if the concentrations of at least two of these, or two of certain combinations thereof, are evaluated analytically. C_T (total CO_2), total alkalinity, CO_2-acidity, CO_3^{2-}-alkalinity and pH are commonly determined in routine water analyses. The measurement of caustic alkalinity and mineral acidity, in addition to the parameters listed, is often useful in the definition of specific systems.

Those parameters determinable by direct analysis are listed in Table 4-6 [7]. From values for any two of the quantities listed, the other parameters and the concentrations of the single ions HCO_3^-, CO_3^{2-}, H^+ and OH^- can be computed. It is evident from the conceptual definitions given in Figure 4-4 that no direct rigorous determination for the concentration of any one of the carbonic species is feasible. For example, acidimetric titration to the CO_2 end point ("methyl orange end point") provides for evaluation of alkalinity rather than of the concentration of HCO_3^-. In waters with pH less than 9, the alkalinity and concentration of HCO_3^- are, of course, practically identical. Similarly, alkalimetric titration to the bicarbonate end point gives CO_2 acidity but not the concentration of H_2CO_3, although these are for all practical purposes the same in waters within the approximate pH range of 5.7 to 7.6.

Precision of Analytic Determinations

Although any two of the quantities listed in Table 4-6 may be used in the evaluation of the others, the question as to the choice of the two parameters whose experimental determination permits the most precise characterization of a carbonate system is significant. No general answer applicable to all natural waters can be given, and experimental expediency must be considered.

Because the sharpness of the end point in an alkalimetric or acidimetric titration is related to the slope, $d(\mathrm{pH})/dC_B$ or $1/\beta$, of the titration curve at the equivalence points, the relative error involved is directly proportional to the buffer intensity at the end point. No inflection point is observed at the carbonate end point ($f = 2$); thus the parameters of acidity and caustic alkalinity, the titrations for which are characterized by this end point, cannot be evaluated by direct titration. Relative errors for the respective analytic determinations are given in Table 4-6, the relative error equations being based on the assumption that the end point can be detected within a certain $\Delta(\mathrm{pH})$ and that the equivalence point pH is known exactly. The latter assumption is, of

[7] W. Weber and W. Stumm, *J. Amer. Water Works Assoc.*, **55**, 1560 (1963).

Table 4-6 Relative Error by Analytic Determinations (after Weber and Stumm [7])

Quantity Determined	Titrant	Equivalence Point Characterized by Equation (Table 4-2)	Relative Error in Determination,† per cent
H^+			$230\Delta(pH)$
C_T			$3\text{--}5$
[Alk]	Strong acid	(12)	$\dfrac{460\Delta(pH)(C_T K_1)^{0.5}}{[\text{Alk}]}$
$[CO_3^{2-} - \text{Alk}]$	Strong acid	(13)	$\dfrac{460\Delta(pH)C_T}{[CO_3^{2-} - \text{Alk}]}\left(\dfrac{K_2}{K_1} + \dfrac{K_W}{C_T K_1}\right)^{0.5}$
Caustic Alk$[OH^- - \text{Alk}]$	Strong acid	(14)	Not feasible‡
[Acy]	Strong base	(14)	Not feasible§
$[CO_2\text{-Acy}]$	Strong base	(13)	$\dfrac{460\Delta(pH)C_T}{[CO_2 - \text{Acy}]}\left(\dfrac{K_2}{K_1} + \dfrac{K_W}{C_T K_1}\right)^{0.5}$
	Na_2CO_3	(13)	$\dfrac{460\Delta(pH)(C_T + [CO_2 - \text{Acy}])}{[CO_2 - \text{Acy}]}\left(\dfrac{K_2}{K_1} + \dfrac{K_W}{K_1(C_T + [CO_2 - \text{Acy}])}\right)^{0.5}$
Mineral Acy $[H^+\text{-Acy}]$	Strong base	(12)	$\dfrac{460\Delta(pH)(C_T K_1)^{0.5}}{[H^+ - \text{Acy}]}$

† Calculation of relative error (E_R, a measure of reproducibility) is defined as:

$$E_R = \frac{\Delta(pH)\beta_{eq}}{C}$$

in which $\Delta(pH)$ is the pH interval within which the equivalence point can be detected; β_{eq} is the buffer capacity at the equivalence point, on the assumption that the titration curve is essentially linear within the critical pH interval; and C is the concentration of the quantity to be determined. Minor terms have been neglected in the derivation of the relative-error expressions given above; hence, they are approximate and valid only for concentrations larger than approximately 10^{-5} M.

Estimated error for the determination of total CO_2 (C_T). The method is based on transferring CO_2 (liberated by acidifying the sample) into $Ba(OH)_2$; the excess of hydroxide is then back titrated with acid.

‡ At this equivalence point no inflection in the titration curve is obtained, and thus titration to the equivalence point pH is accompanied by a prohibitively large error. In many waters a satisfactory acidimetric titration is possible after addition of $BaCl_2$ ($C_{BaCl_2} > C_T$). The titration is carried out in the presence of precipitated $BaCO_3$ with a satisfactory inflection point observed at a pH of approximately 9.

§ The considerations pertinent in this instance are the same as those cited in the preceding footnote; an alkalimetric titration in the presence of $BaCl_2$ may give satisfactory results.

Reproduced with permission from American Water Works Association.

course, not always realistic. In visual titrations with an indicator the relative error might be calculated by assuming that the equivalence point pH can be detected with an uncertainty of 0.4 pH unit. If a pH meter or a screened indicator is used, the detectable visual pH change may be of the order of 0.1–0.2 pH unity. Not all the analytic determinations will be discussed in detail; however, a few essential points not covered in Standard Methods [8] are given below.

Alkalinity

The operational end point of an acidimetric titration for alkalinity generally is not in exact accord with the calculated equivalence point pH, as CO_2 is lost to the atmosphere during titration. The quantity of CO_2 lost in the course of a titration naturally depends on the amount of agitation and on the degree of oversaturation attained at low pH values. It has been the authors' experience, however, that relative CO_2 losses are negligible if smooth stirring, such as is obtainable with a magnetic stirrer, is maintained throughout the course of the titration. Under such circumstances the values for the end point pH at room temperature are, in accord with [(12), Table 4-2], about 0.4 pH units lower than the values quoted in Standard Methods [8]. An independent check on the proper end point pH can be obtained by conductimetric titration of alkalinity with a mineral acid or by a Gran plot discussed below.

As the end point and the buffer capacity at the end point depend on the concentration of CO_2 at this point in a titration, the sharpness of the end point can be improved if CO_2 is continuously removed (e.g., by scrubbing with CO_2-free gas) from the solution during the titration. With decreasing CO_2 the end point pH is progressively shifted toward pH 7 where the sharpness of the end point is the same as in the titration of a strong base with a strong acid.

Alkalinity can be determined fairly precisely even without these precautions. For example, an alkalinity of 10^{-3} equivalents per liter^{-1} (50 mg liter^{-1} as $CaCO_3$) can be determined with a relative error of approximately $\pm 3.5\%$ if the end point is recognizable within 0.3 pH unit of the true equivalence point. For the same precision of end point detection the relative error is ten times as great for a hundredfold dilution. In waters in which low alkalinities (or mineral acidities) have to be determined (for example, rain water, demineralized water and steam condensate), the end point recognition must be more precise in order to obtain results with small relative error. A procedure developed by Gran [9], and Larson and Henley [10], discussed by Thomas [11],

[8] *Standard Methods for the Examination of Water and Waste Water*, American Public Health Association, New York, 1960.
[9] G. Gran, *Analyst*, **77**, 661 (1952).
[10] T. E. Larson and L. Henley, *Anal. Chem.*, **27**, 851 (1965).
[11] J. F. Thomas and J. J. Lynch, *J. Amer. Water Works Assoc.*, **52**, 255 (1960).

and more recently applied for seawater by Dyrssen [12] is based on the principle that added increments of mineral acid linearly increase hydrogen ion concentration.

A detailed derivation of Gran's graphical method of end point location is given in an Appendix to this chapter. A significant advantage of the Gran plot is that it does not matter whether the "pH" values are defined in terms of the IUPAC–NBS convention or in terms of $-\log [H^+]$ units. Figure 4-9 gives an example (from Dyrssen and Sillén [13]) of an emf titration of seawater. This figure illustrates that a single acidimetric titration can give quite accurate data on $[CO_3^{2-}\text{-Alk}]$, [Alk] and C_T.

MINERAL ACIDITY. Mineral acidity, exhibited by waters containing acids stronger than $H_2CO_3^*$, and alkalinity are in a sense complementary parameters since $[H^+\text{-acidity}] + [\text{alkalinity}] = 0$. Many acid waters contain metal ions capable of neutralizing titrant bases. This particular interference frequently can be avoided by adding complex-formers such as fluoride.

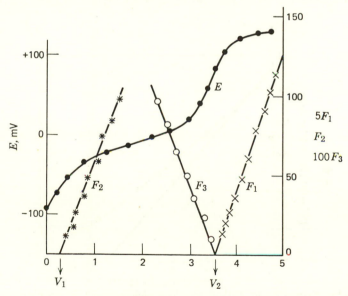

Figure 4-9   emf titration curve for 154 g of seawater with V ml (0.1000 M HCl + 0.4483 M NaCl); characterization of equivalence points by Gran method. F_1, F_2 and F_3 are Gran functions (see Appendix of Chapter 4) for finding the equivalence points V_1 and V_2 corresponding to $f = 1$ and $f = 0$, respectively.

[12] D. Dyrssen, *Acta Chem. Scand.*, **19**, 1265 (1965).
[13] D. Dyrssen and L. G. Sillén, *Tellus*, **19**, 110 (1967).

On occasion it is also possible to remove interfering metal ions by using ion-exchange resins in the sodium form; however, cation exchange resins do not always yield quantitative exchanges with multivalent ions, and it is frequently preferable to obtain the value of mineral acidity by a Gran procedure. The procedure, similar to that described for alkalinity, is based on the fact that in the presence of low concentrations of metal ions or other weak acids, H^+-acidity is titrated preferentially in low-pH regions by initial increments of titrant. Thus the change in $[H^+]$ during the initial stages of an alkalimetric titration of a solution containing metal ions and H^+-acidity is closely proportional to the volume of titrant added.

Experimental data for titrations of 100-ml aliquots of 10^{-2} M HCl and 10^{-2} M HCl $+$ 10^{-3} M Fe(ClO$_4$)$_3$ with 10^{-1} M NaOH are presented in Figure 4-10.

CO_2-*Acidity and* $CO_3{}^{2-}$-*Alkalinity*

There is no direct way to determine rigorously by alkalimetric or acidimetric titration the individual values of $[H_2CO_3^*]$ or $[CO_3^{2-}]$. For many

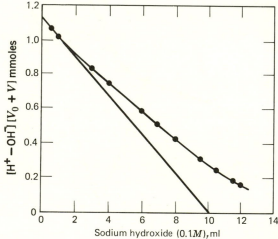

Figure 4-10 Extrapolation method for determining the equivalence point for H^+-acidity in the presence of metal ions or other weak acids. The curve broken by solid circles is for 1×10^{-2} M Fe(ClO$_4$)$_3$. The equivalence point for the HCl alone can be evaluated by extrapolation to $[H^+ - OH^-][V_0 + V] = 0$, because H^+ is titrated preferentially in the initial portion of the titration of the mixed solution. The value of 10 ml of NaOH determined by extrapolation of the data for the combined solution is in good agreement with the value of 9.9 ml obtained from titration of a pure solution of HCl.

waters without mineral acidity and with pH values below 7.6, CO_2-acidity is practically identical to $[H_2CO_3^*]$. Similarly, for waters not containing caustic alkalinity and with pH values above 9, the CO_3-alkalinity is nearly equal to $[CO_3^{2-}]$. Within the pH range 7.6–9, neither CO_2-acidity nor CO_3^{2-}-alkalinity can be interpreted in terms of $[H_2CO_3^*]$ or $[CO_3^{2-}]$, respectively. For example, a water at pH 8 with a CO_2 acidity of 10^{-3} equivalent liter^{-1} determined by titration with strong base actually contains 1.28×10^{-3} M $H_2CO_3^*$. Identification of CO_2-acidity with $[H_2CO_3^*]$ in this water would lead to an underestimation of $[H_2CO_3^*]$ of nearly 30%.

In most natural waters the determination of CO_2-acidity is only slightly less precise than an alkalinity determination. For example, CO_2-acidity can be determined in a water containing 10^{-3} equivalent liter^{-1} of alkalinity and 5×10^{-4} equivalent liter^{-1} of CO_2-acidity with a relative error of approximately 4% provided the end point can be defined within 0.3 pH unit of the true equivalence point pH; however, the error increases quite rapidly as the ratio of CO_2-acidity to alkalinity decreases.

The relative error in an alkalimetric titration becomes larger if Na_2CO_3 instead of NaOH is used as a titrant. The use of soda leads to a reduced slope at the equivalence point because of increased buffer intensity (Figure 4-7).

In titrations of waters of high hardness and alkalinity, $CaCO_3$ saturation is frequently exceeded prior to the end point (pH about 8.3, $f = 1$). If $CaCO_3$ precipitates before the end point is reached, positive errors occur because of the reaction $Ca^{2+} + HCO_3^- = CaCO_3(s) + H^+$. The possibility of this is more pronounced if Na_2CO_3 is used as the titrant. There are, however, various means to circumvent this interference, such as the addition of a cation-exchange resin in the sodium form.

Other frequently significant errors in CO_2 determinations occur because of CO_2 loss to the atmosphere during sampling and transfer. With many natural waters it is most expedient to determine pH and water temperature in the field and, as alkalinity is not changed by loss of CO_2, this quantity can be determined after samples have been transported to the laboratory. (Also if $CaCO_3$ precipitates when the samples are stirred, CO_2 can be added prior to the alkalinity determination to redissolve the precipitate.) The free CO_2 (or CO_2-acidity) and CO_3^{2-} (or CO_3^{2-}-alkalinity) can then be computed from the measured values of pH and alkalinity with the appropriate constants — that is, those valid at the temperature measured in the field and corrected for activity. If a pH meter with a precision of 0.05 pH unit is used, the relative error in $[H^+]$ amounts to about 10%. As $[CO_2]$ is a function of $[H^+]$, values computed from measurements of the latter include relative errors of at least a similar order.

The characterization of a carbonate system by alkalinity and pH has the advantage of expediency and convenience. In most waters, however, precision is not entirely satisfactory; a more precise characterization of the carbonate system is possible if, in addition to alkalinity, the CO_2-acidity or the CO_3^{2-}-alkalinity is determined. This determination, equivalent to a C_T determination, is obtained directly from a complete titration curve (Figure 4-4). If there is interference by substantial amounts of other acids or bases, C_T can also be determined manometrically from the quantity of CO_2 released from the solution after acidification.

Aqueous CO_2 can be determined rather precisely by equilibrating a small volume of a carrier gas with the solution. The CO_2 content in the gas phase can be determined by infrared spectrophotometry, and the $H_2CO_3^*$ content can be calculated (see Example 4-3).

4-7   The Equilibrium Constants

Tables 4-7, 4-8 and 4-9 give the equilibrium constants for various temperatures ($P = 1$ atm). These constants are needed to evaluate the distribution of carbonate species in a quantitative analytical way. Data for equilibrium constants are given for different activity scales (see Section 3-4). For dealing with fresh water systems the infinite dilution activity scale might be most useful, while in seawater the seawater-medium scale is more appropriate.

Table 4-7   Equilibrium Constant for CO_2-Solubility

Equilibrium: $CO_2(g) + aq = H_2CO_3^*$
Henry's law constant: $K = H_2CO_3^*/p_{CO_2}$ (M atm^{-1})

Temp., °C	$\rightarrow 0$	Medium, 1 M NaClO$_4$	Seawater, 19% Cl$^-$
	$-\log K$	$-\log {}^cK$	$-\log {}^cK$
0	1.11†		1.19†
5	1.19†		1.27†
10	1.27†		1.34†
15	1.32†		1.41†
20	1.41†		1.47†
25	1.47†‡	1.51§	1.53†
30	1.53†		1.58†
35	—	1.59§	
40	1.64‡		
50	1.72‡		

† Values based on data taken from Bohr and evaluated by K. Buch, *Meeresforschung*, 1951.
‡ A. J. Ellis, *Amer. J. Sci.*, **257**, 217 (1959).
§ G. Nilsson, T. Rengemo and L. G. Sillén, *Acta Chem. Scand.*, **12**, 878 (1958).

Table 4-8   First Acidity Constant: $H_2CO_3^* = HCO_3^- + H^+$

$K_1 = \dfrac{\{H^+\}\{HCO_3^-\}}{\{H_2CO_3^*\}}$	$K_1' = \dfrac{\{H^+\}[HCO_{3T}^-]}{[H_2CO_3^*]}$	$^cK_1 = \dfrac{[H^+][HCO_{3T}^-]}{[H_2CO_3^*]}$	

	Medium			
Temp., °C	$\to 0$	Seawater, 19‰ Cl$^-$	Seawater	1 M NaClO$_4$
	$-\log K_1$	$-\log K_1'$	$-\log {}^cK_1$	
0	6.579†	6.15‡		
5	6.517†	6.11‡	6.01#	
10	6.464†	6.08‡		
14	—	—	6.02††	
15	6.419†	6.05‡		
20	6.381†	6.02‡		
22	—	6.00§	5.89§	
25	6.352†	6.00,‡ 6.09‖		6.04‡‡
30	6.327†	5.98‡		
35	6.309†	5.97‡		
40	6.298†			
50	6.285†			

† H. S. Harned and R. Davies, Jr., *J. Amer. Chem. Soc.*, **65**, 2030 (1943).

‡ After Lyman (1956), quoted in G. Skirrow, *Chemical Oceanography*, Vol. I, J. P. Riley and G. Skirrow, Eds., Academic Press, New York, 1965, p. 651.

§ A. Distèche and S. Distèche, *J. Electrochem. Soc.*, **114**, 330 (1967).

‖ Calculated as $\log (K_1/f_{HCO_3^-})$ from $f_{HCO_3^-}$ as determination by A. Berner, *Geochim. Cosmochim. Acta*, **29**, 947 (1965).

D. Dyrssen and L. G. Sillén, *Tellus*, **19**, 810 (1967).

†† D. Dyrssen, *Acta Chem. Scand.*, **19**, 1265 (1965).

‡‡ M. Frydman, G. N. Nilsson, T. Rengemo and L. G. Sillén, *Acta Chem. Scand.*, **12**, 878 (1958).

FRESH WATER. In fresh water with $I < 10^{-2}$ M, equilibrium constants valid for infinite dilution may be used. These constants must be corrected with the help of the Debye-Hückel limiting law; the Güntelberg approximation is especially useful. Larson and Buswell [14] have recommended the following salinity corrections

$$pK_1' = pK_1 - \frac{0.5\sqrt{I}}{(1 + 1.4\sqrt{I})} \tag{21}$$

$$pK_2' = pK_2 - \frac{2\sqrt{I}}{(1 + 1.4\sqrt{I})} \tag{22}$$

where pK and pK' are defined as in Table 4-8. In a fresh water whose detailed composition is not known an approximate ionic strength can be estimated

[14] T. E. Larson and A. M. Buswell, *J. Amer. Water Works Assoc.*, **34**, 1667 (1942).

Table 4-9   Second Acidity Constant: $HCO_3^- = H^+ + CO_3^{2-}$

$$K_2 = \frac{\{H^+\}\{CO_3^{2-}\}}{\{HCO_3^-\}} \qquad K_2' = \frac{\{H^+\}[CO_{3_T}^{2-}]}{[HCO_{3_T}^-]} \qquad {}^cK_2 = \frac{[H^+][CO_{3_T}^{2-}]}{[HCO_{3_T}^-]}$$

	Medium			
Temp., °C	$\to 0$	Seawater	0.75 M NaCl	1 M KClO$_4$
	$-\log K_2$	$-\log K_2'$	$-\log K_2'$	$-\log {}^cK_2$
0	10.625†	9.40‡		
5	10.557†	9.34‡		
10	10.490†	9.28‡		
15	10.430†	9.23‡		
20	10.377†	9.17‡		
22	—	9.12§	9.49§	
25	10.329†	9.10‡		9.57‖
30	10.290†	9.02‡		
35	10.250†	8.95‡		
40	10.220†	—		
50	10.172	—		

† H. S. Harned and S. R. Scholes, *J. Amer. Chem. Soc.*, **63**, 1706 (1941).
‡ After Lyman, quoted in G. Skirrow, *Chemical Oceanography*, Vol. I, J. P. Riley and G. Skirrow, Eds., Academic Press, New York, p. 651.
§ A. Distèche and S. Distèche, *J. Electrochem. Soc.*, **114**, 330 (1967).
‖ M. Frydman, G. N. Nilsson, T. Rengemo and L. G. Sillén, *Acta Chem. Scand.*, **12**, 878 (1958).

from the total hardness, H ($H \simeq [Ca^{2+}] + [Mg^{2+}]$) ($M$), and the alkalinity (eq liter^{-1})

$$I \simeq 4H - \tfrac{1}{2}[\text{Alk}] \qquad (23)$$

or from total dissolved solids, S (mg liter^{-1})

$$I \simeq S \times 2.5 \times 10^{-5} \qquad (24)$$

SEAWATER. In seawater precise equilibrium calculations should be made, preferably on the constant medium activity scale. Since the oceans contain an ionic medium of essentially constant composition, and since carbonate and borate species are present as minor constituents (in comparison to other ions), seawater may be defined as the ionic medium. As pointed out by Sillén, equilibrium constants on an ionic medium scale are thermodynamically as well defined as those in the infinite dilution scale.

Even if we adopt seawater as a constant medium, different equilibrium constants have to be defined for different definitions of "pH." If pH is defined operationally in terms of a cell and if the measurement is made by comparing the solution with a standard buffer (NBS), we assume pH $\approx$ p^{a}H $\approx -\log\{H^{+}\}$ (IUPAC, NBS) and equilibria defined by mixed constants (K_1', K_2', in Tables 4-8 and 4-9) may be employed. On the other hand, if one attempts to measure pH in terms of $-\log[H^{+}]$, that is, by calibrating the glass electrode with a dilute solution of a strong acid of known $[H^{+}]$ in the same ionic medium, equilibria defined by $^{c}K_1$ or $^{c}K_2$ should be employed.

Complex or Ion Pair Formation with Sea Water Medium Ions

HCO_3^{-} or CO_3^{-} ions may form complexes with the medium ions (e.g., $MgCO_3$, $NaCO_3^{-}$, $CaCO_3$, $MgHCO_3^{+}$). In the constant ionic medium method, K' and ^{c}K are defined in terms of species that include an unknown number of medium ions, for example,

$$[CO_{3_T}^{2-}] = [CO_3^{2-}] + [MgCO_3] + [CaCO_3] + [NaCO_3^{-}] \quad (25)$$

This is operationally of advantage because in an equilibrium computation, let us say the estimation of $[H_2CO_3^{*}]$ from pH and [Alk], one does not have to consider the complex formation equilibria of HCO_3^{-} and CO_3^{2-} with Ca^{2+}, Mg^{2+} and Na^{+}. It may be necessary to limit the validity of a K' or ^{c}K value to a certain pH range.

That the activity coefficient for CO_3^{2-} as computed from $^{c}K_2$ (seawater) is much lower than the activity coefficient as computed from Debye-Hückel formulas suggests that such complex or ion-pair formation really exists in seawater. Garrels and Thompson [15] have formulated a model for the distribution of major dissolved ions and complex species in seawater. This model will be discussed further in Section 6 of Chapter 6. A comparison of the K_2' values (Table 4-9) determined by Distèche and Distèche [16] for seawater and a 0.75 M NaCl medium (which is approximately equivalent to the ionic strength of seawater) indicates that Mg^{2+} and Ca^{2+} (in addition to Na^{+}) must be involved primarily in complex formation.

Complex and ion-pair formation is of less importance in most fresh waters. Considering the stability constants given by Garrels and Thompson [15], we calculate that waters with $[Ca^{2+}] \leq 2 \times 10^{-3}\ M$ and $[Alk] \leq 4 \times 10^{-3}$ eq liter^{-1} do not form bicarbonato or carbonato complexes of Ca^{2+} exceeding 5% of the total. Carbonato magnesium complexes, however, can occur in Mg^{2+}-rich waters ($Mg^{2+} > 10^{-3}\ M$) of high pH ([Alk] $> 4 \times 10^{-3}$ eq liter^{-1} and pH > 8).

[15] R. M. Garrels and M. E. Thompson, *Amer. J. Sci.*, **260**, 57 (1962).
[16] A. Distèche and S. Distèche, *J. Electrochem. Soc.*, **114**, 330 (1967).

Pressure Dependence

The effect of pressure on the equilibrium constant can be found with the thermodynamic relation

$$\left(\frac{\partial \ln K}{\partial P}\right)_{T,m} = -\frac{\Delta V}{RT} \tag{26}$$

where ΔV (cm^3 mole^{-1}) is the algebraic difference between the total partial molar volumes of products and reactants. For small changes in pressure it is sufficient to regard ΔV as constant and to use (26) in its integrated form

$$\log \frac{K_{P_1}}{K_{P_2}} = -\frac{\Delta V(P_2 - P_1)}{2.3RT} \tag{27}$$

In fresh water systems pressures rarely exceed 30 atm. For most acids $-\Delta V$ is between 10 and 30 cm^3 mole^{-1}, hence the difference in log K does not exceed one or two hundredths of a unit. In recent determinations (Distèche and Distèche [16]) volume changes, $-\Delta V$, of 25.4 and 25.6 cm^3 mole^{-1} have been obtained at 22°C and $I = 0$ for the first and second protolysis reactions of $H_2CO_3^*$. For seawater we encounter pressures of a few hundred atmospheres. In such systems application of (26) and (27) is rendered difficult. At high pressure in seawater $H_2CO_3^*$ and HCO_3^- protolyze to a different extent than at zero ionic strength. Because complex formation and ion-pair equilibria are also pressure dependent, the change in p^cK or pK' values might be quite different from changes in pK values. Distèche and Distèche have obtained a decrease in p^cK_1 and p^cK_2 upon a pressure increase of 1000 bar (1013 atm) of 0.34 and 0.20, respectively. The pressure dependence of p^cK_2 for seawater medium, for example, corresponds to an apparent volume change of -10.6 cm^3 mole^{-1} instead of -25.6 cm^3 mole^{-1} at $I = 0$. Using the experimentally determined values, at the "average depth" of the ocean ($P = 200$ atm) log K_1 and log K_2 values will increase by approximately 0.07 and 0.04 units, respectively.

Although equilibrium constants and, as a consequence, activities or concentrations of individual species are dependent upon pressure and temperature, it is appropriate to emphasize once more that, despite these changes, capacity parameters such as C_T, [Alk] and [Acy] are conservative, that is, independent of pressure, temperature and ionic strength.

4-8   Kinetic Considerations

Ionization equilibria in the dissolved carbonate system are established very rapidly. Somewhat slower (seconds) is the attainment of equilibrium in the hydration or dehydration reaction of CO_2

$$CO_2(aq) + H_2O \rightleftharpoons H_2CO_3 \tag{28}$$

Many natural waters are not in equilibrium with the atmosphere, partially because of unfavorable mixing conditions but primarily because of the slowness of the gas transfer reaction; this transfer is frequently slower than reactions that produce or consume CO_2 in the aqueous phase [respiration, photosynthesis, precipitation and dissolution reactions — e.g., $Ca^{2+} + 2HCO_3^- = CaCO_3 + CO_2 + H_2O$ — mineral alterations — e.g., $NaAlSi_3O_8(s)$ $H_2CO_3^* + 9/2H_2O = Na^+ + HCO_3^- + 2H_4SiO_4 + 1/2Al_2Si_2O_5(OH)_4(s)$.

KINETICS OF THE HYDRATION OF CO_2. As we have seen before, the composite acidity constant of $H_2CO_3^*$ ($= [CO_2] + [H_2CO_3]$) is much smaller than the acidity constant of true H_2CO_3. Kern [17] has reviewed comprehensively the kinetics of the CO_2 hydration and of the H_2CO_3 dehydration reaction. The dehydration reaction

$$H_2CO_3 \underset{k''}{\overset{k'}{\rightleftharpoons}} CO_2 + H_2O \tag{29}$$

is first order with respect to $[H_2CO_3]$ and has a first-order rate constant of $k' = 10$ to $20\ sec^{-1}$ (20–25°C). Its activation energy is ca. 16 kcal mole^{-1}. Similarly the rate of hydration of CO_2 is first order with respect to dissolved $[CO_2]$ and has a rate constant of $k'' = 0.0025$ to $0.03\ sec^{-1}$ (20–25°C). The activation energy is approximately 15 kcal mole^{-1}.

The ratio of these velocity constants permits the estimation of the equilibrium of (30) [cf. (1) and (2) of Table 4-2]:

$$K' = \frac{([CO_2] + [H_2CO_3])}{[H_2CO_3]} = \frac{[H_2CO_3^*]}{[H_2CO_3]} = \frac{K_{H_2CO_3}}{K_{H_2CO_3^*}} = \frac{k'}{k''} \tag{30}$$

K' has a value of about 400–670 at 20–25°C.

Superimposed on these first-order reactions are the processes

$$HCO_3^- \underset{k''''}{\overset{k'''}{\rightleftharpoons}} CO_2 + OH^- \tag{31}$$

with approximate constants $k''' = 2 \times 10^{-4}\ sec^{-1}$ (25°C); and, $k'''' = 8.5 \times 10^3$ liter mole^{-1} sec^{-1}. Processes 31 are kinetically insignificant at pH values < 8. Reactions 29 are subject to catalysis by bases and the enzyme carbonic anhydrase. Considering the order of magnitude of the reaction rate constants, it is obvious that not more than a few minutes are necessary to establish the hydration equilibrium.

[17] D. M. Kern, *J. Chem. Educ.*, **37**, 14 (1960).

CO_2 *Transfer at the Gas–Liquid Interface*

In transferring CO_2 molecules from the atmosphere to an aqueous solution the molecules must be transported through the gas phase and then through a nonturbulent liquid layer adjacent to the surface before they are distributed in the bulk solution phase and before they become partially hydrated. Of these different steps the transfer through the liquid layer adjacent to the surface is slowest and rate determining. This is also valid for the desorption of CO_2 from the solution into the gas phase.

The basic equation (Fick's first law) for transfer across a plane surface is

$$F = \frac{dn}{Adt} = D\frac{\Delta C}{\Delta x} \tag{32}$$

where F is the flux in moles (n) per unit area A (cm^2) per unit of time t (sec). D is the diffusion coefficient (cm^2 sec^{-1}) for the region of thickness Δx (cm) and ΔC is the concentration difference (moles cm^{-3}) across this region. Equation 32 may be written alternatively and more generally as

$$F = K\Delta C = \frac{1}{R_L}\Delta C \tag{33}$$

where K (cm sec^{-1}) is the transfer coefficient, that is, the velocity (cm sec^{-1}) with which the gas molecules traverse the unstirred layer, and R_L ($= 1/K$) (sec cm^{-1}) may be understood as the resistance to transfer in this layer. For the hypothetical system of a completely stagnant solution, R_L can be calculated from fundamental principles (semi-infinite linear diffusion), but in real systems R_L is a function of the hydrodynamics of the system and is strongly influenced by surface-active impurities. For natural waters R_L is of the order of 10^2 to 10^3 sec cm^{-1}. Similar values have been found for other gases. The data of Redfield [18] for O_2 can be interpreted (Eriksson [19]) in terms of an average R_L value corresponding to 1.7×10^2 sec cm^{-1}, whereas Kanwisher's [20] results give R_L values between 2.5×10^2 and 10^3 sec cm^{-1}.

CO_2-EXCHANGE RATES IN A STEADY STATE SYSTEM. The net transfer rate of CO_2 depends on $\Delta C = C_s - C_0$, where C_s is the saturation concentration of aqueous CO_2 at the surface and C_0 is the concentration of CO_2 in the bulk. Thus the net rate of accumulation of CO_2 depends on the difference in the partial pressure of CO_2 in solution and CO_2 in the gas phase. By making

[18] A. C. Redfield, *J. Marine Res.*, **8**, 347 (1948).
[19] E. Eriksson, in *Oceanography*, M. Sears, Ed., American Association for the Advancement of Science, Washington, D.C., 1961, p. 411.
[20] J. Kanwisher, *Deep-Sea Res.*, **10**, 195 (1963).

$C_0 = 0$, we can obtain the CO_2 exchange rate, F_0, that is, the flux of CO_2 in either direction

$$F_0 = \frac{1}{R_L} C_s \qquad (34)$$

Using a representative R_L value of 2×10^2 sec cm^{-1} and a saturation concentration of aqueous CO_2, $C_s = 10^{-8}$ mole cm^{-3} (which corresponds at 25°C to CO_2 at equilibrium with $p_{CO_2} = 3 \times 10^{-4}$ atm) we obtain an exchange rate of 5×10^{-11} moles cm^{-2} sec^{-1} or 15 moles m^{-2} year^{-1}. The residence time of CO_2 in the atmosphere, τ_{CO_2} is obtained from the quotient of the total mass of CO_2 in the atmosphere per unit area and the exchange rate

$$\tau_{CO_2} = \frac{m_{CO_2} \text{ (atm)}}{F_0} \qquad (35)$$

Since there is about 100 moles of CO_2 above each m^2, a residence time of $\tau_{CO_2} = 7$ years is obtained. Similar residence times have been established on the basis of C-14 data [21].

The question may be asked, why is C_s in (34) taken as $[CO_2(aq)]$ rather than $C_T(aq)$? In seawater $C_T \simeq 150[CO_2(aq)]$; correspondingly, F_0 would become 150 times larger than in the previously considered case. The resulting residence time of ~ 0.05 years would, however, be far too small. The slow conversion of CO_2 into H_2CO_3 before ionization precedes a relatively instantaneous exchange of C between H_2CO_3 and HCO_3^-; thus, the exchange of CO_2 as the slow step and not the cycle of HCO_3^- and CO_3^{2-} is important for the ocean–atmosphere CO_2 exchange.

Appendix   Gran Titration

Consider Figure 4-11 and the following definitions.

Symbols: v_0, original volume of sample

v, volume of strong acid added

$\bar{c}_A$, molarity of strong acid

H_2C, HC^- and $C^{2-} = H_2CO_3^*$, HCO_3^- and CO_3^{2-}, respectively

Capacities: $C_T = [H_2C] + [HC^-] + [C^{2-}]$ $\qquad (1)$

$[Alk] = [HC^-] + 2[C^{2-}] + [OH^-] - [H^+]$ $\qquad (2)$

[21] W. S. Broecker, in *The Sea*, M. N. Hill, Ed., Vol. II, Wiley-Interscience, New York, 1963, p.88.

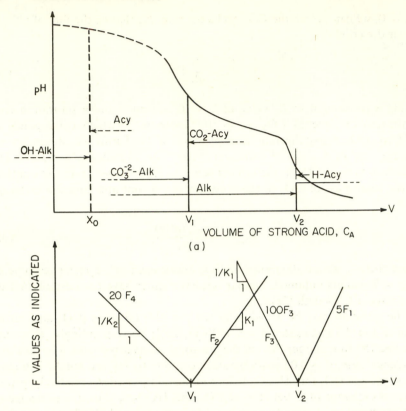

Figure 4-11 Sketch of an acidimetric titration curve (*a*). In *b* the results of *a* are plotted in terms of Gran functions, *F*. The *F* values are defined by (18) and (21)–(24) of the Appendix. x_0, v_1 and v_2 are the volumes of strong acid corresponding to the equivalence points $f = 2$, $f = 1$ and $f = 0$, respectively.

$$[H^+\text{-Acy}] = [H^+] - [HC^-] - 2[C^{2-}] - [OH^-] \tag{3}$$

$$[CO_2\text{-Acy}] = [H_2C] + [H^+] - [C^{2-}] - [OH^-] \tag{4}$$

$$[\text{Acy}] = 2[H_2C] + [HC^-] + [H^+] - [OH^-] \tag{5}$$

$$[CO_3{}^{2-}\text{-Alk}] = [C^{2-}] + [OH^-] - [H^+] - [H_2C] \tag{6}$$

$$[OH^-\text{-Alk}] = [OH^-] - [HC^-] - 2[H_2C] - [H^+] \tag{7}$$

At the various equivalence points the following equalities exist:

$$v_0[\text{Alk}] = v_2\bar{c}_A \tag{8}$$

$$v_0[CO_3{}^{2-}\text{-Alk}] = v_1\bar{c}_A \tag{9}$$

$$v_0[OH^-\text{-Alk}] = x_0\bar{c}_A \tag{10}$$

$$v_0C_T = (v_2 - v_1)\bar{c}_A \tag{11}$$

At any point in the titration curve the following equalities exist:

$$(v_0 + v)[\text{H}^+\text{-Acy}] = (v - v_2)\bar{c}_A \tag{12}$$

$$(v_0 + v)[\text{CO}_2\text{-Acy}] = (v - v_1)\bar{c}_A \tag{13}$$

$$(v_0 + v)[\text{Alk}] = (v_2 - v)\bar{c}_A \tag{14}$$

$$(v_0 + v)[\text{Acy}] = (v - x_0)\bar{c}_A \tag{15}$$

$$(v_0 + v)[\text{CO}_3{}^{2-}\text{-Alk}] = (v_1 - v)\bar{c}_A \tag{16}$$

I. Beyond v_2, (12) can be simplified to:

$$(v_0 + v)[\text{H}^+] \simeq (v - v_2)\bar{c}_A \tag{17}$$

because for $v > v_2$

$$[\text{H}^+] \gg [\text{HC}^-] + 2[\text{C}^{2-}] + [\text{OH}^-]$$

The equivalence point v_2 can be obtained by plotting the left-hand side of (17) versus v.

Instead of $[\text{H}^+]$ we may write $10^{-\text{"pH"}}$. Independent of any pH or activity conventions, operationally one can define "pH" $= \text{pH}° - \log[\text{H}^+]$ where pH_0 is a constant. If the pH-meter reading is made in volts one can also use: $[\text{H}^+] = 10^{-\text{"pH"}} = 10^{(E-e)/k}$ because $E = e + k \log[\text{H}^+]$ where $k = RT/2.3F$ (where $F = $ faraday). Rewriting (17):

$$F_1 = (v_0 + v)10^{-\text{"pH"}} \simeq (v - v_2)\bar{c}_A \tag{18}$$

where $F_1 = 0$ for $v = v_2$.

II. Between v_1 and v_2, (13) and (14) can be simplified because in this range $[\text{H}_2\text{C}] \gg [\text{H}^+] - [\text{C}^{2-}] - [\text{OH}^-]$ and $[\text{HC}^-] \gg [\text{C}^{2-}] + [\text{OH}^-] - [\text{H}^+]$ Accordingly, instead of (13) and (14), respectively,

$$(v_0 + v)[\text{H}_2\text{C}] \simeq (v - v_1)\bar{c}_A \tag{19}$$

and

$$(v_0 + v)[\text{HC}^-] \simeq (v_2 - v)\bar{c}_A \tag{20}$$

Substituting $[\text{H}_2\text{C}] = (1/K_1)[\text{H}^+][\text{HC}^-]$ in (19) and combining with (20) gives:

$$F_2 = (v_2 - v)10^{-\text{"pH"}} = (v - v_1)K_1 \tag{21}$$

$F_2 = 0$ for $v = v_1$; the slope of F_2 versus v being K_1.

III. Equation (21) can be rearranged:

$$F_3 = \frac{v_2 - v}{K_1} \simeq (v - v_1)10^{\text{"pH"}} \tag{22}$$

F_3 gives an additional check on v_2, the slope being $-1/K_1$.

IV. Similarly, other linear functions can be derived; for example, a combination of (20) with (16) (the latter being simplified because $[C^{2-}] > [OH^-] - [H^+] - [H_2C]$ in the titration range where $v < v_1$) gives

$$F_4 = (v_2 - v)10^{\text{"pH"}} = \frac{1}{K_2}(v_1 - v) \tag{23}$$

or

$$F_5 = (v_1 - v)10^{-\text{"pH"}} = K_2(v_2 - v) \tag{24}$$

For plotting, F_1, F_2, etc., may be multiplied by convenient scale factors. The same principles apply for alkalimetric titration in the carbonate or any other proton transfer system.

PROBLEMS

4-1 Does the alkalinity of a natural water (isolated from its surroundings) (a) increase, (b) decrease, or (c) stay constant upon addition of small quantities to the following:

(i) HCl

(ii) NaOH

(iii) Na_2CO_3

(iv) $NaHCO_3$

(v) CO_2

(vi) $AlCl_3$

(vii) Na_2SO_4

4-2 The following short method for the determination of alkalinity has been proposed: Add a known quantity of standard mineral acid and measure the final pH. (Mineral acid quantity should preferably be such that final pH is between 3 and 4.3.) Present the theory and show how you compute the alkalinity.

4-3 A water containing 1.0×10^{-4} moles CO_2 liter^{-1} and having an alkalinity of 2.5×10^{-4} equivalents liter^{-1} has a pH of 6.7. The pH is to be raised to pH 8.3 with NaOH.

(i) How many moles of NaOH per liter of water are needed for this pH adjustment? ($pK_1 = 6.3$, $pK_2 = 10.3$)

(ii) How many moles of lime ($Ca(OH)_2$) would be required?

4-4 An industrial waste from a metals industry contains approximately 5×10^{-3} M H_2SO_4. Before being discharged into the stream, the water is diluted with tap water in order to raise the pH. The tap water has the following composition: pH = 6.5, alkalinity = 2×10^{-3} eq liter^{-1}. What dilution is necessary to raise the pH to approximately 4.3?

4-5 S. L. Rettig and B. F. Jones [*U.S. Geol. Surv. Paper* **501D**, D134 (1964)] designed a manometric method for determining alkalinity. Increments of strong acid (H_2SO_4) are added to the solution in a closed system. After each addition the pressure (predominantly p_{CO_2}) is measured.

(i) Develop an equation that describes the essential features of the partial pressure of CO_2 as a function of g (equivalent fraction of titrant added) and give a semiquantitative plot.

(ii) Discuss the advantages and disadvantages of the method.

4-6 A cascade aerator operating on a well-water will reduce the dissolved CO_2 content from 45 to 18 mg liter $^{-1}$ in one pass. The atmospheric saturation value of CO_2 may be assumed to be 0.5 mg liter $^{-1}$. Laboratory tests indicate that the gas transfer coefficient for H_2S is about 80% that for CO_2. Assume that enough H_2S will be present in the air around the aerator to give a saturation value of 0.1 mg liter $^{-1}$. Estimate the effluent concentration of H_2S if the well water contains 12 mg liter $^{-1}$ of H_2S.

4-7 How is the relative distribution of $H_2CO_3^*$, HCO_3^- and CO_3^- in a solution with [Alk] $= 2.5 \times 10^{-3}$ eq liter $^{-1}$ affected by variation in partial pressure of CO_2? Plot distribution versus p_{CO_2}.

4-8 Compute the pH variation resulting from isothermal evaporation (25°C) of an incipiently 10^{-5} M $NaHCO_3$ solution that remains in equilibrium with the partial pressure of CO_2 of 3×10^{-4} atm.

READING SUGGESTIONS

I. Carbonate Equilibria

Deffeyes, K. S., "Carbonate Equilibria: A Graphic and Algebraic Approach," *Limnol. Oceanog.*, **10**, 412 (1965).

Garrels, R. M., and C. Christ, *Minerals, Solutions and Equilibria*, Harper and Row, New York, 1965. Treatment of carbonate equilibria in Chapter 3.

Hömig, H. E., *Physikochemische Grundlagen der Speisewasserchemie*, W. Classen, Essen, 1963. Gives a comprehensive and rigorous treatment of carbonate equilibria as related to water treatment.

Larson, T. E., and A. M. Buswell, "Calcium Carbonate Saturation Index and Alkalinity Interpretations," *J. Amer. Water Works Assoc.*, **34**, 1664 (1942). Discusses carbonate equilibria and their temperature and activity dependence.

Park, P. K., "Oceanic CO_2 System," *Limnol. Oceanog.*, **14**, 179 (1969).

Skirrow, G., "The Dissolved Gases — Carbon Dioxide," in *Chemical Oceanography*, J. P. Riley and G. Skirrow, Eds., Academic Press, New York, 1965. Comprehensive discussion of carbonate equilibria in seawater.

II. Carbon Dioxide in Natural Waters and in the Atmosphere; Geochemical Considerations

Dyrssen, D., and L. G. Sillén, "Alkalinity and Total Carbonate in Seawater. A Plea for p-T-independent Data," *Tellus*, **19**, 110 (1967). It is shown by practical example how Gran's graphical method can be applied for determining [Alk] and C_T.

Hutchinson, G. E., "The Biochemistry of the Terrestrial Atmosphere," in *The Earth as a Planet*, G. Kuiper, Ed., Univ. Chicago Press, Chicago, 1954, pp. 371–427.

Plass, G. N., "Carbon Dioxide and Climate," *Sci. Amer.*, 3–9 (July 1959).

Revelle, R., and H. E. Suess, "CO_2 Exchange between Atmosphere and Ocean and the Question of an Increase of Atmospheric CO_2 during the Past Decades," *Tellus*, **9**, 18–27 (1957).

Rubey, W. W., "Geological History of Seawater: An Attempt to State the Problem," *Bull. Geol. Soc.*, **62**, 1111 (1951). Considers the role of volatile materials from the earth's interior.

Teal, J. M., and J. Kanwisher, "The Use of p_{CO_2} for the Calculation of Biological Production, with Examples from Waters Off Massachusetts," *J. Marine Res.*, **24**, 5 (1966).

ANSWERS TO PROBLEMS

4-1 Decrease for (i) and (vi); increase for (ii), (iii) and (iv); no change for (v) and (vii).

4-2 $\bar{C}_A[V/(V_0 + V)] - [H^+] \simeq [Alk]$ (V_0 = original volume of sample, V = volume of strong acid added, $\bar{C}_A$ molarity of strong acid). The strong acid which has been added is equivalent to the alkalinity originally contained in the sample plus the mineral acidity that remains after the acid was added. The mineral acidity can be estimated by measuring pH, because $[H\text{-}Acy] \approx [H^+]$.

4-3 (i) 1.0×10^{-4}; (ii) 0.5×10^{-4}. Note that pH = 8.3 corresponds to the equivalence point ($f = 1$). Cf. Figure 4-1 (point y).

4-4 Fivefold dilution. Note that pH = 4.3 corresponds to the equivalence point ($f = 0$, x in Figure 4-1) for the titration of alkalinity. The problem is equivalent to the titration of [Alk] of tap water with the acid of the waste water.

4-6 6 mg liter^{-1} H$_2$S. The rate of decrease in the concentration of the gas will be proportional to the oversaturation, that is, $-(C_0 - C_T)/(C_0 - C_s) = \exp(-kt)$.

4-7 Use (11) and solve by trial and error. Perhaps more conveniently diagrams such as Figure 4-3 may be constructed in order to compute the equilibrium concentrations.

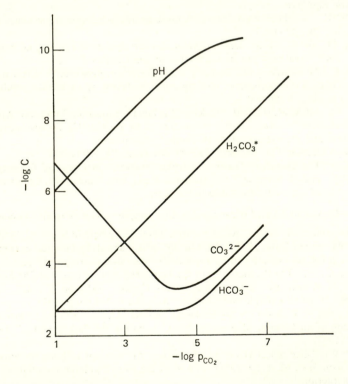

4-8 Use (11) or read result from Figure 4-3. For $[Na^+] = 10^{-5} M$, pH = 6.3; $[Na^+] = 10^{-3} M$, pH = 8.3; $[Na^+] = 10^{-2} M$, pH = 9.2; $[Na^+] = 10^{-1} M$, pH = 9.9.

5 Precipitation and Dissolution

5-1 Introduction

The hydrological cycle interacts with the cycle of rocks. Minerals dissolve in or react with the water. Under different physico-chemical conditions minerals are precipitated and accumulate on the ocean floor and in the sediments of rivers and lakes. Dissolution and precipitation reactions impart to the water constituents which modify its chemical properties. Natural waters vary in chemical composition; consideration of solubility relations aids in the understanding of these variations. This chapter sets forth principles on reactions between solids and water. Here again the most common basis is a consideration of the equilibrium relations.

Dissolution or precipitation reactions are generally slower than reactions among dissolved species, but it is quite difficult to generalize about rates of precipitation and dissolution. There is a lack of data concerning most geochemically important solid–solution reactions, so that kinetic factors cannot be assessed easily. Frequently the solid phase formed incipiently is metastable with respect to a thermodynamically more stable solid phase. Examples are provided by the occurrence under certain conditions of aragonite instead of stable calcite or by the quartz oversaturation of most natural waters. This oversaturation occurs because the rate of attainment of equilibrium between silicic acid and quartz is extremely slow.

The solubilities of many inorganic salts increase with temperature, but a number of compounds of interest in natural waters ($CaCO_3$, $CaSO_4$) decrease in solubility with increase in temperature. Pressure dependence of solubility is very slight but must be considered for the extreme pressures encountered at ocean depths. For example the solubility product of $CaCO_3$ will increase by approximately 0.2 logarithmic units for a pressure of 200 atm.†

HETEROGENEOUS EQUILIBRIA. The extent of the dissolution or precipitation reaction for systems that attain equilibrium can be estimated by considering the equilibrium constants. A simple example is the solubility of silica

$$SiO_2(s) + 2H_2O \underset{\text{precipitation}}{\overset{\text{dissolution}}{\rightleftharpoons}} H_4SiO_4 \tag{1}$$

† $(\partial \log K_{s0}/\partial P)_T = -(\Delta V/RT) \ln 10$ where ΔV, the change in molar volume for the reaction $CaCO_3(s) = Ca^{2+} + CO_3^{2-}$ is ca. -50 ml.

The solubility equilibrium for pure $SiO_2(s)$ is defined by

$$K_{s0} = \{H_4SiO_4\} \tag{2}$$

Equation 2 is obeyed regardless of whether H_4SiO_4 has been added to the solution or comes from the dissolution of $SiO_2(s)$.†

If a solution contains an activity of orthosilicic acid larger than K_{s0}, the solution is oversaturated; from a thermodynamic point of view $SiO_2(s)$ will precipitate. On the other hand, if $\{H_4SiO_4\}$ is smaller than K_{s0}, the solid phase dissolves until $\{H_4SiO_4\} = K_{s0}$. It is obvious that the concentration of dissolved SiO_2 does not depend on the quantity of $SiO_2(s)$ in contact with the solution. The equilibrium condition of (2) is meaningful only in the context of a heterogeneous aqueous solution–solid equilibrium in which the activity of the pure solid phase is constant and can thus by convention be set equal to unity.

Reaction 1 may be compared with the application of the mass law to the following simple heterogeneous equilibria

$$H_2O(l) = H_2O(g); \qquad K = p_{H_2O} \tag{3}$$

$$CaCO_3(s) = CaO(s) + CO_2(g); \quad K = p_{CO_2} \tag{4}$$

$$MgCO_3(s) + 3H_2O(g) = MgCO_3 \cdot 3H_2O(s); \quad K = p_{H_2O}^{-3} \tag{5}$$

in which the activity of a phase involved in the reaction is constant. The vapor pressure of pure water (Reaction 3) ($\{H_2O(l)\} = 1$), however, is higher than the vapor pressure of a solution ($\{H_2O(l)\} < 1$). Thus, in a more general way, the equilibrium of (3) can be given by

$$K = \frac{p_{H_2O}}{\{H_2O(l)\}} \tag{6}$$

where it is understood that in the limiting case of infinite dilution (pure solvent), $\{H_2O(l)\} = 1$. In an analogous way, a solid solution, for example, a solid in which certain lattice sites are occupied by foreign atoms or ions, has a smaller activity than the pure solid. Thus, in a most general way, we can write for the equilibrium of (1):

$$K = \frac{\{H_4SiO_4\}}{\{SiO_2(s)\}} \tag{7}$$

The reference state again is taken as an infinitely dilute solid solution, that is, for the pure solid $SiO_2(s)$, $\{SiO_2(s)\} = 1$. As long as the dissolution reaction is considered a physical process, the solubility may be characterized by the quantity of material that is soluble in a given volume of solvent. Frequently,

† Orthosilicic acid can be formulated as H_4SiO_4 or $Si(OH)_4$.

however, the dissolution reaction is a heterogeneous chemical reaction; then a simple solubility datum is not sufficient to define the solution–solid equilibrium. It is necessary to characterize the solubility by a solubility product; with an equilibrium constant one can characterize the solubility as well as predict how solution variables change the solubility.

In a general way for an electrolyte that dissolves in water according to the reaction

$$A_mB_n(s) \rightleftharpoons mA^{+n}(aq) + nB^{-m}(aq) \tag{8}$$

the equilibrium condition is

$$\{A_mB_n(s)\} = \{A^{+n}(aq)\}^m\{B^{-m}(aq)\}^n \tag{9}$$

The conventional solubility expression

$$K_{s0} = \{A^{+n}(aq)\}^m\{B^{-m}(aq)\}^n \tag{10}$$

results if the activity of the pure solid phase is set equal to unity and if the common standard state convention for aqueous solutions is adopted. Only in some cases can the solubility of a salt be calculated from its solubility product alone. Generally one deals with the solubility of a salt in solutions that contain a common ion, that is, an ion that also exists in the ionic lattice of the solid salt.

Example 5-1 *Solubility of Sulfates, Chlorides, Fluorides and Chromates.* Characterize the solubility of the following salts as a function of the concentration of the common anion from the respective solubility products (K_{s0}):

$$CaSO_4 (pK_{s0} = 4.6), \quad SrSO_4 (6.2), \quad BaSO_4 (9.7), \quad AgCl (10.0),$$
$$PbCl_2 (4.8), \quad Ag_2CrO_4 (12.0), \quad MgF_2 (8.1), \quad CaF_2 (10.3)$$

Figure 5-1 gives a graphical representation of the solubility product, where the log of the metal ion concentration is plotted as a function of the $-\log$ of the common anion, for example,

$$\log [Ca^{2+}] = -pK_{s0(CaSO_4)} + pSO_4^{2-} \tag{i}$$
$$\log [Mg^{2+}] = -pK_{s0(MgF_2)} + 2pF^- \tag{ii}$$
$$\log [Ag^+] = -\tfrac{1}{2}pK_{s0(Ag_2CrO_4)} + \tfrac{1}{2}pCrO_4^{2-} \tag{iii}$$

Correspondingly, in Figure 5-1 the log $[Me^{z+}]$ lines have slopes of 1, 2 and $\tfrac{1}{2}$ and intercepts of $-pK_{s0}$, $-pK_{s0}$ and $-\tfrac{1}{2}pK_{s0}$ for salts of the 1:1, 1:2 and 2:1 type, respectively.

The cations and anions of these salts do not undergo protolysis reactions to any appreciable extent in solutions that are near neutrality. Furthermore, complex formation (or ion-pair binding) between cation and anion may be assumed to be negligible as long as free metal ion and free anion concentration is small (ca. $< 10^{-1.5} M$). Hence the solubility is characterized by the

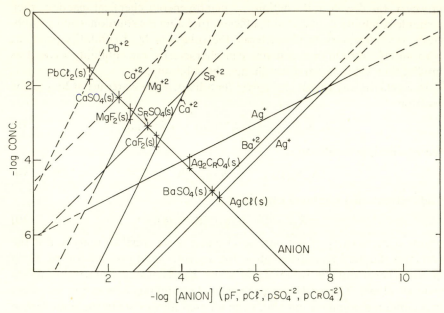

Figure 5-1 Solubility of "simple" salts as a function of the common anion concentration (Example 5-1). The cations and anions of these salts do not protolyze in the neutral pH range. The equilibrium solubility is given by the metal ion concentration. At high anion or cation concentration, complex formation or ion pair binding becomes possible (dashed lines).

If the salt is dissolved in pure water (or in an inert electrolyte), the solubility is defined by the electroneutrality

$$z[Me^{z+}] = n[anion^{n-}]$$

If $z = n$ (e.g., $BaSO_4$), the solubility is given by the intercept (+). If $z \neq n$, the electroneutrality condition is fulfilled at a point slightly displaced from the intersection (‡).

metal ion concentration. (Dashed lines in Figure 5-1 indicate where the foregoing assumptions are no longer valid.)

In the absence of a common anion, for example, if the salt is dissolved in pure water, the solubility is given by the electroneutrality requirement. Considering that $[H^+] \simeq [OH^-]$ we have, for example,

for $CaSO_4(s)$: $[Ca^{2+}] = [SO_4^{2-}]$ (iv)

for $MgF_2(s)$: $2[Mg^{2+}] = [F^-]$ (v)

for $Ag_2CrO_4(s)$: $[Ag^+] = 2[CrO_4^{2-}]$ (vi)

Hence the solubility of a 1:1 salt in pure water is defined by the intersection of the lines of log $[Me^{2+}]$ and log $[anion]$. For a salt of the 1:2 and 2:1 type the solubility is slightly displaced from the intersection in such a way that

$\log [\text{Me}^{2+}] = +0.3 + \log [\text{anion}]$ and $\log [\text{Me}^{2+}] = -0.3 + \log [\text{anion}]$, respectively.

However, the case in which the solubility of a solid can be calculated from the known analytical concentration of added components and from the solubility product alone is very seldom encountered. Ions that have dissolved from a crystalline lattice frequently undergo chemical reactions in solution and therefore other equilibria in addition to the solubility product have to be considered. The reaction of the salt cation or anion with water to undergo acid–base reactions is very common. Furthermore, complex formation of salt cation and salt anion with each other and with one of the constituents of the solution has to be considered. For example, the solubility of FeS(s) in a sulfide-containing aqueous solution depends, in addition to the solubility equilibrium, on acid–base equilibria of the cation (e.g., $\text{Fe}^{2+} + \text{H}_2\text{O} = \text{FeOH}^+ + \text{H}^+$) and of the anion (e.g., $\text{S}^{2-} + \text{H}_2\text{O} = \text{HS}^- + \text{OH}^-$ and $\text{HS}^- + \text{H}_2\text{O} = \text{H}_2\text{S} + \text{OH}^-$) as well as on equilibria describing complex formation (e.g., formation of FeHS^+ or FeS_2^{2-}).

DATA FOR SOLUBILITY CONSTANTS. Available compilations of solubility products [1–3] illustrate that values given by different authors for the same solubility products often differ markedly. Differences of a few orders of magnitude are not uncommon. For example, data for the solubility product of FePO$_4$(s) vary over a range of 10^{13}. There are various reasons for such discrepancies: (a) The formation of a sparingly soluble phase and its equilibrium with the solution is a more complicated process than equilibration reactions in a homogeneous solution phase. (b) The composition and properties, that is, reactivity, of the solids vary for different modifications of the same compound or for different active forms of the same modification. (c) Species influencing the solubility equilibrium (e.g., species formed by complex formation) are overlooked.

PRECIPITATION. An appreciation of the various types of precipitates that may be formed and an understanding of the changes which the precipitates undergo in aging are prerequisite for understanding and interpreting solubility equilibrium constants. A lucid treatment of these considerations, especially emphasizing precipitation and dissolution of metal hydroxides, has been given by Feitknecht [4] and by Feitknecht and Schindler [3].

[1] L. G. Sillén and A. E. Martell, *Stability Constants of Metal–Ion Complexes*, Special Publ., No. 17, The Chemical Society, London, 1964.
[2] W. M. Latimer, *Oxidation Potentials*, Prentice-Hall, New York, 1952.
[3] W. Feitknecht and P. Schindler, *Solubility Constants of Metal Oxides, Metal Hydroxides and Metal Hydroxide Salts in Aqueous Solution*, Butterworths, London, 1963.
[4] W. Feitknecht, *15th International Congress of Pure and Applied Chemistry, Lisbon, 1956*, Vol. III.

An "active" form of the compound, that is, a very fine crystalline precipitate with disordered lattice, is generally formed incipiently from strongly oversaturated solutions. Such an active precipitate may persist in metastable equilibrium with the solution and may convert ("age") only slowly into a more stable and "inactive" form. Measurements of the solubility of "active" forms give solubility products that are higher than those of the inactive forms. "Inactive" solid phases with ordered crystals are also formed from solutions that are only slightly oversaturated.

Hydroxides and sulfides often occur in amorphous and several crystalline modifications. Amorphous solids may be "active" or "inactive." As Feitknecht and Schindler [3] point out, initially formed amorphous precipitates or active forms of unstable crystalline modifications may undergo two kinds of changes during aging. Either the active form of the unstable modification becomes inactive or a more stable modification is formed. With amorphous compounds deactivation may be accompanied by condensation [e.g., $Fe(OH)_3$]. If a metal oxide is more stable than the primarily precipitated hydroxide, dehydration may occur. When several of the processes mentioned take place together, nonhomogeneous solids are formed upon aging. In dissolution experiments with such nonhomogeneous solids, the more active components are dissolved more readily, that is, the solubility may depend upon the quantity of the solid present. Feitknecht and Schindler give the following example:

$$(amorph)Fe(OH)_3(active) \begin{cases} \nearrow (amorph)FeO_{n/2}(OH)_{3-n}(inactive) \\ \longrightarrow \alpha\text{-}FeOOH(active) \\ \searrow \alpha\text{-}Fe_2O_3 \end{cases} \qquad (11)$$

In determining solubility equilibrium constants, many investigators have been motivated by a need to gain information that was pertinent primarily for relatively short-term conditions (minutes to hours) typically encountered in the laboratory. In operations of analytical chemistry, for example, precipitates are frequently formed from strongly oversaturated solutions; the conditions of precipitation of the incipient active compound rather than the dissolution of the aged inactive solid are often of primary interest. Most solubility products measured in such cases refer to the most active component.† On the other hand, in dealing with heterogeneous equilibria of natural water systems, the more stable and inactive solids are frequently more pertinent.‡ Aging

† Strictly speaking, solubility products for active solid compounds are, because of their time dependence, not equilibrium constants; they are of operational value to estimate the conditions (pMe, pH, etc.) under which precipitation occurs.
‡ A recent critical review [5] on thermodynamic properties of minerals is very useful.
[5] R. A. Robie and D. R. Waldbaum, *Geol. Survey Bull.* **1259** (1968).

often continues for geological time-spans. Furthermore, the solid phase has frequently been formed in nature under conditions of slight supersaturation. Solubility constants determined under conditions where the solid has been identified by X-ray diffraction are especially valuable.

5-2   The Solubility of Oxides, Hydroxides and Carbonates

Many minerals with which water comes into contact are oxides, hydroxides, carbonates and hydroxide-carbonates. The same ligands (hydroxides and carbonates) are dissolved constituents of all natural waters. The solid and solute chemical species under consideration belong to the ternary system $Me^{n+}-H_2O-CO_2$. In this section a discussion of solubility equilibria is given. Complications arising in treating such solubility equilibria from nonpure solid phases (solid solutions) and from the possible occurrence of carbonate, mixed carbonato-hydroxo and polynuclear complexes are neglected, but will be taken up in later sections (5-5 and 6-6). In this section we restrict ourselves to applying operational equilibrium constants (K' or ^cK).

Oxides and Hydroxides

If a pure solid oxide or hydroxide is in equilibrium with free ions in solution, for example,

$$Me(OH)_2(s) = Me^{2+} + 2OH^- \tag{12}$$

$$MeO(s) + H_2O = Me^{2+} + 2OH^- \tag{13}$$

the conventional solubility product is given by

$$^cK_{s0} = [Me^{2+}][OH^-]^2 \text{ (mole}^3 \text{ liter}^{-3}) \tag{14}$$

The subscript zero indicates that the equilibrium of the solid with the simple (uncomplexed) species Me^{2+} and OH^- is considered.† Sometimes it is more appropriate to express the solubility in terms of reaction with protons, since the equilibrium concentrations of OH^- ions may be extremely small, for example,

$$Me(OH)_2(s) + 2H^+ = Me^{2+} + 2H_2O \tag{15}$$

$$MeO(s) + 2H^+ = Me^{2+} + H_2O \tag{16}$$

Then the solubility equilibrium can be characterized by

$$^{c*}K_{s0} = \frac{[Me^{2+}]}{[H^+]^2} \text{ [mole}^{-1} \text{ liter]} \tag{17}$$

† The complete dissolution reaction considers the water that participates in the dissolution reaction: $Me(OH)_2(s) + (x + zy)H_2O(l) = Me(H_2O)_x^{2+} + z(OH^-)(H_2O)_y$.

Table 5-1 Constants for Solubility Equilibria of Oxides, Hydroxides, Carbonates and Hydroxide Carbonates

Reaction	Symbol for Equilibrium Constant†	log K (25°C)	I
I. Oxides and Hydroxides			
$H_2O(l) = H^+ + OH^-$	K_W	−14.00	0
$(am)Fe(OH)_3(s) = Fe^{3+} + 3OH^-$	K_{s0}	−13.77	1 M NaClO$_4$
$(am)Fe(OH)_3(s) = FeOH^{2+} + 2OH^-$	K_{s0}	−38.7	3 M NaClO$_4$
$(am)Fe(OH)_3(s) = Fe(OH)_2^+ + OH^-$	K_{s1}	−27.5	3 M NaClO$_4$
$(am)Fe(OH)_3(s) + OH^- = Fe(OH)_4^-$	K_{s2}	−16.6	3 M NaClO$_4$
$2(am)Fe(OH)_3(s) = Fe_2(OH)_4^{4+} + 4OH^-$	K_{s4}	−4.5	3 M NaClO$_4$
$(am)FeOOH(s) + 3H^+ = Fe^{3+} + 2H_2O$	K_{s22}	−51.9	3 M NaClO$_4$
α-$FeOOH(s) + 3H^+ = Fe^{3+} + 2H_2O$	$*K_{s0}$	3.55	3 M NaClO$_4$
α-$Al(OH)_3$(gibbsite) $+ 3H^+ = Al^{3+} + 3H_2O$	$*K_{s0}$	1.6	3 M NaClO$_4$
γ-$Al(OH)_3$(bayerite) $+ 3H^+ = Al^{3+} + 3H_2O$	$*K_{s0}$	8.2	0
$(am)Al(OH)_3(s) + 3H^+ = Al^{3+} + 3H_2O$	$*K_{s0}$	9.0	0
$Al^{3+} + 4OH^- = Al(OH)_4^-$	$*K_{s0}$	10.8	0
$CuO(s) + 2H^+ = Cu^{2+} + H_2O$	K_4	32.5	0
$Cu^{2+} + OH^- = CuOH^+$	$*K_{s0}$	7.65	0
$2Cu^{2+} + 2OH^- = Cu_2(OH)_2^{2+}$	K_1	6.0 (18°C)	0
$Cu^{2+} + 3OH^- = Cu(OH)_3^-$	K_{22}	17.0 (18°C)	0
$Cu^{2+} + 4OH^- = Cu(OH)_4^{2-}$	K_3	15.2	0
$ZnO(s) + 2H^+ = Zn^{2+} + H_2O$	K_4	16.1	0
$Zn^{2+} + OH^- = ZnOH^-$	$*K_{s0}$	11.18	0
$Zn^{2+} + 3OH^- = Zn(OH)_3^-$	K_1	5.04	0
$Zn^{2+} + 4OH^- = Zn(OH)_4^{2-}$	K_3	13.9	0
$Cd(OH)_2(s) + 2H^+ = Cd^{2+} + 2H_2O$	K_4	15.1	0
$Cd^{2+} + OH^- = CdOH^+$	$*K_{s0}$	13.61	0
$Mn(OH)_2(s) = Mn^{2+} + 2OH^-$	K_1	3.8	1 M LiClO$_4$
$Mn(OH)_2(s) + OH^- = Mn(OH)_3^-$	K_{s0}	−12.8	0
$Fe(OH)_2$(active) $= Fe^{2+} + 2OH^-$	K_{s3}	−5.0	0
$Fe(OH)_2$(inactive) $= Fe^{2+} + 2OH^-$	K_{s0}	−14.0	0
$Fe(OH)_2$(inactive) $+ OH^- = Fe(OH)_3^-$	K_{s0}	−14.5 (−15.1)	0
	K_{s3}	−5.5	0

Reaction	Constant	Value	Conditions
$Mg(OH)_2$(active) $= Mg^{2+} + 2OH^-$	K_{s0}	-9.2	0
$Mg(OH)_2$(brucite) $= Mg^{2+} + 2OH^-$	K_{s0}	-11.6	0
$Mg^{2+} + OH^- = MgOH^+$	K_1	2.6	0
$Ca(OH)_2(s) = Ca^{2+} + 2OH^-$	K_{s0}	-5.43	0
$Ca(OH)_2(s) = CaOH^+ + OH^-$	K_{s1}	-4.03	0

II. Carbonates and Hydroxide Carbonates

Reaction	Constant	Value	Conditions
$CO_2(g) + H_2O = H^+ + HCO_3^-$	K_{p1}	-7.82	0
		-7.5	Seawater
		-7.3	5°C, 200 atm, seawater
$HCO_3^- = H^+ + CO_3^{2-}$	(K_2)	-10.33	0
		-9.0	Seawater
		-9.1	5°C, 200 atm, seawater
$CaCO_3$(calcite) $= Ca^{2+} + CO_3^{2-}$	K_{s0}	-8.35	0
		-6.2	Seawater
$CaCO_3$(aragonite) $= Ca^{2+} + CO_3^{2-}$	K_{s0}	-8.22	0
$SrCO_3(s) = Sr^{2+} + CO_2^{2-}$	K_{s0}	-9.03	0
		-6.8	Seawater
$ZnCO_3(s) + 2H^+ = Zn^{2+} + H_2O + CO_2(g)$	$*K_{ps0}$	7.95	0
$Zn(OH)_{1.2}(CO_3)_{0.4}(s) + 2H^+ = Zn^{2+} + H_2O + CO_2(g)$	$*K_{ps0}$	9.8	0
$Cu(OH)(CO_3)_{0.5}(s) + 2H^+ = Cu^{2+} + \frac{3}{2}H_2O + \frac{1}{2}CO_2(g)$	$*K_{ps0}$	7.08	0
$Cu(OH)_{0.67}(CO_3)_{0.67}(s) + 2H^+ = Cu^{2+} + \frac{4}{3}H_2O + \frac{2}{3}CO_2(g)$	$*K_{ps0}$	7.08	0
$MgCO_3$(magnesite) $= Mg^{2+} + CO_3^{2-}$	K_{s0}	-4.9	0
$MgCO_3$(nesquehonite) $= Mg^{2+} + CO_3^{2-}$	K_{s0}	-5.4	0
$Mg_4(CO_3)_3(OH)_2 \cdot 3H_2O$(hydromagnesite) $= 4Mg^{2+} + 3CO_3^{2-} + 2OH^-$	K_{s0}	29.5	0
$CaMg(CO_3)_2$(dolomite) $= Ca^{2+} + Mg^{2+} + 2CO_3^{2-}$	K_{s0}	-16.7	0
$FeCO_3$(siderite) $= Fe^{2+} + CO_3^{2-}$	K_{s0}	-10.4	0
$CdCO_3(s) + 2H^+ = Cd^{2+} + H_2O + CO_2(g)$	K_{ps0}	6.44	1 M NaClO₄

The constants given here are taken from quotations or selections in (a) L. G. Sillén and A. E. Martell, *Stability Constants of Metal–Ion Complexes*, Special Publ., No. 17, The Chemical Society, London, 1964; (b) W. Feitknecht and P. Schindler, *Solubility Constants of Metal Oxides, Metal Hydroxides and Metal Hydroxide Salts in Aqueous Solutions*, Butterworths, London, 1963; and (c) P. Schindler, "Heterogeneous Equilibria Involving Oxides, Hydroxides, Carbonates and Hydroxide Carbonates," in *Equilibrium Concepts in Natural Water Systems*, Advances in Chemistry Series, No. 67, American Chemical Society, Washington, D.C., 1967, p. 196. Unless otherwise specified a pressure of 1 atm is assumed.

† Most of the symbols used for the equilibrium constants are those given in *Stability Constants of Metal–Ion Complexes* (see Table 5-2).

Table 5.1   Continued

III. CaCO$_3$ Solubility Equilibrium

Reaction	$-\log K$ (infinite dilution) at						$-\log {}^cK$ (seawater)[†]	
	5°C	10°C	15°C	20°C	25°C	40°C	5	25
$CaCO_3(s) = Ca^{2+} + CO_3^{2-}$	8.09	8.15	8.22	8.28	8.34	8.51	5.67	6.19
$CaCO_3(s) + H^+ = HCO_3^- + Ca^{2+}$	−2.47	−2.34	−2.21	−2.10	−1.99	−1.71	−3.51	−2.77
$H_2CO_3^* = H^+ + HCO_3^-$	6.52	6.46	6.42	6.38	6.35	6.30	6.15	5.99
$CO_2(g) + H_2O = H_2CO_3^*$	1.20	1.27	1.34	1.41	1.47	1.64	1.27	1.53
$HCO_3^- = H^+ + CO_3^{2-}$	10.56	10.49	10.43	10.38	10.33	10.22	9.18	8.96

† Quoted from L. G. Sillén, in *Oceanography*, M. Sears, Ed., American Association for the Advancement of Science, Washington, D.C., 1961.

or for a solubility equilibrium with a trivalent metal ion

$$FeOOH(s) + 3H^+ = Fe^{3+} + 2H_2O \tag{18}$$

by

$$^{c*}K_{s0} = \frac{[Fe^{3+}]}{[H^+]^3} \; [mole^{-2} \; liter^2] \tag{19}$$

The definitions for the solubility equilibrium contained in (14) and (17) are interrelated. For $MeO_{z/2}$ or $Me(OH)_z$ the following general equation obtains

$$^{c*}K_{s0} = {}^{c}K_{s0}/K_w{}^z \tag{20}$$

where K_w is the ion product of water. A few representative solubility products are given in Table 5-1. From these equilibrium constants, $[Me^{z+}]$ in equilibrium with the pure solid phase can be computed readily as a function of pOH or pH. Especially convenient is the graphical representation in a log $[Me^{z+}]$–pH diagram for which the relationship of (21) can be derived

$$\log [Me^{z+}] = \log {}^{c*}K_{s0} - z\mathrm{pH}$$
$$\log [Me^{z+}] = \log {}^{c}K_{s0} + z p K_w - z\mathrm{pH} \tag{21}$$

Equation 21 is plotted for a few oxides or hydroxides in Figure 5-2. Obviously, log $[Me^{z+}]$ plots linearly as a function of pH with a slope of $-z$ ($d \log [Me^{z+}]/d\mathrm{pH} = -z$) and an intercept at log $[Me] = 0$, with the value pH $= -(1/z)\mathrm{p}^{c*}K_{s0}$.

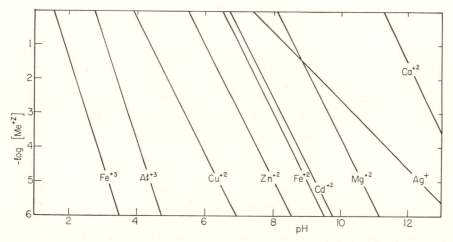

Figure 5-2 Solubility of oxides and hydroxides. Free metal ion concentration in equilibrium with solid oxides or hydroxides. The occurrence of hydroxo metal complexes must be considered for evaluation of complete solubility.

The relations depicted in Figure 5-2 or characterized by (14) do not fully describe the solubility of oxides or hydroxides. We have to consider that the solid can be in equilibrium with hydroxo metal–ion complexes† $[Me(OH)_n]^{z-n}$.

We already have encountered such hydroxo complexes as conjugate (Brønsted) bases of aquo metal ions, for example,

$$Zn^{2+} + H_2O = ZnOH^+ + H^+; \quad *K_1 \qquad (22)$$

or

$$FeOH^{2+} + H_2O = Fe(OH)_2^+ + H^+; \quad *K_2 \qquad (23)$$

We can characterize the solubility of the metal oxide or hydroxide, Me_T, by

$$Me_T = [Me^{2+}] + \sum_1^n [Me(OH)_n^{z-n}] \qquad (24)$$

Figure 5-3 gives examples for the solubility of amorphous $Fe(OH)_3$, ZnO and CuO.

Example 5-2 *Graphical Representation of Solubility of* ZnO(s). The construction of a logarithmic diagram will be illustrated for the solubility equilibrium of ZnO. It is convenient to write all the possible reactions for solid phase with solute species, for example,

$$ZnO(s) + 2H^+ = Zn^{2+} + H_2O; \qquad \log *K_{s0} = 11.2 \ (25°C) \quad (i)$$
$$ZnO(s) + H^+ = ZnOH^+; \qquad \log *K_{s1} = 2.2 \qquad (ii)$$
$$ZnO(s) + 2H_2O = Zn(OH)_3^- + H^+; \qquad \log *K_{s3} = -16.9 \qquad (iii)$$
$$ZnO(s) + 3H_2O = Zn(OH)_4^{2-} + 2H^+; \quad \log *K_{s4} = -29.7 \qquad (iv)$$

These constants given here are equivalent to those given in Table 5-1. The equilibrium constant of (ii) obtains from the following combination

$$ZnO(s) + 2H^+ = Zn^{2+} + H_2O; \quad \log *K_{s0} = 11.2 \qquad (v)$$
$$Zn^{2+} + OH^- = ZnOH^+; \qquad \log K_1 = 5.0 \qquad (vi)$$
$$H_2O = H^+ + OH^-; \qquad \log K_w = -14 \qquad (vii)$$

$$ZnO(s) + H^+ = ZnOH^+; \qquad \log *K_{s1} = 2.2 \qquad (ii)$$

In accordance with (i–iv) the lines characterizing the logarithmic concentrations of Zn^{2+}, $Zn(OH)^+$, $Zn(OH)_3^-$ and $Zn(OH)_4^{2-}$ as a function of pH

---

† More generally and more exactly, polynuclear complexes $[Me_m(OH)_n]^{zm-n}$, and complexes with other ligands, L^{y-}, in the solution, $[Me_pL_p]^{(zp-yq)}$, and mixed complexes with OH^- and L^{y-} have to be considered too. In the examples discussed here, the possible occurrence of these more complicated species has little effect upon the solubility characteristics (see Section 6-6).

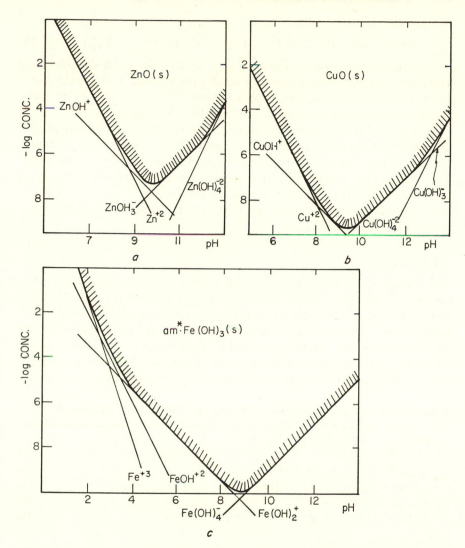

Figure 5-3 Solubility of amorphous $Fe(OH)_3$, ZnO and CuO. The construction of the diagram is explained in Example 5-2. The possible occurrence of polynuclear complexes, for example, $Fe_2(OH)_2^{4+}$, $Cu_2(OH)_2^{2+}$, has been ignored. Such complexes do not change the solubility characteristics markedly for the solids considered here.

have slopes of -2, -1, $+1$ and $+2$, respectively. The intercepts are also defined readily; for example, for $Zn(OH)^+$: $\log [ZnOH^+] = 0$ when $pH = -p*K_{s1}$ (ii); or $\log [ZnOH^+] = \log [Zn^{2+}]$ when $pH = p*K_1$ (22). Summing up all the soluble zinc species gives the line that surrounds the shaded area.

Numerically, $[Zn(II)_T]$ is given by

$$Zn(II)_T = {}^*K_{s0}[H^+]^2 + {}^*K_{s1}[H^+] + {}^*K_{s3}[H^+]^{-1} + {}^*K_{s4}[H^+]^{-2} \quad (25)$$

As Figure 5-3 illustrates in a general way, solid oxides and hydroxides have amphoteric characteristics; they can react with protons or hydroxide ions. There is a pH value for which the solubility is at a minimum.† In more alkaline or more acidic pH regions, the solubility becomes larger.

At this point it is necessary to remind ourselves that the solubility predictions based on solubility products given in Table 5-1 or depicted in Figures 5-1 and 5-2 are valid only for solutions in which, under the appropriate condition, these metal oxides or hydroxides are thermodynamically stable or at least metastable. The affinity of ligands other than OH^- (i.e., S^{2-}, CO_3^{2-}, PO_4^{3-}) present in natural waters may preclude the existence of stable oxides or hydroxides (Section 5-3).

Carbonates

In the $Me^{z+}-H_2O-CO_2$ system in the presence of the earth's atmosphere, carbonates are frequently more stable than oxides or hydroxides as solid phases. Thus in natural water systems the concentration of some metal ions is controlled by the solubility of metal carbonates. The equilibrium constants used to characterize solubility equilibria of carbonates are summarized in Table 5-2. The various solubility expressions [(6–11), Table 5-2] are interrelated and can all be expressed in terms of the conventional solubility

Table 5-2   Equilibrium Constants Defining the Solubility of Carbonates

I. Equilibrium Among Solutes and $CO_2(g)$

$H_2O = H^+ + OH^-$	K_W	(1)
$CO_2(g) + H_2O = H_2CO_3^*(aq)$	K_H	(2)
$CO_2(g) + H_2O = HCO_3^- + H^+$	$K_{p1} = K_H K_1$	(3)
$H_2CO_3^* = HCO_3^- + H^+$	K_1	(4)
$HCO_3^- = CO_3^{2-} + H^+$	K_2	(5)

II. Solid–Solutions Equilibria

$MeCO_3(s) = Me^{2+} + CO_3^{2-}$	K_{s0}	(6)
$MeCO_3(s) + H^+ = Me^{2+} + HCO_3^-$	${}^*K_s = K_{s0}K_2^{-1}$	(7)
$MeCO_3(s) + H_2CO_3^* = Me^{2+} + 2HCO_3^-$	${}^+K_{s0} = K_{s0}K_1K_2^{-1}$	(8)
$MeCO_3(s) + H_2O + CO_2(g) = Me^{2+} + 2HCO_3^-$	${}^+K_{ps0} = {}^+K_{s0}K_H$	(9)
$MeCO_3(s) + 2H^+ = Me^{2+} + CO_2(g) + H_2O$	${}^*K_{ps0} = K_{s0}K_2^{-1}K_{p1}^{-1}$	(10)
$MeCO_3(s) + 2H^+ = Me^{2+} + H_2CO_3^*$	${}^*K_{s0} = K_{s0}K_2^{-1}K_1^{-1}$	(11)

† Solubilities near this minimum usually cannot be determined experimentally. They are obtained by extrapolating from measurements in pH regions where solubilities are higher.

product K_{s0}. A listing of the different formulations should indicate merely that the solubility can be characterized by different experimental variables. For example, we can fully define a solubility equilibrium with a solid carbonate by p_{CO_2}, $[Me^{2+}]$ and $[H^+]$ [(10), Table 5-2]; by p_{CO_2}, $[Me^{2+}]$ and $[HCO_3^-]$ [(9), Table 5-2] or by $[H^+]$, $[Me^{2+}]$ and $[HCO_3^-]$ [(7), Table 5-2]. Parameters such as these are more accessible to direct analytic determination than $[CO_3^{2-}]$. Table 5-1 lists solubility products for a few metal carbonates.

In treating heterogeneous equilibria of the system Me^{2+}–CO_2–H_2O it is important to distinguish two cases: (a) systems that are closed to the atmosphere (we consider only the solid phase and the solution phase, that is, we treat $H_2CO_3^*$ as a nonvolatile acid) and (b) systems that include a (CO_2-containing) gas phase in addition to the solid and solution phase. For each of the two cases a few representative models will be discussed (Table 5-3). Because of its significance in natural water systems, calcite will be used as an example of the solid phase.

Model I. System Closed to Atmosphere

(*a*) DISSOLUTION OF $CaCO_3(s)$ IN PURE WATER. The following solute species will be encountered: Ca^{2+}, $H_2CO_3^*$, HCO_3^-, CO_3^{2-}, H^+ and OH^-. Since we have six unknowns, at a given pressure and temperature we need six equations to define the solution composition. Four mass laws interrelate the equilibrium concentrations of the solutes [first and second acidity constant of $H_2CO_3^*$, ion product of water and solubility product of $CaCO_3(s)$]. An additional equation obtains if one considers that all Ca^{2+} that becomes dissolved must equal in concentration the sum of the dissolved carbonic species

$$[Ca^{2+}] = C_T \tag{25}$$

Furthermore the solution must fulfill the condition of electroneutrality

$$2[Ca^{2+}] + [H^+] = [HCO_3^-] + 2[CO_3^{2-}] + [OH^-] \tag{26}$$

The set of six equations has to be solved simultaneously. It may be convenient to start with the solubility product

$$[Ca^{2+}] = \frac{K_{s0}}{[CO_3^{2-}]} \tag{27}$$

Using ionization fractions (Table 4-2)

$$[CO_3^{2-}] = C_T\alpha_2; \quad [HCO_3^-] = C_T\alpha_1; \quad [H_2CO_3^*] = C_T\alpha_0$$

Equation 27, rewritten as $[Ca^{2+}] = K_{s0}/C_T\alpha_2$, can be combined with (25) to give:

$$[Ca^{2+}] = C_T = \left(\frac{K_{s0}}{\alpha_2}\right)^{0.5} \tag{28}$$

Table 5-3 Solubility Equilibria in the System CaO–CO$_2$–H$_2$O

Components in Addition to CaO, CO$_2$ and H$_2$O	Selected Variables Necessary to Define System in Addition to Pressure and Temperature	Equation Defining Composition	Composition, $P = 1$ atm, 25°C
Model 1 System Closed to Atmosphere (Phases: Calcite and Aqueous Solution)			
A 0	$C_T = [\text{Ca}^{2+}]$	(29)	$pH = 9.9$, $p\text{Ca} = 3.9$, $p\text{HCO}_3^- = 4.05$, $p\text{CO}_3^{2-} = 4.4$, $p\text{Alk} = 3.62$
B Acid or base	$C_T = [\text{Ca}^{2+}]$ $C_B - C_A$		
C Acid or base	$C_T = \text{constant}$	(30) (34)	Figure 5-4 Figure 5-5
Model 2 System in Equilibrium with Gas Phase (Phases: Calcite, Aqueous Solution and CO$_2$(g))			
A 0	0†	(39)	$pH = 8.4$, $p\text{Ca} = 3.3$, $p\text{HCO}_3^- = 3.0$, $p\text{CO}_3^{2-} = 5.0$, $p\text{H}_2\text{CO}_3^* = 5.0$, $p\text{Alk} = 3.0$
B Acid or base	$C_B - C_A$	(40)	Figure 5-6

† Instead of selecting P (total pressure) as a variable, it is more convenient to select p_{CO_2} (partial pressure of CO$_2$) as a variable.

This can be substituted into the charge condition of (26)

$$\left(\frac{K_{s0}}{\alpha_2}\right)^{0.5}(2 - \alpha_1 - 2\alpha_2) + [H^+] - \frac{K_w}{[H^+]} = 0 \tag{29}$$

Equation 29 can be solved by trial and error. The result is given in Table 5-3.

(*b*) CaCO$_3$ PLUS ACID OR BASE. The addition of acid C_A or base C_B will change the pH and the solubility relations. The problem is analogous to the alkalimetric or acidimetric titration of a CaCO$_3$ suspension. The addition of acid (i.e., $C_A = [HCl]$) or base ($C_B = [NaOH]$) will shift the charge balance (26) to the following:

$$C_A - C_B = \left(\frac{K_{s0}}{\alpha_2}\right)^{0.5}(2 - \alpha_1 - 2\alpha_2) + [H^+] - \frac{K_w}{[H^+]} \tag{30}$$

With the help of this equation it is possible to compute the quantity of acid or base, C_A or C_B, needed per liter of solution [which remains in equilibrium with CaCO$_3$(s)] to reach a given pH value (Figure 5-4*a*). For each pH value

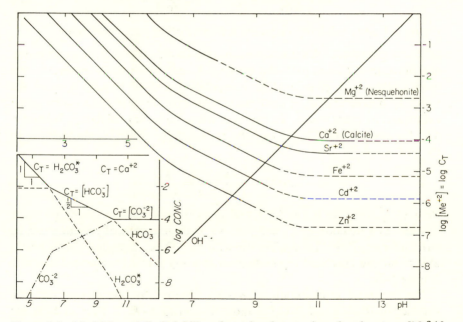

Figure 5-4   Model Ia and Ib. Solubility of metal carbonates in a closed system: [Me^{2+}] = C_T. The insert gives the essential features for the construction of the diagram for CaCO$_3$(s) and gives equilibrium concentrations of all the carbonate species. A suspension of pure MeCO$_3$(s) ($C_B - C_A = 0$) is characterized by the intersection of [OH$^-$] and [Me^{2+}] = C_T. Dashed portions of the curves indicate conditions under which MeCO$_3$(s) is not thermodynamically stable.

the solubility of $CaCO_3(s)$, for example, $[Ca^{2+}]$ or C_T, can be calculated with (28), which is still valid under our assumptions. Then the equilibrium concentrations of the other solutes can be readily obtained.

It is convenient to plot (28) graphically in a logarithmic concentration–pH diagram. The construction of the plot is facilitated by considering that the following conditions exist in the various pH regions:

for $pH > pK_2$:

$$d \log \frac{\alpha_2}{d\text{pH}} = 0; \qquad d \log \frac{C_T}{d\text{pH}} = 0; \qquad -\log C_T = \tfrac{1}{2}pK_{s0} \qquad (31)$$

for $pK_1 < pH < pK_2$:

$$d \log \frac{\alpha_2}{d\text{pH}} = +1; \qquad d \log \frac{C_T}{d\text{pH}} = -\tfrac{1}{2} \qquad (32)$$

for $pH < pK_1$:

$$d \log \frac{\alpha_2}{d\text{pH}} = +2; \qquad d \log \frac{C_T}{d\text{pH}} = -1 \qquad (33)$$

Figure 5-4 gives solubility diagrams for some other metal carbonates.

(c) $C_T = $ CONSTANT. What is the maximum soluble metal ion concentration as a function of C_T and pH? This example is analytically quite important. We have a water of a given analytic composition and we inquire whether the water is oversaturated or undersaturated with respect to a solid metal carbonate. In other words, we compute the equilibrium solubility. Numerically, for any pH and C_T, the equilibrium solubility must be maintained. In the case of calcite

$$[Ca^{2+}] = \frac{K_{s0}}{[CO_3{}^{2-}]} = \frac{K_{s0}}{C_T \alpha_2} \qquad (34)$$

Because α_2 is known for any pH, (34) gives the equilibrium saturation value of Ca^{2+} as a function of C_T and pH. An analogous type of equation can be written for any $[Me^{2+}]$ in equilibrium with $MeCO_3(s)$. Equation (34) is amenable to simple graphical representation in a $\log [Me^{2+}]$ versus pH diagram (Figure 5-5).

The graphical representation becomes obvious by considering that the product of $[Ca^{2+}]$ and $[CO_3{}^{2-}]$ must be constant (K_{s0}). Thus at high pH (pH $> pK_2$) where the $\log [CO_3{}^{2-}]$ line has a slope of zero, the $\log [Ca^{2+}]$ line must also have a slope of zero. Here the saturation concentration $[Ca^{2+}]$ must be equal to $K_{s0}/[CO_3{}^{2-}]$. In the region where $pK_1 < pH < pK_2$, $\log [CO_3{}^{2-}]$ has a slope of $+1$; correspondingly, $\log [Ca^{2+}]$ must have a slope of -1. At pH $< pK_1$, $\log [CO_3{}^{2-}]$ has a slope of $+2$; in order to

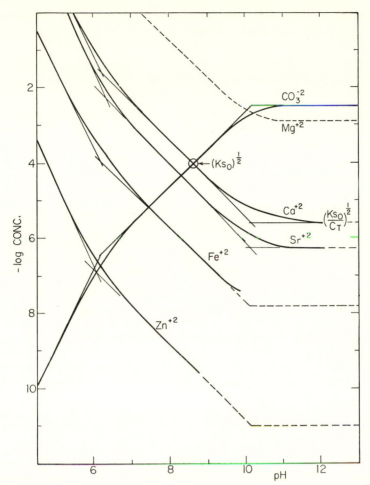

Figure 5-5 Model Ic, solubility of $MeCO_3(s)$ for a closed system with C_T = constant. The diagram gives the maximum soluble $[Me^{2+}]$ as a function of pH for a given C_T. Dashed portions of the curves indicate conditions under which $MeCO_3(s)$ is not thermodynamically stable.

maintain constancy of the product $[Ca^{2+}][CO_3^{2-}]$, $\log [Ca^{2+}]$ must have a slope of -2. Figure 5-5 shows the equilibrium saturation values of a few metal ions with respect to their carbonates for $C_T = 10^{-2.5} M$. Such a diagram is well suited for comparing the solubility of various metal carbonates and their pH dependence. The relations characterized by (34) are valid only for conditions (pH, C_T, temp.) under which the corresponding solid metal carbonates are stable or metastable. Criteria for determining the most stable solid phase will be discussed in Section 5-3.

Model II. System in Equilibrium with $CO_2(g)$

(a) $CaCO_3(s)$–CO_2–H_2O. We open the system previously discussed to the atmosphere. Specifically, we prepare a solution by adding $CaCO_3(s)$ (calcite) to pure H_2O and expose this solution to a gas phase containing CO_2. This model is representative of conditions encountered typically in fresh waters. For our example we select a partial pressure of CO_2 corresponding to that of the atmosphere ($-\log p_{CO_2} = 3.5$). Because of the additional phase we can write, in addition to the four mass laws, an independent relationship for the solubility of CO_2 in the aqueous solution (e.g., Henry's law). Furthermore, an electroneutrality condition can be formulated. If we specify the temperature and a partial pressure of CO_2, the system is completely defined.

Because of the equilibrium with CO_2 in the gas phase, $[Ca^{2+}]$ is no longer equal to C_T, but the same electroneutrality condition (26) pertains:

$$2[Ca^{2+}] + [H^+] = C_T(\alpha_1 + 2\alpha_2) + [OH^-] \tag{35}$$

Furthermore $[Ca^{2+}]$ can be expressed as a function of C_T and $[H^+]$ and C_T is defined by p_{CO_2}

$$[Ca^{2+}] = \frac{K_{s0}}{C_T\alpha_2} \tag{36}$$

$$C_T = \frac{K_H p_{CO_2}}{\alpha_0} \tag{37}$$

$$[CO_3{}^{2-}] = \frac{K_H p_{CO_2}\alpha_2}{\alpha_0} \tag{38}$$

$$[Ca^{2+}] = \frac{(K_{s0}/K_H p_{CO_2})\alpha_0}{\alpha_2} \tag{39}$$

With the substitutions of (37) and (38), (35) can now be solved for $[H^+]$. The result is given in Table 5-3. Comparing this result with that of Model Ia, we see that the influence of atmospheric CO_2 has depressed the pH markedly and that $[Ca^{2+}]$ and $[Alk]$ have been raised to values very representative of those in natural waters.

(b) $CaCO_3$–H_2O–CO_2 PLUS ACID OR BASE. pH changes can occur upon addition of acid or bases (e.g., addition of wastes, dissolution of volcanic volatile compounds like HCl, biological reactions) to the model discussed before. The electroneutrality condition can be formulated most generally by considering that the charge balance of (35) has been shifted because of the addition of C_A (acid) or C_B (base):

$$C_B + 2[Ca^{2+}] + [H^+] = C_T(\alpha_1 + 2\alpha_2) + [OH^-] + C_A \qquad (40)$$

Since $[Ca^{2+}]$ and C_T in this equation can again be expressed as a function of p_{CO_2} and $[H^+]$, the "titration curve" of a $CaCO_3$ suspension that remains in equilibrium with $CO_2(g)$ can be computed.

Equation 39 can be plotted graphically (Figure 5-6). The distribution of the dissolved carbonate species at a given temperature is defined entirely by p_{CO_2}.

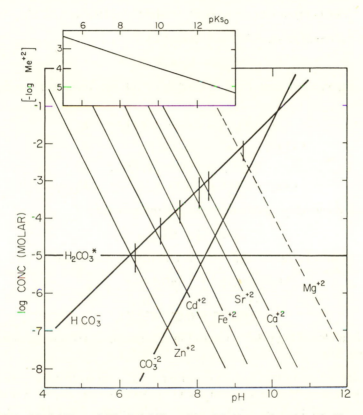

Figure 5-6   Model IIa and IIb. Solubility of $MeCO_3(s)$ as a function of pH at constant p_{CO_2}. $-\log p_{CO_2} = 3.5$ (corresponding to partial pressure of CO_2 in atmosphere). If no excess acid or base is added ($C_B - C_A = 0$), the equilibrium composition of the solution is given by the electroneutrality condition, $2[Me^{2+}] \simeq [HCO_3^-]$. This condition is indicated by a vertical dash slightly displaced from the intersection of $[Me^{2+}]$ with $[HCO_3^-]$. The insert gives $-\log[Me^{2+}]$ for pure $MeCO_3(s)$ suspensions in equilibrium with $p_{CO_2} = 10^{-3.5}$ atm as a function of pK_{so}.

The line for $[Mg^{2+}]$ has been calculated for nesquehonite, $MgCO_3 \cdot 3H_2O$, which, however, is not a stable phase under these conditions.

Hence, Figure 5-6 is essentially the superposition of (40) upon Figure 4-3. The log $[Ca^{2+}]$ line has a slope of $+2$ with respect to pH. Differentiating (39) with respect to pH and considering that $d \log (\alpha_0/\alpha_2)/pH = +2$, leads to a slope of $+2$ with respect to pH for the log $[Ca^{2+}]$ line in the diagram.

Example 5-3   *Solubility of Pure* $MeCO_3(s)$ *Suspensions.*   Derive an equation that shows how the solubility of various bivalent metal carbonates varies with their solubility product.

Consider the reaction

$$MeCO_3(s) + CO_2(g) + H_2O = Me^{2+} + 2HCO_3{}^-; \quad {}^+K_{ps0} \qquad (i)$$

We can represent its equilibrium constant by

$$ {}^+K_{ps0} = K_{s0}K_1K_HK_2{}^{-1} \qquad (ii)$$

This becomes evident from the addition of the following equilibrium reactions

$$MeCO_3(s) = Me^{2+} + CO_3{}^{2-}; \quad K_{s0} \qquad (iii)$$

$$CO_2(g) + H_2O = H_2CO_3^*; \quad K_H \qquad (iv)$$

$$H_2CO_3^* = H^+ + HCO_3{}^-; \quad K_1 \qquad (v)$$

$$CO_3{}^{2-} + H^+ = HCO_3{}^-; \quad K_2{}^{-1} \qquad (vi)$$

The electroneutrality equation can be approximated by

$$2[Me^{2+}] \simeq [HCO_3{}^-] \qquad (vii)$$

which can now be substituted into the equilibrium expression of (i)

$$\frac{[Me^{2+}][HCO_3{}^-]^2}{p_{CO_2}} \simeq \frac{4[Me^{2+}]^3}{p_{CO_2}} \simeq {}^+K_{ps0} \qquad (viii)$$

$$[Me^{2+}] \simeq 0.63\,{}^+K_{ps0}^{1/3}p_{CO_2}^{1/3} \qquad (ix)$$

Hence in a plot of log $[Me^{2+}]$ versus log K_{s0} a slope of $1/3$ obtains (see insert in Figure 5-6).

HYDROXIDE CARBONATES. Some metal ions form solid hydroxide carbonates. Examples are hydromagnesite $Mg_4(CO_3)_3(OH)_2 \cdot 3H_2O$, azurite $Cu_3(OH)_2$-$(CO_3)_2$, malachite $Cu_2(OH)_2CO_3$ and hydrozincite $Zn_5(OH)_6(CO_3)_2$. In evaluating solubility characteristics of such hydroxide carbonates the same principles apply as were discussed before. For example, the solubility of hydromagnesite can be characterized by the reaction

$$Mg_4(CO_3)_3(OH)_2 \cdot 3H_2O = 4Mg^{2+} + 3CO_3{}^{2-} + 2OH^- \qquad (41)$$

where

$$ {}^cK_{s0} = [Mg^{2+}]^4[CO_3{}^{2-}]^3[OH^-]^2 \qquad (42)$$

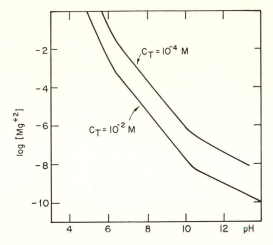

Figure 5-7   Solubility of Hydromagnesite, $Mg_4(CO_3)_3(OH)_2 \cdot 3H_2O(s)$.

$[Mg^{2+}]$ in equilibrium with hydromagnesite is given by

$$[Mg^{2+}] = \frac{^c K_{s0}^{1/4} K_W^{1/2}}{C_T^{3/4} \alpha_2^{3/4} [H^+]^{1/2}} \tag{43}$$

The dependence of $[Mg^{2+}]$ on pH for a given C_T is represented in Figure 5-7. As (43) suggests, the slopes of the log $[Mg^{2+}]$ versus pH line are -2, $-\frac{5}{4}$ and $-\frac{1}{2}$, for pH $<$ pK_1, p$K_1 <$ pH $<$ pK_2 and pH $>$ pK_2, respectively. In Example 5-6, the solubility of hydromagnesite is compared with that of other compounds in the system $Mg–CO_2–H_2O$.

5-3   The Stability of Hydroxides, Carbonates, Hydroxide Carbonates and Sulfates

In the previous sections we have applied equilibrium constants for heterogeneous equilibria in a rather formal way. Thermodynamically meaningful conclusions are justified only if, under the specified conditions (concentration, pH, temperature, pressure), the solutes are in equilibrium with that solid phase for which the mass law relationship has been formulated or if under the specified conditions the assumed solid phase is really stable or at least metastable. It remains to be illustrated how we can establish which phase predominates under a given set of conditions.

Which Solid Phase Controls the Solubility of Fe(II)?

In order to exemplify how to find out which solid predominates as a stable phase for selected conditions, we may consider the solubility of Fe(II) in a carbonate-bearing water of low redox potential ($P = 1$ atm, 25°C, I = 6 × 10^{-3} M). We may first consider the solubility products of $Fe(OH)_2(s)$ and of $FeCO_3(s)$

$$K_{s0}(Fe(OH)_2) \simeq 10^{-14.7} \text{ (moles}^3 \text{ liter}^{-3})$$

$$K_{s0}(FeCO_3) \simeq 10^{-10.7} \text{ (moles}^2 \text{ liter}^{-2})$$

The numerical values of these equilibrium constants cannot be compared directly. Note that the constants have different units. It would be incorrect and misleading to infer from the numerical values of the solubility products that $Fe(OH)_2(s)$ is less soluble than $FeCO_3(s)$. It is appropriate rather to inquire which solid controls the solubility for a given set of conditions (P, T, pH, C_T, [Alk] or p_{CO_2}), that is, gives the smallest concentration of soluble Fe(II).

Example 5-4   *Control of Solubility.*   Is it $FeCO_3(s)$ or $Fe(OH)_2(s)$ that controls the solubility of Fe(II) in anoxic water of [Alk] = 10^{-4} eq liter^{-1} and pH = 6.8? We estimate maximum soluble [Fe(II)] by considering the solubility equilibrium (25°C) with (i) $FeCO_3(s)$ as well as that with (ii) $Fe(OH)_2(s)$.

(i) Assuming solubility equilibrium with $FeCO_3(s)$

$$FeCO_3(s) = Fe^{2+} + CO_3^{2-}; \quad \log {}^cK_{s0} \simeq -10.4 \qquad (i)$$

$$H^+ + CO_3^{2-} = HCO_3^-; \quad \log {}^cK_2 \simeq +10.1 \qquad (ii)$$

$$\overline{FeCO_3(s) + H^+ = Fe^{2+} + HCO_3^-; \quad \log {}^{c*}K_s \approx -0.3 \qquad (iii)}$$

Correspondingly, $\log [Fe^{2+}] = \log {}^{c*}K_s - pH - \log [HCO_3^-]$; and since at this pH, $[Fe^{2+}] \simeq [Fe(II)]$ and $[HCO_3^-] \simeq [Alk]$, we obtain $\log [Fe(II)] = -3.1$.

(ii) Assuming solubility equilibrium with $Fe(OH)_2(s)$

$$Fe(OH)_2(s) = Fe^{2+} + 2OH^-; \quad \log {}^cK_{s0} = -14.5 \qquad (iv)$$

$$2H^+ + 2OH^- = 2H_2O; \quad -2 \log {}^cK_W = +27.8 \qquad (v)$$

$$\overline{Fe(OH)_2(s) + 2H^+ = Fe^{2+} + 2H_2O; \quad \log {}^{c*}K_{s0} = +13.3 \qquad (vi)}$$

Thus, $\log [Fe^{2+}] = \log {}^*K_{s0} - 2pH$; and $\log [Fe(II)] \simeq -0.3$. Because $[Fe^{2+}]$ (or [Fe(II)]) is smaller for hypothetical equilibrium with $FeCO_3(s)$ than with $Fe(OH)_2$, siderite [$FeCO_3(s)$] is more stable than $Fe(OH)_2(s)$.

SOLUBILITY AND PREDOMINANCE DIAGRAMS. From thermodynamic information, diagrams can be constructed which circumscribe the stability boundaries of the solid phases. Depending on the variables used, different kinds of predominance diagrams can be constructed.

For systems closed to the atmosphere, a *solubility diagram* (e.g., log [Fe^{2+}], log [H$^+$], log C_T) can conveniently illustrate the conditions under which a particular solid phase predominates. Figure 5-8 gives a solubility diagram for Fe(II) considering FeCO$_3$(s) and Fe(OH)$_2$(s) as possible solid phases. The construction of the diagram consists essentially in a superposition of a pH-dependent Fe(OH)$_2$(s) solubility diagram (Figure 5-2) and a pH- and C_T-dependent FeCO$_3$(s) solubility diagram (Figure 5-5). For a given [H$^+$] and C_T the solid compound giving the smaller Fe^{2+} is more stable. Thus, for the conditions in Figure 5-8a, FeCO$_3$(s) dictates the maximum concentration of Fe(II) below pH values of approximately 10, and Fe(OH)$_2$ limits soluble iron above pH 10.

The same information can be gained from an *activity ratio diagram*. The construction is very simple and is illustrated in Figure 5-8b. We again choose pH as a master variable and make our calculation for a given C_T. In this figure we plot the ratios between the activities of the various soluble and solid species as a function of pH. In our case one of the species {Fe^{2+}} (or [Fe^{2+}]) is chosen as a reference state. Thus the ordinate values are log {A_i}/{Fe^{2+}}. Because the diagram gives activities on a relative scale (relative to [Fe^{2+}]) we treat the activities of solid phases formally in the same way as the activities (or concentrations) of the solutes. Two equations determine the ratios {Fe(OH)$_2$}(s)/{Fe^{2+}} and {FeCO$_3$(s)}/{Fe^{2+}}. For the solubility of Fe(OH)$_2$(s) we have

$$\frac{\{Fe^{2+}\}}{\{H^+\}^2\{Fe(OH)_2(s)\}} = {}^*K_{s0_{Fe(OH)2}} \tag{44}$$

Thus, log ({Fe(OH)$_2$(s)}/{Fe^{2+}}) plots as a function of pH as a straight line with a slope of +2:

$$\log \frac{\{Fe(OH)_2(s)\}}{\{Fe^{2+}\}} = p^*K_{s0_{Fe(OH)2}} + 2pH \tag{45}$$

and an intercept pH = $\frac{1}{2}p^*K_{s0_{Fe(OH)2}}$.

For the solubility of FeCO$_3$

$$\frac{\{Fe^{2+}\}\{CO_3^{2-}\}}{\{FeCO_3(s)\}} = K_{s0_{FeCO3}} \tag{46}$$

Because {CO$_3^{2-}$} = $C_T\alpha_2$, (46) can be rearranged to

$$\log \frac{\{FeCO_3(s)\}}{\{Fe^{2+}\}} = pK_{s0_{FeCO3}} + \log C_T + \log \alpha_2 \tag{47}$$

Equation 47 can be plotted considering that log α_2 = 0 at pH > pK_2 and that $d(\log \alpha_2)/d$pH is +1 or +2 in the pH regions pK_1 < pH < pK_2 and pH < pK_1, respectively.

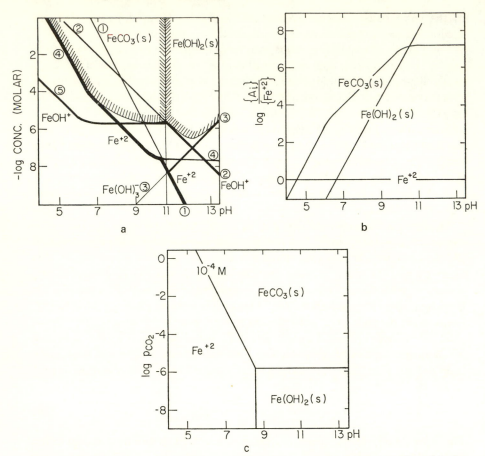

Figure 5-8 Stability of $Fe(OH)_2(s)$ and $FeCO_2(s)$ (siderite) (25°C; $I = 0$). With the help of such diagrams it is possible to evaluate the conditions (pH, C_T, $[Fe^{2+}]$, p_{CO_2}) under which a solid phase [$Fe(OH)_2(s)$ or $FeCO_3(s)$] is stable.

a: Solubility diagram of Fe(II) in a $C_T = 10^{-3}$ M carbonate system. The numbers on the curves refer to equations describing the respective equilibria in this system as follows:

1. $Fe(OH)_2(s) = Fe^{2+} + 2OH^-$ $K_{s0} = 2 \times 10^{-15}$
2. $Fe(OH)_2(s) = [Fe(OH)]^+ + OH^-$ $K_{s1} = 4 \times 10^{-10}$
3. $Fe(OH)_2(s) + OH^- = [Fe(OH)_3]^-$ $K_{s3} = 8.3 \times 10^{-6}$
4. $FeCO_3(s) = Fe^{2+} + CO_3^{2-}$ $K_{s0} = 2.1 \times 10^{-11}$
5. $FeCO_3(s) + OH^- = [Fe(OH)]^+ + CO_3^{2-}$ $K_{s1} = 0.1 \times 10^{-5}$

b: Activity ratio diagram for $C_T = 10^{-3}$ M. The same conclusions as under *a* apply: above pH $\simeq$ 10, Fe(OH)(s) has the highest relative activity; it can precipitate as a pure phase. Below pH $\simeq$ 10, $FeCO_3$ becomes more stable than $Fe(OH)_2$ and controls the solubility of Fe(II).

c: $\log p_{CO_2}$–pH predominance diagram. $FeCO_3(s)$ is the stable phase at p_{CO_2} larger than 10^{-6} atm, that is, in atmosphere.

At any pH the ordinate values in the activity ratio diagram give the activities for the various species on a relative scale. Thus in Figure 5-8 at pH $= 12, Fe(OH)_2(s)$ has the highest relative activity. This solid phase will precipitate at this pH; as a pure solid its activity will be unity and $FeCO_3(s)$ must have an activity of much less than unity and cannot exist as a pure solid phase. Figure 5-8 shows that, for $C_T = 10^{-3} M$, $FeCO_3(s)$ is stable below ca. pH $= 10$.

For *open systems* it is convenient to select $\log p_{CO_2}$ and $-\log [H^+]$ as variables. An assumption must then be made about the concentrations of the solute. The computation of the straight lines in Figure 5-8c is based on the following equations. For the coexistence of Fe^{2+} and $Fe(OH)_2(s)$

$$\log {}^*K_{s0_{Fe(OH)_2}} - 2pH + pFe^{2+} = 0. \tag{48}$$

The thermodynamic coexistence of Fe^{2+} and $FeCO_3(s)$ as a function of pH and p_{CO_2} is expressed, perhaps most conveniently, by considering the solubility equilibrium in the form of the reaction

$$FeCO_3(s) + 2H^+ = Fe^{2+} + CO_2(g) + H_2O; \quad {}^*K_{ps0_{FeCO_3}}$$
$$\log p_{CO_2} = \log {}^*K_{ps0_{FeCO_3}} - 2pH + pFe^{2+} \tag{49}$$

The coexistence of $FeCO_3(s)$ and $Fe(OH)_2$ is given by the equilibrium

$$FeCO_3(s) + H_2O = Fe(OH)_2(s) + CO_2(g) \tag{50}$$

where [compare (48) and (49)]

$$K = \frac{{}^*K_{ps0_{FeCO_3}}}{{}^*K_{s0_{Fe(OH)_2}}} = p_{CO_2} \tag{51}$$

The equilibrium partial pressure of CO_2 for (51) is approximately 10^{-6} atm. At p_{CO_2} higher than 10^{-6} atm, for example, in systems exposed to the atmosphere, $Fe(OH)_2$ is not stable and will be converted to $FeCO_3(s)$. In Figure 5-8a the enhancement of the solubility caused by the formation of $FeOH^+$ ($= 10^{-4} M$) also has been considered.

Similar predominance diagrams for the systems $Zn^{2+}-CO_2-H_2O$ and $Cu^{2+}-CO_2-H_2O$ are given in Figure 5-9. In equilibrium with air ($p_{CO_2} = 10^{-3.5}$ atm), according to these diagrams, hydrozincite $[Zn_5(OH)_6(CO_3)_2]$ is more stable than $Zn(OH)_2(s)$† and tenorite (CuO) is the stable solid copper salt.

Example 5-5   *Free Energy of Formation of a Metal Carbonate.*   The Gibbs free energy of formation of a metal carbonate can be computed from the free energy of formation of MeO(s) and $CO_2(g)$ and the free energy of the reaction

† ZnO(s), however, is more stable than hydrozincite at this p_{CO_2}.

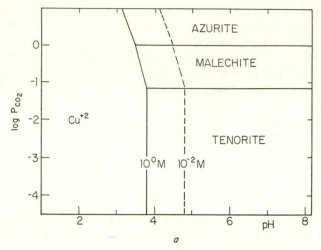

Corrosion Sci, 7, 629 (1967)
R. Grauer & W. Feitknecht

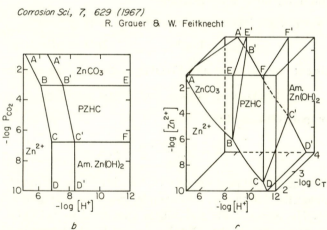

Figure 5-9 Predominance diagrams for the systems $Cu^{2+}-CO_2-H_2O$ and $Zn^{2+}-CO_2$ $-H_2O$. *a*: $\log p_{CO_2}$–pH diagram for Cu(II) ($I = 0$, 25°C) (see P. Schindler, 1967).

b: $\log p_{CO_2}$–pH diagram for Zn(II) ($I = 0.2$, 25°C) (taken from R. Grauer and W. Feitknecht, 1967). $[Zn^{2+}] = 10^{-1}$ and 10^{-4} M along the lines $ABCD$ and $A'B'C'D'$, respectively. PZHC = hydrozincite.

c: $\log p_{CO_2} = pH - \log [Zn^{2+}]$ predominance diagram (taken from R. Grauer and W. Feitnecht, 1967).

Reproduced with permission from Pergamon Press.

MeO(s) + CO_2(g) = MeCO$_3$(s). Estimate a free energy of CdCO$_3$(s) from the following information (at 25°C) [cf. Gamsjäger, Stuber and Schindler, *Helv. Chim. Acta*, **48**, 723 (1965)]:

$$C(s) + O_2(g) = CO_2(g); \quad \Delta G_1^\circ = -94,260 \text{ (cal)} \quad \text{(i)}$$

$$Cd(s) + \tfrac{1}{2}O_2(g) = CdO(s); \quad \Delta H_2^\circ = -62,360 \text{ (cal)}$$

$$\Delta S_2^\circ = -23.7 \text{ (cal deg}^{-1}) \quad \text{(ii)}$$

$$CdO + CO_2(g) = CdCO_3(s); \quad \Delta G_3^\circ = -12,145 \text{ (cal)} \quad \text{(iii)}$$

For (ii), ΔG_2° can be computed from ΔH_2° and $T\Delta S_2^\circ$. The former has been determined calorimetrically from the heat of combustion; the latter has been estimated from specific heat data (C_p). Using $\Delta G^\circ = \Delta H^\circ - T\Delta S^\circ$, we obtain $\Delta G_2^\circ = -62,360 + 7,080 = -55,280$ (cal). Summing up (i), (ii) and (iii), one obtains

$$Cd(s) + C(s) + \tfrac{3}{2}O_2(g) = CdCO_3(s); \quad \Delta G^\circ = -161,690 \text{ (cal)} \quad \text{(iv)}$$

Note: Gamsjäger, Stuber and Schindler determined the solubility of CdCO$_3$(s) and evaluated the free energy of formation from the following cycle

$C(s) + O_2(g) = CO_2(g)$	$\Delta G_1^\circ = -94,260$ cal	(i)
$H_2(g) + \tfrac{1}{2}O_2(g) = H_2O(g)$	$\Delta G_5^\circ = -54,635$ cal	(v)
$H_2O(g) = H_2O(l)_I$	$\Delta G_6^\circ = -2,130$ cal	(vi)
$Cd_I^{2+} + CO_2(g) + H_2O(l)_I = CdCO_3(s) + 2H_I^+$	$\Delta G_7^{\circ\prime} = RT \ln {}^{c*}K_{ps0}$	
	$= 8,830$ cal	(vii)
$Cd(s) + 2H_I^+ = Cd_I^{2+} + H_2(g)$	$\Delta G_8^{\circ\prime} = 2fE_{\text{cell(I)}}^{\circ\prime}$	
	$= -18,980$ cal	(viii)

$$Cd(s) + C(s) + \tfrac{3}{2}O_2(g) = CdCO_3(s) \qquad \Delta G_4^\circ = -161,170 \text{ cal} \quad \text{(iv)}$$
$$(\Delta G_4^\circ = \Delta G_1^\circ + \Delta G_5^\circ + \Delta G_6^\circ + \Delta G_7^{\circ\prime} + \Delta G_8^{\circ\prime})$$

(Solute species present in the ionic medium of ionic strength I are indicated by suffix I.) The equilibria of (vii) and (viii) have been determined by solubility measurements ($^{c*}K_{ps0} = [H^+]/[Cd^{2+}]p_{CO_2}$) and from emf measurements (electrochemical cell), respectively, in solutions of constant ionic medium (3 M NaClO$_4$); the free energy of evaporation of water [reaction (vi)] is also given for the same ionic medium. Note that ΔG_4°, the free energy of formation, is unaffected by the ionic medium.

Example 5-6   *The Stability of Carbonates in the System* Mg–CO$_2$–H$_2$O. Below are listed standard free energies of formation, $\bar{G}_f^\circ$, of Mg^{2+}-bearing minerals; these are based on data derived by Langmuir [D. Langmuir, *J. Geol.*,

73, 730 (1965)], while $\bar{G}_f^\circ$ values for solutes were taken from Latimer [2]:

	$\bar{G}_f^\circ$, kcal mole^{-1}
$Mg(OH)_2(s)$, brucite	-200.0
$MgCO_3(s)$, magnesite	$-241.9\dagger$
$MgCO_3 \cdot 3H_2O(s)$, nesquehonite	$-412.7\dagger$
$Mg_4(CO_3)_3(OH)_2 \cdot 3H_2O(s)$, hydromagnesite	-1100.1
$Mg^{2+}(aq)$	-109.0
$CO_3^{2-}(aq)$	-126.2
$OH^-(aq)$	-37.6
$H_2O(l)$	-126.2

† These values are at variance with observations of others who claim that magnesite is more stable than nesquehonite [pK_{so} (magnesite) > pK_{so}(nesquehonite)].

On the basis of these data, establish stability domains as a function of variables C_T, pH and p_{CO_2} for the minerals listed.

We first compute equilibrium constants using the relationships $\Delta G^\circ = -RT \ln K$ and obtain (at 25°C)

	ΔG°, kcal mole^{-1}	$-\log K_{s0}$
$Mg(OH)_2(s) = Mg^{2+} + 2OH^-$ brucite	-15.8	11.6
$MgCO_3(s) = Mg^{2+} + CO_3^{2-}$ magnesite	$+6.7$	4.9
$MgCO_3(s) \cdot 3H_2O(s) = Mg^{2+} + CO_3^{2-} + 3H_2O$ nesquehonite	$+7.4$	5.4
$Mg_4(CO_3)_3(OH)_2 \cdot 3H_2O = 4Mg^{2+} + 3CO_3^{2-} + 2OH^-$ hydromagnesite $\quad + 3H_2O$	$+40.2$	29.5

In order to obtain a first survey on the stability relationships, we construct an activity ratio diagram using Mg^{2+} as a reference state. The equations are

For brucite:

$$\log \frac{\{Mg(OH)_2(s)\}}{\{Mg^{2+}\}} = pK_{s0} - 2pK_w + 2pH$$

$$= -16.4 + 2pH \tag{i}$$

For magnesite:

$$\log \frac{\{magnesite(s)\}}{\{Mg^{2+}\}} = pK_{s0} + \log C_T + \log \alpha_2 \tag{ii}$$

$$= 4.9 + \log C_T + \log \alpha_2$$

For nesquehonite:

$$\log \frac{\{\text{nesquehonite(s)}\}}{\{Mg^{2+}\}} = pK_{s0} + \log C_T + \log \alpha_2$$

$$= 5.4 + \log C_T + \log \alpha_2 \qquad \text{(iii)}$$

For hydromagnesite†:

$$\log \frac{\{\text{hydromagnesite(s)}\}^{\frac{1}{4}}}{\{Mg^{2+}\}} = \tfrac{1}{4}pK_{s0} - \tfrac{1}{2}pK_W + \tfrac{3}{4}\log C_T + \tfrac{3}{4}\log \alpha_2 + \tfrac{1}{2}pH$$

$$= 0.4 + \tfrac{3}{4}\log C_T + \tfrac{3}{4}\log \alpha_2 + \tfrac{1}{2}pH \qquad \text{(iv)}$$

These equations are plotted in Figure 5-10a for an assumed value of $\log C_T = -2.5$. It is convenient to start to plot these equations at high pH values where $\log \alpha_2 = 0$. In the pH region $pK_1 < pH < pK_2$, $d\log \alpha_2/dpH$ has a slope of $+1$. Our activity ratio diagram postulates that for $\log C_T = -2.5$, nesquehonite is stable below pH $\simeq 7.3$; brucite becomes stable above pH $\simeq 9.7$. In the intermediate pH range (7.3 to 9.7) hydromagnesite is the stable phase.

In Figure 5-10b a solubility diagram is plotted for $-\log C_T$ of 2.5. Equations i to iv with unit activity for the solid phase can be used.

For an open system, activity ratio or solubility diagrams for given p_{CO_2} values can be constructed. Figure 5-10c gives $\log p_{CO_2}$–pH predominance diagrams for $pMg^{2+} = 0$ and 3.0. The dissolution reaction may be rearranged to give mass law expressions with p_{CO_2} and $[H^+]$, for example:

For nesquehonite:

$$MgCO_3 \cdot 3H_2O(s) = Mg^{2+} + CO_3^{2-} + 3H_2O; \qquad \log K_{s0} = -5.4$$
$$CO_3^{2-} + 2H^+ = H_2CO_3^*; \qquad -\log (K_1K_2) = +16.6$$
$$H_2CO_3^* = CO_2(g) + H_2O; \qquad -\log K_H = +1.5$$

$$\overline{MgCO_3 \cdot 3H_2O(s) + 2H^+ = Mg^{2+} + CO_2(g) + 3H_2O; \qquad \log {}^*K_{ps0} = +12.7}$$

Similarly one derives for hydromagnesite:

$$Mg_4(CO_3)_3(OH)_2 3H_2O + 8H^+ = 4Mg^{2+} + 3CO_3^{2-} + 5H_2O;$$
$$\log {}^*K_{ps0} = +52.8$$

According to Figure 5-10c we infer that at 25°C both nesquehonite and brucite are unstable if brought in contact with humid air; hydromagnesite is the stable phase under these conditions.

Mixed Carbonates; Dolomite

Two mixed carbonates of Ca^{2+} and Mg^{2+} occur in nature: dolomite $CaMg(CO_3)_2(s)$ and huntite $CaMg_3(CO_3)_4(s)$. Dolomite constitutes a large

† $\{\text{hydromagnesite(s)}\}^{\frac{1}{4}} = \{Mg(CO_3)_{0.75}(OH)_{0.5} \cdot 0.75H_2O(s)\}$.

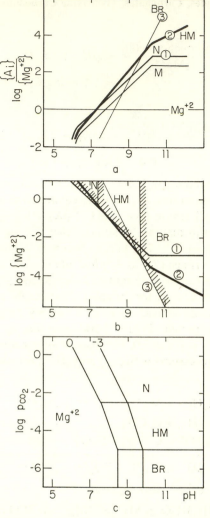

Figure 5-10 Stability in the system $Mg^{2+}-CO_2-H_2O$ ($I = 0$, 25°C) (see Example 5-6). Br = brucite, $Mg(OH)_2(s)$; HM = hydromagnesite, $Mg_4(CO_3)_3(OH)_23H_2O(s)$; N = nesquehonite, $MgCO_33H_2O(s)$; M = magnesite, $MgCO_3(s)$.

a: Activity ratio diagram for $-\log C_T = 2.5$. Equations defining relative activity for A_i = M, N(s)(1), HM(s)(2), Br(s)(3). Stable phases are:

$$N(s):\ pH < 7.4$$
$$HM(s):\ pH = 7.4-9.7$$
$$Br(s):\ pH > 9.7$$

b: Solubility ($-\log\{Mg^{2+}\}$ versus pH) diagram for $-\log C_T = 2.5$. The numbers have the same meaning as in *a*. As *a* suggests, Br(s), HM(s) and N(s) control the maximum soluble of $\{Mg^{2+}\}$ in the high pH, the alkaline and the neutral pH range, respectively.

c: Predominance diagram for $\log\{Mg^{2+}\} = 0$ and -3. Thermodynamically, N(s) is only stable at high p_{CO_2}. Br(s) can exist only at very low p_{CO_2}; if exposed to air it should carbonize to HM(s).

192

fraction of the total quantity of carbonate rocks. The conditions under which dolomite precipitates in nature are not well understood. Attempts to precipitate a dolomite phase from oversaturated solutions under atmospheric conditions have been unsuccessful [6]. Dolomite may be formed by the reaction of Mg^{2+}-rich waters with carbonate sediments (Deffeyes et al. [7]). Investigations by Peterson, Bien and Berner [8] have shown that calcium-rich dolomite is being formed at a very slow rate (growth rate of crystals of the order of hundreds of Angstrøms per thousand years) at the present time in Deep Spring Lake, California. The lake, in an arid intermontane basin, maintains only a small brine body during summer; its water contains much carbonate and sulfate and has a pH of about 9.5 to 10.

Although precipitation of dolomite from most natural waters would appear to be unimportant as a controlling factor in carbonate equilibria, the extensive areas underlain by dolomite bed rock suggest that the solubility characteristics are not unimportant [9]. Thus water in equilibrium with dolomite rocks may have a composition quite different from that of waters in equilibrium with $CaCO_3$.

THE SOLUBILITY OF DOLOMITE. The lack of understanding of the dolomite precipitation process is reflected in the discrepancy of solubility products reported by different investigators. Published figures range from $10^{-16.5}$ to $10^{-19.5}$. As mentioned before, solubility equilibrium can be reached (under atmospheric conditions) only from undersaturation. The time of approaching equilibrium is unknown. Thus it is very difficult to ascertain equilibrium in laboratory experiments.

Nature, however, has provided us with a long-term solubility experiment. As shown by Hsu [10], the well waters of Florida show a constant ratio of magnesium to calcium ($[Ca^{2+}]/[Mg^{2+}] = 0.8 \pm 0.1$). Hsu points out that the tendency for subsurface waters to have such a nearly constant magnesium–calcium ratio suggests that waters in porous dolomitic limestones might have equilibrated with both the calcite and dolomite phases.

The solubility product of dolomite is given by

$$K_{s0} = \{Ca^{2+}\}\{Mg^{2+}\}\{CO_3{}^{2-}\}^2 \qquad (52)$$

For the reaction

$$2CaCO_3(s)(\text{calcite}) + Mg^{2+} = CaMg(CO_3)_2(s) + Ca^{2+} \qquad (53)$$

[6] D. L. Graf and J. R. Goldsmith, *J. Geol.*, **64**, 173 (1956).
[7] K. S. Deffeyes et al., *Science*, **143**, 687 (1964).
[8] M. N. A. Peterson, G. S. Bien and R. A. Berner, *J. Geophys. Res.*, **68**, 6493 (1963).
[9] O. P. Bricker and R. M. Garrels, in *Principles and Applications of Water Chemistry*, S. D. Faust and J. V. Hunter, Eds., Wiley, New York, 1967, p. 449.
[10] K. J. Hsu, *J. Hydrol.*, **1**, 288 (1963).

the equilibrium constant, K, is defined in terms of the activity ratio of Ca^{2+} and Mg^{2+} and the K_{so} values of dolomite and calcite:

$$K = \frac{K^2_{so_{CaCO3}}}{K_{so_{dolomite}}} = \frac{\{Ca^{2+}\}}{\{Mg^{2+}\}} \tag{54}$$

For solution in contact and equilibrium with dolomite and calcite, the activity ratio is a constant at any temperature and pressure. In the Floridian aquifer waters the concentration ratio $(f_{Ca^{2+}}/f_{Mg^{2+}} \simeq 1)$ remains nearly constant even though $[Ca^{2+}]$ and $[Mg^{2+}]$ vary from less than 10^{-3} to 10^{-2} M. With an average value of $[Mg^{2+}]/[Ca^{2+}]$ of 0.78 and $K_{so_{calcite}} = 5 \times 10^{-9}$ (25°C), a $K_{so_{dolomite}} = 2.0 \times 10^{-17}$ can be calculated [10]. The same value (25°C, $P = 1$ atm) has been obtained from laboratory investigations [11]. With this constant and the constants given earlier, a predominance diagram for the system Mg^{2+}, Ca^{2+}, H_2O, CO_2 can be constructed. The diagram shown in Figure 5-11 is similar to that given by Langmuir. Obviously, dolomite is thermodynamically the stable phase in seawater of average salinity and average calcium to magnesium ratio. But despite an abundance of dolomite in ancient sedimentary rocks, no unequivocal proof of recent dolomite in seawater of typical composition has been given.

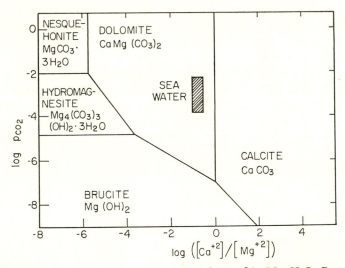

Figure 5-11 Stability relations in the system $Ca^{2+}-Mg^{2+}-CO_2-H_2O$. Based on free energy values given by Langmuir (25°C). Seawater is within the thermodynamic predominance of dolomite as a soluble phase, but no convincing evidence for the formation of recent dolomite in seawater of normal salinity has been demonstrated.

[11] D. Langmuir, Ph.D. thesis, Harvard University, Cambridge, Mass., 1964.

Dolomites found in nature seldom have exact stoichiometric composition and are frequently structurally rich in calcium (protodolomite). Dolomite, as well as calcite, has a tendency to form solid solutions with many metal ions. Calcite has a tendency to accommodate Mg^{2+} in its structure to form *magnesian calcite*. Kinetically, the deposition of magnesian calcite may be more favorable than the deposition of dolomite.

Example 5-7 *Solubility of Dolomite.* Consider fresh water containing 5×10^{-4} M Mg^{2+} and total carbonate carbon, $C_T = 2 \times 10^{-3}$ M. Assuming a closed system, define the pH range in which the solubility of Ca^{2+} should be dominated by dolomite solubility.

If the solubilities of dolomite and calcite ($25°C$, $P = 1$ atm) are $K_{so} = 2 \times 10^{-17}$ and 5×10^{-9}, respectively, the pH at the dolomite–calcite predominance boundary is defined by $[Ca^{2+}]/[Mg^{2+}] = 1.25$. This corresponds to a pH where $[Ca^{2+}] = 6.25 \times 10^{-4}$ M $= K_{so}/C_T\alpha_2$, that is, where $-\log \alpha_2 = 2.4$. Thus at pH values above pH $= 7.8$ dolomite should be more stable that calcite. A graphical representation is given in Figure 5-12.

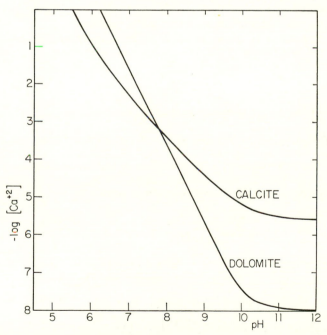

Figure 5-12 Comparison of the stability of calcite and dolomite (Example 5-7). Equilibrium solubility of calcite and dolomite for $Mg^{2+} = 5 \times 10^{-4}$ and $C_T = 2 \times 10^{-3}$ M. ($25°C$, $I = 0$). Dolomite becomes more stable than calcite at pH values above 7.8.

Sulfides

The principles considered so far for hydroxides and carbonates can be applied to salts containing other anions. Of special importance are the sulfides. Solubility of sulfides, however, is frequently complicated by the formation of thio-complexes.

Example 5-8 *Solubility of α-CdS.* 1. Estimate the solubility of α-CdS at pH = 4.5 and $p_{H_2S} = 1$ atm (25°C).

2. Construct a predominance diagram for the system $Cd^{2+}-H_2S-CO_2-H_2O$ with $p_{CO_2} = 10^{-3.5}$ atm; the following information [W. Kraft, H. Gamsjäger and E. Schwarz-Bergkampf, *Monatsh. Chem.*, **97**, 1134 (1966)] is available for $I = 1$ M NaClO₄:

$$\log {}^{c*}K_{ps0(CdS)} = -5.8; \qquad \log {}^{c*}K_{ps0(CdCO_3)} = 6.44$$
$$\log {}^{c}K_{p12(H_2S)} = -1.05\dagger; \qquad \log {}^{c}K_{1(H_2S)} = -6.90$$
$$\log {}^{c}K_{2(HS^-)} = -14.0; \qquad \log {}^{c}K_{p12(CO_2)} = -1.51\dagger$$
$$\log {}^{c}K_{1(H_2CO_3^*)} = -6.04; \qquad \log {}^{c}K_{(HCO_3^-)} = -9.57$$

It is assumed (but not confirmed experimentally) that under the conditions specified, no thio- or hydroxo-complexes are formed.

1. The solubility is given by $[Cd^{2+}]$ in equilibrium in α-CdS(s) and $p_{H_2S} = 1$.

$${}^{c*}K_{ps0} = \frac{[Cd^{2+}]p_{H_2S}}{[H^+]^2}$$

Hence at pH = 4.5, $[Cd^{2+}] = 10^{-14.8}$ M.

2. The predominance diagram is characterized mainly by the reaction

$$CdCO_3(s) + H_2S(g) = CdS(s) + H_2O(l) + CO_2(g) \qquad (i)$$

The equilibrium constant of this reaction is given by

$$\log \frac{{}^{c*}K_{ps0}(CdCO_3)}{{}^{c*}K_{ps0}(CdS)} = 12.24 = \log \frac{p_{CO_2}}{p_{H_2S}} \qquad (ii)$$

The coexistence of Cd^{2+} with CdS(s) and with CdCO₃(s), respectively, is given by the equilibria defined by the respective ${}^{c*}K_{ps0}$ values:

$$6.44 = \log [Cd^{2+}] + 2pH + \log p_{CO_2} \qquad (iii)$$
$$-5.8 = \log [Cd^{2+}] + 2pH + \log p_{H_2S} \qquad (iv)$$

† K_{p12} is Henry's law constant for the dissolution of the gas. The constants for CO₂, $H_2CO_3^*$ and HCO₃⁻ are from M. Frydman et al., *Acta Chem. Scand.*, **12**, 878 (1958).

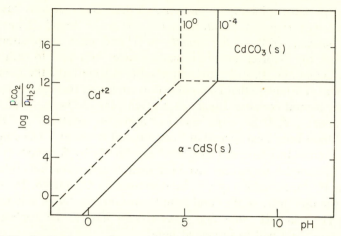

Figure 5-13 Predominance diagram in the system $Cd^{2+}-H_2S(g)-CO_2(g)-H_2O$ (see Example 5-8). $CdCO_3(s)$ becomes unstable with respect to $CdS(s)$ at extremely small partial pressures of H_2S. Based on data by Kraft, Gamsjäger and Schwarz-Bergkampf for $I = 1\ M\ (NaClO_4)$ and 25°C. A fixed p_{CO_2} of $10^{-3.5}$ atm and $[Cd^{2+}] = 10^{-4}$ and $= 10°$ are assumed.

For a fixed $\log p_{CO_2} = -3.5$, these equations become

$$\log [Cd^{2+}] = 9.94 - 2pH$$

$$\log [Cd^{2+}] = -2.3 - 2pH + \log \frac{p_{CO_2}}{p_{H_2S}} \tag{v}$$

These relations are plotted in Figure 5-13 for the two conditions $\log [Cd^{2+}] = -4$ and 0. Extremely small partial pressures of H_2S are sufficient to convert $CdCO_3$ into CdS. As pointed out by Kraft, Gamsjäger and Schwarz-Bergkampf, natural greenocite (α-CdS) is quite resistant to weathering and carbonation.

5-4   The Conditional Solubility Product

We have already illustrated that the solubility of many salts is not governed by the solubility product alone because other equilibria besides the solubility equilibrium occur in solution. The cation of the salt can react with water to produce hydroxo complexes and the anion may also undergo acid–base reactions. A related phenomenon is that the cation and anion of the salt can form complexes with each other or with other species present in solution. All these side reactions tend to remove the ions of the slightly soluble salt from the solution and, thus, to increase the solubility.

In order to make solubility predictions, all these side reactions have to be considered. The required equilibrium data may be found frequently in *Stability Constants of Metal–Ion Complexes* [1] and other compilations [2, 3, 5]. However, the involvement of so many species may lead to rather complicated calculations. In order to simplify the calculations, the concept of "conditional constants," that is, equilibrium constants that hold only under given experimental conditions (e.g., at a given pH) has been introduced by Schwarzenbach [12]. Other terms such as "apparent constant" or "effective constant" have been used instead of "conditional constant." The concept of conditional constants has been discussed extensively by Ringbom [13].

Effects of complex formation involving ligands and metal ions other than OH^- and H^+ upon solubility will be discussed in Chapter 6. In order to illustrate the usefulness of the conditional constant in treating side reactions involving H^+ and OH^- ions and to clarify its nature, the application to solubility equilibria will be discussed here.

SOLUBILITY PRODUCT VALID AT A GIVEN pH.   Consider the solubility equilibrium

$$M_nA_m(s) = mA^{-n} + nM^{+m} \tag{55}$$

with the solubility product

$$K_{s0} = [A^{-n}]^m[M^{+m}]^n \tag{56}$$

The conditional solubility product is defined by the equation

$$P_s = A_T{}^m M_T{}^n = \frac{K_{s0}}{\alpha_M{}^n \alpha_A{}^m} \tag{57}$$

where M_T is the total concentration of the metal and A_T the total concentration of the anion of slightly soluble salt, irrespective of the form in which each is present, for example (omitting charges)

$$M_T = [M] + [MOH] + [M(OH)_2] + \cdots + [M(OH)_n] \tag{58}$$

$$B_T = [A] + [HA] + [H_2A] + \cdots + [H_nA] \tag{59}$$

The α values are similar to those defined previously for acid-base equilibria

$$\alpha_M = \frac{[M]}{M_T} \tag{60}$$

$$\alpha_A = \frac{[A]}{A_T} \tag{61}$$

[12] G. Schwarzenbach, *Complexometric Titrations*, Interscience, New York, 1957.
[13] A. Ringbom, *Complexation in Analytical Chemistry*, Wiley-Interscience, New York, 1963.

They are functions of equilibruim constants, for example,

$$\alpha_M = \frac{[M]}{[M] + [MOH] + [M(OH)_2] + \cdots + [M(OH)_n]}$$

$$= \left(1 + \frac{*K_1}{[H^+]} + \frac{*K_1 *K_2}{[H^+]^2} + \cdots + \frac{*K_1 *K_2 \cdots *K_n}{[H^+]^n}\right)^{-1} \quad (62)$$

$$\alpha_A = \frac{[A]}{[A] + [HA] + [H_2A] + \cdots + [H_nA]}$$

$$= \left(1 + \frac{[H^+]}{K_n} + \frac{[H^+]^2}{K_n K_{n-1}} + \cdots + \frac{[H^+]^n}{K_n K_{n-1} \cdots K}\right)^{-1} \quad (63)$$

where $*K_1, *K_2, \ldots$ and $K_1, K_2, \ldots$ are acidity constants of the metal ion (hydrolysis constants) and acidity constants of the base, respectively. It is convenient to represent α values as a function of pH in graphs or tables. For the α values known for any pH of interest the conditional solubility product (Equation 57) is readily obtained. The conditional constant gives the relationship between the quantities that are of direct interest; in other words, a complicated solubility equilibrium may be reduced to a form analogous to that encountered when the cation and anion do not undergo any chemical side reactions (Example 5-1). The following simple examples illustrate the principle. In these examples, a conventional approach may not be any more troublesome numerically than the use of conditional constants, but with the concept of conditional constants even very complicated systems may be attacked more readily.

Example 5-9   *The Solubility of* $MgNH_4PO_4(s)$. In which pH range is the precipitation of $MgNH_4PO_4$ possible from a water containing Mg_T $(= [Mg^{2+}] + [MgOH^+]) = 10^{-2}\,M$, $N_T\,(= [NH_4^+] + [NH_3]) = 10^{-3}\,M$ and $P_T\,(= [H_3PO_4] + [H_2PO_4^-] + [HPO_4^{2-}] + [PO_4^{3-}]) = 10^{-4}\,M$? The conditional solubility product of $MgNH_4PO_4(s)$ is defined by

$$P_s = \frac{K_{s0\,MgNH4PO4}}{\alpha_{Mg}\alpha_N\alpha_S} = Mg_T \times N_T \times P_T$$

The following constants are available ($25°C$, $I = 0$)

$$\log K_{s0\,MgNH4PO4} = -12.6$$
$$\log *K_{1(\text{hydrolysis of } Mg^{+2})} = -11.4$$
$$\log K_{1(NH_4^+)} = -9.24$$
$$\log K_{1(H_3PO_4)} = -2.1$$
$$\log K_{2(H_2PO_4^-)} = -7.2$$
$$\log K_{3(HPO_4^{2-})} = -12.3$$

We first use these K values without correcting for salinity. (But for a more rigorous answer the K values must be converted into ^cK values.)

The α values are defined by

$$\alpha_{Mg} = \left(1 + \frac{*K_1}{[H^+]}\right)^{-1}$$

$$\alpha_N = \left(1 + \frac{K_{1(NH_4^+)}}{[H^+]}\right)^{-1}$$

$$\alpha_P = \left(1 + \frac{[H^+]}{K_3} + \frac{[H^+]^2}{K_2K_3} + \frac{[H^+]^3}{K_1K_2K_3}\right)^{-1}$$

The α values are plotted as a function of pH in Figure 5-14a. The conditional solubility product reaches its minimum at pH $= 10.7$ $[= \frac{1}{2}pK_{3(H_2PO_4^-)} + pK_{(NH_4^+)}]$. Thus precipitation of $MgNH_4PO_4(s)$ is favored in alkaline solutions. The conditional solubility product can now be compared with the product of the actual concentrations, $Q_{sT} = N_T \times P_T \times Mg_T = 10^{-9}$. As Figure 5-14$b$ shows, only within the pH range 9–12 is $Q_{sT} > P_s$; in principle, a precipitation is possible. However, because the difference between pQ_{sT} and pP_s is quite small (< 0.6) no efficient precipitation appears possible. Furthermore the effect of ionic strength has been ignored. Using a Güntelberg approximation and an ionic strength of $I = 0.1$, pK_{so} (hence p^cP_{sT}) becomes smaller by ca. 1.6 units; the solution actually does not become oversaturated with respect to $MgNH_4PO_4(s)$.

Example 5-10   *Solubility in the System* $Zn^{2+}-CO_2-H_2O$. Calculate conditional solubility constants for the solubility of $ZnCO_3(s)$ and hydrozincite, $Zn_5(OH)_6(CO_3)_2(s)$. The following equilibrium constants are available $(I = 0.2, 25°C)$:

$$Zn(OH)_2(amorph) + 2H^+ = Zn^{2+} + 2H_2O; \qquad \log {}^{c*}K_{so} = 12.7 \qquad (i)$$

$$ZnO(s) + 2H^+ = Zn^{2+} + H_2O; \qquad \log {}^{c*}K_{so} = 11.4 \qquad (ii)$$

$$ZnCO_3(s) + 2H^+ = Zn^{2+} + H_2CO_3^*; \qquad \log {}^{c*}K_{so} = 6.7 \qquad (iii)$$

$$Zn(OH)_{1.2}(CO_3)_{0.4}(s) + 2H^+$$
$$= Zn^{2+} + 0.4H_2CO_3^* + 1.2H_2O; \quad \log {}^{c*}K_{so} = 9.4 \qquad (iv)$$

$$Zn^{2+} + H_2O = ZnOH^+ + H^+; \qquad \log {}^{c*}K_1 = -9.4 \qquad (v)$$

$$Zn^{2+} + 3H_2O = Zn(OH)_3^- + 3H^+; \qquad \log {}^{c*}\beta_3 = -28.2 \qquad (vi)$$

$$Zn^{2+} + 4H_2O = Zn(OH)_4^{2-} + 4H^+; \qquad \log {}^{c*}\beta_4 = -40.5 \qquad (vii)$$

We may define the following α values:

$$\alpha_{Zn} = \frac{[Zn^{2+}]}{Zn_T} = \left(1 + \frac{{}^{c*}K_1}{[H^+]} + \frac{{}^{c*}\beta_3}{[H^+]^3} + \frac{{}^{c*}\beta_4}{[H^+]^4}\right)^{-1} \qquad (viii)$$

$$\alpha_0 = \frac{[H_2CO_3^*]}{C_T} = \left(1 + \frac{K_1}{[H^+]} + \frac{K_1K_2}{[H^+]^2}\right)^{-1} \tag{ix}$$

Thus the conditional solubility products for (iii), $ZnCO_3(ZC)$, and (iv), hydrozincite (HZ), are

$$P_{s(ZC)} = Zn_T \times C_T = \frac{c^*K_{s0_{ZC}}[H^+]^2}{\alpha_{Zn}\alpha_0} \tag{x}$$

or

$$pP_{s(ZC)} = -6.7 + 2pH + \log \alpha_{Zn} + \log \alpha_0 \tag{xi}$$

$$P_{s(HZ)} = Zn_T \times C_T^{0.4} = \frac{c^*K_{s0_{HZ}}[H^+]^2}{\alpha_{Zn}\alpha_0^{0.4}} \tag{xii}$$

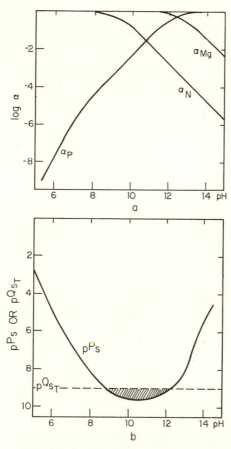

Figure 5-14   Solubility of $MgNH_4PO_4$ (Example 5-9). The conditional solubility product $P_s = P_T \times N_T \times Mg_T$ (b) is readily calculated from the α values (a). Minimum solubility at pH $\simeq$ 10.7.

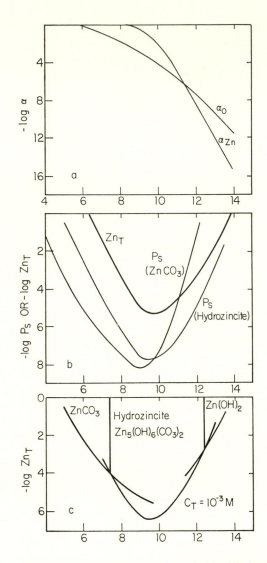

Figure 5-15   Conditional solubility products of $ZnCO_3(s)$ and hydrozincite $(Zn_4(OH)_6 \cdot (CO_3)_2(s))$.

  a: α_{Zn} and α_0 versus pH.
  b: The conditional solubility products and largest possible Zn_T.
  c: Maximum soluble Zn_T in closed system ($C_T = 10^{-3}\ M$).

or

$$pP_{s(\text{HZ})} = -9.4 + 2pH + \log \alpha_{\text{Zn}} + 0.4 \log \alpha_0 \qquad \text{(xiii)}$$

$\log \alpha$ and pP values are plotted in Figure 5-15.

There is an additional constraint: Because the solubility of $Zn(OH)_2(s)$ (or the more stable $ZnO(s)$) must not be exceeded, Zn_T in (viii), (x) and (xii) cannot assume any arbitrary value:

$$Zn_T \leq \frac{{}^*K_{s0\,\text{Zn(OH)}_2}[H^+]^2}{\alpha_{\text{Zn}}} \qquad \text{(xiv)}$$

Equation xiv derives from (i) and (viii). Figure 5-15b gives a plot of the largest possible concentration of Zn_T as a function of pH.

The use of the conditional constants may be illustrated by computing with the help of (x), (xii) and (xiv) the maximum total zinc concentration that can be maintained in a carbonate-bearing water ($C_T = 10^{-3} M$) without becoming oversaturated with respect to zinc carbonate, hydrozincite or amorphous $Zn(OH)_2$. The result is given in Figure 5-15c. The conclusions are similar to those given earlier. The stability boundary between $Zn(OH)_2$ and hydrozincite depends on which polymorphous form of $Zn(OH)_2(s)$ is assumed. Considering a more stable (and less soluble) form such as $ZnO(s)$ will shift the boundary to a less alkaline value.

5-5   The Activity of the Solid Phase

If a solid and a solution are in equilibrium, all components occurring in both phases have, according to the Gibbs equilibrium condition, the same chemical potential in both phases. The solubility of a salt depends on the activity of the solid phase. There are various factors that affect the activity of the solid phase: (a) the lattice energy, (b) the degree of hydration, (c) solid solution formation and (d) the free energy of the surface.

In defining a solid–solution equilibrium, the pure solid phase is taken as a reference state and its activity is, because of its constancy, set equal to unity. This means that the activity of the solid phase is implicitly contained in the solubility equilibrium constant, and this can thus be used to deduce information about the constitution of the solid phase.

Polymorphism

A chemical substance can exist in more than one crystalline form. These polymorphic forms are also called allotropic modifications. Except at a transition temperature only one form of crystal is thermodynamically stable for a given chemical composition of a system. The metastable modifications are more soluble than the stable one; for example, the solubilities of quartz and of amorphous silica in H_2O (25°C) are approximately 2×10^{-4} M and 2×10^{-3} M, respectively, as orthosilicic acid. Obviously quartz is the stable solid phase; its free energy of formation is ca. 1.36 kcal smaller (more negative) than that of amorphous SiO_2

$$SiO_2(\text{amorph}) = SiO_2(\text{quartz}); \quad \log K = 1.0; \quad \Delta G° = -1.36 \text{ kcal} \quad (64)$$

Which of these solid phases shall we consider in making "equilibrium" calculations? At low temperature and pressure amorphous silica precipitates from oversaturated silica solutions; the rate of quartz formation is extremely slow and cannot be observed under these conditions. Obviously it would be wrong to assume that only the thermodynamically stable modification prevails as the solid phase in the system under consideration. It is necessary to adduce some information on the kinetics of the solid–solution interactions and to have, if possible, some information on the type of solid phases that are in contact with the actual solution or with the natural water in order to postulate which solid phase may persist in metastable existence for the periods of time under consideration. The existence of a metastable modification for geological time spans is not uncommon.

There are at least five polymorphic forms of $CaCO_3$ which occur in nature. Calcite and aragonite are especially important. At $P = 1$ atm and 25°C, calcite is less soluble than aragonite and hence is the stable form thermodynamically. Nevertheless, recent carbonate sediments are composed to a large extent of aragonite (Bricker and Garrels [9]). Aragonite is being deposited in vast amounts under conditions of temperature and pressure where calcite is the thermodynamically stable phase (Cloud [14]). Furthermore, aragonite is the dominant form of $CaCO_3$ deposited by organisms. Organisms can also form calcite in which part of the Ca^{2+} is substituted by Mg^{2+} (magnesian calcite).

The solid phase that precipitates first will be governed by kinetic factors. Trace elements can influence the kinetics significantly. Aragonite has usually a higher Sr^{2+} content than calcite; the inference has been made that the presence of Sr^{2+} in the water favors the precipitation of aragonite and retards or "prevents" the conversion into calcite.

With respect to $CaCO_3$, seawater appears to behave quite differently from fresh water, and Mg^{2+} has been suggested as responsible for the anomalous behavior of $CaCO_3$ in seawater (Pytkowicz [15]). Mg^{2+} apparently is incorporated into the surface layer of the mineral, causing the apparent (and metastable) solubility to increase.

[14] P. E. Cloud, "Carbonate Deposition West of Andros Island, Bahamas," *U.S. Geol. Surv. Profess. Papers*, **350** (1962).
[15] R. M. Pytkowicz, *J. Geol.*, **73**, 196 (1965).

Organisms always have the capability of carrying out endergonic reactions either coupled with exergonic ones or driven by the energy of photosynthesis. Thus organisms can excrete or produce solid phases from solutions undersaturated with respect to these solid phases.

VARIOUS HYDRATES OF A SOLID.   In a similar way as different allotropic modifications have different activities, the various hydrates of a solid have different solubilities, and the one with the lowest solubility at a given temperature, pressure and water activity (e.g., vapor pressure) is the most stable one. The degree of hydration can have a pronounced effect on the rate of precipitation or on the rate of dissolution. Thus, it is not necessarily the most stable hydrate of a solid which is kinetically most persistent within a certain observation period.

Solid Solution Formation

The solids occurring in nature are seldom pure solid phases. Isomorphous replacement by a foreign constituent in the crystalline lattice is an important factor by which the activity of the solid phase may be decreased. If the solids are homogeneous, that is, contain no concentration gradient, one speaks of homogeneous solid solutions. The thermodynamics of solid solution formation has been discussed by Vaslow and Boyd [16] for solid solutions formed by $AgCl(s)$ and $AgBr(s)$.

To express theoretically the relationship involved we consider a heterogeneous system where $AgBr(s)$ as solute becomes dissolved in solid $AgCl$ as solvent. This corresponds to the reaction that takes place if $AgCl(s)$ is shaken with a solution containing Br^-. The reaction might formally be characterized by the equilibrium

$$AgCl(s) + Br^- = AgBr(s) + Cl^- \tag{65}$$

The equilibrium constant for (65), that is, the distribution constant D, corresponds to the quotient of the solubility product constants of $AgCl(s)$ and $AgBr(s)$

$$\frac{\{Ag^+\}\{Cl^-\}}{\{AgCl(s)\}} = K_{s0_{AgCl}} \tag{66}$$

$$\frac{\{Ag^+\}\{Br^-\}}{\{AgBr(s)\}} = K_{s0_{AgBr}} \tag{67}$$

$$\frac{\{Cl^-\}\{AgBr(s)\}}{\{Br^-\}\{AgCl(s)\}} = \frac{K_{s0_{AgCl}}}{K_{s0_{AgBr}}} = D \tag{68}$$

[16] F. Vaslow and G. E. Boyd, *J. Amer. Chem. Soc.*, **74**, 4691 (1952).

The activity ratio of the solids may be replaced by the ratio of the mole fractions ($X_{AgCl} = n_{AgCl}/n_{AgCl} + n_{AgBr}$) multiplied by activity coefficients:

$$\frac{\{Cl^-\}}{\{Br^-\}} \frac{X_{AgBr}.f_{AgBr}}{X_{AgCl} f_{AgCl}} = \frac{K_{s0_{AgBr}}}{K_{s0_{AgCl}}}$$

or   (69)

$$\frac{X_{AgBr}}{X_{AgCl}} = \frac{K_{s0_{AgCl}}}{K_{s0_{AgBr}}} \frac{\{Br^-\} f_{AgCl}}{\{Cl^-\} f_{AgBr}}$$

According to this equation the extent of dissolution of Br^- in solid AgCl (X_{AgBr}/X_{AgCl}) is a function of (a) the solubility product ratio of AgCl to AgBr; (b) the solution composition, that is, the activity ratio of Br^- to Cl^-; and (c) a solid solution factor, given by the ratio of the activity coefficients of the solid solution components (f_{AgCl}/f_{AgBr}).

As a first approximation, we may assume that f_{AgCl}/f_{AgBr} is equal to unity (ideal solid solution) and that the activity ratio of the species in the fluid may be replaced by the concentration ratio

$$\frac{[Cl^-]}{[Br^-]} \frac{X_{AgBr}}{X_{AgCl}} \approx \frac{K_{s0_{AgCl}}}{K_{s0_{AgBr}}} = D \qquad (70)$$

The qualitative significance of solid solution formation can be demonstrated with the help of this simplified equation, using the following numerical example.

Consider a solid solution of 10% AgBr in AgCl (90%) which is in equilibrium with Cl^- and Br^-; the composition of the suspension is:

Aqueous Phase	Solid Phase
$[Cl^-] = 10^{-3.0}\ M$	$X_{AgCl} = 0.9$
$[Br^-] = 10^{-6.55}\ M$	$X_{AgBr} = 0.1$
$[Ag^+] = 10^{-6.8}\ M$	

In accordance with (70), $D = 400$ ($\simeq$ quotient of the K_{s0} values at 25°C; $pK_{s0_{AgCl}} = 9.7$ and $pK_{s0_{AgBr}} = 12.3$).

The composition of the equilibrium mixture shows that Br^- has been enriched significantly in the solid phase in comparison to the liquid phase ($D > 1$). If one considered the concentrations of aqueous $[Br^-]$ and $[Ag^+]$, one would infer, by neglecting to consider the presence of a solid solution phase, that the solution is undersaturated with respect to AgBr ($[Ag^+][Br^-]/K_{s0_{AgBr}} = 0.1$). Because the aqueous solution is in equilibrium with a solid solution, however, the aqueous solution is saturated with Br^-. Although the solubility of the salt that represents the major component of the solid phase is only slightly affected by the formation of solid solutions, the solubility of the minor component is appreciably reduced. The observed occurrence of certain metal ions in sediments formed from solutions that appear to be formally (in the absence of any consideration of solid solution formation) unsaturated with respect to the impurity can, in many cases, be explained by solid solution formation.

Usually, however, the distribution coefficients determined experimentally are not equal to the ratios of the solubility product because the ratio of the activity coefficients of the constituents in the solid phase must not be assumed to be equal to 1. A comparison of solubility product quotients and actually observed D values, given in Table 5-4, illustrates that activity coefficients in the solid phase may differ markedly from 1. Let us consider, for example, the coprecipitation of $MnCO_3$ in calcite. Assuming that the ratio of the activity coefficients in the aqueous solution is close to unity, the equilibrium distribution may be formulated as [cf. (68)]

$$\frac{K_{s0_{CaCO_3}}}{K_{s0_{MnCO_3}}} = D = \frac{[Ca^{2+}]}{[Mn^{2+}]} \frac{X_{MnCO_3} f_{MnCO_3}}{X_{CaCO_3} f_{CaCO_3}} \tag{71}$$

$$D = D_{obs} \frac{f_{MnCO_3}}{f_{CaCO_3}} \tag{72}$$

The solubility product quotient at 25°C ($pK_{s0_{MnCO_3}} = 11.09$, $pK_{s0_{CaCO_3}} = 8.37$) can now be compared with an experimental value of D. The data of Bodine, Holland and Borcsik [17] (see also Figure 5-16) give $D_{obs} = 17.4$ (25°C).

Table 5-4 Solid Solution Formation

Solid Solution	T, °C	D_{obs}†	D‡	Ref.
AgBr in AgCl	30	211.4	315.7	[a]
$Ra(IO_3)_2$ in $Ba(IO_3)_2$	25	1.42	1.32	[b]
$RaSO_4$ in $BaSO_4$	20	1.8	5.9	[c]
$MnCO_3$ in calcite	250	17.4	525	[d]
$SrCO_3$ in calcite	250	0.13	5	[d]

† For solid solution of B in A, [A(s) + B(aq) = B(s) + A(aq)]:

$$D_{obs} = \frac{[A(aq)] X_B}{[B(aq)] X_A}$$

‡ D = quotient of solubility products, $D = K_{s0(A)}/K_{s0(B)}$.

[a] F. Vaslow and G. E. Boyd, *J. Amer. Chem. Soc.*, **75**, 4691 (1952).
[b] A. Polessitsky and A. Karataewa, *Acta Physicochim.* (*U.R.S.S.*), **8**, 259 (1938) (quoted as cited in ref. a).
[c] O. Hahn, *Applied Radiochemistry*, Cornell University Press, Ithaca, N.Y., 1933, p. 88.
[d] M. W. Bodine, H. D. Holland and M. Borcsik, *Symposium on Problems of Post-magnetic Ore Deposition*, Vol. II, Prague, 1965, p. 401.

[17] M. W. Bodine, H. D. Holland and M. Borcsik, *Symposium on Problems of Post-magnative Ore Deposition*, Vol. II, Prague, 1965, p. 407.

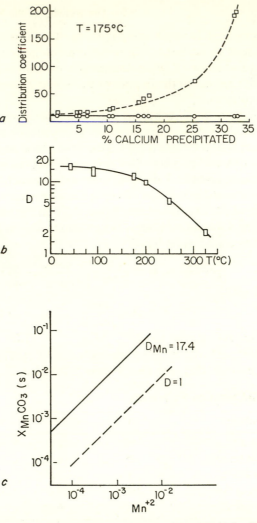

Figure 5-16 Solid solution formation; dissolution of $MnCO_3(s)$ in $CaCO_3(s)$ (calcite). a and b from Bodine, Holland, and Borcsik [17]. In a the distribution coefficient is plotted against per cent calcium precipitated. D has been calculated for homogeneous solid solution formation (squares)

$$D = \left(\frac{m_{Mn^{2+}}}{m_{Ca^{2+}}}\right)_{solid} \times \frac{[Ca^{2+}]}{[Mn^{2+}]}$$

and for heterogeneous solid solution formation (circles)

$$D = \frac{\log \dfrac{[Mn^{2+}]_0}{[Mn^{2+}]_f}}{\log \dfrac{[Ca^{2+}]_0}{[Ca^{2+}]_f}}$$

In b the distribution coefficient is plotted as a function of temperature. c schematically shows the enrichment of the calcite in manganese relative to the composition in the aqueous solution at 25°C.

Because D is smaller than K_0, the solid solution factor acts to lower the solution of $MnCO_3$ in $CaCO_3$ significantly from that expected if an ideal mixture had been formed. Assuming that in dilute solid solutions (X_{MnCO_3} very small) the activity coefficient of the solvent is close to unity ($X_{CaCO_3}f_{CaCO_3} \simeq 1$), an activity coefficient for the solute is calculated to be $f_{MnCO_3} = 31$. Qualitatively, such a high activity coefficient reflects a condition similar to that of a gas dissolved in a concentrated electrolyte solution where the gas, also characterized by an activity coefficient larger than unity, is "salted out" from the solution. Thermodynamically, the activity coefficient of a solid solute is related to the decrease in partial molar free energy resulting from the transfer of one mole of solute from a large quantity of an ideal solid solution to the real solid solution of the same mole fraction (Vaslow and Boyd [16]). This excess free energy of solid solution will be small if an ion of similar size and bond type, forming crystals with similar lattice parameters, becomes isomorphously substituted.

Example 5-11   *Solid–Solution of* Sr^{2+} *in Calcite.*   It has been proposed (e.g., Schindler [18]) that a solid solution of Sr^{2+} in calcite might control the solubility of Sr^{2+} in the ocean. Estimate the composition of the solid solution phase (X_{SrCO_3}). The following information is available: the solubilities of $CaCO_3(s)$ calcite and $SrCO_3(s)$ in seawater at 25°C are characterized by $p^cK_{s0} = 6.1$ and 6.8, respectively. The equilibrium concentration of CO_3^{2-} is $[CO_3^{2-}] = 10^{-3.6}\ M$. The actual concentration of Sr^{2+} in seawater is $[Sr^{2+}] \simeq 10^{-4}\ M$. According to the value quoted in Table 5-4,

$$([Ca^{2+}]/[Sr^{2+}])(X_{SrCO_3}/X_{CaCO_3}) = 0.13\ (25°C)$$

Assuming a saturation equilibrium of seawater with Sr^{2+}–calcite, the equilibrium concentrations would be $[Ca^{2+}] \simeq 10^{-2.5}\ (= {}^cK_{s0}/[CO_3^{2-}])$ and $[Sr^{2+}] = 10^{-4}\ M$ (= actual concentration). Thus $X_{SrCO_3}/X_{CaCO_3} = 0.004$ and $X_{CaCO_3} = 0.996$.

It may be noted that, since the distribution coefficient is smaller than unity, the solid phase becomes depleted in strontium relative to the concentration in the aqueous solution. The small value of D may be interpreted in terms of a high activity coefficient of strontium in the solid phase $f_{SrCO_3} \approx 38$. If the strontium were in equilibrium with strontianite, $[Sr^{2+}] \simeq 10^{-3.2}\ M$, that is, its concentration would be more than six times larger than at saturation with $Ca_{0.996}Sr_{0.004}CO_3(s)$. This is an illustration of the consequence of solid solution formation where with $X_{CaCO_3}f_{CaCO_3} \simeq 1$:

$$[Sr^{2+}] = \frac{X_{SrCO_3}f_{SrCO_3}{}^cK_{s0SrCO_3}}{[CO_3^{2-}]}$$

[18] P. Schindler, in *Equilibrium Concepts in Natural Waters*, Advances in Chemistry Series, No. 67, American Chemical Society, Washington, D.C., 1967, p. 196.

that is, the solubility of a constituent is greatly reduced when it becomes a minor constituent of a solid solution phase.

Besides homogeneous solid solutions, *heterogeneous arrangement* of foreign ions within the lattice is possible. While homogeneous solid solutions represent a state of true thermodynamic equilibrium, heterogeneous solid solutions can persist in metastable equilibrium with the aqueous solution. Heterogeneous solid solutions may form in such a way that each crystal layer as it forms is in distribution equilibrium with the particular concentration of the aqueous solution existing at that time (Doerner and Hoskins [19]; Gordon, Salutsky and Willard [20]). Correspondingly, there will be a concentration gradient in the solid phase from the center to the periphery. Such a gradient results from very slow diffusion within the solid phase. Following the treatment given by Doerner and Hoskins and by Gordon, Salutsky and Willard, the distribution equilibrium for the reaction

$$CaCO_3(s) + Mn^{2+} = MnCO_3(s) + Ca^{2+} \qquad (73)$$

is written as in (68), but we consider that the crystal surface is in equilibrium with the solution:

$$\left(\frac{[MnCO_3]}{[CaCO_3]}\right)_{\text{crystal surface}} \times \frac{[Ca^{2+}]}{[Mn^{2+}]} = D' \qquad (74)$$

If $d[MnCO_3]$ and $d[CaCO_3]$, the increments of $MnCO_3$ and $CaCO_3$ deposited in the crystal surface layer, are proportional to their respective solution concentrations, (75) obtains

$$\frac{d[MnCO_3]}{d[CaCO_3]} = D' \frac{[Mn^{2+}]_0 - [Mn^{2+}]}{[Ca^{2+}]_0 - [Ca^{2+}]} \qquad (75)$$

or after rearrangement

$$\frac{d[MnCO_3]}{[Mn^{2+}]_0 - [Mn^{2+}]} = D' \frac{d[CaCO_3]}{[Ca^{2+}]_0 - [Ca^{2+}]} \qquad (76)$$

where $[Ca^{2+}]_0$ and $[Mn^{2+}]_0$ represent initial concentrations in the aqueous solution. Integration of (76) leads to

$$\log \frac{[Mn^{2+}]_0}{[Mn^{2+}]_f} = D' \log \frac{[Ca^{2+}]_0}{[Ca^{2+}]_f} \qquad (77)$$

where $[Mn^{2+}]_f$ and $[Ca^{2+}]_f$ represent final concentrations in the aqueous solutions. Figure 5-16a illustrates that experimentally determined distribution

[19] H. A. Doerner and W. M. Hoskins, *J. Amer. Chem. Soc.*, **47**, 662 (1925).
[20] L. Gordon, M. L. Salutsky and H. H. Willard, *Precipitation from Homogeneous Solution*, Wiley, New York, 1959.

coefficients for the coprecipitation of Mn(II) in calcite give constant values for heterogeneous solid-solution formation only.

Most of the distribution coefficients measured to date on a variety of relatively insoluble solids are characterized by the Doerner-Hoskins relation. This relationship is usually obeyed with crystals that have been precipitated from homogeneous solution or under conditions similar to those encountered in precipitation from homogeneous solution (Gordon, Salutsky and Willard [20]). If the precipitation occurs in such a way that the aqueous phase remains as homogeneous as possible and the precipitant ion is generated gradually throughout the solution, large and well-formed crystals likened to the structure of an onion [21] are obtained. Each infinitesimal crystal layer is equivalent to a shell of an onion. As each layer is deposited, there is insufficient time for reaction between solution and crystal surface before the solid becomes coated with succeeding layers. Kinetic factors make the metastable persistence of such compounds possible for relatively long — often for geological time spans.

MAGNESIAN CALCITE. Of special significance is the substitution of magnesium for calcium in the calcite structure. Although no complete solid solution series is possible at low temperature, analyses of natural calcites known to have been formed at low temperatures show $MgCO_3$ contents as high as 18 mole % (Bricker and Garrels [9]). Many marine organisms secrete hard parts of magnesian calcite.

As pointed out by Bricker and Garrels, magnesian calcites are metastable under earth surface conditions and lose Mg^{2+} within geologically short time spans. Calcites from mesozoic and older rocks are very low in Mg^{2+} except under unusual circumstances (Goldsmith et al. [22]). Partial substitution of Mg^{2+} in calcite enhances the solubility. Although no true metastable equilibrium solubility can be established, Chave et al. [23] have measured the ultimate pH of suspensions of magnesian calcites with various Mg^{2+} content exposed to a CO_2 pressure of 1 atm. The solubility of calcite is compared with that of aragonite and magnesian calcite in Figure 5-17, which has been calculated by Bricker and Garrels [9] from the data of Chave et al. [23]. As the magnesian calcite is dissolved, Mg^{2+} becomes enriched in the solution and a purer calcite is precipitated $[(Ca, Mg)CO_3(s) = CaCO_3(s) + Mg^{2+}]$ (incongruent dissolution). Because the apparent activity coefficient of $MgCO_3$ in calcite (f_{MnCO_3}) is smaller than unity, the aqueous solution becomes enriched in Mg^{2+} during the dissolution. As Figure 5-17 suggests, solutions having contact with magnesian calcite may have Mg^{2+}/Ca^{2+} ratios higher

[21] O. Hahn, *Applied Radiochemistry*, Cornell Univ. Press, Ithaca, N.Y., 1936.
[22] J. R. Goldsmith et al., *Geochim. Cosmochim. Acta*, **7**, 212 (1955).
[23] K. E. Chave et al., *Science*, **137**, 33 (1962).

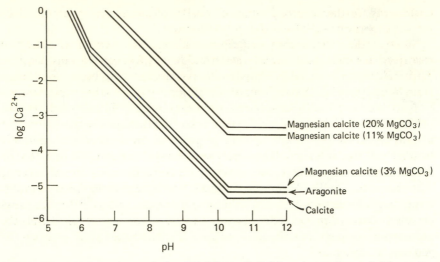

Figure 5-17 Comparison of the Solubility of Calcite, Aragonite and Magnesium Calcite (from Bricker and Garrels, [9]. Solubility of Ca^{2+} as a function of pH for $C_T = 10^{-3}\ M$ (25°C). The solubility data have been computed from results obtained by Chave et al. [23]. The data do not represent a true (reversible) equilibrium solubility. Reproduced with permission from J. Wiley and Sons.

than that of the solid phase and may contain $[Ca^{2+}]$ greater than predicted from the solubility of pure calcite.

OCCLUSION AND SURFACE ADSORPTION. In occlusion and surface adsorption the impurities become trapped in the solids or adsorbed on the surface under *nonequilibrium conditions*. The extent of occlusion depends upon the rate of precipitation. In adsorption the microcomponent is adsorbed or undergoes chemical reaction (surface complex formation) on the solid. Occasionally it is possible to distinguish between solid solution formation and incorporation of foreign ions by occlusion or adsorption by the use of X-ray diffraction patterns. Although the method is not very sensitive, solid compound formation will cause a change in the lattice dimension; the latter is proportional to the concentration of the solute. No observable changes in the X-ray diffraction pattern result from surface adsorption.†

† More recently H. D. Winland [J. Sediment. Petrol., *39*, 1529 (1969)] has given a different interpretation based on the stability of $CaCO_3$ polymorphs. Because of the impurity of polymorphs that exist in equilibrium with seawater, aragonite, as defined by partition coefficients, has the lower free energy of formation and should be considered the thermodynamically stable form of $CaCO_3$ in shallow marine environments.

Solubility of Fine Particles

Finely divided solids have a greater solubility than large crystals. As a consequence small crystals are thermodynamically less stable and should recrystallize into large ones. For particles smaller than about 1 μ or of specific surface area greater than a few square meters per gram, surface energy may become sufficiently large to influence surface properties. Similarly the free energy of a solid may be influenced by lattice defects such as dislocations and other surface heterogeneities.

The change in the free energy ΔG involved in subdividing a coarse solid suspended in aqueous solution into a finely divided one of molar surface S is given by

$$\Delta G = \tfrac{2}{3}\bar{\gamma}S \tag{78}$$

where $\bar{\gamma}$ is the mean free surface energy (interfacial tension) of the solid–liquid interface.

Equation 78 can be derived [24] from the thermodynamic statement that at constant temperature and pressure and assuming only one "mean" type of surface

$$dG = \mu_0 dn + \bar{\gamma} ds \tag{79}$$

or

$$\mu = \mu_0 + \frac{\bar{\gamma} ds}{dn} \tag{80}$$

This can be rewritten as

$$\mu = \mu_0 + \frac{M}{\rho} \bar{\gamma} \frac{ds}{dv} \tag{81}$$

where M = formula weight, ρ = density, n = number of moles and s = surface area of a single particle.

Because the surface and the volume of a single particle of given shape are $s = kd^2$ and $v = ld^3$,

$$\frac{ds}{dv} = \frac{2s}{3v} \tag{82}$$

Since the molar surface is $S = Ns$, and the molar volume $V = Nv = M/\rho$, where N is particles per mole, (81) can be rewritten as

$$\mu = \mu_0 + \tfrac{2}{3}\bar{\gamma}S \tag{83}$$

[24] B. V. Enüstün and J. Turkevich, *J. Amer. Chem. Soc.*, **82**, 4502 (1960).

or in terms of Equation 78

$$\frac{d \ln K_{s0}}{dS} = \frac{2}{3} \frac{\bar{\gamma}}{RT} \qquad (84)$$

or

$$\log K_{s0_{(S)}} = \log K_{s0_{(S=0)}} + \frac{\frac{2}{3}\bar{\gamma}}{2.3RT} S \qquad (85)$$

The specific surface effect can also be expressed by substituting $S = M\alpha/\rho d$ where α is a geometric factor which depends on the shape of the crystals ($\alpha = k/l$).

Recently Schindler [25] and Schindler et al. [26] have investigated the effect of particle size and molar surface on the solubilities of ZnO, $Cu(OH)_2$ and CuO. An example of their data is given in Figure 5-18 where the solubility constant of ZnO is plotted, in accordance with (85), semilogarithmically as a function of the molar surface. The mean surface tension $\bar{\gamma}$ can be computed from the slope ($\bar{\gamma}$ for interface CuO–0.2 M $NaClO_4 = 770 \pm 300$ erg cm^{-2}). Table 5-5 compares $\bar{\gamma}$ for a few solid–liquid interfaces.

METASTABILITY AND PARTICLE SIZE. The particle size may play a significant role in the inversion from one polymorphous form to another. Following

Table 5-5   Surface Tension of Solid–Liquid Interfaces

Interface	$\bar{\gamma}$, ergs cm^{-2}, 25°C	Ref.
SiO_2(amorphous)–H_2O	46	[a]
$BaSO_4$–H_2O	123	[b]
$SrSO_4$–H_2O	86	[b]
$PbCO_2$–H_2O	112	[b]
$SrCO_3$–H_2O	92	[b]
$(NH_2)CH_2COOH$(glycine)–H_2O	29	[c]
ZnO–0.2 M $NaClO_4$	770	[d]
CuO–0.2 M $NaClO_4$	690	[d]
$Cu(OH)_2$–0.2 M $NaClO_4$	410	[d]

[a] G. B. Alexander, *J. Phys. Chem.*, **61**, 1563 (1957).
[b] A. G. Walton, *Microchim. Acta*, **3**, 422 (1963).
[c] H. H. Lo, M.S. thesis, Case Institute of Technology, 1965, quoted from A. G. Walton, *Science*, **148**, 601 (1965).
[d] P. Schindler, H. Althaus, F. Hofer and W. Minder, *Helv. Chim. Acta*, **48**, 1204 (1965).

[25] P. Schindler, in *Equilibrium Concepts in Natural Water Systems*, Advances in Chemistry Series, No. 67, American Chemical Society, Washington, D.C., 1967, p. 196.
[26] P. Schindler, H. Althaus, F. Hofer and W. Minder, *Helv. Chim. Acta*, **48**, 1204 (1965).

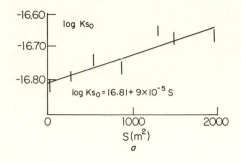

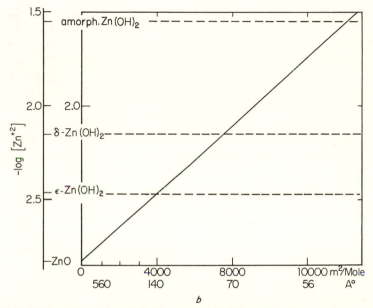

Figure 5-18   Effect of particle subdivision upon solubility of ZnO.

a: Effect of molar surface upon solubility product of ZnO (25°C, $I = 0$). (From Schindler, 1967.)

b: Comparison of the solubility (pH = 7) of ZnO, as it is influenced by molar surface or particle diameter, with the solubility of polymorphous modifications of $Zn(OH)_2$ (see scale on the left). (From data by Schindler et al., 1965.) ZnO is the most stable of the modifications, but if its molar surface area is larger than 4000 m² ($d < 140$ Å) it becomes less stable than (coarse) ϵ-$Zn(OH)_2$.

Reproduced with permission from American Chemical Society.

arguments presented by Schindler [25] and Schindler et al. [26] we may consider the reaction

$$Cu(OH)_2(s) = CuO(s) + H_2O(l) \qquad (86)$$

Schindler et al. determined the solubility constants and the influence of molar surface upon solubility:

$Cu(OH)_2$: $\log {}^*K_{so} = 8.92 + 4.8 \times 10^{-5}S$; $\bar{\gamma} = 410 \pm 130 \text{ erg cm}^{-2}$

CuO: $\log {}^*K_{so} = 7.89 + 8 \times 10^{-5}S$; $\bar{\gamma} = 690 \pm 150 \text{ erg cm}^{-2}$

Thus the solubility of $Cu(OH)_2$ is approximately 10 times larger than that of CuO. The inversion of $Cu(OH)_2$ into CuO (reaction 86) should occur exergonically. However, if CuO is very finely divided ($S_{CuO} > 12,000 \text{ m}^2$) it becomes less stable than coarse $Cu(OH)_2$ ($S_{Cu(OH)_2} = 0$):

$$\Delta G(\text{cal}) = -1400 + [0.109S_{CuO} - 0.065S_{Cu(OH)_2}] \qquad (87)$$

Figure 5-19 plots the enhancement of the solubility of CuO and $Cu(OH)_2$ (pH = 7.0) with an increase in molar surface. We may recall that the solid phase which gives the smaller solubility is the thermodynamically more stable phase. As Figure 5-19 illustrates, at sufficiently large subdivision of the solids ($S_{CuO} = S_{Cu(OH)_2} > 31,000 \text{ m}^2$; $d \simeq 40 \text{ Å}$) $Cu(OH)_2$(s) may become more stable (less soluble) than CuO(s); this results from the larger interfacial energy of the oxide as compared to the hydroxide. As pointed out by Schindler et al. [26], this dependence of stability (solubility) upon particle size may be one of the reasons why crystal nuclei of $Cu(OH)_2$ are formed incipiently when Cu(II) is precipitated; upon subsequent crystal growth, and corresponding decrease

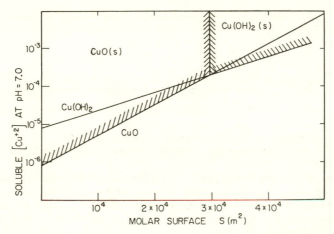

Figure 5-19 Influence of molar surface upon solubility of CuO and of $Cu(OH)_2$ at pH = 7.0. From data on solubility constants and surface tensions by Schindler et al., 1965. (The relations depicted have been validated experimentally only for $S < 10^4 \text{ m}^2$.) The figure suggests that $Cu(OH)_2$(s) becomes more stable than CuO(s) for very finely divided CuO crystals ($S > 3 \times 10^4 \text{ m}^2$, $d < 40 \text{ Å}$). Plausibly, in precipitating Cu(II), $Cu(OH)_2$(s) may first be precipitated (d = very small), but CuO(s) becomes more stable than $Cu(OH)_2$ upon growth of the crystals, and an inversion of $Cu(OH)_2$ into the more stable phase becomes possible.

in S, CuO becomes more stable and $Cu(OH)_2$ is inverted into $CuO(s)$. These arguments are schematic and simplified. The assumption that $\bar{\gamma}$ remains independent of surface area probably breaks down when we deal with nuclei consisting of a very small number of ions or molecules. Furthermore, the relations given are valid only for particles of uniform size.

5-6 The Effect of Inert Electrolyte Upon Solubility; Nonideality Corrections

If we consider a solid–solution equilibrium, any decrease in the activity of the solute species is accompanied by an increase in solubility. Hence, in dilute solutions ($I < 1$) the solubility of a salt is enhanced in the presence of inert electrolytes. In more concentrated solutions activity coefficients may become larger than 1 and in very concentrated electrolytes ($I > 1$) the solubility of a salt may become smaller again.

In this section we discuss briefly how to apply nonideality corrections using the Debye-Hückel equation and how to convert activities from the infinite dilution activity scale to the constant ionic medium activity scale without making nonthermodynamic assumptions. But we should keep in mind that the nonideality effects are frequently small in comparison to the other uncertainties involved in the calculations of solid-solution equilibrium. Frequently the solubility constants are not known with sufficient accuracy and side reactions are disregarded or unknown. If values for the same constant reported by different authors vary by a few orders of magnitude and we have no basis to select critically the "best" value, it makes little sense to apply salinity corrections. In dealing with solid–solution equilibria there are many causes in addition to effects by inert electrolytes which may limit the accuracy and the applicability of equilibrium calculations. Despite such limitations, and even if we ignore corrections for nonideality, equilibrium calculations may be useful as an aid to appreciate the effect of pertinent variables and may help in the interpretation of observed facts.

Solubility and Debye-Hückel Theory

The solubility product of equilibrium $M_nN_m(s) = nM + mN$ can be formulated

$$K_{s0} = [M]^n[N]^m f_M^n f_N^m \tag{88}$$

By using the extended Debye-Hückel law, (88) can be rearranged to

$$\log K_{s0} = \log {}^cK_{s0} - (nz_M^2 + mz_N^2)\left(\frac{A\sqrt{I}}{1 + Ba\sqrt{I}}\right) \tag{89}$$

In dealing with media containing mixed electrolytes it may be appropriate to set A = 0.5 and Ba = 1 in (89); then

$$pK_{s0} = p^c K_{s0} + (nz_M^2 + mz_N^2)\left(\frac{0.5\sqrt{I}}{1 + \sqrt{I}}\right) \qquad (90)$$

Instead of (90), we may use the Davies equation:

$$pK_{s0} = p^c K_{s0} + (nz_M^2 + mz_N^2)\left(\frac{0.5\sqrt{I}}{1 + \sqrt{I}} - 0.2I\right) \qquad (91)$$

Both (90) and (91) illustrate that the solubility increases with increasing I and that the effect is especially pronounced for salts containing multivalent ions [e.g., the correction factor is nine times larger for $FePO_4(s)$ than for $AgCl(s)$].

ELUCIDATION OF STANDARD FREE ENERGY OF FORMATION FROM MEASUREMENTS IN AN IONIC MEDIUM. The experimental evaluation of a solubility constant obviously has to be carried out in solutions of finite concentration. Most solubility (and other equilibrium) constants are reported as so-called thermodynamic constants; the experimental constants are extrapolated to the infinite dilution (zero ionic strength) or the constants are corrected to $I = 0$ by the application of a theoretical or empirical formula (usually the Debye-Hückel theory).

Within the last 25 years many solution chemists have preferred to determine so-called "stoichiometric" constants; these are measured in the presence of a large excess of inert electrolyte (e.g., KNO_3 or $NaClO_4$). In this way a constant ionic environment can be maintained even if one component of the solution is varied; hence ionic strength and all activity coefficients remain constant even if equilibria are shifted with respect to other more dilute solution conditions. As pointed out in Chapter 3, concentration and activity become identical, that is, analytically determined concentration equals potentiometrically determined concentration (or activity) if the same ionic medium is used as a standard. These stoichiometric solubility constants can be determined quite accurately; thermodynamically they are equally well defined as the so-called "thermodynamic" equilibrium constants, valid at infinite dilution. One serious drawback is that the stoichiometric constants are strictly valid only for the specified ionic medium. The standard free energy associated with a solubility equilibrium valid at a given ionic strength, $\Delta G^{\circ\prime}$, is shifted by a constant from the standard free energy defined by the conventional thermodynamic scale, ΔG°.

As Schindler [27] has shown, the free energy of formation of a compound can be obtained from characteristic stoichiometric equilibrium constants without making any nonthermodynamic assumptions. Let us consider, for

[27] P. Schindler, *Helv. Chim. Acta*, **42**, 577 (1959).

example (see Table 5-6), the determination of the free energy of formation of zinc oxide or zinc hydroxide.

The standard free energy of formation of ZnO or $Zn(OH)_2$ [$\Delta G°$ of (5) or (6) of Table 5-6] can be evaluated from measurements carried out with solutions of a particular ionic medium. Note that $\Delta G°$ is unaffected by the choice of the ionic medium.

Values for $\Delta G_{1a}^{°\prime}$ or $\Delta G_{1b}^{°\prime}$ are obtained from the corresponding solubility constants ($^{c*}K_{s0}$). $\Delta G_2^{°\prime}$ results from emf measurements of a cell,† $Pt(H_2)/$ solution of ionic strength I, Zn^{2+}/Zn, where the solution contains the same ionic medium used for the determination of the solubility constant. Excellent literature data are available for $\Delta G_3°$ and $\Delta G_4^{°\prime}$. Example 5-5 illustrates the evaluation of the free energy of formation of a metal carbonate.

The Conversion of $^{c}K_s$ into K_s

Such thermodynamic cycles can be used to convert stoichiometric constants into thermodynamic ones and vice versa (Schindler [25]). The procedure is illustrated in Example 5-12 in which the following equation, generally valid for oxides and carbonates of bivalent metals, is derived:

$$RT \ln \frac{^*K_{s0}}{^{c*}K_{s0}} = RT \ln \frac{p_{H_2O(I)}}{p_{H_2O(I=0)}} + 2f(E_{cell(I)}° - E_{cell(I=0)}°) \qquad (92)$$

Equation 92 interrelates the four basic quantities $^{c*}K_{s0}$, $^*K_{s0}$, $E_{(I=0)}°$ and $E_{(I)}°$.

Table 5-6 Evaluation of Standard Free Energy of Formation from Heterogeneous Equilibria in Constant Ionic Medium†

Reaction	Free Energies	
$Zn^{2+}(aq)_{(I)} + H_2O_{(I)} = ZnO(s) + 2H^+_{(I)}$	$\Delta G_{1a}^{°\prime} = RT \ln {}^{c*}K_{s0(ZnO)}$	(1a)
$Zn^{2+}(aq)_{(I)} + 2H_2O_{(I)} = Zn(OH)_2(s) + 2H^+_{(I)}$	$\Delta G_{1b}^{°\prime} = RT \ln {}^{c*}K_{s0(Zn(OH)_2)}$	(1b)
$Zn(s) + 2H^+(aq)_{(I)} = Zn^{2+}_{(I)} + H_2(g)$	$\Delta G_2^{°\prime} = 2fE_{cell(I)}°$	(2)‡
$H_2(g) + \frac{1}{2}O_2(g) = H_2O(g)$	$\Delta G_3° = -54,635$ cal	(3)
$H_2O(g) = H_2O(l)_{(I)}$	$\Delta G_4^{°\prime} = RT \ln p_{H_2O(I)}$	(4)
$Zn(s) + \frac{1}{2}O_2(g) = ZnO(s)$	$\Delta G° = \Delta G_{1a}° + \Delta G_2° + \Delta G_3° + \Delta G_4°$	(5)§
$Zn(s) + H_2(g) + O_2(s) = Zn(OH)_2(s)$	$\Delta G° = \Delta G_{1b}° + \Delta G_2°$ $+ 2(\Delta G_3° + \Delta G_4°)$	(6)§

† Cf. P. Schindler, H. Althaus and W. Feitknecht, *Helv. Chim. Acta*, **47**, 982 (1964).

‡ $E_{cell(I)}°$ corresponds to the standard emf value at a given ionic strength; f = Faraday.

§ $\Delta G°$ for ZnO(s) or $Zn(OH)_2(s)$ is obtained from measurements in solutions of constant ionic medium (I).

† Schindler, et al. [26] used zinc amalgam instead of Zn; this difference (smaller than 10^{-5} V) can be disregarded for the calculation of $\Delta G_2^{°\prime}$.

Example 5-12   *The Conversion of °K into K.* Compute $*K_{ps0}$ (25°C) for $ZnCO_3(s)$ from $^{c*}K_{ps0} = 8.17 \pm 0.05$ ($I = 0.2$ M $NaClO_4$, 25°C) (cf. Schindler [25]). We first calculate the free energy of formation of $ZnCO_3$ by building up $ZnCO_3(s)$ from elements through reactions where equilibrium constants (K or E_{cell}) have been determined for an ionic medium of 0.2 M $NaClO_4$:

$$Zn(s) + 2H_{(I)}^+ = Zn_{(I)}^{2+} + H_2(g); \qquad \Delta G_1^{\circ\prime} = 2fE_{cell(I)}^\circ \qquad (i)$$

$$C(s) + O_2(g) = CO_2(g); \qquad \Delta G_2^\circ \qquad (ii)$$

$$H_2(g) + \tfrac{1}{2}O_2(g) = H_2O(g); \qquad \Delta G_3^\circ \qquad (iii)$$

$$H_2O(g) = H_2O(l)_{(I)}; \qquad \Delta G_4^{\circ\prime} = RT \ln p_{H_2O(I)} \qquad (iv)$$

$$Zn_{(I)}^{+2} + CO_2(g) + H_2O_{(I)} = ZnCO_3(s) \dotplus 2H_{(I)}^+; \quad \Delta G_5^{\circ\prime} = RT \ln {}^{c*}K_{ps0} \qquad (v)$$

$$Zn(s) + C(s) + \tfrac{3}{2}O_2(g) = ZnCO_3(s); \qquad \Delta G_6^\circ = \Delta G_1^{\circ\prime} + \Delta G_2^\circ + \Delta G_3^\circ$$
$$+ \Delta G_4^{\circ\prime} + \Delta G_5^{\circ\prime} \quad (vi)$$

With ΔG_6°, the free energy of formation of $ZnCO_3$, $*K_{ps0}$ (valid at $I = 0$) can be computed if reliable values for the standard free energy of formation of $Zn^{2+}(aq)_{(I=0)}$, $CO_2(g)$ and $H_2O(l)_{(I=0)}$ are available. We can also close the cycle and "break down" $ZnCO_3(s)$ into the elements through reactions where equilibria are considered at infinite dilution:

$$ZnCO_3(s) + 2H_{(I=0)}^+ = Zn_{(I=0)}^{2+} + CO_2(g)$$
$$+ H_2O_{(I=0)}; \quad \Delta G_7^\circ = -RT \ln *K_{ps0} \qquad (vii)$$

$$Zn_{(I=0)}^{+2} + H_2(g) = Zn(s) + 2H_{(I=0)}^+; \qquad \Delta G_8^\circ = -2fE_{cell(I=0)} \qquad (viii)$$

$$CO_2(g) = C(s) + O_2(g); \qquad -\Delta G_2^\circ \qquad (ix)$$

$$H_2O_{(I=0)} = H_2O(g); \qquad \Delta G_{11}^\circ = -RT \ln p_{H_2O(I=0)} \qquad (x)$$

$$H_2O(g) = H_2(g) + \tfrac{1}{2}O_2(g); \qquad -\Delta G_3^\circ \qquad (xi)$$

$$ZnCO_3(s) = Zn(s) + C(s) \qquad \Delta G_{13}^\circ = \Delta G_7^\circ + \Delta G_8^\circ - \Delta G_2^\circ$$
$$+ \tfrac{3}{2}O_2(g); \qquad\qquad + \Delta G_{11}^\circ - \Delta G_3^\circ \quad (xii)$$

Because the sum of all the free energy contributions of (i) to (v) and (vii) to (xi) must equal zero,

$$RT \ln \frac{*K_{ps0}}{{}^{c*}K_{ps0}} = RT \ln \frac{p_{H_2O(I)}}{p_{H_2O(I=0)}} + 2f(E_{cell(I)}^\circ - E_{cell(I=0)}^\circ) \qquad (xiii)$$

Schindler found $E_{cell(I=0.2)}^\circ = -769.3 \pm 0.3$ mV. The standard potential, however, is $E_{cell(I=0)}^\circ = -762.8$ mV. Because the first term on the right-hand side of (xii) is essentially zero, we obtain

$$\log \frac{*K_{ps0}}{{}^{c*}K_{ps0}} \simeq \frac{-0.0065}{0.02958} = -0.22 \quad \text{or} \quad \log *K_{ps0} = 7.95 \, (\pm \, 0.05)$$

Schindler also shows that this value is in agreement with that obtained from converting the thermodynamic constant into the stoichiometric one using the Davies equation.

$$\log {}^{c*}K_{ps0} = \log {}^*K_{ps0} + \frac{\sqrt{I}}{1 + \sqrt{I}} - 0.3(I) \qquad \text{(xiv)}$$

By using (xiv), the constant valid at infinite dilution becomes $\log {}^*K_{ps0} = 7.92 \pm 0.05$. .

5-7   The Coexistence of Phases in Equilibrium

We illustrated earlier that in equilibrium calculations the elementary algebraic rule, according to which a set of n unknowns can be resolved if they are connected by n independent equations, can be used to calculate the number of independent variables needed to define an equilibrium system under specified conditions; for example, we considered that in a homogeneous phase carbonate solution, in addition to temperature and pressure, two independent variables, such as [Alk] and pH or C_T and $[CO_3^{2-}]$, are necessary in order to define completely the composition of the solution. Similarly, for a ternary $CaO–CO_2–H_2O$ system it was shown that p_{CO_2} and T might be freely chosen in order to fix the composition of an equilibrium mixture containing $CaCO_3(s)$, aqueous solution and $CO_2(g)$. Considerations such as these are statements similar to those given more formally by the Gibbs phase rule. One object of the phase rule is to determine the conditions of reproducibility of the equilibrium system so that it can be copied at will except as to the size and shape.

The Phase Rule

The phase rule is often stated in the form

$$F = C + 2 - P \qquad (93)$$

where F is the number of degrees of freedom, that is, the independent internal variables, C is the number of components and P is the number of phases present in the system and in equilibrium with each other. A *phase* is a domain with uniform composition and properties. Such specific properties as density, specific heat and compressibility of a phase are the same throughout its extent. Phases are, for example, a gas, a gaseous mixture, a homogeneous liquid solution, a uniform solid substance, a solid solution. The number of *components* C is defined as the minimum number of substances which we must bring together in order to duplicate the system under consideration. Frequently a variety of choices of components may be made but it is only important to find the least number of substances capable of expressing the composition of the mixture. For example, a system containing solid CaO and $CaCO_3$ in equilibrium with an aqueous solution and a partial pressure of CO_2 may be copied by mixing either set of the following components: CaO, CO_2, H_2O; $Ca(OH)_2$, H_2CO_3, H_2O; or $CaCO_3$, CaO, H_2O. Thus, $CaCO_3$ is equivalent to $CaO + CO_2$, or the carbon dioxide gas phase is equivalent to $CaCO_3 - CaO$.

The number of *degrees of freedom* or the *variance* of a system is the number of variable factors (independent internal variables) such as temperature, pressure, partial pressure, concentration, mole fraction for which one can choose arbitrary values. For a system in equilibrium, not more arbitrary values than F internal variables can be selected.

DERIVATION. The phase rule can be derived from exact definitions of thermodynamic functions. Only an outline of such a derivation is given here; rigorous derivations will be found in thermodynamic textbooks.

In general the number of independent variables F must be equal to the number of variables U minus the number of independent equations I:

$$F = U - I \qquad (94)$$

In any phase, the composition is defined by $C - 1$ concentration terms. For example, $C - 1$ mole fractions (X) define the composition because one of the mole fractions is given by the condition that the sum of the mole fractions is unity. Hence, in a system of P phases the internal variables U are, besides pressure and temperature, $C - 1$ concentration terms in each phase

$$U = P(C - 1) + 2 \qquad (95)$$

For a system in equilibrium at constant temperature and pressure, the chemical potential for any given component has the same value in every phase. For each unique phase equilibrium we may write one independent equation

$$I = C(P - 1) \qquad (96)$$

Thus the number of variables remaining undetermined according to (94) for the system at equilibrium is

$$F = P(C - 1) + 2 - C(P - 1) = C - P + 2 \qquad (97)$$

SIMPLE APPLICATIONS. The phase rule is a useful tool for organizing equilibrium models. In considering such a model it is expedient to examine the variables and the relations connecting them. A knowledge of the variance (F) is of importance in assessing whether or not sufficient information is available for a definite numerical solution of an equilibrium problem; by assigning appropriate values to independent variables the variance can be reduced. Furthermore, the rule illustrates in a lucid way that systems that are quite different in character may behave in a similar manner.

Some of the most simple applications of the phase rule can perhaps be explained by reference to the examples listed in Table 5-7.

All systems of one component (Example 1) can be perfectly defined by giving values to, at most, two variable factors, usually P and T. If any two phases can coexist in equilibrium in a system of $C = 1$, only one independent variable, T or P, can be selected. At a given temperature the pressure has a definite value: $P = f(T)$. An invariant system ($C = 1$) is characterized by a triple point in a P, T diagram; both P and T are fixed if three phases coexist. The phase rule predicts that in a one-component system four phases (e.g., two crystalline solid phases such as ice I and ice III together with liquid water and water vapor) cannot be in equilibrium with each other.

As a further illustration, consider Example 3 of Table 5-7. A gas dissolving in a liquid is a bivariant system. If T is specified the equilibrium solubility of the gas in the liquid still varies with the partial pressure of the gas (Henry's law).

Example 4 illustrates a two-component hydrate where, for $P = 3$, at a given temperature the vapor pressure of the system must be constant.

In Examples 5 and 6 the reactions

$$CaCO_3(s) = CaO(s) + CO_2(g) \tag{98}$$
$$3Fe_2O_3(s) = 2Fe_3O_4(s) + \tfrac{1}{2}O_2(g) \tag{99}$$

are compared. Because the solids in (99) interact to form a solid solution, the system is bivariant at equilibrium. The system of (98), however, has only one degree of freedom because the two solids do not interact and form two solid phases.

Example 7 of Table 5-7 illustrates that one degree of freedom is lost for every new phase that is in thermodynamic coexistence. An equilibrium system containing calcite, aqueous solution and $CO_2(g)$ is bivariant; if we fix the partial pressure of CO_2 at a given temperature the composition of the solution (pH, C_T, [Alk], etc.) is constant and remains unchanged with isothermal evaporation or dilution. Even the addition of the base $Ca(OH)_2$ does not change the composition of the system as long as equilibrium between the three phases is presumed to exist. Hence, in such regard the system behaves as a pH-stat or a pCa-stat; the buffer intensity,

$$\beta^{Ca(OH)_2}_{CaCO_3(s),aq,CO_2(g)}$$

is infinite. If we "make" four phases [$Ca(OH)_2(s)$, calcite, aqueous solution, $CO_2(g)$] with the same three components, only one degree of freedom remains.

Table 5-7 Phase Rule Relationships

Components		Phases			Variables	
No. C	Type	No. P	Type, Examples	No. of Independent Variables	Examples of Internal Variables	Examples of Independent Variables
(1) 1	H_2O	1	Liquid	2	P, T	P, T
		2	Ice, liquid	1	P, T	P or T
		2	Liquid, vapor	1	P, T, p_{H_2O}	T or p†
		3	Ice, liquid vapor	0	—	—
(2) 2	SiO_2, H_2O	1	Aqueous solution	3	$[H_4SiO_4], P, T$	$[H_4SiO_4], P, T$
		2	Quartz, aqueous solution	2	$[H_4SiO_4], P, T$	P, T
		3	Quartz, cristobalite, aqueous solution	1	$[H_4SiO_4], P, T$	T or P
(3) 2	H_2O, CO_2	2	Aqueous solution, $CO_2(g)$	2	$T, P, p_{H_2O}, p_{CO_2},$ $[CO_2], [H_2CO_3]$	T and p_{CO_2}†
		3	Ice, solution, $CO_2(g)$	1	$T, P, p_{H_2O}, p_{CO_2},$ $[CO_2], [H_2CO_3]$	P or p_{CO_2}
(4) 2	$CuSO_4$, H_2O	2	$CuSO_4(s)$, $H_2O(g)$	2	P, T, p_{H_2O}	T, p_{H_2O}
		2	$CuSO_4 \cdot H_2O(s)$, $H_2O(g)$	2	P, T, p_{H_2O}	T, p_{H_2O}
		3	$CuSO_4(s)$, $CuSO_4 \cdot H_2O(s)$, $H_2O(g)$	1	P, T, p_{H_2O}	T

(5)	2	CaO, CO$_2$; or CaCO$_3$, CO$_2$; or CaCO$_3$, CaO	3	CaCO$_3$(s), CaO(s), CO$_2$(g)	1	T, p_{CO_2}, P	T
(6)	2	Fe$_3$O$_4$, O$_2$; or Fe$_2$O$_3$, O$_2$; or Fe$_3$O$_4$, Fe$_2$O$_3$	2	Solid solution of Fe$_2$O$_3$ in Fe$_3$O$_4$, O$_2$(g)	2	$T, p_{O_2}, P, x_{Fe_3O_4}, x_{Fe_2O_3}$	T, p_{O_2}
(7)	3	CaO, CO$_2$, H$_2$O; or Ca(OH)$_2$, CaCO$_3$, H$_2$O; or CaCO$_3$, CaO, H$_2$O	1	Aqueous solution	4	$P, T, [Ca^{2+}], [Alk], [H^+], [CO_3^{2-}], C_T$	$T, P, [Alk], pH$
			2	Calcite, aqueous solution	3	$P, T, [Ca^{2+}], [Alk], [H^+], [CO_3^{2-}], C_T$	T, P, C_T; or $T, P,$ Alk
			3	Calcite, aqueous solution, CO$_2$(g)	2	$P, T, [Ca^{2+}], [Alk], [H^+], [CO_3^{2-}], C_T, p_{CO_2}$	T, p_{CO_2}; or T, P
			4	Calcite, Ca(OH)$_2$(s), aqueous solution, CO$_2$(g)	1	$P, T, [Ca^{2+}], [Alk], [H^+], [CO_3^{2-}], C_T$	T
			5	Calcite, Ca(OH)$_2$(s), ice, aqueous solution, CO$_2$(g)	0	$P, T, [Ca^{2+}], [Alk], [H^+], [CO_3^{2-}], C_T$	—

† Note in examples (1) and (3) that $P = p_{H_2O}$ and $P = p_{CO_2} + p_{H_2O}$, respectively.

As long as all four phases coexist in equilibrium at a given temperature, the p_{CO_2} remains constant and is independent of the quantities of the two solids.

Example 5-13   *Salt Deposition in Oceans.*   A large volume of an aqueous solution containing K^+, Mg^{2+}, Na^+, Cl^- and SO_4^{2-} is concentrated by isothermal evaporation. Find the maximum number of solid phases that can be in equilibrium with the solution. It can be assumed that the temperature of evaporation does not happen to be an exact transition point. (Cf. Van't Hoff, *Zur Bildung der Ozeanischen Salzablagerungen*, Braunschweig, 1905–1909; this problem is taken from L. G. Sillén, P. W. Lange and C. O. Gabrielson, *Problems in Physical Chemistry*, Prentice-Hall, Englewood Cliffs, N.J., 1952.) An aqueous solution that contains a different cations and b different anions contains $(a + b)$ components (H_2O and $a + b - 1$ salts). In our example, $C = 5$ (e.g., H_2O and Na_2SO_4, K_2SO_4, $MgCl_2$, KCl). Because $F = 2(P, T)$ there must be $P = 5$ phases. Accordingly, a maximum number of four solid phases can be in equilibrium with the solution.

In constructing a thermodynamic model for a natural water system, we usually incorporate the components and specify the phases to be included. In a stepwise construction the addition of each new component ($\Delta C = 1$) must be accompanied by either a new phase ($\Delta P = 1$) or a new internal variable ($\Delta F = 1$). The creation of a new phase in a system of a given number of components reduces the degree of freedom by one; more phases demand fewer assumptions for solving an equilibrium model. The maintenance of constant concentration conditions in an aqueous solution can be accomplished in principle by the coexistence in equilibrium of a sufficient number of phases. Hence, the solid phases comprised in the lithosphere are of particular significance in natural water models. Although there might not be true equilibrium in any actual aquatic system, the composition of the lithosphere represents a significant regulatory factor for the composition of the hydrosphere.

It is appropriate to mention here the work of Sillén [28], who epitomized the application of equilibrium models for portraying the prominent features of the seawater system. His analysis, which will be discussed in some detail later (Chapter 8), indicates that heterogeneous equilibria of solid minerals comprise the principal buffer systems in oceanic waters; according to such a model the CO_2 content of the oceans and of the atmosphere is controlled by equilibria with solids at the sediment–water interface.

LOCAL EQUILIBRIUM. If we consider a large volume of sediments or of rock assemblage, we will find phases that are thermodynamically incompatible.

[28] L. G. Sillén, in *Oceanography*, M. Sears, Ed., American Association for the Advancement of Science, Washington, D.C., 1961, p. 549.

But, as pointed out by Thompson [29], it is generally possible to regard any part of such a system as substantially in internal equilibrium if that part is made sufficiently small. With respect to phase equilibrium the requirement appears to be met if no mutually incompatible phases were in actual contact and if phases of variable composition were to show only continuous variation from point to point.

CONGRUENT AND INCONGRUENT SOLUBILITY. If we add $MgCO_3(s)$(magnesite) to distilled water (25°C), this solid will dissolve incipiently; but upon further addition, $Mg(OH)_2(s)$ will precipitate because under these conditions $Mg(OH)_2$ will be less soluble than $MgCO_3(s)$ ($pK_{s0(Mg(OH)_2)} = 10.9$; $pK_{s0(MgCO_2)} = 4.9$). Such a dissolution behavior is referred to as incongruent solubility. The term incongruent solution is generally used if a mineral upon dissolution reacts to form a new solid; or if a saturated solution, in the presence of solid phases with which it is in equilibrium, reacts upon isothermal evaporation with a change in the solid phases. In natural environments incongruent solubility is probably more prevalent than congruent solution.

From a point of view of the phase rule, the dissolution of a simple salt can be interpreted as a system of three components containing not more than two phases (aqueous solution and solid phase; absence of the vapor phase). Hence at a given pressure and temperature, one degree of freedom remains, that is, the concentration of the solution in equilibrium with the solid can undergo change. Ricci [30] has given a lucid and quantitative treatment of incongruent solubility of normal salts. In Figure 5-20 the congruent dissolution behavior of $CaCO_3$ is contrasted with the incongruent dissolution characteristics of $MgCO_3$. A solution of the salt in pure H_2O lies on the diagonal line. The maximum solution concentration of the salt that can be achieved is given by C_{max}. This concentration cannot be exceeded (under conditions of equilibrium). In the case of the $Mg(OH)_2–CO_2–H_2O$ system, however, $Mg(OH)_2$ is precipitated. Such a "hydrolytic precipitation" occurs if the solubility of the base C_{max} is smaller than the solubility of the salt S. Thus, congruence and incongruence is characterized by $C_{max} > S$ and $C_{max} < S_{inc}$, respectively.

5-8   Crystal Formation; The Initiation and Production of the Solid Phase

The formation of a solid phase by precipitation from solution, usually takes place in three steps.

[29] J. B. Thompson, "Local Equilibrium in Metasomatic Processes" in *Researches in Geochemistry*, P. H. Abelson, Ed., Wiley, New York, 1959.
[30] J. E. Ricci, *Hydrogen Ion Concentration*, Princeton Univ. Press, Princeton, N.J., 1952.

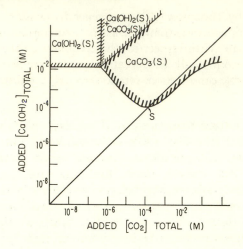

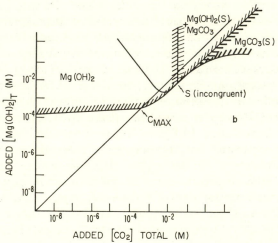

Figure 5-20 Congruent and incongruent solubility of $CaCO_3(s)$ and $MgCO_3(s)$.

1. The interaction between ions or molecules leads to the formation of a cluster.

$$X + X \rightleftharpoons X_2$$
$$X_2 + X \rightleftharpoons X_3$$
$$X_{j-1} + X \rightleftharpoons X_j \text{ (cluster)} \tag{100}$$
$$\text{Nucleation: } X_j + X \rightleftharpoons X_{j+1} \text{ (nucleus)} \tag{101}$$

Nucleation corresponds to the formation of the new centers from which spontaneous growth can occur. The nucleation process determines the size and the size distribution of crystals produced.

2. Subsequently, material is deposited on these nuclei

$$X_j + X \rightleftharpoons \text{crystal growth} \tag{102}$$

and crystallites are being formed (crystal growth).

3. Large crystals may eventually be formed from fine crystallites by a process called ripening.

A better insight into the mechanisms of the individual steps in the formation of crystals would be of great help in explaining the creation and transformation of sedimentary deposits. Valuable reviews are available on the principles of nucleation of crystals and the kinetics of precipitation and crystal growth [31–35]. Only a few important considerations are summarized here in order to illustrate the wide scope of questions to be answered in order to predict rates and mechanisms of precipitation in natural systems.

Nucleation

If one gradually increases the concentration of a solution, exceeding the solubility product with respect to a solid phase, the new phase is not formed until a certain degree of supersaturation has been achieved. Stable nuclei can only be formed after an activation energy barrier has been surmounted. In Section 5-5 we have seen that small crystallites are more soluble than large crystals; hence the energy barrier is related to the additional free energy needed to form the more soluble nuclei and is due to the surface energy of these small particles. Equation 85 can be rewritten for an aggregate of spherical shape

$$kT \ln \frac{a}{a_0} = \frac{2V\bar{\gamma}}{r} \tag{103}$$

where a/a_0 is the concentration (or activity) ratio (supersaturation) with a_0 the theoretical solubility of a large crystal. V is the "molecular" volume, $\bar{\gamma}$ the interfacial energy and r the radius. Once nuclei of critical size X_j [in (100)] have been formed, crystallization is spontaneous.

As discussed before (Section 5-5), incipiently formed nuclei might be of a different polymorphous form than the final crystals. If the nucleus is smaller than one unit cell, the growing crystallite produced initially is most likely amorphous; substances with large unit cell tend to be precipitated initially as

[31] A. E. Nielsen, *Kinetics of Precipitation*, Macmillan, New York, 1964.
[32] G. H. Nancollas and N. Purdie, *Quart. Rev.* (London), **18**, 1 (1964).
[33] B. Chalmers, *Principles of Solidification*, Wiley, New York, 1964.
[34] A. G. Walton, *The Formation and Properties of Precipitates*, Wiley-Interscience, New York, 1967.
[35] A. G. Walton, *Science*, **148**, 601 (1965).

an amorphous phase. Walton et al. [36] have shown, for example, that calcium phosphate nucleated from solution at high pH is an amorphous or soft metastable material with a stoichiometric composition of $[Ca^{2+}]/[P] = 1.5$ which ultimately converts into apatite.

The free energy of the formation of a nucleus, ΔG_j, consists essentially of work required to create a surface and of energy gained (volume free energy) from making bonds. For a spherical nucleus, the first quantity is given by $4\pi r^2 \bar{\gamma}$ and the second quantity (always negative) can be expressed as $jkT \ln (a/a_0)$, where j is the number of molecules or ions in the nucleus, or expressed in terms of volume, $j = 4\pi r^3/3V$. Hence the free energy of nucleus formation may be written as

$$\Delta G_j = \frac{-4\pi r^3}{3V} kT \ln \frac{a}{a_0} + 4\pi r^2 \bar{\gamma} \tag{104}$$

In Figure 5-21a, ΔG_j is plotted as a function of r for assumed constant values† of V and $\bar{\gamma}$ for a few values of a/a_0. Obviously the activation energy ΔG_a decreases with increasing supersaturation as does the size of the nucleus r_j. ΔG_a can be calculated readily by inserting into (104) r_j as obtained from (103).

$$r_j = \frac{2\bar{\gamma}V}{kT \ln (a/a_0)} \tag{105}$$

for which

$$\Delta G_a = \frac{16\pi \bar{\gamma}^3 V^2}{3[kT \ln (a/a_0)]^2} \tag{106}$$

Equations 105 and 106 are also obtainable by finding the maximum for ΔG_j in (104), that is, by letting $d(\Delta G_j)/dr = 0$.

The rate at which nuclei form J may be represented according to conventional rate theory as

$$J = A \exp \frac{-\Delta G_a}{kT} \tag{107}$$

where A is a factor related to the efficiency of collisions of ions or molecules. Accordingly, the rate of nucleation is controlled by the interfacial energy, the supersaturation, the collision frequency efficiency and the temperature. For given values of T, A and $\bar{\gamma}$, the nucleation rate J (nuclei formed $cm^{-3} sec^{-1}$) can be calculated as a function of a/a_0 (Figure 5-21b). J is critically dependent upon the supersaturation. For example, using the conditions specified for Figure 5-21a, one can calculate that nucleation is almost instantaneous at $a/a_0 = 100$ (10^5 nuclei per $cm^{-3} sec^{-1}$) while (homogeneous) nucleation

[36] A. G. Walton, W. J. Bodin, H. Furedi, and A. Schwartz, *Can. J. Chem.*, **45**, 2095 (1967).
† $\bar{\gamma}$ is not necessarily independent of cluster size.

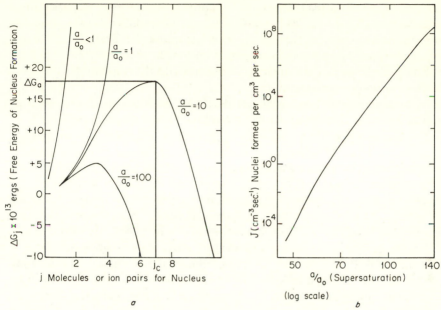

Figure 5-21   Nucleation. The energy barrier and the nucleation rate depend critically on the supersaturation.

a: Free energy of formation of a spherical nucleus as a function of its size, calculated for different supersaturations (a/a_0) [cf. (104)].

b: Double logarithmic plot of nucleation rate versus supersaturation (a/a_0) calculated with (106) and (107).

The curves have been calculated for the following assumptions: $\bar{\gamma} = 100$ ergs cm^{-2}; "molecular" volume, $V = 3 \times 10^{-23}$ cm^3; collision frequency efficiency [Eq. (107)], $A = 10^{30}$ cm^{-3} sec^{-1}.

should not occur even within geological time spans for a tenfold super-saturation (for $a/a_0 = 10$, one nucleus cm^{-3} within 10^{70} sec). It is appropriate to speak of a critical supersaturation that must be exceeded before nucleation will occur.

The theory given so far applies if supersaturation is built up homogeneously and under conditions of steady state (stationary distribution of clusters). Experimentally, nucleation data may often be fitted in a plot of log J versus log c in accordance with the semi-theoretical equation

$$J = knC^n = \frac{d[X_j]}{dt} \tag{108}$$

where C is the initial concentration. As Figure 5-21*b* shows, log J plots nearly rectilinearly as a function of log (a/a_0); thus, if the concentration is not markedly changed as a result of nucleation ($c \approx a$), (108) is not in disaccord with classical nucleation theory (107). According to the theory of Christiansen and Nielsen [37], n in (108) is characteristic of the number of ions needed to form the critical cluster, that is, $n = j$. While in the chemical theory the size of the nucleus depends on the supersaturation, in the "constant number" theory [37] the size of the nucleus is constant and the nucleation rate is less dependent upon supersaturaton.

According to (108) experimental data can be fitted by a plot of $\log t$ (induction period) versus $\log c$ characterized by a slope of $n - 1$:

$$k = tc^{(n-1)} \tag{109}$$

The induction period t may be defined as the time necessary for incorporation of a constant fraction of the solute into crystallites; experimentally it is the time lapse until the onset of precipitation is observed. Instead of c, the ion product or its appropriate root, for example, $([Ca^{2+}][F^-]^2)^{1/3}$ for CaF_2, may be plotted. Note that the nucleus does not arise by a simultaneous collision of all the individual ions (or molecules) of which it is composed. If $n = 4$, the nucleus A_4B_4 is formed by stepwise bimolecular additions, but the addition of an additional ion to A_4B_4 may be interpreted to result in a fourth-order dependence on the concentration because all the preceding steps are considered to be in equilibrium.

Pytkowicz [15] has investigated the nucleation of calcium carbonate in artificial and natural seawater. He suggests from an extrapolation of his results that inorganic precipitation of $CaCO_3$ from average seawater would, under conditions of apparently homogeneous nucleation, require ca. 100,000 years. In Mg^{2+}-free seawater nucleation is considerably faster and less dependent upon supersaturation. Qualitatively this inhibition may be explained [15] by considering that in seawater, where $[Mg^{2+}]$ is five times larger than $[Ca^{2+}]$, collisions that form MgCa-carbonate clusters are more probable than collisions that form $CaCO_3$ clusters.

Heterogeneous nucleation, however, is the predominant formation process for crystals in natural waters. In a similar way as catalysts reduce the activation energy, foreign solids may catalyze the nucleation process by reducing the energy barrier. If the surface of the heteronucleus matches well with the crystal, the interfacial energy between the two solids, $\bar{\gamma}_2$, is smaller than the interfacial energy between the crystal and the solution, $\bar{\gamma}$, and nucleation may take place at a lower concentration on a heteronucleus than in a homogeneous solution. The matching between surfaces is more a matter of agreement in lattice type and atomic distances than chemical similarity [31]. Favorable precipitation sites in the solid substrate will be those where strong adsorption occurs, particularly if bonding with the substrate is possible [34]. The critical

[37] J. H. Christiansen and A. E. Nielsen, *Acta Chem. Scand.*, **5**, 673 (1951).

supersaturation will decrease with increasing degree of lattice matching between nucleus and substrate [38, 39]. In the extreme case substrate and nucleus are identical and $\bar{\gamma}_2$ [and thus in turn the second term in (104)] becomes zero.

Phase changes in natural waters are almost invariably initiated by heterogeneous solid substrates. Inorganic crystals, skeletal particles, clays, sand and biocolloids might serve as heteronuclei. At the same time trace quantities of surfactants and other material that may become adsorbed at the active sites on potential substrate particles might inhibit or prevent the onset of nucleation. An example of heterogeneous nucleation is provided by the nucleation of apatite on calcite surfaces (Figure 10-17).

Crystal Growth

The growth of crystals occurs in successive reaction steps: (a) the transport of solute to the crystal solution interface, (b) the adsorption of solute at the surface and (c) the incorporation of the crystal constituents into the lattice. Furthermore as a back reaction [4] the crystal dissolution must be considered. Growth kinetics depends on the rate-limiting step; essentially two limiting theories on growth kinetics have been established: (a) diffusion controlled growth and (b) interface controlled growth.

Diffusion controlled growth (or dissolution) can be characterized by

$$\frac{dc}{dt} \simeq -ks(c - c_0) \quad (110)$$

where c_0 is the equilibrium concentration, that is, the concentration in the immediate proximity of the crystal surface, s, the surface area. The rate constant k depends on the diffusion coefficient and the extent of turbulence in solution.

Interface controlled growth does not depend on the turbulence. The results can be fitted generally to rate laws of the type

$$\frac{dc}{dt} = k's(c - c_0)^n \quad (111)$$

where n is the order of reaction. n has been determined for silver chloride ($n = 2$), silver chromate ($n = 3$), magnesium oxalate ($n = 2$) and potassium chloride ($n = 1$). While dissolution reactions are nearly always diffusion controlled, growth of crystals appears to be frequently controlled by interfacial processes such as adsorption or dislocation steps.

[38] D. Turnbull and B. Vonnegut, *Ind. Eng. Chem.*, **44**, 1292 (1952).
[39] J. B. Newkirk and D. Turnbull, *J. Appl. Phys.*, **26**, 579 (1955).

"CRYSTAL POISONS." Because the rate-determining step is frequently controlled at the interface, small amounts of soluble foreign constituents may alter markedly the growth rate of crystals and their morphology. A few examples are given in Section 6-6. The retarding effect of substances that become adsorbed may be explained as being due primarily to the obstruction by adsorbed molecules to the deposition of lattice ions. Davies and Nancollas [40] have shown, for some cases, that the rate constant for crystal growth is reduced by an amount reflecting the extent of adsorption.

GROWTH OF CONCRETIONS. In sedimentary rocks we commonly find concretions, that is, material formed by deposition of a precipitate, such as calcite or siderite, around a nucleus of some particular mineral grain or fossil. The origin of most concretions is not known, but as mentioned by Berner [41], their often-spherical shape with concentric internal structure suggests diffusion as an important factor affecting growth. The rate of growth, if diffusion controlled, is readily amenable to mathematical treatment. Berner [41] has provided some idea of the time scale involved in growth of postdepositional concretions. His calculations illustrate, for example, that for a typically slowly flowing groundwater, with a supersaturation in $CaCO_3$ of 10^{-4} M (assumed to be constant), the time of growth of calcite concretions ranges from 2500 yr for concretions of 1-cm radius to 212,000 yr for those of 5-cm radius. Hence concretion growth, if diffusion and convection is the rate-controlling step, is relatively rapid when considered on the scale of geologic time.

In qualitative analytical chemistry inorganic solutes are distinguished from each other by a separation scheme based on the selectivity of precipitation reactions. In natural waters certain minerals are being dissolved while others are being formed. Under suitable conditions a cluster of ions or molecules selects from a great variety of species the appropriate constituents to form particular crystals. The birth of a crystal and its growth is an impressive example of nature's selectivity.

READING SUGGESTIONS

Butler, J. N., *Ionic Equilibrium, A Mathematical Approach*, Addison-Wesley, Reading, Mass., 1964.

Case, L. O., "The Phase Rule in Analytical Chemistry," in *Treatise on Analytical Chemistry*, Part I, Volume 2, I. M. Kolthoff and P. J. Elving, Eds., Wiley-Interscience, New York, 1963, p. 957.

[40] C. W. Davies and G. H. Nancollas, *Trans. Faraday Soc.*, **51**, 818, 823 (1955).
[41] R. A. Berner, *Geochem. Cosmochim. Acta*, **32**, 477 (1968).

Garrels, R. M., M. E. Thompson and R. Siever, "Solubility of Some Carbonates at 25°C and One Atmosphere Total Pressure," *Amer. J. Sci.*, **258**, 402 (1960). Contains a discussion on calcite–dolomite equilibria.

Leussing, D. L., "Solubility," in *Treatise on Analytical Chemistry*, Part I, Volume 1, I. M. Kolthoff and P. J. Elving, Eds., Wiley-Interscience, New York, 1963, p. 675.

Nancollas, G. H., and N. Purdie, "The Kinetics of Crystal Growth," *Quart. Rev.* (London), **18**, 1 (1964). Concise review in nucleation and crystal growth.

Robie, R. A., and D. R. Waldbaum, "Thermodynamic Properties of Minerals and Related Substances at 298.15°K and One Atmosphere Pressure and at Higher Temperatures," *Geol. Survey Bull.*, **1259** (1968). Contains a critical summary of the available thermodynamic data for minerals and related substances in a convenient form for the use of earth scientists.

Schindler, P. W., "Heterogeneous Equilibria Involving Oxides, Hydroxides, Carbonates and Hydroxide Carbonates," in *Equilibria Concepts in Natural Water Systems*, Adv. Chem. Series, No. 67, American Chemical Society, Washington, D.C., 1967, p. 196. Lucid discussion of evaluation of solubility data emphasizing metastable equilibria with polymorphous modifications.

Applications of Precipitation and Dissolution Reactions in Water Treatment

Fair, G. M., J. C. Geyer and D. A. Okun, chapters 29 and 30 in *Water Supply and Waste Water Disposal*, Volume 2, Wiley, New York, 1968.

Hömig, H. E., *Physikochemische Grundlagen der Speisewasserchemie*, Classen, Essen, 1963. Includes excellent discussion on solubility relations at higher temperatures (boilers).

PROBLEMS

5-1 (i) How much Fe^{2+} (expressed in moles liter^{-1}) could be present in a 10^{-2} M $NaHCO_3$ solution without causing precipitation of $FeCO_3$? The solubility product of $FeCO_3$ is $K_s = 10^{-10.7}$.

(ii) How would this maximum soluble Fe^{2+} concentration change upon lowering the pH of that solution by 1 unit?

5-2 C. V. Cole [*Soil Sci.*, **83**, 141 (1956)] established the following equation for soil water:

$$\log [Ca^{2+}] + 2pH = \text{const.} - \log p_{CO_2}$$

Under what conditions does this equation hold? Express the constant in terms of known equilibrium constants.

5-3 How much acid or base has to be added to a saturated solution of $CaCO_3$ to adjust the soluble Ca^{2+}-level to

(i) 10^{-3} M

(ii) 5×10^{-4} M

if the solution is shielded from the atmosphere?

5-4 Tillmans (1907) made investigations on the solubility of marble ($CaCO_3$) by calcium bicarbonate solutions containing different amounts of $H_2CO_3^*$. He shook his solutions for 10 days in closed flasks with marble chips. After this period equilibrium was attained. He determined experimentally the equilibrium concentrations of the $H_2CO_3^*$ and the alkalinity. For all waters that showed pH values below 8.5, after the saturation with $CaCO_3$, he found the following empirical relation: $[H_2CO_3^*] = K[Alk]^3$.

(i) Show that this relationship can be derived by mass law considerations.

(ii) Express K of the above equation in terms of K_1 = first acidity constant of $H_2CO_3^*$, K_2 = second acidity constant of $H_2CO_3^*$, K_{so} = solubility product of $CaCO_3$.

5-5 10^{-3} mole of $Mg(OH)_2(s)$ is added to 1 liter of water. The system is then exposed to a partial pressure of CO_2 of 10^{-1} atm.

(i) Which solid phase precipitates?

(ii) What is the pH of the solution?

(Free energy data or solubility constants are quoted in Example 5-6.)

5-6 This problem deals with surveys in anoxic basins [Richards et al., *Limnol. Oceanog.*, 10, R197 (1965)]. Compute the solubility product of FeS from the figure given below. The computed slope of the line is $[Fe^{2+}][S^{2-}] = 5.3 \times 10^{-18}$. 18°C. $I = 0.06$.

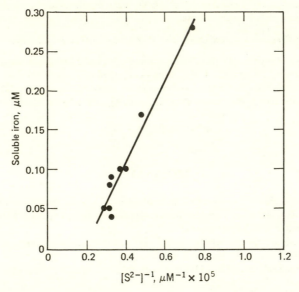

5-7 Some solid $CaCO_3$ is in equilibrium with its saturated solution. How will the amount of Ca^{2+} in the solution be affected (increase, decrease or no effect) by adding small amounts of the following:

(i) KOH   (v) $FeCl_3$

(ii) $CaCl_2$ (vi) Na_2CO_3

(iii) $(NH_4)_2SO_4$ (vii) H_2O

(iv) Sodium metaphosphate $(NaPO_3)_n$

5-8 Consider the system containing the phases $Ca_{10}(PO_4)_6(OH)_2(s)$, $CaCO_3(s)$, $CaHPO_4(s)$, aqueous solution and gas $(CO_2(g))$ in equilibrium at a temperature of 10°C.

(a) Give suitable components and degrees of freedom for this system.

(b) Under conditions of equilibrium does $[Ca^{2+}]$ increase, decrease or stay constant upon addition of small quantities of the following: (i) H_3PO_4; (ii) CO_2, (iii) NaOH, (iv) $Ca(OH)_2$, (v) HCl, (vi) H_2O?

ANSWERS TO PROBLEMS

5-1 (i) $10^{-6.8}$; (ii) $10^{-5.8}$

5-2 Reaction: $CaCO_3(s) + 2H^+ = Ca^{2+} + CO_2(g) + H_2O$. Const. $= *K_{ps0} = K_{s0}K_2^{-1}K_H^{-1}K_1^{-1}$ [compare (10) of Table 5-2].

5-3 (i) 1×10^{-3} mole liter^{-1} acid; (ii) 5×10^{-4} mole liter^{-1} acid [by use of (30)].

5-4 Reaction: $CaCO_3(s) + H_2CO_3^* = Ca^{2+} + 2HCO_3^-$; charge balance: $2[Ca^{2+}] \approx [HCO_3^-]$; $[Alk] \approx [HCO_3^-]$. $[HCO_3^-]^3/[H_2CO_3^*] = 2K_{s0}K_1K_2^{-1} = K$.

5-5 No precipitation, pH $= 6.1$.

5-6 $^cK_{s0} = 5.3 \times 10^{-18}$; $pK_{s0} \approx p^cK_{s0} + 4\sqrt{I}/(1 + \sqrt{I}) \approx 18.1$. (The value given by the *National Bureau of Standards Circular 500* for 25°C and $I = 0$ is $pK_{s0} = 17.3$.)

5-7 Increase for (ii), (iii), (v); decrease for (i), (iv), (vi); no change for (vii).

5-8 (a) CO_2, H_2O, P_2O_5, CaO; 5 phases, hence, one degree of freedom, but specification of temperature makes the system nonvariant.

(b) No change for (i), (ii), (iv) and (vi) (because these represent additions of components); decrease for (ii); increase for (v).

6   Metal Ions in Aqueous Solution: Aspects of Coordination Chemistry

6-1   Introduction

All chemical reactions have one common denominator: the atoms, molecules or ions involved tend to improve the stability of the electrons in their outer shell. In a broad classification of chemical reactions we distinguish between two general groups of reactions by which atoms achieve such stabilization. (a) Redox processes, in which the oxidation states of the participating atoms change, and (b) reactions in which the coordinative relationships are changed. What do we mean by a change in coordinative relations? The coordinative relations are changed if the coordinative partner is changed or if the coordination number† of the participating atoms is changed. This may be illustrated by the following examples.

1. If an acid is introduced into water

$$HClO + H_2O = H_3O^+ + ClO^-$$

the coordinative partner of the hydrogen ion (which has a coordination number of one) is changed from ClO to H_2O.

2. The precipitation that frequently occurs in the reaction of a metal ion with a base

$$Mg \cdot aq^{2+} + 2OH^- = Mg(OH)_2(s) + aq$$

can be interpreted in terms of a reaction in which the coordinative relations are changed, in the sense that a three-dimensional lattice is formed in which each metal ion is surrounded by and coordinatively "saturated" by the appropriate number of bases.

3. Metal ions can also react with bases without formation of precipitates in reactions such as

$$Cu \cdot aq^{2+} + 4NH_3 = [Cu(NH_3)_4]^{2+} + aq$$

---

† The coordination number is indicative of the structure and specifies the number of nearest neighbors (ligand atoms) of a particular atom.

238

In this simple classification of reactions no distinction needs to be made between acid–base, precipitation and complex formation reactions; they are all coordinative reactions and hence phenomenologically and conceptually similar.

DEFINITIONS. In the following, any combination of cations with molecules or anions containing free pairs of electrons (bases) is called coordination (or complex formation) and can be either electrostatic or covalent, or a mixture of both. The metal cation will be called the *central atom*, and the anions or molecules with which it forms a coordination compound will be referred to as *ligands*. If the ligand is composed of several atoms, the one responsible for the basic or nucleophilic nature of the ligand is called the ligand atom. If a base contains more than one ligand atom, and thus can occupy more than one coordination position in the complex, it is referred to as a *multidentate* complex former. Ligands occupying one, two, three, etc., positions are referred to as unidentate, bidentate, tridentate, etc. Typical examples are oxalate and ethylendiamine as bidentate ligands, citrate as a tridentate ligand, ethylene-diamine tetraacetate (EDTA) as hexadentate ligand. Complex formation with multidentate ligands is called *chelation*, and the complexes are called chelates. The most obvious feature of a chelate is the formation of a ring. For example, in the reaction between glycine and $Cu \cdot aq^{2+}$, a chelate

with two rings, each of five members, is formed. Glycine is a bidentate ligand; O— and N— are the donor atoms. If there is more than one metal atom (central atom) in a complex, we speak about multi- or *polynuclear complexes*.

One essential distinction between a proton complex and a metal complex is that the *coordination number* of protons is different from that of metal ions. The coordination number of the proton is 1 (in hydrogen bonding, H^+ can also exhibit a coordination number of 2). Most metal cations exhibit an even coordination number of 2, 4, 6 and occasionally 8. In complexes of coordination number 2, the ligands and the central ion are linearly arranged. If the coordination number is 4, the ligand atoms surround the central ion either in a square planar or in a tetrahedral configuration. If the coordination number is 6, the ligands occupy the corners of an octahedron, in the center of which stands the central atom.

6-2   Protons and Metal Ions

In all solution environments the bare metal ions are in continuous search of a partner. All metal cations in water are hydrated, that is, they form aquo complexes. The coordination reactions in which metal cations participate in aqueous solutions are exchange reactions with the coordinated water molecules exchanged for some preferred ligands. The barest of the metal cations is the free hydrogen ion, the proton. Hence in some regards there is little difference in principle between a free metal ion and a proton.

BRÖNSTED ACIDITY AND LEWIS ACIDITY. In Figure 6-1 alkalimetric titration curves for the reaction of phosphoric acid and $Fe(H_2O)_6^{3+}$, respectively, with a base (OH^- ion) are compared. Millimolar solutions of H_3PO_4 and ferric perchlorate have a similar pH value. Both acids ($Fe \cdot aq^{3+}$ and H_3PO_4) are multiprotic acids, that is, they can transfer more than one proton.

In Figure 6-2 the titration of H_3O^+ with ammonia is compared with the titration of $Cu \cdot aq^{2+}$ with ammonia. pH and pCu ($= -\log [Cu \cdot aq^{2+}]$) are plotted as a function of the base added. In both cases "neutralization curves" are observed. In the case of the H_3O^+–NH_3 reaction a pronounced pH jump occurs at the equivalence point. The pCu jump is less pronounced in the $Cu \cdot aq^{2+}$–NH_3 reaction because NH_3 is bound to the Cu^{2+} ion in a stepwise consecutive way ($CuNH_3^{2+}$, $Cu(NH_3)_2^{2+}$, $Cu(NH_3)_3^{2+}$, $Cu(NH_3)_4^{2+}$,

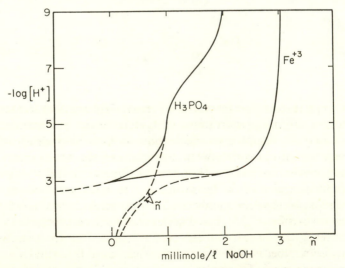

Figure 6-1   Alkalimetric titration of $10^{-3} M$ H_3PO_4 and $10^{-3} M$ $Fe \cdot aq^{3+}$. Both H_3PO_4 and $Fe \cdot aq^{3+}$ are multiprotic Brönsted acids. Millimolar solutions of H_3PO_4 and $Fe(ClO_4)_3$ have similar pH values.

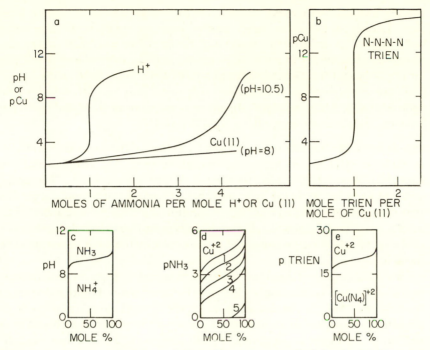

Figure 6-2   Titration of H_3O^+ and $Cu \cdot aq^{2+}$ with ammonia (*a*) and with Tetramine (trien) (*b*). Equilibrium diagrams for the distribution of NH_3–NH_4^+ (*c*) of the amino copper(II) complexes (*d*), and of Cu^{2+}, Cu-trien (*e*). The similarity of titrating H^+ with a base and titrating a metal ion with a base (Lewis acid–base interaction) is obvious. Both neutralization reactions are used analytically for the determination of acids and metal ions. A pH or pMe indicator or indicator electrodes (glass electrode for H^+ and copper electrode for Cu^{2+}) can be used for the end point indication.

$Cu(NH_3)_5^{2+})$ (Figure 2*d*). If, however, four NH_3 molecules are packaged together into one single molecule such as trien (triethylenetetramine, H_2N—CH_2—CH_2—NH—CH_2—CH_2—NH—CH_2—CH_2—NH_2) a 1:1 Cu-trien complex is formed and a simple titration curve with a very pronounced pCu jump is observed at the equivalence point (Figure 6-2*b*). In this case the Cu-trien equilibrium (Figure 6-2*e*) is as simple as the H^+–NH_3 equilibrium (Figure 6-2*c*). Such neutralization reactions are exploited analytically for the determination of acids or metal ions; a hydrogen ion electrode (glass electrode) and a metal ion-sensitive electrode (e.g., a copper electrode for Cu^{2+}), or a pH or pMe indicator, are used as sensors for H^+ and Me^{+n}, respectively.

The examples given illustrate the phenomenological similarity between the "neutralization" of H^+ with bases and that of metal ions with complex formers. The bases, molecules or ions that can "neutralize" H^+ or metal ions possess free pairs of electrons. Acids are proton donors according to Brön-sted. Lewis, on the other hand, has proposed a much more generalized definition of an acid in the sense that he does not attribute acidity to a particular element but to a unique electronic arrangement: the availability of an empty orbital for the acceptance of a pair of electrons. Such acidic or acid-analogue properties are possessed by H^+, metal ions and other Lewis acids such as $SOCl_2$, $AlCl_3$, SO_2, BF_3. In aqueous solutions protons and metal ions compete with each other for the available bases.

THE ACIDITY OF THE METAL IONS. It is frequently difficult to determine the number of H_2O molecules in the hydration shell, but many metal ions coordinate 4 or 6 H_2O molecules per ion. Water is a weak acid. The acidity of the H_2O molecules in the hydration shell of a metal ion is much larger than that of water. As pointed out before (Section 3-2) this enhancement of the acidity of the coordinated water may be interpreted qualitatively as the result of the repulsion of the protons of H_2O molecules by the positive charge of the metal ion.

Hence the acidity of aquo metal ions is expected to increase with decrease of the radius and an increase of charge of the central ion. Figure 6-3 attempts to illustrate how the oxidation state of the central atom determines the predominant species (aquo, hydroxo, hydroxo-oxo and oxo complexes) in the pH range of aqueous solutions. Metal ions with $z = +1$ are generally coordinated with H_2O atoms. Most bivalent metal ions are also coordinated with water up to pH values of 6 to 12. Most trivalent metal ions are already coordinated with OH^- ions within the pH range of natural waters. For $z = +4$ the aquo ions have become too acidic and are out of the accessible pH range of aqueous solutions with few exceptions, for example, Th(IV). There O^{2-} already begins to appear as a ligand, for example, for C(IV) where we have oxo-hydroxo complexes, $H_2CO_3 = CO(OH)_2$ or $HCO_3^- = CO_2(OH)^-$, in the pH range 4.5 to 10; above pH $= 10$, O^{2-} becomes the exclusive ligand (CO_3^{2-}). With even higher oxidation states of the central atom, hydroxo complexes can only occur at very low pH values. The scheme given in Figure 6-3 represents an oversimplification. For every oxidation state, a distribution of acidity according to the ionic radius exists: thus the acidity, as indicated by the pK values given in parentheses, increases in the series of the following aquo ions of $z = +2$.

$$Ba^{2+} (14.0), \quad Ca^{2+} (13.3), \quad Mg^{2+} (12.2), \quad Be^{2+} (5.7)\dagger$$

The electrostatic rules given are quite useful, but the picture is still rather oversimplified because other factors related to the electron distribution are

---

† $Be(aq)^{2+}$ may in fact only form polynuclear hydrolysis products [e.g., $Be_3(OH)_3^{3+}$] but on a comparable basis it is more acidic than $Mg(aq)^{2+}$.

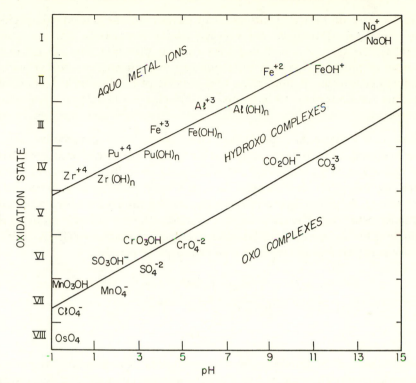

Figure 6-3   Predominant pH range for the occurrence of aquo, hydroxo, hydroxo-oxo and oxo complexes for various oxidation states. The scheme attempts to show a useful generalization, but many elements cannot be properly placed in this simplified diagram because other factors, such as radius and those related to electron distribution, have to be considered in interpreting the acidity of metal ions. A similar diagram has been given by Jørgensen (1963).

involved. For example, $Cd \cdot aq^{2+}$ is a stronger acid than $Ca \cdot aq^{2+}$ although these ions have the same charge to radius ratio.

Hydrolysis of Metal Ions

Hydrolysis is a somewhat antiquated term which, according to the Brönsted theory, is no longer necessary in order to describe the proton transfer from an acid to water or the proton transfer from water to a base. Unfortunately the term is still used, especially if the acid is an aquo metal ion.

More than 30 years ago Brönsted postulated that multivalent metal ions participate in a series of consecutive proton transfers:

$$Fe(H_2O)_6{}^{3+} = Fe(H_2O)_5OH^{2+} + H^+ = Fe(H_2O)_4(OH)_2{}^+ + 2H^+$$
$$= Fe(OH)_3(H_2O)_3(s) + 3H^+ = Fe(OH)_4(H_2O)_2{}^- + 4H^+$$

In the case of Fe(III), hydrolysis can go beyond the uncharged species $Fe(OH)_3(H_2O)_3(s)$ to form anions such as the ferrate(III) ion, probably $[Fe(OH)_4 \cdot 2H_2O]^-$. All hydrated ions can, in principle, donate a larger number of protons than that corresponding to their charge and can form anionic hydroxo-metal complexes, but, because of the limited pH range of aqueous solutions, not all elements can exist as anionic hydroxo or oxo complexes.

POLYNUCLEAR HYDROXO COMPLEXES. The scheme of a consecutive stepwise hydroxide binding is too simple. Although the hydrolysis products listed for hydrolysis of $Fe \cdot aq^{3+}$ are all known and identified, the intermediate steps are frequently complicated. In a few cases, the main products are monomeric. Polymeric hydrolysis species (isopolycations) have been reported for most metal ions. Thus the existence of multinuclear hydrolysis products is a rather general phenomenon. The hydrolyzed species such as $Fe(H_2O)_5OH^{2+}$ can be considered to dimerize by a condensation process

$$2Fe(H_2O)_5OH^{2+} = [(H_2O)_4Fe \underset{OH}{\overset{OH}{\diagup \diagdown}} Fe(H_2O)_4]^{4+} + 2H_2O$$

The existence of the dimer has been corroborated experimentally by potentiometric, spectrophotometric and magnetochemical methods. The dimer may undergo additional hydrolytic reactions which could provide additional hydroxo groups which then could form more bridges. The terms "ol" and "oxo" are often used in referring to the —OH— and —O— bridges. A sequence of such hydrolytic and condensation reactions, sometimes called olation and oxolation,† leads, under conditions of oversaturation with respect to the (usually very insoluble) metal hydroxide, to the formation of colloidal hydroxo polymers and ultimately to the formation of precipitates. In the pH range lower than the zero point of charge of the metal hydroxide precipitate, positively charged metal hydroxo polymers prevail. In solutions more alkaline than the zero point of charge, anionic hydroxo complexes (isopolyanions) and negatively charged colloids exist. Although multinuclear complexes have been recognized for many years for a few hydrolysis systems such as Cr(III) and Be(II) and for anions of Cr(VI), Si(IV), Mo(VI) and V(V), more recent studies have shown that multinuclear hydrolysis products of metallic cations are of almost universal occurrence in the water solvent system. Table 6-1 gives an illustration for some of the hydrolysis species

† Olation may be followed by oxolation, a process in which the bridging OH group is converted to a bridging O group.

Table 6-1 Multinuclear, Hydroxo and Oxo Complexes†

Type‡	Metals Believed to Form Such Complexes§
A. Cationic Complexes	

Me—OH—Me 　　　　Be(II), Mg(II), Zn(II), Cd(II), UO_2^{2+}

$$Me\underset{OH}{\overset{OH}{\diamond}}Me$$ 　　　　Cu(II), Fe(III), Hg(II), Sc(II), Sn(II)

$$Me\underset{OH}{\overset{OH}{\diamond}}Me\underset{OH}{\overset{OH}{\diamond}}Me$$ 　　　　Hg(II), Sn(II), Pb(II), Sc(III)

$$Me\underset{OH}{\overset{OH}{\diamond}}Me\underset{OH}{\overset{OH}{\diamond}}Me\cdots$$ 　　　　Sc(III), In(III)

$(Me(OH)_2)_n$

(cube structure) 　　　　Pb(II)

$(Me_4(OH)_4)\|$

Varied 　　　　$Be_3(OH)_3^{3+}$, $Bi_6(OH)_{12}^{6+}$, $Pb_6(OH)_8^{4+}$, $Al_7(OH)_{17}^{4+}$, $Al_{13}(OH)_{34}^{5+}$, $Mo_7O_{24}^{6-}$, $V_{10}O_{28}^{6-}$

B. Oxo Complexes of Metals, Metalloids and Nonmetals

Type	Examples
O_3XOXO_3	$Cr_2O_7^{2-}$, $S_2O_7^{2-}$, $P_2O_7^{4-}$
$O_3X(O_4X)XO_3$	$P_3O_{10}^{5-}$
$(XO_3)_n$	$(PO_3)_n^{n-}$, $(SiO_3)_n^{2n-}$, $CrO_3(s)$
$(X_2O_5)_n$	$P_2O_5(s)$, $(Si_2O_5)_n^{n-}$
$(XO_2)_n$	$SiO_2(s)$

† Modified from P. Schindler, personal communication, 1968.

‡ Charges are omitted; the structural arrangement given, although plausible, is hypothetical.

§ The list is not complete; many species that have been claimed by various authors are omitted. More detailed information is obtainable from L. G. Sillén and A. E. Martell, *Stability Constants of Metal–Ion Complexes*, Special Publ., No. 17, The Chemical Society, London, 1964. Many of the species given here may not be thermodynamically stable.

‖ Such a structure has been proposed by Esval and Tyree (see S. Y. Tyree, "The Nature of Inorganic Solute Species in Water," in *Equilibrium Concepts in Natural Water Systems*, Advances in Chemistry Series, No. 67, American Chemical Society, Washington, D.C., 1967, p. 183.

reported for various metal ions of interest in water. Additional information and corresponding equilibrium constants may be found in *Stability Constants of Metal–Ion Complexes*. Table 6-1 is not comprehensive and further developments in the near future may lead to corrections.

6-3   The Stability of Hydrolysis Species

The establishment of hydrolysis equilibria is frequently very fast, as long as the hydrolysis species are simple. Some metal ions like Cr(III), Co(III), Pt(II) and ruthenium, however, form their complexes very slowly. Polynuclear complexes are frequently formed rather slowly. Many of these polynuclear hydroxo complexes are kinetic intermediates in the slow transition from free metal ions to solid precipitates and are thus thermodynamically unstable. Some metal ion solutions "age," that is, they change their composition over periods of weeks because of slow structural transformations of the isopolyions. Such nonequilibrium conditions can frequently be recognized if the properties of metal ion solutions (electrode potentials, spectra, conductivity, light scattering, coagulation effects, sedimentation rates, etc.) depend on the history of the solution preparation.

Hydrolysis equilibria can be interpreted in a meaningful way if the solutions are not oversaturated with respect to the solid hydroxide or oxide. Occasionally it is desirable to extend equilibrium calculations into the region of oversaturation; but quantitative interpretations on the species distribution must not be made unless metastable supersaturation can be demonstrated to exist. Most hydrolysis equilibrium constants have been determined in the presence of a swamping "inert" electrolyte of constant ionic strength ($I = 0.1$, 1 or 3). As we have seen before, the formation of hydroxo species can be formulated in terms of acid–base equilibria. The formulation of equilibria of hydrolysis reactions is in agreement with that generally used for complex formation equilibria (see Table 6-2).

The following rules can be established:

1. The tendency of metal ion solutions to protolyze (hydrolyze) increases with dilution and with decreasing (H^+).

2. The fraction of polynuclear complexes in a solution decreases on dilution.

The first rule can be illustrated by comparing the equilibria ($I = 0$, 25°C)

$$Mg^{2+} + H_2O = MgOH^+ + H^+; \quad \log {}^*K_1 = -11.4 \qquad (1)$$

$$Cu^{2+} + H_2O = CuOH^+ + H^+; \quad \log {}^*K_1 = -6.0 \qquad (2)$$

Table 6-2   Formulation of Stability Constants†

I. *Mononuclear Complexes*
  (a) Addition of ligand

$$M \xrightarrow[K_1]{L} ML \xrightarrow[K_2]{L} ML_2 \cdots \xrightarrow[K_i]{L} ML_i \cdots \xrightarrow[K_n]{L} ML_n$$
$$\xrightarrow{\quad\quad \beta_2 \quad\quad}$$
$$\xrightarrow{\quad\quad\quad \beta_i \quad\quad\quad}$$
$$\xrightarrow{\quad\quad\quad\quad \beta_n \quad\quad\quad\quad}$$

$$K_i = \frac{[ML_i]}{[ML_{(i-1)}][L]} \tag{1}$$

$$\beta_i = \frac{[ML_i]}{[M][L]^i} \tag{2}$$

  (b) Addition of protonated ligands

$$M \xrightarrow[*K_1]{HL} ML \xrightarrow[*K_2]{HL} ML_2 \cdots \xrightarrow[*K_3]{HL} ML_i \cdots \xrightarrow[*K_n]{HL} ML_n$$
$$\xrightarrow{\quad\quad *\beta_2 \quad\quad}$$
$$\xrightarrow{\quad\quad\quad *\beta_i \quad\quad\quad}$$
$$\xrightarrow{\quad\quad\quad\quad *\beta_n \quad\quad\quad\quad}$$

$$*K_i = \frac{[ML_i][H^+]}{[ML_{(i-1)}][HL]} \tag{3}$$

$$*\beta_i = \frac{[ML_i][H^+]^i}{[M][HL]^i} \tag{4}$$

II. *Polynuclear Complexes*

  In β_{nm} and $*\beta_{nm}$ the subscripts n and m denote the composition of the complex $M_m L_n$ formed. [If $m = 1$, the second subscript $(= 1)$ is omitted.]

$$\beta_{nm} = \frac{[M_m L_n]}{[M]^m [L]^n} \tag{5}$$

$$*\beta_{nm} = \frac{[M_m L_n][H^+]^n}{[M]^m [HL]^n} \tag{6}$$

---

† The same notation as that used in L. G. Sillén and A. E. Martell, *Stability Constants of Metal–Ion Complexes*, Special Publ., No. 17, The Chemical Society, London, 1964, is used.

At great dilution (pH $\rightarrow$ 7), most of the Cu(II) of a pure Cu-salt solution [e.g., $Cu(ClO_4)_2$] will occur as a hydroxo complex

$$\alpha_{CuOH^+} = \frac{[CuOH^+]}{Cu_T} = \left(1 + \frac{[H^+]}{*K_1}\right)^{-1} = 0.91 \tag{3}$$

On the other hand, because of the small acidity of Mg^{2+}, even at infinite dilution the fraction of hydrolyzed Mg^{2+} ions of a solution of an Mg^{2+}-salt is very small:

$$\alpha_{MgOH^+} = \frac{[MgOH^+]}{Mg_T} = \left(1 + \frac{[H^+]}{*K_1}\right)^{-1} = 0.0001 \qquad (4)$$

Accordingly, only the salt solutions of sufficiently acid metal ions which fulfill the condition [1]

$$p*K_1 < \tfrac{1}{2}pK_w \qquad \text{or} \qquad p*\beta_n < \frac{n}{2}pK_w \qquad (5)$$

where $*\beta_n$ is the cumulative acidity constant (see Table 6-2), undergo substantial hydrolysis upon dilution. The progressive hydrolysis upon dilution is the reason that some metal salt solutions tend to precipitate upon dilution. (See Example 6-2.)

MONONUCLEAR WALL. If hydrolysis leads to mononuclear and polynuclear hydroxo complexes, it can be shown that mononuclear species prevail beyond a certain dilution. If we consider, for example, the dimerization of $CuOH^+$

$$2CuOH^+ = Cu_2(OH)_2^{2+}; \quad \log *K_{22} = 1.5 \qquad (6)$$

it is apparent from the dimensions of the equilibrium constant ($conc^{-1}$) that the dimerization is concentration dependent. Thus for a Cu(II) system where $Cu_T = [Cu^{2+}] + [Cu(OH)^+] + 2[Cu_2(OH)_2^{2+}]$, (6) can be formulated as

$$\frac{[Cu_2(OH)_2^{2+}]}{[CuOH^+]^2} = \frac{[Cu_2(OH)_2^{2+}]}{(Cu_T - [Cu^{2+}] - 2[Cu_2(OH)_2^{2+}])^2} = *K_{22} \qquad (7)$$

and it becomes obvious that $[Cu_2(OH)_2^{2+}]$ is dependent upon Cu_T. With the help of (7) and (2) for each pH, the mononuclear wall (e.g., Cu_T for $[Cu_{dimer}] = 1/100[Cu_{monomer}]$ can be calculated (compare Examples 6-1 and 6-3). As pointed out before, for many metals the polynuclear species are formed only under conditions of oversaturation with respect to the metal hydroxide or metal oxide and are thus not stable thermodynamically, for example,

$$Cu_2(OH)_2^{2+} = Cu^{2+}(aq) + Cu(OH)_2(s), \quad \Delta G° = -2.6 \text{ kcal} \qquad (8)$$

As shown by Schindler [1], multinuclear hydrolysis species usually are not observed during dissolution of the most stable modification of the solid hydroxide or oxide; they are formed, however, by oversaturating a solution with respect to the solid phase. Such polynuclear species, even if thermodynamically unstable, are of significance in natural water systems. Many multinuclear hydroxo complexes may persist as metastable species for years.

---

[1] P. Schindler, private communication.

Quantitative application of known hydrolysis equilibria is illustrated in the next two examples.

Example 6-1   *The Hydrolysis of* Iron(III).   The addition of $Fe(ClO_4)_3$ to H_2O may lead to the following soluble species: Fe^{3+}, $Fe(OH)^{2+}$, $Fe(OH)_2{}^{1+}$, $Fe(OH)_4{}^-$ and $Fe_2(OH)_2{}^{4+}$.
Compute the equilibrium composition of:

1. A homogeneous solution to which $10^{-4}\ M$ $(10^{-2}\ M)$ of iron(III) has been added and the pH adjusted within the range 1–4.5 with acid or base;
2. An iron(III) solution in equilibrium with amorphous ferric hydroxide.

The following equilibrium constants are available $I = 3(NaClO_4)$ (25°C).

$$Fe^{3+} + H_2O = FeOH^{2+} + H^+; \qquad \log {}^*K_1 = -3.05 \qquad \text{(i)}$$
$$Fe^{3+} + 2H_2O = Fe(OH)_2{}^+ + 2H^+; \qquad \log {}^*\beta_2 = -6.31 \qquad \text{(ii)}$$
$$2Fe^{3+} + 2H_2O = Fe_2(OH)_2{}^{4+} + 2H^+; \quad \log {}^*\beta_{22} = -2.91 \qquad \text{(iii)}$$
$$Fe(OH)_3(s) + 3H^+ = Fe^{3+} + 3H_2O; \qquad \log {}^*K_{s0} = 3.96 \qquad \text{(iv)}$$
$$Fe(OH)_3(s) + H_2O = Fe(OH)_4{}^- + H^+; \qquad \log {}^*K_{s4} = -18.7 \qquad \text{(v)}$$

1. In the *homogeneous system*, the concentration condition [(vi) or (vii)] must be fulfilled.

$$Fe_T = [Fe^{3+}] + [FeOH^{2+}] + [Fe(OH)_2{}^+] + 2[Fe_2(OH)_2{}^{4+}] \qquad \text{(vi)}$$

$$Fe_T = [Fe^{3+}]\left(1 + \frac{{}^*K_1}{[H^+]} + \frac{{}^*\beta_2}{[H^+]^2} + \frac{2[Fe^{3+}]{}^*\beta_{22}}{[H^+]^2}\right) \qquad \text{(vii)}$$

As with other polyprotic acids we may define successive distribution coefficients: $\alpha_0 = [Fe^{3+}]/Fe_T$, $\alpha_1 = [FeOH^{2+}]/Fe_T$, $\alpha_2 = [Fe(OH)_2{}^+]/Fe_T$ and $\alpha_{22} = 2[Fe_2(OH)_2{}^{4+}]/Fe_T$. [$\alpha_{22}$ gives the fraction of iron(III) present in the form of the dimer.]
Inspecting (vii) we note that the last term, proportional to the polymer concentration, is an implicit function of the concentration of iron(III). α_0 may be defined with the help of (i)–(iii) and (vii):

$$\alpha_0 = \left(1 + \frac{{}^*K_1}{[H^+]} + \frac{{}^*\beta_2}{[H^+]^2} + \frac{2Fe_T\alpha_0{}^*\beta_{22}}{[H^+]^2}\right)^{-1} \qquad \text{(viii)}$$

or

$$\frac{\alpha_0{}^2 2Fe_T{}^*\beta_{22}}{[H^+]^2} + \alpha_0\left(1 + \frac{{}^*K_1}{[H^+]} + \frac{{}^*\beta_2}{[H^+]^2}\right) - 1 = 0 \qquad \text{(ix)}$$

Equation ix is written in the form corresponding to the sum $\alpha_{22} + \alpha_0 + \alpha_1 + \alpha_2 - 1 = 0$. Accordingly, the remaining distribution coefficients are defined by

$$\alpha_{22} = \frac{\alpha_0{}^2 2 Fe_T \beta_{22}}{[H^+]^2} \qquad \text{(x)}$$

$$\alpha_1 = \frac{\alpha_0 {}^* K_1}{[H^+]} \qquad \text{(xi)}$$

$$\alpha_2 = \frac{\alpha_0 {}^* \beta_2}{[H^+]^2} \qquad \text{(xii)}$$

Now the computation can be carried out readily, starting with the quadratic equation ix, where we compute α_0 for a given Fe_T and for varying $[H^+]$. The results are plotted in Figure 6-4. It is obvious that the extent of hydrolysis depends on pH and Fe_T. Comparing the distribution diagrams for $Fe_T = 10^{-4} M$ and $10^{-2} M$ at a given pH, we note that the fraction of the dimer is concentration dependent. It may also be noted that α_0 is a measure of the relative extent of complex formation by OH^- ions:

$$\log \alpha_0 = pFe_T - pFe^{3+} = \Delta pFe \qquad \text{(xiii)}$$

The higher ΔpM, the better the metal ion is complexed.

2. The species distribution in the *heterogeneous system* can be calculated by considering (iv) in addition to the hydrolysis equilibria. By combining (i)–(iii) with (iv), we obtain

$$\log [FeOH^{2+}] = \log {}^* K_{s0} + \log {}^* K_1 + 2 \log [H^+] \qquad \text{(xiv)}$$
$$\log [Fe(OH)_2{}^+] = \log {}^* K_{s0} + \log \beta_2 + \log [H^+] \qquad \text{(xv)}$$
$$\log [Fe_2(OH)_2{}^{4+}] = 2 \log {}^* K_{s0} + \log {}^* \beta_{22} + 4 \log [H^+] \qquad \text{(xvi)}$$

together with (xvii) and (xviii) which follow from (iv) and (v)

$$\log [Fe^{3+}] = \log {}^* K_{s0} + 3 \log [H^+] \qquad \text{(xvii)}$$
$$\log [Fe(OH)_4{}^-] = \log {}^* K_{s4} - \log [H^+] \qquad \text{(xviii)}$$

In a double logarithmic diagram, (xiv)–(xviii) can be plotted as straight lines with well-defined slopes of $+4$, $+3$, $+2$, $+1$ and -1, respectively (Figure 5-3).

Example 6-2 *Dilution of a Ferric Salt Solution.* Estimate the species distribution of a pure ferric perchlorate solution (1 M) as a function of dilution. (Perchlorate is chosen as an anion since it does not appear to form complexes with Fe^{3+}.) Variation in activity coefficients caused by dilution may be ignored; this corresponds to diluting with an inert salt of constant I medium.

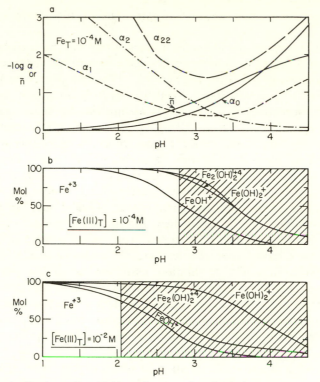

Figure 6-4 Distribution of Fe(III) species as a function of pH (Example 6-1). The extent of hydrolysis depends on pH and on the total Fe(III) concentration.

a: plots of the average ligand number $\bar{n}$ and the distribution coefficients α, respectively, as a function of pH for a 10^{-4} M Fe(III) solution. $\bar{n}$ is experimentally accessible from [H^+] measurements in alkalimetric titrations; α_1 can be determined independently from potentiometric measurements (ferro–ferri cell) or from spectrophotometric or magneto-chemical measurements. Either one or both parameters can be used to arrive at equilibrium constants.

Distribution diagrams for the various Fe(III) species are depicted in *b* and *c* using representative equilibrium constants. In the shaded areas the solution becomes over-saturated with respect to $Fe(OH)_3(s)$ ($K_s = 10^{-38}$). Additional polynuclear hydrolysis species occur as kinetic intermediates in the usually slow transition to $Fe(OH)_3(s)$ in this pH range.

The problem consists essentially in simultaneously solving (i)–(iii) and (vi) of Example 6-1 together with the proton condition

$$[H^+] = [FeOH^{2+}] + 2[Fe(OH)_2{}^+] + 2[Fe_2(OH)_2{}^{4+}] + [OH^-] \qquad (i)$$

Equation i also follows from the electroneutrality condition of the solution, considering that $[ClO_4{}^-] = 3Fe_T$. The existence of $Fe(OH)_4{}^-$ may be ignored since it does not occur above concentrations of $10^{-9}\,M$ in solutions of pH < 7.

There are various operational approaches that can be used to solve the requisite five equations simultaneously. One convenient approach starts by making a guess of $[H^+]$ for a given Fe_T. [This guess may be based on a tentative calculation assuming that only one hydrolysis reaction predominates at a given concentration; e.g., one may assume that for concentrated Fe(III) solutions hydrolysis to the dimer (Eq. iii of Example 6-1) determines $[H^+]$; while for very dilute solutions $[H^+]$ is controlled by hydrolysis to $Fe(OH)_2{}^+$ (Eq. ii of Example 6-1).] With this guess and with the aid of Equation ix of Example 6-1 we calculate a tentative value of α_0. Then tentative values of α_1, α_2 and α_{22} and thus tentative concentrations for the individual species are obtained. The adequacy of the guess is then checked with the help of the proton condition (Eq. i); with a new and improved guess and subsequent iteration, convergence can be obtained. The results, plotted in Figure 6-5, illustrate the dilution rules given earlier. The extent of hydrolysis increases upon dilution. In a $10^{-3}\,M$ solution, assuming metastable oversaturation, approximately 30% of the iron is present as free ferric ion; this fraction is reduced to 4% in a $10^{-4}\,Fe_T$ solution. The data given show that because of pH increase, the proportion of the dimer increases upon dilution up to about $10^{-2}\,M\,Fe_T$ but on further dilution the fraction of Fe(III) present as a dimer decreases.

Figure 6-5 also shows the dilution at which precipitation of amorphous $Fe(OH)_3$ occurs if equilibrium is attained. Accordingly, only concentrated $(10^{-1}\,M$ and stronger) Fe(III) solutions with pH values below 2 are thermodynamically stable with respect to solid amorphous ferric hydroxide. [Even such solutions, however, are not stable with respect to more stable solid modifications, i.e., α-FeOOH (goethite).] Diluting an unacidified Fe(III) salt solution leads to precipitation of $Fe(OH)_3(s)$. For example, a $10^{-3}\,M$ $Fe(ClO_4)_3$ solution changes its color upon standing; continuously changing spectra (UV and visible) are observed. Within days such solutions become turbid and within weeks precipitates of $Fe(OH)_3$ can be observed. Extremely large dilutions of a Fe(III) salt $(Fe_T < 10^{-10}\,M)$ are necessary to maintain iron(III) in solution. As Figure 6-5 suggests, the species that predominates at such dilutions (pH $\approx$ 7) is $Fe(OH)_2{}^+$.

The Formation Curve

Figure 6-1 illustrates that aquo metal ions can be titrated alkalimetrically. The morphology of the titration curve frequently can give valuable insight into the type of species encountered during the titration. In Section 3-8 it was

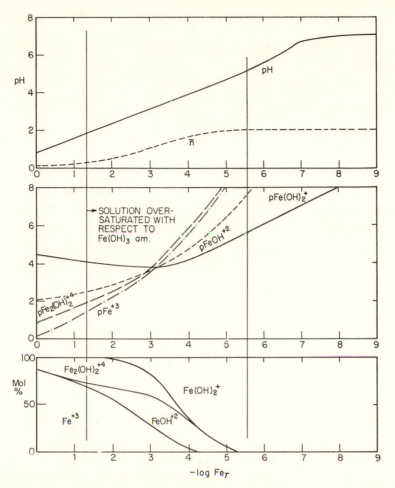

Figure 6-5   Dilution of $Fe(ClO_4)_3$. 1 M $Fe(ClO_4)_3$ is diluted. The extent of hydrolysis is, for example, illustrated by $\bar{n}$ [OH^- bound per Fe(III)]. Fe(III) solutions are not stable with respect to solid amorphous ferric hydroxide over most of the pH range (pH > 1.3).

shown that the alkalimetric titration curve of a polyprotic acid, for instance H_3PO_4, is given by

$$f = \frac{C_B}{P_T} = \frac{[H_2PO_4^-] + 2[HPO_4^{2-}] + 3[PO_4^{3-}]}{P_T} + \frac{[OH^-] - [H^+]}{P_T} \tag{9}$$

$$= \alpha_1 + 2\alpha_2 + 3\alpha_3 + \frac{[OH^-] - [H^+]}{P_T} \tag{10}$$

Similarly, the alkalimetric titration of aquo ferric ion can be characterized by

$$f = \frac{C_B}{\text{Fe}_T} = \frac{[\text{FeOH}^{2+}] + 2[\text{Fe(OH)}_2{}^+] + 2[\text{Fe}_2(\text{OH})_2{}^{4+}] + 4[\text{Fe(OH)}_4{}^-]}{\text{Fe}_T}$$

$$+ \frac{[\text{OH}^-] - [\text{H}^+]}{\text{Fe}_T} \tag{11a}$$

$$= \alpha_1 + 2\alpha_2 + \alpha_{22} + 4\alpha_4 + \frac{[\text{OH}^-] - [\text{H}^+]}{\text{Fe}_T} \tag{11b}$$

The α values are now defined as in Example 6-1 (note that $\alpha_{22} = 2[\text{Fe}_2(\text{OH})_2{}^{4+}]/$ Fe_T). The first term on the right-hand side of (11a) is a measure of the average number of hydroxide ions bound per iron(III) atom; it is called the *ligand number* or the *formation function*, $\bar{n}$

$$\bar{n} = \frac{[\text{FeOH}^{2+}] + 2[\text{Fe(OH)}_2{}^+] + 2[\text{Fe}_2(\text{OH})_2{}^{4+}] + 4[\text{Fe(OH)}_4{}^-]}{\text{Fe}_T}$$

$$= \alpha_1 + 2\alpha_2 + \alpha_{22} + 4\alpha_4 \tag{12}$$

The ligand number, generally applicable to any ligand, is usually plotted as a function of the logarithm of the free ligand concentration; hence, as a function of log $[\text{OH}^-]$ or $-\log [\text{H}^+]$ in the case of hydrolysis equilibria. In the case of H_3PO_4 (Eqs. 9, 10) $\bar{n}$ may be interpreted similarly as the average number of protons deficient per P-atom with respect to H_3PO_4. (In interpreting the solution composition, proton deficiency is mathematically equivalent to hydroxide excess.) In Figure 6-1 the ligand number is plotted as a function of $-\log [\text{H}^+]$, and in Figure 6-5 as a function of Fe_T. It is apparent from (11a) that the ligand number can be obtained directly from potentiometric measurements. For the evaluation of equilibrium constants, attempts are made to interpret data of $\bar{n}$ obtained from measurements over a wide range of Me_T and $[\text{H}^+]$ in terms of probable hydrolysis species. A convenient feature of the formation curve of mononuclear systems is that it depends solely on the ligand concentration and not on the total concentration of the metal ion. The existence of polynuclear complexes is detected most readily if the variation in Me_T causes shifts in the formation curve.

The ligand number is also a convenient parameter under nonequilibrium conditions. In a $\text{Me}^{z+}–\text{H}_2\text{O}$ system, $\bar{n}$ also gives the average charge q of the hydrolysis species

$$q = z - \bar{n} \tag{13}$$

BUFFER INTENSITY. The buffer intensity, $\beta = dC_B/d\text{pH}$, is related to the slope of the formation curve of a hydrolysis system. For the iron(III) system we can formulate [cf. (11)]

$$\beta = \frac{dC_B}{d\text{pH}}$$

$$= \frac{d[\text{FeOH}_2{}^+] + 2d[\text{Fe(OH)}_2{}^+] + 2d[\text{Fe}_2(\text{OH})_2{}^{4+}] + 4d[\text{Fe(OH)}_4{}^-]}{d\text{pH}}$$

$$+ \frac{d[\text{OH}^-] - d[\text{H}^+]}{d\text{pH}} \tag{14}$$

Substituting (12) into (14) gives

$$\beta = \frac{dC_B}{d\text{pH}} = \text{Fe}_T \frac{d\bar{n}}{d\text{pH}} + \frac{d[\text{OH}^-] - d[\text{H}^+]}{d\text{pH}} \tag{15}$$

where the last term is equal to $2.3([\text{H}^+] + [\text{OH}^-])$. If $[\text{H}^+]$ and $[\text{OH}^-]$ are small in comparison to the total concentration (Fe_T), the slope of the formation curve ($\bar{n}$ versus $-\log[\text{H}^+]$) of a hydrolysis system is a measure of the buffer intensity. Each hydrolysis species contributes to the buffer intensity. Frequently no steps appear in the formation curve because successive acidity (hydrolysis) constants are very close together; therefore a rather large buffer intensity that remains relatively uniform over a wide pH range may be imparted to the solution by metal ion hydrolysis (see Figure 6-4a).

Hydrous Metal Oxides

It has been pointed out that the formation of a precipitate can often be considered the final stage in the formation of polynuclear complexes. Aggregates of ions which form the building stones in the lattice are produced in the solution, and these aggregates combine with other ions to form neutral compounds.

In concurrence with the scheme depicted in Figure 6-3, there is a continuous transition between metallic and metalloid behavior. This may also be made evident by comparing the metal hydroxides in a section of the periodic table (Figure 6-6). Many multivalent hydrous oxides are amphoteric because of the acid–base equilibria involved in the hydrolysis reactions of aquo-metal ions. Thus *hydrogen ions and hydroxide ions* are primarily *the potential determining ions* for hydrous oxide precipitates. Alkalimetric or acidimetric titration curves for hydrous metal oxides, that is, formation curves for heterogeneous systems, provide a quantitative explanation for the manner in which the charge of the hydrous oxide depends on the pH of the medium. The amphoteric behavior of solid metal hydroxides becomes evident from such titration curves. From an operational point of view such hydrous oxides can be compared with amphoteric polyelectrolytes and can be considered as

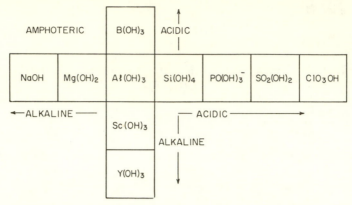

Figure 6-6 Acidity and basicity of some metal hydroxides.

hydrated solid electrolytes, frequently possessing a variable space lattice in which the proportion of different ions, cations as well as anions, is variable within the limits of electrical neutrality of the solid. These hydroxides show a strong tendency to interact specifically with anions as well as with cations. Interactions with anions (ligand exchange or anion exchange) predominate under conditions where the metal oxide is positively charged, that is, at pH values below the isoelectric point. At pH values higher than the isoelectric point the relative number of extra coordinated OH^- ions (or of hydroxo groups that have dissociated H^+ ions) will increase with increased concentration of base. Under such conditions the solid phase is capable of interacting with cations (cation exchange) (Section 9-5).

6-4   Colloid-Chemical Properties of Hydroxo Metal Complexes

It has been known for a long time, but not sufficiently appreciated, that metal ion hydrolysis species are strongly adsorbed at solid–solution interfaces [2]. This tendency to be adsorbed is especially pronounced for polynuclear poly-hydroxo species. No adequate theory for this enhanced adsorption by hydrolysis is available, but a few qualitative reasons can be given [3]. (a) Hydrolyzed species are larger and less hydrated than nonhydrolyzed species. (b) The enhancement of adsorption is apparently due to the presence of a coordinated OH^- group. Simple OH^- ions are bound at many solid surfaces and are frequently potential determining ions; hydroxo metal complexes may

[2] W. Stumm and J. J. Morgan, *J. Amer. Water Works Assoc.*, **54**, 971 (1962).
[3] W. Stumm and C. R. O'Melia, *J. Amer. Water Works Assoc.*, **60**, 514 (1968).

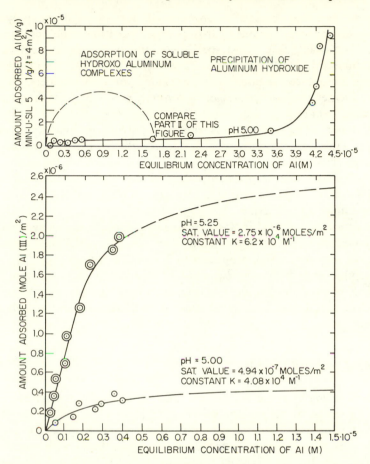

Figure 6-7 Adsorption of hydrolyzed Al(III) on colloidal silica, fitted by Langmuir isotherms [from H. Hahn and W. Stumm, *Advan. Chem. Ser.*, **79**, 91 (1968)]. Reproduced with permission from American Chemical Society.

similarly or to an even larger extent be adsorbed to the solid surface [4]. Alternatively, the replacement of an aquo group by a hydroxo group in the coordination sheath of a metal atom may render the complex more hydrophobic by reducing the interaction between the central metal atom and the remaining aquo groups. This reduction in solvent–hydroxo complex interaction might then in turn enhance the formation of covalent bonds between the metal atom and specific sites on the solid surface by reducing the energy necessary to displace water molecules from the coordination sheath. Finally,

[4] C. R. O'Melia and W. Stumm, *J. Colloid Interfac. Sci.*, **23**, 437 (1967).

adsorption becomes especially pronounced for polyhydroxo–polymetal species because more than one hydroxide group per "molecule" can become attached at the interface. Most species containing OH-groups, cationic as well as anionic hydroxo complexes, have been observed to be adsorbed at solid–liquid interfaces. In addition to metal ion hydrolysis products, one may cite polysilicates, polyphosphates and heteropoly ions (e.g., phosphotungstate and silicotungstate).

It is interesting to note that adsorption of polyhydroxo metal complexes can occur against electrostatic repulsion forces, that is, the chemical adsorption energy can outweigh the electrochemical work involved in the adsorption. Figure 6-7 shows some experimental data in the adsorption of hydrolyzed Al(III) [solutions oversaturated with respect to $Al(OH)_3(s)$]. The data can be fitted by Langmuir isotherms.

Coagulation as practiced in water treatment is brought about by metal ion hydrolysis species and not by free multivalent metal ions [5]. These hydrolysis species also can play a significant role in flotation processes [6].

Adsorption of hydrolysis species must be assumed to occur on most interfaces and is thus of considerable significance in the transformations of multivalent metal ions and in the fate of radionuclides in natural water systems.

6-5   Metal Ions and Ligands

Considerable emphasis has been placed thus far on hydroxo complexes; this is amply justified by the ubiquitousness of OH^- in water and by the strong affinity of many metal ions to OH^-. But other proton acceptors can serve as electron pair donors and thus coordinate as Lewis bases with metal ions. There is no single generally valid correlation between the stability of proton complexes (reciprocal of protolysis constant) and the stability of metal complexes, but some weak bases such as ClO_4^- and NO_3^- have very little tendency to form metal complexes.

A *and* B *Metal Cations*

Preference for one type of ligand over another is a function of the cation in aqueous solutions. Hence, the combination of ligand and cations can become highly selective. Ahrland, Clatt and Davies [7] divided metal ions into two

[5] E. Matijevic et al., *J. Phys. Chem.*, **65**, 826 (1961).
[6] M. C. Fuerstenau, P. Somasundaran and D. W. Fuerstenau, *Trans. Inst. Mining Met.*, **74**, 381 (1965).
[7] S. Ahrland, S. J. Clatt and W. R. Davies, *Quart. Rev.* (London), **12**, 265 (1958).

categories, depending on whether the metal ions formed their most stable complexes with the first ligand atom of each periodic group (F, O, N) or with a later member of the group (I, S, P). As Table 6-3 shows, this classification into A and B metal cations is reflected by the number of electrons in the outer shell. Class A metal cations have the electron configuration of inert

Table 6-3   Classification of Metal Ions and Other Lewis Acids

A-Metal Cations	Transition Metal Cations	B-Metal Cations
Electron configuration of inert gas. Low polarizability. "*Hard* Spheres" (H^+), Li^+, Na^+, K^+, Be^{2+}, Mg^{2+}, Ca^{2+}, Sr^{2+}, Al^{3+}, Sc^{3+}, La^{3+}, Si^{4+}, Ti^{4+}, Zr^{4+}, Th^{4+}	1–9 outer shell electrons; spherically not symmetrical V^{2+}, Cr^{2+}, Mn^{2+}, Fe^{2+}, Co^{2+}, Ni^{2+}, Cu^{2+}, Ti^{3+}, V^{3+}, Cr^{3+}, Mn^{3+}, Fe^{3+}, Co^{3+}	Electron number corresponds to Ni^0, Pd^0 and Pt^0 (10 or 12 outer shell electrons). Low electronegativity, high polarizability. "*Soft* Spheres" Cu^+, Ag^+, Au^+, Te^+, Ga^+, Zn^{2+}, Cd^{2+}, Hg^{2+}, Pb^{2+}, Sn^{2+}, Tl^{3+}, Au^{3+}, In^{3+}, Bi^{3+}

According to Pearson's [8] *Hard and Soft Acids*

Hard Acids	Borderline	Soft Acids
All A-metal cations plus Cr^{3+}, Mn^{3+}, Fe^{3+}, Co^{3+}, UO^{2+}, VO^{2+} as well as species like BF_3, BCl_3, SO_3, RSO_2^+, RPO_2^+, CO_2, RCO^+, R_3C^+	All bivalent transition metal cations plus Zn^{2+}, Pb^{2+}, Bi^{3+} SO_2, NO^+, $B(CH_3)_3$	All B-metal cations minus Zn^{2+}, Pb^{2+}, Bi^{3+} All metal atoms, bulk metals I_2, Br_2, ICN, I^+, Br^+

Preference for Ligand Atom

$N \gg P$	$P \gg N$
$O \gg S$	$S \gg O$
$F \gg Cl$	$I \gg F$

Qualitative Generalizations on Stability Sequence

Cations	Cations
Stability $\simeq$ prop. $\dfrac{\text{Charge}}{\text{Radius}}$	Irving-Williams Order $Mn^{2+} < Fe^{2+} < Co^{2+} < Ni^{2+} < Cu^{2+} > Zn^{2+}$
Ligands	Ligands
$F > O > N = Cl > Br > I > S$ $OH^- > RO^- > RCO_2^-$ $CO_3^{2-} \gg NO_3^-$ $PO_4^{3-} \gg SO_4^{2-} \gg ClO_4^-$	$S > I > Br > Cl = N > 0 > F$

[8] R. G. Pearson, *J. Amer. Chem. Soc.*, 85, 3533 (1963).

gases. These ions may be visualized to be of spherical symmetry and their electron sheaths are not readily deformed under the influence of electric fields, such as those produced by adjacent charged ions (low polarizability). Metaphorically, one may think of *hard spheres.* Class B metal cations, on the other hand, have a more readily deformable electron sheath (high polarizability) and may be visualized as *soft spheres.*

Metal cations in class A form complexes preferentially with the fluoride ion and ligands having oxygen as a donor atom. Water is more strongly attracted to these metals than ammonia or cyanide. No sulfides (precipitates or complexes) are formed, since OH^- ions are bound before HS^- or S^{2-}. Chloro or iodo complexes are weak and occur most readily in acid solutions. The univalent alkali ions essentially form complexes only with water (some weak complexes of Li^+ and Na^+ with chelating agents such as EDTA and polyphosphates are known). Chelating agents containing only nitrogen or sulfur as ligand atoms do not coordinate with A cations to form complexes of appreciable stability. Class A metal cations tend to form difficultly soluble precipitates with OH^-, CO_3^{2-} and PO_4^{3-}; no reaction occurs with sulfur and nitrogen donors (addition of NH_3 alkali sulfides or alkali cyanides produces solid hydroxides).

With class A metals a simple electrostatic picture of the binding of cation and ligand may be a satisfactory explanation of complex stability; for example, the stability increases rapidly with an increase in charge on the metal ion and those with the smallest radius form the most stable complexes. Some stability sequences are indicated in Table 6-3.

On the other hand, class B metal ions coordinate preferentially with bases containing I, S, P or N as donor atoms. Thus, metal ions in this class may bind ammonia stronger than water, CN^- in preference to OH^- and form more stable I^- or Cl^- complexes than F^- complexes. These metal cations, as well as transition metal cations (Table 6-3), form insoluble sulfides and soluble complexes with S^{2-} and HS^-.

In this group electrostatic forces do not appear to be of primary importance because neither the charge nor the size of interacting ions is entirely decisive for the stability sequence. Noncolored components often yield a colored compound, thus indicating a significant deformation of the electron sheath. Hence, in addition to coulombic forces, other types of interactions caused by nonelectrovalent forces must be considered. These other types of interactions have been interpreted in terms of quantum mechanics. In a simplified picture we may visualize bond formation as resulting from the sharing of an electronic pair between the central atom and the ligand (covalent bond). The tendency toward complex formation increases with the capability of the cation to take up electrons (ionization potential of the metal) and with decreasing electronegativity, that is, with increasing tendency of the ligand to donate electrons.

In the series F, O, N, Cl, Br, I, S, the electronegativity decreases from left to right while the stability of complexes with B cations increases. There are further factors such as steric hindrance and entropy effects; thus, stability sequences with cations of the B group are often irregular.

TRANSITION METAL CATIONS have between zero and ten d-electrons (Table 6-3). For these cations a reasonably well-established rule on the sequence of complex stability, the Irving-Williams order [9], is valid. According to this rule the stability of complexes increases in the series $Mn^{2+} < Fe^{2+} < Co^{2+} < Ni^{2+} < Cu^{2+} > Zn^{2+}$. An example is given in Figure 6-8. As a first approximation one might argue qualitatively that the electrovalent behavior of the bivalent transition metal cations remains almost constant, but that the non-electrovalent behavior changes markedly.

The Irving-Williams order is usually explained with the aid of *crystal field theory*. The crystal field is the electric field acting at the central metal ion due to the attached groups or ligands. This field affects the energy of electrons, particularly those in the d-orbitals of the central ion. If d-electrons can preferentially occupy lower energy d-orbitals, the complex becomes more stable than otherwise by an amount called the crystal field stabilization

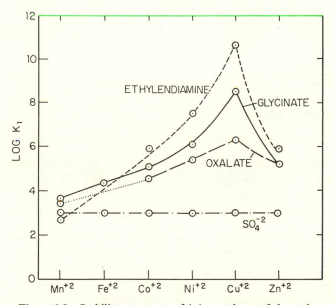

Figure 6-8 Stability constants of 1:1 complexes of d-metals.

[9] H. Irving and R. J. P. Williams, *J. Chem. Soc.*, **1953**, 3192.

energy (CFSE). The ions Mn^{2+} (5 d-electrons) and Zn^{2+} (10 d-electrons) cannot show any crystal field stabilization; but CFSE increases progressively from Fe^{2+} (6 electrons) to Cu^{2+} (10 electrons).

SOFT AND HARD ACIDS AND BASES. The division of metal ions into class A, class B and transition ions† represents a convenient classification scheme. One might wish to have such a classification not only for compounds but also for reactions. Complex formation reactions have been interpreted in terms of Lewis acid–base interaction. The relative binding strength of an electron pair donor (Lewis basicity) depends on the nature of the electron pair acceptor (acidity). On the basis of the behavior of A metal cations and B metal cations discussed above, two broad categories of reactions can be established:

1. The reaction of an A cation with a ligand that attaches preferentially to A ions ($F^- \gg Cl^-$; $OH^- > NH_3$).

2. The reaction of a B cation with a ligand that has a strong tendency to share electrons with B ions ($I^- > Cl^- > F^-$; $NH_3 > OH^- > H_2O$).

In Pearson's concept of soft and hard acids and bases (SHAB concept), the reactions listed under (1) are classified as hard acid–hard base interactions, while the typical reactions of B cations fall into the category of soft acid–soft base interactions. The classification of hard and soft acids can be extended to species other than metal ions (Table 6-3). It is beyond the scope of this discussion to elaborate on various ramifications of the SHAB concept, but some useful qualitative trends, usable for qualitatively predicting chemical reactions and relative sequence in the stability of compounds, are indicated in Table 6-3. However, it is not possible to set up a single scale of strengths for Lewis acids or Lewis bases.

Ion Pairs and Complexes

Principally two types of complex species can be distinguished:

1. The metal ion or the ligand or both retain the coordinated water in forming the complex compound, that is, the metal ion and the base are separated by one or more water molecules.

2. The interacting ligand is immediately adjacent to the metal cation.

In the first case one speaks of ion pairs; in the latter case of complexes. Most methods of determining stability constants do not distinguish between the two types of associations. In some cases a distinction is possible through kinetic or spectrophotometric investigation. Kinetically, when true complexes are being formed, a dehydration step must precede the association reaction.

† In a more detailed classification we also have to consider the rare-earth ions (with 1–13 electrons in excess of xenon) and the more complicated actinide series (elements 89 to 103).

Association accompanied by changes in the absorbance of visible light is indicative of complex formation reactions as such, whereas the formation of ion pairs may be accompanied by changes in the UV region.

Simple electrostatic models which consider coulombic forces interacting between the ions over appropriate distances permit the estimation of stability constants of ion pairs. Such calculations give the following range of stability constants $(25°C, I = 0)$:

> For ion pairs with opposite charge of 1, $\log K \simeq 0-1$
> For ion pairs with opposite charge of 2, $\log K \simeq 1.5-2.2$
> For ion pairs with opposite charge of 3, $\log K \simeq 2.8-4.0$

The formation of ion pairs is characterized by a positive ΔH and a positive ΔS, that is, it is caused by a gain in entropy.

The fact that stability constants of bivalent cations with SO_4^{2-} are relatively constant is indicative that the association with SO_4^{2-} occurs without complete displacement of the hydration sheath.

Example 6-3   *The Stability of the* $MgCO_3$ *Complex.*   To what extent is $MgCO_3(aq)$ a significant species in a solution of pH = 8.0 containing $Mg_T = 10^{-2}\ M$ and $[Alk] = 10^{-2}\ eq\ liter^{-1}$, $K_1 = 10^{2.2}$ $(I = 0.1, 25°C)$.

The stability of $MgCO_3(aq)$ is defined by

$$\frac{[MgCO_3(aq)]}{[Mg^{2+}][CO_3^{2-}]} = K_1 = 10^{2.2} \tag{i}$$

Furthermore (since $[MgOH^+]$ is negligible at pH 8) Mg_T is given by

$$Mg_T = [Mg^{2+}](1 + K_1[CO_3^{2-}]) \tag{ii}$$

At pH = 8, $C_T \simeq [Alk]$ and C_T is given by

$$C_T = [HCO_3^-] + [CO_3^{2-}] + [MgCO_3(aq)] \tag{iii}$$

Assuming that in (iii) the last term is negligible in comparison to the first two terms, we estimate $[CO_3^{2-}]$. Using $p^cK_2 \simeq 10.0$, we obtain $[CO_3^{2-}] \approx 10^{-4}$ M. Using this value in (ii) and (i), an equilibrium concentration of $[MgCO_3(aq)] \simeq 1.6 \times 10^{-4}\ M$ obtains, that is, approximately 1.6% of Mg_T in solution is present as carbonato complex. (The assumptions made in arriving at this result were justified.)

6-6   Complex Formation and the Solubility of Solids

We have seen that precipitation can be interpreted in terms of formation of polynuclear complexes. As has been illustrated for OH^- as a ligand (see Figure 5-3), the presence of metal complexing species will affect the solubility of slightly soluble metal salts. It is well known that the presence of foreign ligands increases the solubility. Thus ammonia increases the solubility of silver halides or citrate and polyphosphates may dissolve $CaCO_3$, but even without foreign ligands complex formation in solution among constituents of the lattice will affect the solubility. Computation of the solubility solely from the solubility product and acid–base equilibria is possible only if the solid phase consists of a salt, that is, if the building stones of the lattice ionize completely in solution. Few of the solid binary compounds actually consist of truly ionic lattices; only then is it possible to calculate the solubility from the free metal ion concentration, $[Me^{+z}]$, in equilibrium with the solid phase

$$Me_T = [Me^{+z}] \tag{16}$$

In Section 5-2, it was shown that the solubility of oxides and hydroxides can be enhanced by the formation of mononuclear hydroxo complexes. The total solubility was given [(13), Section 5-2] as

$$Me_T = [Me^{+z}] + \sum_1^n [Me(OH)_n^{z-n}] \tag{17}$$

This equation can now be generalized if we consider the possibility of complex formation in solution with any ligand L or its protonated form H_iL. Then the total solubility includes, in addition to the free metal ion concentration, the sum of all the soluble metal complexes:

$$Me_T = \sum^m [Me_mL_nH_i] \tag{18}$$

where all values of m, n and i have to be considered for the summation in (18). As pointed out by Schwarzenbach and Widmer [10], if a compound Me_mL_n has a very small solubility product K_{s0}, it must be expected that molecular associates between the metal ion and the ligand exist in solution as stable complexes. This is understandable if we consider that the same forces that result in slightly soluble lattices are operative in the formation of soluble complexes. Metal ions of the class B cations (soft acids) are especially capable of forming covalent bonds (σ and π bonds). The presence of covalent bonds is frequently evident when the color of the solids (green copper hydroxide carbonate, yellow AgI, black Ag_2S) is not a composite of the colors of the ions. The following examples will illustrate the significance of complex formation in enhancing the solubility of solids.

The first example is based on a well-documented case history of the solubility of Ag_2S given by Schwarzenbach and Widmer. Ag^+, as a typical B

[10] G. Schwarzenbach and M. Widmer, *Helv. Chim. Acta*, **46**, 2613 (1963); **49**, 111 (1966).

cation ("soft acid"), has a strong tendency to coordinate with sulfur donor atoms. In this instance the concentration of free Ag^+ and the solubility is particularly illustrative. A second example examines the effect of carbonate complexing upon the solubility of Cu(II).

Example 6-4 *The Solubility of* Ag_2S. 1. Compute from the information given in Table 6-4 the concentration of free Ag^+ in a sulfide solution of 0.1 M total sulfur concentration $(S_T = [H_2S] + [HS^-] + [S^{2-}] = 0.1\ M)$ and pH $= 13$.

Considering the mass laws of (1), (2) and (3), the free Ag^+ concentration can be computed by

$$\log [Ag^+] = \tfrac{1}{2}(\log K_{s_0} - \log S_T - \log \alpha_2) \tag{i}$$

where α_2 is defined as usual:

$$\alpha_2 = \frac{[S^{2-}]}{S_T} = \left(1 + \frac{[H^+]}{K_{HS}} + \frac{[H^+]^2}{K_{H_2S}K_{HS}}\right)^{-1}$$

Ignoring the fact that the solution has a slightly higher I than that for which the constants have been determined, one obtains for pH $= 13$ a $\log \alpha = -1$ and, correspondingly, *pAg $= 23.85$.*

This result corresponds to *one silver ion per liter*. Somewhat naively one might be inclined to ask the "statistical" question: where do we find the single Ag^+ if we subdivide a liter of saturated solution into ten 100-ml portions. Interestingly enough, a silver electrode that can sense silver ions potentiometrically will detect pAg $= 23.85$ in each one of the 100-ml portions. Although a silver electrode is extremely sensitive to free Ag^+, it could not specifically respond to solitary silver ions. The electrode can, of course, function only because larger concentrations of Ag(I) are present in the solution of thio silver(I) complexes. We would reach the same conclusions by making a solubility determination using a radiochemical method and a Ag_2S

Table 6-4 Equilibria Pertinent in Effecting the Solubility of Ag_2S†

$Ag_2S(s) = 2Ag^+ + S^{2-}$	$\log K_{s0} = -49.7$	(1)
$H_2S(aq) = H^+ + HS^-$	$\log K_{H2S} = -6.68$	(2)
$HS^- = H^+ + S^{2-}$	$\log K_{HS} = -14.0$	(3)
$Ag^+ + SH^- = AgSH$	$\log K_1 = 13.3$	(4)
$AgSH + SH^- = Ag(SH)_2^-$	$\log K_2 = 3.87$	(5)
$2Ag(SH)_2^- = HSAgSAgSH^{2-} + H_2S(aq)$	$\log K = 3.2$	(6)
$Ag_2S(s) + 2HS^- = HSAgSAgSH^{2-}$	$\log K_{s3} = -4.82$	(7)‡

† Cf. Schwarzenbach and Widmer [10]. $I = 0.1$ (NaClO$_4$), 20°C.
‡ Equilibrium constant obtained from (1) and (3)–(6).

that is tagged with radioactive Ag^+. Indeed, radiochemically we would detect a solubility of approximately 10^{-8} M. Accordingly the real solubility of Ag_2S under the conditions specified is approximately 14 orders of magnitude larger than free $[Ag^+]$.

2. Consider all the equilibria given in Table 6-4 and estimate as a function of pH the concentration of all the species in equilibrium with a saturated solution of $Ag_2S(s)$ in 0.02 M S_T.

Total dissolved silver is given by

$$Ag_T = [Ag^+] + [AgSH] + [Ag(SH)_2^-] + 2[Ag_2S_3H_2^{2-}] \qquad \text{(ii)}$$

where $[Ag^+]$ is defined by

$$[Ag^+] = \left(\frac{K_{so}}{S_T\alpha_2}\right)^{\frac{1}{2}} \qquad \text{(i)}$$

and the concentrations of AgSH and $Ag(SH)_2^-$ can be compared from (3–5) of Table 6-4

$$[AgSH] = [Ag^+]K_1S_T\alpha_1 \qquad \text{(iii)}$$

and

$$[Ag(SH)_2^-] = [Ag^+]K_1K_2S_T^2\alpha_1^2 \qquad \text{(iv)}$$

where

$$\alpha_1 = \frac{[HS^-]}{S_T} = \left(\frac{[H^+]}{K_{H_2S}} + 1 + \frac{K_{HS}}{[H^+]}\right)^{-1}$$

The concentration of $HS–AgS–AgSH^{2-}$ is computed most readily from (7) of Table 6-4

$$[Ag_2S_3H_2^{2-}] = K_{s_3}S_T^2\alpha_1^2 \qquad \text{(v)}$$

Summing up the terms in (ii), for total soluble silver as a function of S_T and an implicit function of $[H^+]$ we obtain

$$Ag_T = \left(\frac{K_{so}}{S_T\alpha_2}\right)^{\frac{1}{2}}(1 + K_1S_T\alpha_1 + K_1K_2S_T^2\alpha_1^2] + 2Ks_3S_T^2\alpha_1^2 \qquad \text{(vi)}$$

(see Figure 6-9). For every pH the individual species can be computed readily by the equations given above. In the pH range below 4, Ag_T consists primarily of AgSH. An inspection of (i) and (iii) shows that $d\log[AgSH]/d$pH = $d\log(\alpha_1/\sqrt{\alpha_2})/d$pH = 0. In the pH range 5–9 the solubility of silver is given by $Ag(SH)_2^-$; its logarithmic concentration plots with a slope of $+1$ and $-1/2$ in the pH range up to and beyond pH = $6.7(pK_{H_2S})$, respectively;

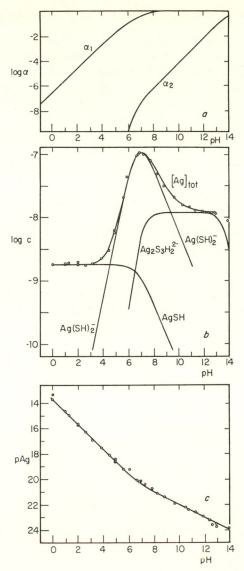

Figure 6-9 The solubility of Ag_2S (see Example 6-4). The solubility is not given by the concentration of free Ag^+; it is determined by the presence of the complexes AgSH, $Ag(SH)_2^-$ and $HSAgSAgSH^{2-}$ (cf. Schwarzenbach and Widmer *Helv. Chim. Acta*, **49**, 111 (1966). Total concentration, Ag_T over precipitated Ag_2S has been measured radiochemically. Free concentration, $[Ag^+]$, has been determined from emf measurements with a silver electrode. The data are valid for a solution containing a total concentration of S_T ($= [H_2S] + [HS^-] + [S^{2-}]$) $= 0.02$ *M* and an ionic strength of $I = 0.1$ ($NaClO_4$) at 20°C.

a: Distribution coefficients of HS^- and S^{2-}, respectively.

b: Soluble silver complexes and total Ag(I) as a function of pH.

c: Free $[Ag^+]$ versus pH.

Reproduced with permission from Schweiz Chemikerverband, Zürich, Switzerland.

these slopes arise because $d\log [\text{Ag(SH)}_2{}^-]/d\text{pH} = d\log(\alpha_1{}^2/\sqrt{\alpha_2})/d\text{pH}$ equals $+1$ and $-1/2$ for the regions $\text{pH} < \text{p}K_{\text{H}_2\text{S}}$ and $\text{pH} > \text{p}K_{\text{H}_2\text{S}}$, respectively. Only at high pH (>9) does the dimer $\text{Ag}_2\text{S}_3\text{H}_2{}^{2-}$ become significant in controlling the solubility; in this pH range $\alpha_1 = 1$ up to pH 14 ($\text{p}K_{\text{HS}}-$) because its concentration does not vary with pH.

Note: Schwarzenbach and Widmer obtained Ag_T experimentally by a radiochemical method. They determined free $[\text{Ag}^+]$ from emf measurements with a silver electrode. The existence of the thio complexes was postulated and their stability calculated from a combination of these measurements. In this context it should be mentioned that it is not possible from such solubility experiments to distinguish mathematically between a species $\text{Ag(SH)}_2{}^-$ and polynuclear species of the type $(\text{Ag}_2\text{S})_x\text{Ag(SH)}_2{}^-$. But, for chemical reasons, the formation of the polynuclear species is improbable, so that x is taken equal to zero.

Example 6-5 *Solubility of* Cu(II) *in Natural Water; Effect of Complexing by Carbonate.*   Estimate the solubility of Cu(II) in a carbonate-bearing water of constant C_T ($C_T = 10^{-2}\ M$) (closed system). The pertinent Cu(II) equilibria are given in Table 6-5.

In order to gain insight into the predominant solid phases and soluble species, it appears expedient to construct first an activity ratio diagram.

Table 6-5   **Cu(II) Equilibria, 25°C** ($I = 0$)

Item	Formula	Log $K\dagger$
1	$\text{CuO(s)} + 2\text{H}^+ = \text{Cu}^{2+} + \text{H}_2\text{O}$ (Tenorite)	7.65
2	$\text{Cu}_2(\text{OH})_2\text{CO}_3(\text{s}) + 4\text{H}^+ = 2\text{Cu}^{2+} + 3\text{H}_2\text{O} + \text{CO}_2(\text{g})$ (Malachite)	14.16
3	$\text{Cu}_3(\text{OH})_2(\text{CO}_3)_2(\text{s}) + 6\text{H}^+ = 3\text{Cu}^{2+} + 4\text{H}_2\text{O} + 2\text{CO}_2(\text{g})$ (Azurite)	21.24
4	$\text{Cu}^{2+} + \text{H}_2\text{O} = \text{CuOH}^+ + \text{H}^+$	-8
5	$2\text{Cu}^{2+} + 2\text{H}_2\text{O} = \text{Cu}_2(\text{OH})_2{}^{2+} + 2\text{H}^+$	-10.95
6	$\text{Cu}^{2+} + \text{CO}_3{}^{2-} = \text{CuCO}_3(\text{aq})$	6.77
7	$\text{Cu}^{2+} + 2\text{CO}_3{}^{2-} = \text{Cu(CO}_3)_2{}^{2-}(\text{aq})$	10.01
8	$\text{CO}_2(\text{g}) + \text{H}_2\text{O} = \text{HCO}_3{}^- + \text{H}^+$	-7.82
9	$\text{Cu}^{2+} + 3\text{H}_2\text{O} = \text{Cu(OH)}_3{}^- + 3\text{H}^+$	-26.3
10	$\text{Cu}^{2+} + 4\text{H}_2\text{O} = \text{Cu(OH)}_2{}^{2-} + 4\text{H}^+$	-39.4

† Given or quoted by P. Schindler, "Heterogeneous Equilibria Involving Oxides, Hydroxides, Carbonates and Hydroxide Carbonates," in *Equilibrium Concepts in Natural Water Systems*, Advances in Chemistry Series, No. 67, American Chemical Society, Washington, D.C., 1967.

Taking $\{Cu^{2+}\}$ as a reference we obtain the following activity ratios for the equilibria given in Table 6-5:

$$\log \frac{\{CuO(s)\}}{\{Cu^{2+}\}} = -7.65 + 2pH \tag{i}$$

$$\log \frac{\{Cu(OH)(CO_3)_{0.5}(s)\}}{\{Cu^{2+}\}} = -3.17 + 1.5\,pH + 0.5 \log C_T + 0.5 \log \alpha_1 \tag{ii}$$

Equation ii results from a combination of (2) and (8) of Table 6-5.
    Similarly combining (3) and (8) gives

$$\log \frac{\{Cu(OH)_{0.67}(CO_3)_{0.67}(s)\}}{\{Cu^{2+}\}} = -1.85 + pH + 0.67 \log C_T + 0.67 \log \alpha_1 \tag{iii}$$

$$\log \frac{\{CuOH^+\}}{\{Cu^{2+}\}} = -8 + pH \tag{iv}$$

$$\log \frac{\{CuCO_3(aq)\}}{\{Cu^{2+}\}} = 6.77 + \log C_T + \log \alpha_2 \tag{v}$$

$$\log \frac{\{Cu(CO_3)_2^{2-}(aq)\}}{\{Cu^{2+}\}} = 10.01 + 2 \log C_T + 2 \log \alpha_2 \tag{vi}$$

$$\log \frac{\{Cu(OH)_3^-\}}{\{Cu^{2+}\}} = -26.3 + 3pH \tag{vii}$$

$$\log \frac{\{Cu(OH)_4^{2-}\}}{\{Cu^{2+}\}} = -39.4 + 4pH \tag{viii}$$

In a double logarithmic diagram, log activity ratio versus pH [(i)–(viii)] can be plotted in straight-line portions having readily defined slopes. Figure 6-10a shows that under the specified conditions malachite and tenorite qualify as stable solid phases; malachite is stable below pH = 7 while tenorite is more stable in the alkaline region. With respect to dissolved species the activity ratio diagram reveals that under the specified conditions the following species predominate: Cu^{2+} up to pH = 6; $CuCO_3(aq)$ in the pH range 6–9.3; $Cu(CO_3)_2^{2-}(aq)$ in the pH range 9.3–10.7; $Cu(OH)_3^-$ and $Cu(OH)_4^{2-}$ above pH 10.7 and 12.9, respectively. With this information a logarithmic solubility diagram can be sketched (Figure 6-10b).

Some Geochemical Implications

The two examples given above underline the need to identify all pertinent species and to consider all reactions and equilibria related to the solubility of

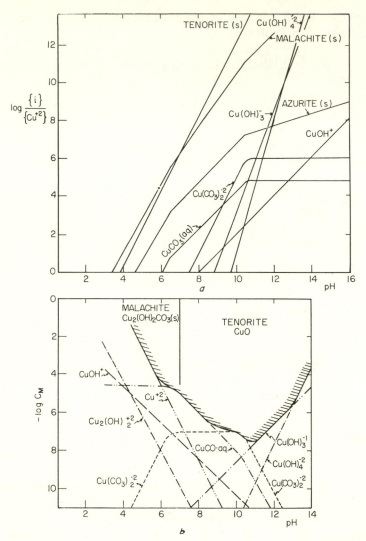

Figure 6-10   Solubility of Cu(II). a: Activity ratio diagram. b: Solubility diagram. The solid line surrounding the shaded area gives the total solubility of Cu(II) which up to pH value of 6.96, is governed by the solubility of malachite [$Cu_2(OH)_2CO_3(s)$]. In the low pH region azurite [$Cu_3(OH)_2(CO_3)_2(s)$] is metastable but may become stable at higher C_T. Above pH 7 the solubility is controlled by the solubility of CuO (tenorite). The predominant species with increasing pH are Cu^{2+}, $CuCO_3(aq)$, $Cu(CO_3)_2^{2-}$ and hydroxo copper(II) anions. $C_T = 10^{-2}\ M$.

a mineral. Calculations from the solubility product alone can often be very misleading; although the solubility products of HgS and PbS may differ by more than 20 orders of magnitude, their solubility does not differ by more than a few orders of magnitude. Complex formation among constituents of the lattice or with constituents of the solution must be taken into account in exploring mechanisms of ore transport by the ore-forming fluids. The observation by Barton [11] that those metal sulfides having the smallest solubility products behave as if they were the most soluble is compatible with the coordination chemical interpretation that the smaller the solubility product, the stronger is the tendency to form soluble associates.

The Influence of Complex Formers on the Crystallization of Solids

Complex formers present in solution may often have no or little effect on the solubility of solids; they may, however, affect the kinetics of nucleation and of growth and dissolution of crystals. Kitano and Hood [12] have studied the effects of various organic complex formers on polymorphic crystal formation of calcium carbonate. Some of their conclusions are the following.

1. Organic material which slowed the rate of calcium carbonate formation because of very *strong* complex formation with calcium ions greatly favored the formation of calcite. Citrate, malate, pyruvate, glycylglycine, and glycogen are of this type.

2. Organic material which affected the rate of calcium carbonate formation *moderately* had a fairly strong influence on the crystal type of calcium carbonate formed. Serine, lactate, glycoprotein, succinate, glutamate, glycine, arginine and taurine are of this type. Their effects were different from each other: glycoprotein favored calcite and vaterite formation; chondroitinsulfate, succinate, lactate and arginine favored calcite formation; taurine favored aragonite formation, glutamate favored vaterite formation, and glycine and serine favored vaterite and aragonite formation.

3. Organic material which had *little* influence on the rate of calcium carbonate formation also had little influence on the crystal type of calcium carbonate. Galactose, dextrose, alanine and acetate are of this type.

4. The rate of calcium carbonate formation decreased and the influence of organic material on the crystal type of calcium carbonate increased with an increase in the concentration of organic material in solution.

[11] P. B. Barton, in *Researches in Geochemistry*, P. H. Abelson, Ed., Wiley, New York, 1959, p. 279.
[12] Y. Kitano and D. W. Hood, *Geochim. Cosmochim. Acta*, **29**, 29 (1965).

6-7   Inorganic Complexes in Natural Waters

The formation constants of hydroxo complexes, especially of class A cations, are large when compared to most other inorganic ligands. This explains why, with most tri- and tetravalent metal ions, the hydroxides or hydrous oxides are the only stable precipitates in the pH range of natural waters.† Insoluble phosphates are formed with Al(III), Fe(III) and Zr(IV) but they are stable only in solutions of low pH. With many bivalent ions, however, CO_3^{2-}, S^{2-}, S_2^{2-} and PO_4^{3-} may successfully compete with OH^- ions to satisfy the coordinative requirements of metal ions. Thus $FeCO_3$, $MnCO_3$, $CaCO_3$ and basic carbonates of Zn(II) are the predominant solid phases in carbonate-bearing water systems. The solubility of Cu^{2+}, Mg^{2+} and Be^{2+}, however, is predominantly controlled by the solubility of the respective hydroxides.

What soluble inorganic complexes do we have to consider in natural waters? At the great dilutions usually encountered, very little complexing occurs. In comparing stability constants for the formation of chloro and fluoro complexes with those of hydroxo complexes [13]

$$K_{\text{MeOH}} = \frac{[\text{MeOH}]}{[\text{Me}][\text{OH}]} \tag{19}$$

$$K_{\text{MeCl}} = \frac{[\text{MeCl}]}{[\text{Me}][\text{Cl}]} \tag{20}$$

$$K_{\text{MeF}} = \frac{[\text{MeF}]}{[\text{Me}][\text{F}]} \tag{21}$$

we establish that in seawater coordination with Cl^- predominates over that with hydroxide only for Ag(I), Hg(II), Au(I) and possibly Cd(II) and Pb(II), that is,

$$\log \frac{K_{\text{MeCl}}}{K_{\text{MeOH}}} > \log \frac{[\text{OH}^-]}{[\text{Cl}^-]} \approx -5.4 \tag{22}$$

and that fluoro magnesium complexes are more stable than hydroxo complexes, that is,

$$\log \frac{K_{\text{MgF}}}{K_{\text{MgOH}}} > \log \frac{[\text{OH}^-]}{[\text{F}^-]} \approx -1.5 \tag{23}$$

---

† Interestingly, Fe(III) can form basic carbonates under suitable conditions. Dvořák, Feitknecht and Georges [*Helv. Chim. Acta*, **52**, 501, 515, 523 (1969)] have prepared amorphous as well as crystalline basic carbonates of Fe(III) [i.e., $(NH_4)_2Fe_2(OH)_4(CO_3)_2 \cdot H_2O$; $(NH_4)_4Fe_2(OH)_4(CO_3)_3 \cdot 3H_2O$].

[13] L. G. Sillén, in *Oceanography*, M. Sears, Ed., American Association for the Advancement of Science, Washington, D.C., 1961, p. 549.

There appear to be no other metal ions existing primarily as fluoro complexes in natural waters.

The Garrels and Thompson Model [14] of Ion Association in Sea Water

We have already pointed out that carbonate, bicarbonate and sulfate ion pairs or complexes might not be expected in most fresh waters ($I < 0.05$) above 1 or 2% of the reported concentrations. (Compare Example 6-4.) Seawater ($I = 0.7$) and brines, however, contain species sufficiently concentrated for the formation of ion pairs or complexes.

Garrels and Thompson have established a seawater model by computing the distribution of the most important dissolved species. Their calculations are based on stability constants and activity coefficients (infinite dliution scale) given in Table 6-6.† The 18 species in their model (K, KSO_4^-, Na^+, $NaCO_3^-$, $NaHCO_3(aq)$, $NaSO_4^-$, Ca^{2+}, $CaHCO_3^+$, $CaCO_3(aq)$, $CaSO_4(aq)$, Mg^{2+}, $MgHCO_3^+$, $MgCO_3(aq)$, $MgSO_4(aq)$, HCO_3^-, CO_3^{2-}, SO_4^{2-} and Cl^-) make up 99.7% of average surface seawater (19‰ chlorinity, 25°C). The distribution of the species, as calculated from the stability constants assumed, is given in Table 6-6. According to this model, the major cations are present as aquo metal ions. However, a significant fraction of bicarbonate, carbonate and sulfate appears to be complex bound to metal ions. According to this model the solubility of solid carbonates is influenced very strongly by the tendency of the carbonate ligand to form Mg and Na complexes [15]. According to this model, Garrels and Christ [16] show that the Mg^{2+} and Na^+ content of seawater is nearly as important as H^+ in controlling $CaCO_3$ composition. They point out that, for example, an organism that excludes Mg^{2+} from its cell fluids would tend to cause precipitation of $CaCO_3$ because of the loss of the complexing power of Mg^{2+} for CO_3^{2-}.

The calculated distribution of species, of course, depends upon the stability constants and activity coefficients used for the calculations. Unfortunately, however, considerable uncertainty exists on the stability of these associates.

---

† The calculations consist essentially in solving 18 equations with 18 unknowns simultaneously. Ten stability constants are available and 8 mass balance relations, for example,

$$[SO_{4_T}] = [SO_4^{2-}] + [MgSO_4(aq)] + [CaSO_4(aq)] + [KSO_4^-] + [NaSO_4^-]$$

for Na_T, Mg_T, Ca_T, K_T, Cl_T, SO_{4_T}, HCO_{3_T}, CO_{3_T}. As Garrels and Thompson point out, the calculation can be simplified if the assumption is made first that the cations are not significantly complexed (e.g., $[Ca^{2+}] = Ca_T$).

[14] R. M. Garrels and M. E. Thompson, *Amer. J. Sci.*, **260**, 57 (1962).
[15] R. M. Garrels, M. E. Thompson and R. Siever, *Amer. J. Sci.*, **259**, 43 (1961).
[16] R. M. Garrels and C. L. Christ, *Solutions, Minerals and Equilibria*, Harper and Row, New York, 1965, p. 107.

Table 6-6   Distribution of Species in Seawater According to Model by Garrels and Thompson [14]

I. Major Dissolved Species in Seawater† (19‰ chlorinity, 25°C, pH = 8.15)

Metal Ion, Me	Conc. Total, molal	Free Ion, %	$MeSO_4$(aq), %	$MeHCO_3$, %	$MeCO_3$(aq), %
Na(I)	0.48	99	1+	—	—
K(I)	0.010	99	1	—	—
Mg(II)	0.054	87	11	1	0.3
Ca(II)	0.010	91	8	1	0.2

Ligand, L	Conc. Total, molal	Free Ion, %	CaL, %	MgL, %	NaL, %	KL, %g
SO_4^{2-}	0.028	54	3	22	21	0.5
HCO_3^-	0.0024	69	4	19	8	—
CO_3^{2-}	0.00027	9	7	67	17	—
Cl^-	0.56	100	—	—	—	—

II. Stability Constants (log K_1 values for $I = 0$, 25°C)

Metal Ion	Ligand				
	OH^-	HCO_3^-	CO_3^{2-}	SO_4^{2-}	Cl^-
H^+	14.0	6.4	10.33	2	—‡
K^+	—	—	—	0.96	—
Na^+	−0.7	−0.25	1.27	0.72	—
Ca^{2+}	1.30	1.26	3.2	2.31	—
Mg^{2+}	2.58	1.16	3.4	2.36	—

III. Activity Coefficients of Individual Species in Seawater ($I = 0.7$, 19‰ chlorinity, 25°C)

	Activity Coefficient
Noncharged species ($NaHCO_3$, $MgCO_3$, $CaCO_3$, $MgSO_4$, $CaSO_4$)	1.13
HCO_3^-, $NaCO_3^-$, $NaSO_4^-$, KSO_4^-, $MgHCO_3^+$, $CaHCO_3^+$	0.68
Na^+	0.76
K^+	0.64
Mg^{2+}	0.36
Ca^{2+}	0.28
Cl^-	0.64
CO_3^{2-}	0.20
SO_4^{2-}	0.12

† Computed on the basis of stability constants and activity coefficients given in Parts II and III, respectively, of this table.

‡ A dash denotes negligible complex formation.

The probable error in determining stability constants and the uncertainty of assigning meaningful activity constants in mixed electrolytes are very large. For example, Raaflaub and Riesen et al. [17] have determined stability constants for $MgCO_3(aq)$ which are one order of magnitude lower than that given by Garrels and Thompson. Calculations using these lower values would result in a species distribution in seawater which is different from that given by Garrels and Thompson, but some other pertinent qualitative features of the model of Garrels and Thompson would still be valid. Similar questions arise concerning the stability of the $CaCO_3(aq)$ species [18].

6-8 Chelates

One of the reasons why metal ions in solution in natural waters are not appreciably complexed by ligands other than H_2O or OH^- is that most bases indigenous to natural waters are unidentate ligands. As we shall see, unidentate ligands form less stable complexes than multidentate ligands, especially in dilute solutions. Carbonate, sulfate and phosphate can, but seldom do, serve as bidentate ligands. The solid carbonates of Ca^{2+}, Mn^{2+}, Fe^{2+}, Co^{2+} and Zn^{2+} have six oxygen atoms around the metal ion, but besides $CaCO_3(aq)$ and $MgCO_3(aq)$ soluble carbonato complexes may exist only at high pH for Be(II), the heavier lanthanides and for Cu(II). Perhaps for steric reasons these anions usually act as unidentate ligands.

As with hydroxide, complexes with other ligands are formed stepwise. In a series such as

$$Cu^{2+}, \; Cu(NH_3)^{2+}, \; Cu(NH_3)_2^{2+}, \; Cu(NH_3)_3^{2+}, \; Cu(NH_3)_4^{2+}, \; Cu(NH_3)_5^{2+}$$

the successive stability constants generally decrease; that is, the Cu^{2+} ion takes up one NH_3 molecule after the other. (There are exceptions to this behavior.) As can be seen from Figure 6-2d, relatively high concentrations of NH_3 are necessary to complex copper(II) effectively and to form a tetramine complex. In natural waters the concentrations of the ligands and the affinity of the ligands to the metal ion, with the exception of H_2O and OH^-, are usually sufficiently small so that at best a one-ligand complex may be formed.

The Chelate Effect

Complexes with monodentate ligands are usually less stable than those with multidentate ligands. More important is the fact that the degree of

[17] J. Raaflaub, *Helv. Chim. Acta*, **43**, 629 (1960). W. Riesen, H. Gamsjäger and P. Schindler, *Chimia*, **23**, 186 (1969).
[18] D. Langmuir, *Geochim. Cosmochim. Acta*, **32**, 835 (1968).

complexation decreases more strongly with dilution for monodentate complexes than for multidentate complexes (chelates). This is illustrated in Figure 6-11, where the degree of complexation is compared as a function of concentration for uni, bi and tetradentate copper(II) amine complexes. Free $[Cu \cdot aq^{2+}]$ is plotted as a function of dilution in the left-hand graph, while the quantitative degree of complexation, as measured by ΔpCu [$= \log (Cu_T/[Cu^{2+}])$] is given in the right-hand graph. It is obvious from this figure that the complexing effect of NH_3 on Cu^{2+} becomes negligible at concentrations that might be encountered in natural water systems. Chelates, however, remain remarkably stable even at very dilute concentrations.

The curves drawn in Figure 6-11 have been calculated on the basis of constants taken from *Stability Constants of Metal–Ion Complexes*. The calculations are essentially the same as those outlined for the hydroxo complex formation; although algebraically simple, they are tedious and time-consuming. In the case of the $Cu–NH_3$ system, the following species have to

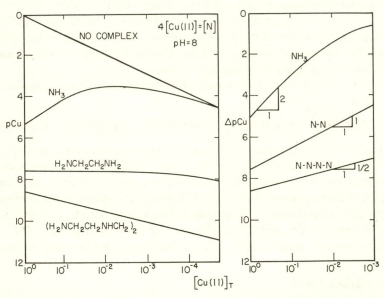

Figure 6-11   The chelate effect on complex formation of $Cu \cdot aq^{2+}$ with monodentate, bidentate and tetradentate amines. pCu is plotted as a function of concentration in the left-hand diagram. On the right the relative degree of complexation as measured by ΔpCu as a function of concentration is depicted. The extent of complexing is larger with chelate complex formers than with unidentate ligands. Unidentate complexes are dissociated in dilute solutions while chelates remains essentially undissociated at great dilutions.

be considered

$$Cu^{2+}, \quad Cu(NH_3)^{2+}, \quad Cu(NH_3)_2^{2+}, \quad Cu(NH_3)_3^{2+}, \quad Cu(NH_3)_4^{2+},$$
$$Cu(NH_3)_5^{2+}, \quad NH_4^+, \quad NH_3$$

Thus, for every $[H^+]$, eight equations have to be solved simultaneously in order to compute the relative concentrations of each species present. Six mass laws (five stability expressions for the five different amine complexes and the acid–base equilibrium of NH_4^+–NH_3) and two concentration conditions make up the eight equations. As concentration conditions one might formulate equations defining Cu_T and NH_{3_T}

$$Cu_T = [Cu^{2+}] + [Cu(NH_3)^{2+}] + [Cu(NH_3)_2^{2+}] + \cdots \tag{24}$$

$$NH_{3_T} = [NH_4^+] + [NH_3] + [Cu(NH_3)^{2+}] + 2[Cu(NH_3)_2^{2+}] + \cdots \tag{25}$$

Guidelines for coping with these and more involved types of calculations have been provided by Ringbom [19], Schwarzenbach [20] and others. Computers are, of course, also very useful [21].

In Figure 6-11 a concentration of complex former equivalent to that of metal ion was considered. ΔpM, of course, increases with increasing concentration of the complex former over the metal. Figure 6-12 shows the effect of various ligands on the complex formation with ferric iron. Here the concentration of the complexing agent is kept at a constant value and in excess of $[Fe(III)_T]$. If the ligand is in large excess over the metal, the quantitative degree of complexation ΔpM is independent of the total metal ion concentration. In this figure we again observe the increase in stability in going from monodentate (F^-, SO_4^{2-}, HPO_4^{2-}) to bidentate (oxalate), to tridentate (citrate) and to hexadentate (EDTA, DCTA) ligands. Figure 6-12 also illustrates that in all aqueous solutions $[H^+]$ and $[OH^-]$ influence markedly the degree of complexation. At low pH, H^+ competes successfully with the metal ions for the ligand. At high pH, OH^- competes successfully with the ligand for the coordinative positions on the metal ion. Furthermore, at low and high pH mixed hydrogen–metal and hydroxide–ligand complexes can be formed. [In the case of EDTA ($= L$), in addition to FeL^- the complexes $FeHL$, $FeOHL^{2-}$ and $Fe(OH)_2L^{3-}$ have to be considered.] Because of the competing influence of H^+ or OH^-, the complexing effect cannot be estimated solely from the stability constants.

The calculations predict that a 10^{-2} M solution of sulfate, fluoride, phosphate, oxalate or citrate can keep 10^{-3} M ferric iron in solution [i.e., prevent

[19] A. Ringbom, *Complexation in Analytical Chemistry*, Wiley-Interscience, New York, 1963.
[20] G. Schwarzenbach, *Complexometric Titrations*, Interscience, New York, 1959.
[21] A. Bard and D. M. King, *J. Chem. Educ.*, **42**, 127 (1965).

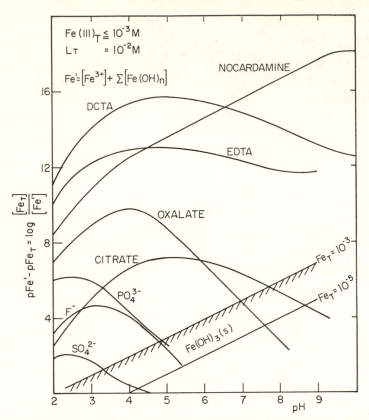

Figure 6-12   Complexing of Fe(III). The degree of complexation is expressed in terms of ΔpFe for various ligands ($10^{-2} M$). The competing effect of H^+ at low pH values and of OH^- at higher pH values explains that effective complexation is strongly dependent on pH. Mono-, di- and tri-dentate ligands ($10^{-2} M$) are not able to keep a $10^{-3} M$ Fe(III) in solution at higher pH values. EDTA is ethylenediaminetetraacetate. DCTA is 1,2-diaminocyclohexane-tetraacetate. Nocordamine is a trihydroxamate (see Figure 6-14).

precipitation of Fe(OH)$_3$(s)] only up to pH values of 3.3, 4.7, 4.8, 6.9 and 7.6, respectively. From another point of view, the pH of ferric iron precipitation will be strongly affected by the presence of coordinating anions; for example, Figure 6-12 illustrates that in the presence of $10^{-2} M$ ligand a $10^{-3} M$ solution of ferric iron will form precipitates around pH 4.5 if the ligand is phosphate (or $H_2PO_4^-$), whereas precipitation will not occur until pH values around 7.5 if the ligand is citrate. The precipitates formed initially in the presence of such ligands are usually nonstoichiometric mixed precipitates [e.g., phosphato-hydroxo iron(III) precipitates].

Figure 6-13 illustrates complexation of Zn(II) by a few chelate formers.

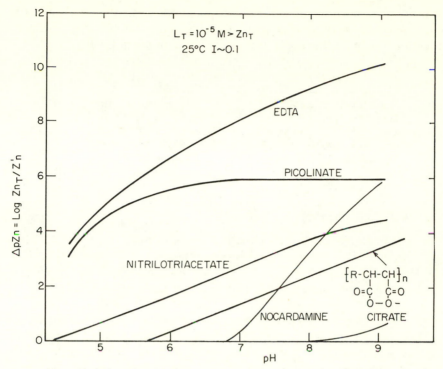

Figure 6-13   Chelation of Zn(II). The extent of complex formation (ΔpZn) as a function of pH is strongly influenced by the acid–base behavior of the metal ion and of the complex-former. In the case of Zn(II), the tendency to form hydroxo species is less within the pH range of interest than in the case of Fe(III).

Example 6-6   *The Stability of Ethylenediamine Complexes of Nickel.* Estimate the equilibrium composition of a $10^{-4} M$ solution of Ni(en)$_3$Cl$_2$ (en = enthylenediamine) whose pH has been adjusted to (a) pH = 7.2 and (b) pH = 3.0. The following constants are available: ($I = 0.1$, $25°C$) $\log \beta_1 = 7.5$, $\log \beta_2 = 13.8$, $\log \beta_3 = 18.3$, $\log K_{H_2en^{2+}} = -7.4$, $\log K_{Hen^+} = -10.1$.

In addition to the five equations defined by these equilibrium constants, three additional equations can be established (for simplicity, charges are omitted)

$$\text{Ni}_T = [\text{Ni}] + [\text{Ni(en)}] + [\text{Ni(en)}_2] + [\text{Ni(en)}_3] \tag{i}$$

$$\text{en}_T = [\text{en}] + [\text{Hen}] + [\text{H}_2\text{en}] + [\text{Ni(en)}] + 2[\text{Ni(en)}_2] + 3[\text{Ni(en)}_3] \tag{ii}$$

Furthermore, for a pure solution of the salt Ni(en)$_3$Cl$_2$ we obtain the concentration condition

$$3\text{Ni}_T = \text{en}_T \tag{iii}$$

Substituting the formulations for the equilibrium constants into (i) and (ii) and calculating by trial and error, we obtain the following approximate equilibrium compositions (per cent contribution in parentheses):

(a) pH = 7.2	(b) pH = 3.0
$[Ni^{2+}] = 2.12 \times 10^{-5}$ (21%)	$[Ni^{2+}] = 9.99 \times 10^{-5}$ (~100%)
$[Ni(en)^{2+}] = 6.35 \times 10^{-5}$ (64%)	$[Ni(en)^{2+}] = 10^{-11.5}$ (0%)
$[Ni(en)_2^{2+}] = 1.20 \times 10^{-5}$ (12%)	$[Ni(en)_2^{2+}] = 10^{-20.2}$ (0%)
$[Ni(en)_3^{2+}] = 0.003 \times 10^{-5}$ (0%)	$[Ni(en)_3^{2+}] = 10^{-30.7}$ (0%)
$[en] = 9.5 \times 10^{-8}$	$[en] = 10^{-15}$ (0%)
$[Hen^+] = 7.6 \times 10^{-5}$	$[H_2en^{2+}] = 2.99 \times 10^{-4}$ (100%)
$[H_2en^{2+}] = 1.2 \times 10^{-4}$	$[Hen^+] = 10^{-7.8}$ (0%)

Note the following:

1. The predominant species in the solution is H_2en^{2+}.
2. Around pH = 3, where there is very little complex formation (less than 10^{-11} M), the predominant species is H_2en^{2+}. Even at the higher pH value (7.2), a significant portion of the metal ion remains uncomplexed.

INCREASE IN ENTROPY PRIMARILY RESPONSIBLE FOR THE CHELATE EFFECT. A chelate ring is more stable than the corresponding complex with unidentate ligands. The enthalpy change resulting from either complex formation, however, is frequently about the same; for example, ΔH for the formation of a diamine complex is approximately equal to ΔH for the formation of an ethylenediamine complex [22]. Hence, essentially the same type of bond occurs in these complexes. That $-\Delta G°$ for the chelate is larger than for the corresponding complex with an unidentate ligand must be accounted for primarily by the fact that the formation of a ring is accompanied by a larger increase in entropy than that encountered in the formation of a nonchelate complex.

Metal Ion Buffers

The analogy between Me^{z+} and H^+ can be extended to the concept of buffers. pH buffers are made by mixing acids and conjugate bases in proper proportions

$$[H^+] = \frac{K[HA]}{[A^-]}$$ (26)

[22] G. Schwarzenbach, in *Advances in Inorganic Chemistry and Radiochemistry*, Volume 3, H. J. Eméleus and A. G. Sharpe, Eds., Academic Press, New York, 1961, p. 257.

Metal ions can be similarly buffered by adding appropriate ligands to the metal ion solution

$$[Me] = \frac{K[MeL]}{[L]} \qquad (27)$$

Such pMe buffers resist a change in $[Me^{2+}]$. It is well known that the living cell controls not only pH but also pCu, pMn, pMg, etc., and that complex formers are used as the buffering component; pMe buffers are convenient tools for investigating phenomena pertaining to metal ions. It is unwise to prepare a pH 6 solution by diluting a concentrated HCl solution, but this mistake is frequently made with metal ion solutions. If, for example, we want to study the toxic effect of Cu^{2+} on algae, it might be more appropriate to prepare a suitable pCu buffer. If a copper salt solution is simply diluted, the concentration (or activity) of the free Cu^{2+} may, because of hydrolysis and adsorption and other side reactions, be entirely different from that calculated by considering the dilution only.

Example 6-7   pCa *Buffer.*   Calculate pCa of a solution of the following composition: EDTA (ethylenediamine tetraacetate) $= Y_T = 1.95 \times 10^{-2}\,M$, $Ca_T = 9.82 \times 10^{-3}$, pH $= 5.13$, $I = 0.1$ (20°C). EDTA is a tetraprotic acid. For the conditions given the four acidity constants are characterized by $pK_1 = 2.0$, $pK_2 = 2.67$, $pK_3 = 6.16$, $pK_4 = 10.26$. Two Ca complexes are formed with EDTA, $CaHY^-$ and CaY^{2-}. Hence we need the stability constants

$$\frac{[CaY^{2-}]}{[Ca^{2+}][Y^{4-}]} = K_{CaY} = 10^{10.6}$$

$$\frac{[CaHY^-]}{[Ca^{2+}][HY^{3-}]} = K_{CaHY} = 10^{3.5}$$

The computation may start by setting up equations for the concentration conditions (for simplicity charges are omitted)

$$
\begin{aligned}
Ca_T &= [Ca] + [CaY] + [CaHY] \\
&= [Ca](1 + [Y]K_{CaY} + [Y][H]K_4^{-1}K_{CaHY}) \\
&= [Ca]\alpha^{-1}{}_{Ca}
\end{aligned}
\qquad (i)
$$

and, where as before,

$$\alpha_{Ca} = \frac{[Ca]}{Ca_T}$$

$$
\begin{aligned}
Y_T &= [H_4Y] + [H_3Y] + [H_2Y] + [HY] + [Y] + [CaY] + [CaHY] \\
&= [Y](\alpha_4')^{-1} + [Ca][Y]K_{CaY} + [Ca][H][Y]K_4^{-1}K_{CaHY}
\end{aligned}
\qquad (ii)
$$

where

$$\alpha_4' = \frac{[Y]}{\displaystyle\sum_{i=0}^{i=4}[H_iY]} = \left(1 + \frac{[H]}{K_4} + \frac{[H]^2}{K_4K_3} + \frac{[H]^3}{K_4K_3K_2} + \frac{[H]^4}{K_4K_3K_2K_1}\right)^{-1} \quad \text{(iii)}$$

α_4' can be plotted most conveniently in a double logarithmic diagram (log α_4' versus $-$log $[H^+]$). Equations i and ii, containing the two unknowns [Y] and [H], are best solved by trial and error. Most conveniently we may start by calculating α_4' from (iii) or taking its value from a graphical plot. Then [Y] may be estimated from (iii) by assuming that $\sum [H_iY] \simeq (Y_T - Ca_T)$.

With this tentative [Y], first values of α_{Ca}^{-1} and [Ca] are estimated, respectively, with the help of (i). We then may check whether the assumption that $\sum [H_iY] = [Y_T - Ca_T]$ was appropriate. Subsequent reiteration gives the result $[Ca] = 4.12 \times 10^{-5}$ (pCa = 4.39) and $[Y] = 6.05 \times 10^{-9}$. For illustration the concentration of all the other species are also given

$[CaY] = 9.66 \times 10^{-3}$; $[CaHY] = 1.09 \times 10^{-4}$; $[H_4Y] = 2.26 \times 10^{-8}$

$[H_3Y] = 3.07 \times 10^{-5}$; $[H_2Y] = 8.8 \times 10^{-3}$; $[HY] = 8.21 \times 10^{-4}$

Note: If $[Ca^{2+}]$ is lost from the solution, that is, by adsorption at the glass wall, the Ca^{2+} ion buffer described will have a tendency to maintain constant pCa. For example, removal from the solution of 2×10^{-5} M Ca^{2+} per liter $(dCa_T = 2 \times 10^{-5}$ M, corresponding to approximately 50% of the free $[Ca^{2+}]$) will change the free $[Ca^{2+}]$ by approximately 4×10^{-7} M $(dpCa = 0.005)$, that is,

$$\beta_{pH, Y_T}^{Ca_T} = \frac{dCa_T}{dpCa} \simeq 4 \times 10^{-3} \text{ (mole per pCa unit)}$$

6-9   Organic Complexes in Natural Waters; Problems of Specificity

The current literature refers frequently to soluble complex-bound metal ions in natural waters. Chelation with organic molecules is frequently postulated to explain the occurrence of metal ions at concentrations higher than those calculated from known solubility product values. The fact that some organic constituents may increase or decrease biological activity (or the rate of uptake of metal ions) has been cited frequently as an indication for the solubilization of useful trace metals or the masking of harmful metals caused by chelation. Similarly the observations that a large fraction of metal species in natural waters are nonextractable with suitable reagents, not adsorbable on selective ion exchangers or nondialysable have been accounted for in terms of complex formation.

Despite these arguments in favor of organic complexation, very little direct proof for the existence of soluble chelates in natural waters is available. It is very difficult to detect soluble chelates in natural waters by any direct method or to isolate them from natural waters, especially with the very small amounts of metal ions that are usually present [23]. In this section we will investigate interactions between metal ions and organic constituents, and inquire as to the type of metal organic associations to be found in natural waters.

Concentrations of dissolved organic matter in natural waters usually range from 0.1 to 10 mg liter^{-1}. The upper concentration is reached in polluted lakes, streams and estuaries. The lower concentration is more typically encountered in unpolluted and nonproductive fresh water and seawater. The organic matter found includes compounds such as amino acids, polysaccharides, amino sugars, fatty acids, organic phosphorus compounds, aromatic compounds containing —OH and —COOH functional groups and porphyrins which contain donor atoms suitable for complex formation. Natural waters contain primarily metals of Group I and Group II of the periodic table, but they also contain trace quantities of all naturally occurring metals and this includes most transition elements.

The Competition by Ca^{2+} and Mg^{2+} Decreases Tendency to Form Soluble Complexes

The ratio of inorganic to organic constituents is very high. How can trace concentrations of organic matter specifically interact with individual metal ions? Does not the presence of Ca^{2+} and Mg^{2+} at concentrations many orders of magnitude larger than potential organic complex forming species blur any complex-forming tendency of the organic functional groups? These questions are well justified because in natural water systems the stability of a complex is influenced not only by H^+ and OH^-, but also by other metal ions, especially Ca^{2+} and Mg^{2+}.

In order to illustrate the problem we select a system that appears to be especially favorable to complex formation with Fe(III). As we have seen in Figure 6-12, EDTA forms rather strong complexes with Fe(III). To such a system we now add Ca^{2+}. Simple complex formation of Fe(III) by EDTA

$$Fe^{3+} + Y^{4-} = FeY^-; \quad \log K_{FeY} = 25.1 \tag{28}$$

is now replaced by

$$CaY^{2-} + Fe^{3+} = FeY^- + Ca^{2+}; \quad \log \frac{K_{FeY}}{K_{CaY}} = 14.4 \tag{29}$$

where $\log K_{CaY} = 10.7$.

[23] E. K. Duursma and W. Sevenhuysen, *Neth. J. Sea Res.*, **3**, 95 (1966).

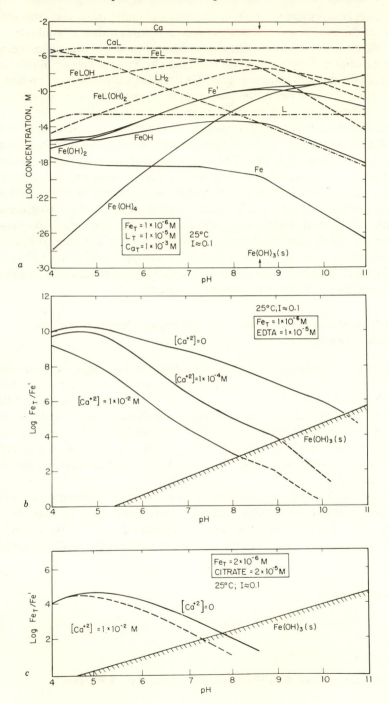

As (29) illustrates, FeY^- cannot be formed to any significant extent if the ratio $[Ca^{2+}]$ to $[Fe^{3+}]$ is sufficiently large. For an exact calculation the hydroxo complexes $FeYOH$ and $FeY(OH)_2$ have to be considered in addition to CaY^{2-} and FeY^-. (Above pH 4 the complexes $CaHY^-$ and $FeHY$ can be neglected.) For a given set of conditions in the system

$$M_1 = Ca^{2+}, \ M_2 = Fe^{3+}, \ M_3 = H^+$$
$$L_1 = EDTA\,(Y^{4-}), \ L_2 = OH^-$$

the calculated equilibrium concentrations of the more important species are plotted in Figure 6-14a. Figure 6-14b illustrates more explicitly how Ca^{2+} depresses the extent of Fe(III)–EDTA complexation. Such a strong complex former, even if available in excess of Fe(III), cannot, in the presence of Ca^{2+}, prevent Fe(III) from precipitating as $Fe(OH)_3(s)$ in alkaline solutions. The competing effect of other cations depends on the relative stability of the complexes that are formed. For example, in the case of $L = citrate^{4-}$, the equilibrium, $CaHL^- + Fe^{3+} = FeHL + Ca^{2+}$, is characterized by an equilibrium constant $K = K_{FeHL}/K_{CaHL}$ of $10^{7.4}$. Correspondingly, the relative effect of Ca^{2+} upon Fe(III) complexing by citrate is much smaller than that encountered with EDTA (Figure 6-14c).

The examples given are not yet very realistic. In nature we encounter a multitude of complex-forming bases, each one most likely present at concentrations smaller than $10^{-6} \ M$ (~ 0.1 mg liter^{-1}) organic matter. Furthermore other competing ions (Mg^{2+}, Al^{3+}, etc.) need to be considered. At the end of this chapter a hypothetical multimetal, multiligand system will be considered.

Complex Formation at Solid–Solution Interfaces

It is difficult to distinguish operationally between dissolved and colloidally dispersed substances. Statements frequently found in the literature which claim, for example, that decaying vegetable matter such as humic acids which

---

Figure 6-14   Competitive effect of Ca^{2+} on complex formation of Fe(III). In most practical systems, ions other than H^+ or OH^- compete as coordination partners, as illustrated here in a "simple" two-metal, two-ligand system:

$$M_1 = Fe^{3+}, \ M_2 = Ca^{2+}, \ (M_3 = H^+)$$
$$L_1 = EDTA \text{ or citrate}, \ L_2 = OH^-$$

a: Ca^{2+}, Fe(III)–EDTA equilibria and the relationship to $Fe(OH)_3(s)$ ($K_{s0} = 10^{-38}$) formation. $Fe(OH)_3(s)$ precipitates at pH > 8.6.

b: Effects of total Ca concentration upon Fe(III)–EDTA equilibrium.

$$Me' = \sum_{0}^{k^\bullet} Me(OH)_n; \ Me_T = Me' + \sum MeHL + \sum MeOHL$$

c: Effect of Ca^{2+} on Fe(III)–citrate equilibrium.

carry carboxyl and hydroxyl functional groups form soluble chelates in natural waters must be taken with reservations. These substances can coordinate and form chelates with iron(III). Such substances undoubtedly can bind ferric iron and at high concentrations are capable of preventing the precipitation of ferric iron. Colored water usually contains higher iron(III) concentrations. It is not yet certain whether such substances can really keep ferric iron in *solution* at the pH values of interest. It might be more probable that the coordinative products formed between the color bases, OH^-, and Fe(III) are insoluble and are present as highly dispersed colloids (lakes†). In a similar way, small quantities of phosphates or polyphosphates, usually present in concentrations far below those required for the formation of soluble complexes, aid markedly in the formation of stable, negatively charged, colloidal dispersions. As we have seen, in the absence of complex formers other than OH^- the solubility of ferric iron cannot exceed concentrations of ca. 10^{-8} M within the pH range 6–9. Since this does not appear to be in accord with the analytical findings for natural waters, it has been suggested by various authors that the iron(III) is frequently present as an organic complex bound to iron. Although this possibility cannot be excluded in all cases, one must be aware that it is analytically very difficult to distinguish between dissolved and suspended iron. Lengweiler, Buser and Feitknecht [24] have shown that with very dilute Fe(III) solutions, containing Fe^{59} as a tracer and brought to a pH between 5 and 10, essentially all the iron hydroxide can be sedimented by ultracentrifugation (93,000 g, 180 min). The size of the $Fe(OH)_3$ particles varies with the pH of the solution. The diameter can be smaller than 100 Å.

Surface complex formation appears to be of significance in accounting for the uptake of certain metal ions on organic aggregates [25]. Alternatively a large variety of dissolved organic substances may be adsorbed by clay minerals. This adsorption on clays may frequently be interpreted as complex formation of the organic bases with exposed octahedral aluminum ions. Typically, the organic anions which are most strongly adsorbed to clays are good complex formers for Al(III) (multivalent phenolates such as pyrogallol, gallic acid, tannic acid alizarine dyes). Furthermore, anion adsorption occurs preferably at the edge surface of the clays where Al ions are exposed. As a result of the surface complex formation, the edge charge reverses and clays disperse (peptize). Interestingly, many interactions between solid surfaces and organic material are from anionic organic bases with negatively charged colloids, for example, humic acid anions on negatively charged silica. As

† In dye chemistry, the term "lake" is used to describe colloidally dispersed colored complexes.

[24] H. Lengweiler, W. Buser and W. Feitknecht, *Helv. Chim. Acta*, **44**, 796 (1961).
[25] A. Siegel and B. Burke, *Deep-Sea Res.*, **12**, 789 (1965).

discussed more fully in Section 9-7, short-range chemical interaction, such as surface complex formation, often outweighs electrostatic repulsion. Interactions between partners carrying charge of like sign may be enhanced by complex formation in the immediate vicinity of the solid surface between the metal ion and the functional group of the organic constituents if either the metal ion or the organic constituent is specifically sorbed to the surface. The following qualitative picture may be visualized.

1. Metal ions accumulate in the double layer and may adsorb at the solid surface carrying a negative surface potential.

2. The approach of the negatively charged organic entity to the negatively charged surface is facilitated.

3. Complex formation between metal ions in the double layer and functional groups of the organic compound result in the formation of RCOO—Me^+ or RSO_3—Me^+ groups that can be anchored on the surface, thus lowering the electrochemical free energy of attachment; correspondingly, the extent of adsorption is increased and the orientation ability of the adsorbed organic compound may be altered [26].

Selectivity in Complex Formation

Compounds containing F and O donor atoms are rather general complex formers. On the other hand, the soft bases, for example, ligands with N, Cl, Br, I and S are more selective; the selectivity increases with decreasing electronegativity of the element. Polyamines and compounds with sulfur are much more selective for B cations and transition metals than polycarboxylic acid. Although different ions may be held with varying degree of strength, inspection of *Stability Constants of Metal–Ion Complexes* shows that no complete specificity for a single metal ion has been achieved with synthetic chelating agents. Thus it is not surprising that of the many valuable metals (e.g., 7×10^7 tons of gold and 7×10^9 tons of copper) known to be dissolved in seawater, none has yet been commercially extracted with the help of a chelating agent [27].

CONCENTRATION OF METAL IONS BY ORGANISMS. Although simple chelating agents cannot concentrate a single metal exclusively, living organisms are almost capable of achieving this result. Some extreme examples are cited by Bayer: vanadium is a million times more concentrated in the blood cells of the tunicate *Phallusia mamillata* than in seawater. Similarly copper is concentrated by a factor of 10^5 in the blood of the cuttlefish *Octopus vulgaris*. It

[26] A. Sommerauer, D. L. Sussman and W. Stumm, *Kolloid-Z. Z. Polym.*, **225**, 147 (1968).
[27] E. Bayer, *Angew. Chem.*, **3**, 325 (1964).

appears that in the organism organic complexing agents are highly selective for one ion. What molecules have such a near-absolute specificity? Because functional groups containing donor atoms are not sufficiently selective, it is necessary to provide molecules with special geometrical arrangements. A simple example is provided by 1,10-phenanthroline, which gives color reactions with Fe(II) and Cu(I). The reaction with Fe(II) can be suppressed by introducing alkyl groups into the 2- and 9 positions, giving a reagent which still reacts with Cu(I) [28]. Introducing bulky constituents into a chelate ligand usually does not aid complex formation but only hinders it from a kinetic point of view; but specificity can be improved by this method [29]. It is conceivable that protein structures contain enclaves into which only special metal ions can fit, and that specificity in enzyme catalysis by metals arises in this way [30]. In line with this argument are the observations made by Bayer and coworkers on the hemocyanin derived from *Octopus vulgaris*. This hemocyanin has a molecular weight of 2.7×10^6. Its composition is that of a protein containing no prosthetic groups, that is, the complex-forming function is part of the molecular protein. Apparently, in addition to the donor atoms (e.g., mercapto groups of the cysteine linked by peptide bonds), the sequence of the links and the steric structure must be responsible for the specificity. Bayer [27] has been able to synthesize macromolecular chelating agents possessing high selectivity for rare-metal ions.

The general problem of the enrichment of metals in the marine biosphere has been treated by Goldberg [31]. He pointed out that, in general, the relative concentration factors closely parallel the order of stability of the complexes formed by metal ions with ligands. Table 6-7, prepared by Goldberg, shows a qualitative correspondence between the enrichments and the "Irving-Williams" order which suggests, according to Goldberg, that the selective and general accumulation processes are most probably initiated by complexing reactions taking place at exposed external surfaces of the organisms.

FERRICHROMES. An interesting model of a complexing system of iron(III) in nature is provided by the ferrichromes. Figure 6-12 shows the complexing tendency of nocardamine for Fe^{3+}. Nocardamine belongs to a group of compounds generally known as ferrichromes but also classified as siderochrome compounds by some workers. These ferrichromes represent a class of naturally occurring heteromeric peptides, containing a trihydroxamate as an iron(III) binding center. Figure 6-15 represents a structure of one of these

[28] H. Irving, E. J. Butler and M. F. Ring, *J. Chem. Soc.*, **1949**, 1489.
[29] I.-B. Eriksson-Quensel and T. Svedberg, *Biol. Bull.*, **71**, 498 (1936).
[30] F. R. N. Gurd, *Chemical Specificity in Biological Interactions*, Academic Press, New York, 1954.
[31] E. D. Goldberg, in *Chemical Oceanography*, J. P. Riley and G. Skirrow, Eds., Academic Press, New York, 1965, p. 163.

Table 6-7 Concentration Factors of Elements in Marine Organisms†

Organisms	Metal									
	Cu	Ni	Pb	Co	Zn	Mn	Mg	Ca	Sr	Ba
Seaweeds (Black and Mitchell, 1952)	—	550– 2000–	—	—	900	—	—	—	23	—
Benthic algae (Lowman, 1964)	400– 90,000	40,000	—	—	—	1000– 30,000	—	—	—	—
Plankton (Nicholls et al., 1960)	—	<20– 8000	30– 12,000	<100 16,000	—	—	—	—	—	—
Marine animals (F. G. Lowan, Personal Communication)	—	3000– 70,000	—	—	—	2000– 10,000	—	—	—	—
Anchovetta (Goldberg, 1962)	80	—	10,000	—	400	1000	0.1	7	8	20
Yellow fin tuna (Goldberg, 1962)	200	—	—	—	700	80	0.2	6	7	2
Skipjack tuna (Goldberg, 1962)	100	50	—	—	500	40	0.3	5	5	3
Sponges (Bowen and Sutton, 1951)	1400	420	—	50	—	—	—	0.07	3.5	—

† E. D. Goldberg, in *Chemical Oceanography*, J. P. Riley and G. Skirrow, Eds., Academic Press, New York, 1965, p. 185. The only intercomparisons that are valid are along horizontal lines (different techniques were used for different organisms). The elements are arranged in the "Irving-Williams" order (see Figure 6-8) to illustrate the relation between concentration factors and complex stability. Reproduced with permission from Academic Press.

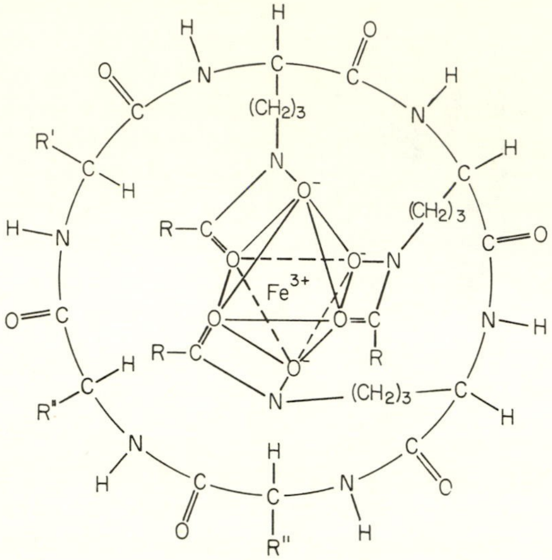

Figure 6-15 Structure of a ferrichrome (desferri-ferrichrome). One of the strongest complex formers presently known for Fe(III). The iron binding center is a trihydroxamate. It has been suggested that such naturally occurring ferrichromes play an important role in the biosynthetic pathways of iron. After Neilands [32]. Reproduced with permission from Birkhäuser Verlag, Basel, Switzerland.

ferrichromes (Desferri-ferrichrome). Polyhydroxamic acid rather specifically forms strong soluble complexes with Fe(III) and Mn(III). Its structure is related to formaldoxime, which can be used analytically for the determination of iron and manganese. The ferrichromes appear to be the strongest iron(III) complex formers presently known. The strong affinity of Fe^{3+} for the trihydroxamate is plausible in view of the tendency of Fe^{3+} to coordinate preferentially with basic oxygen donor atoms. The binding center (Figure 6-15) consists of an octahedral arrangement of oxygen ligand atoms, thus satisfying all the coordinative requirements of Fe^{3+}. These natural trihydroxamates bind iron so highly that iron remains complex bound in solution even at great dilutions and high pH. Furthermore, competition of earth alkali ions with iron(III) appears to be rather weak, thus making the ferrichromes relatively specific for ferric iron.

Ferrichromes appear to be widely distributed in microorganisms. Fungi have been used for the routine preparation of ferrichromes in the laboratory [32]. Ferrichrome has been suggested as a cofactor in microbial iron metabolism. According to Neilands, the ferric ion, once coordinated as soluble

[32] J. B. Neilands, in *Essays in Coordination Chemistry*, W. Schneider, R. Gut and G. Anderegg, Eds., Birkhäuser, Basel, 1964, p. 222.

trihydroxamate, can be transported to or into the cell and donated to the iron enzymes. Neilands has also shown that many microorganisms are capable of augmenting the bisynthesis of metal-free trihydroxamic acids during iron deficiency and he infers that this may represent an evolutionary invention whereby the organism protects itself against iron deprivation.

It is of course not yet known to what extent in natural water systems this special ligand serves the double function of dissolving iron and specifically donating it to biosynthetic pathways at the point of demand for iron, but it certainly makes an elegant model.

A Hypothetical Multimetal, Multiligand System for Major and Minor Metal Ions and Ligands

In order to form a more comprehensive picture of the possible chelating influence of natural organic compounds in natural waters, it is necessary to consider a system that contains, besides the major and minor ions typically found in natural waters, a variety of organic substances individually present at small concentrations ($C \leq 10^{-6} M$). Because of lack of information, the organic substances considered in the hypothetical system are not realistically representative of all detailed properties of natural waters. The organic substances selected for the model include one amino acid (glycine), citrate as a representative of a substance of importance in the biochemical pathway, salicylate which contains an arrangement of functional groups (carboxylic acid and aromatic OH) similar to that of humic acids and other natural color compounds, nocardamine (a ferrichrome) as a rather specific and strong complex former for Fe(III) and nitrilotriacetate (NTA) which is a very strong but rather unspecific chelating agent.

The system (Table 6-8) essentially consists of a mixture of nine metal ions and nine ligands, each one added at a fixed total (analytical) concentration. For convenience a fixed pH of 8.0 (e.g., controlled by other chemical equilibria at the sediment water interface), a temperature of 25°C and a constant ionic strength of 0.5 M has been assumed.

Computation of Equilibrium Composition

Within the given nine-metal, nine-ligand matrix approximately 150 metal–ligand species [$M_i H_j L_k (OH)_l$, where i, j, k or $l \geq 0$], for which stability constants are known, exist. Thus approximately 150 nonlinear equations with an equal number of unknowns have to be solved simultaneously. Furthermore, for each combination we need to check whether the solubility with respect to possible solid phases has been exceeded. After we know which solid phases exist, the concentrations of the solution phase have to be recomputed.

Table 6-8 A Hypothetical 9-Metal, 9-Ligand System †

Metal	Ligand										
	SO$_4$ −1.6	CO$_3$ −3.7	F −4.2	PO$_4$ −5.7	NTA −6.0	NOC −7.0	GLY −7.5	SAL −7.0	CIT −6.5	OH	Free M
Ca −2.0	−3.2	−4.7	−6.0	−8.0 *Solid*	−6.3	−20.2	−10.7	—	−6.9	−7.1	−2.03
Mg −1.3	−2.5	−4.1	−4.5	−7.5	−6.6	−16.8	−9.6	—	−6.9	−5.2	−1.33
Sr −3.8	−5.0	—	—	−10.4	−9.5	−21.4	−12.5	—	−9.4	−9.2	−3.83
Fe(III) −6.0	−17.4	—	−17.2	−18.6	−8.9	−7.0	−17.3	−14.0	−9.3	*Solid*	−18.0
Mn(II) −6.7	−7.9	−8.9	—	—	−10.0	−16.3	−14.0	−13.3	−11.7	−8.0	−6.73
Cu(II) −7.0	−10.3	−9.6	−12.9	—	−7.2	−16.1	−10.3	−11.1	−7.5	−9.9	−9.2
Zn −6.5	−7.8	—	−10.5	—	−6.9	−16.0	−11.3	−12.3	−11.7	−9.1	−6.8
Cd −9.0	−9.9	−12.4	—	−17.8	−9.7	−21.8	−14.1	−16.1	−13.0	−8.3	−9.16
Ni −7.0	−8.9	—	−11.7	—	−7.1	−17.1	−11.3	−13.3	−9.8	−10.9	−7.8
Free L	−1.67	−5.2	−4.5	−10.6	−10.7	−21.5	−9.3	−12.5	−16.3	−8.2	
LH$_n$	−7.7	−4.3	−9.5	−7.1	−8.9	−17.6	−7.5	−7.0	−8.3	−6.0	

† NTA = nitrilotriacetate, NOC = nocardamine, GLY = glycine, SAL = salicylate, CIT = citrate. Numbers along top indicate logs of *total* ligand concentration, numbers along left-hand side indicate logs of *total* metal ion concentration. The numbers in the M-L matrix are the logs of the *total* ML complex (including H or OH forms). "Solid" indicates that a stable solid phase can be formed by reaction between the metal and ligand under the conditions assumed. Dashes indicate that no species was considered in making the computation. LH$_n$ refers to the sum of the protonated ligand concentrations. The computations were made using an iterative, fast-converging computer program developed for the purpose by Morel and Morgan [33].

Obviously these computations become extremely tedious unless we are assisted by a suitable computer program. The computations for the system of Table 6-8 were made using an iterative, fast-converging computer program developed for the purpose by Morel [33].

The literature on computation of chemical equilibria is an extensive one. (See references at end of chapter.) However, many papers on that subject have such a general approach that it is difficult — for a nonspecialist at least — to derive any practical application. Other authors are generally interested in solving so narrow a problem that their methods cannot be generalized and used for other problems. The results of Table 6-8 has been obtained by solving the system of nonlinear equations using an iterative process based on the Newton method.

THE RESULT. The system considered in Table 6-8 has some, but not all, the chemical properties of seawater; it is quasi-realistic and serves primarily for illustrative purposes. Some of the equilibrium species and the corresponding equilibrium constants used for the calculation (selected from *Stability Constants of Metal–Ion Complexes*) have been overlooked. Blank spaces in Table 6-8 indicate that no species was considered in making the computation. The following results are of particular interest.

Two solid phases are found under equilibrium conditions, namely, apatite $[Ca_5(PO_4)_3(OH)]$ and ferric oxide $[Fe(OH)_3$ or $FeOOH]$. Of the nine metal ions, six are present predominantly as free aquo metal ions, Cu(II) and Cd(II) were present primarily as NTA complexes, while soluble iron(III) exists almost exclusively as a nocardamine (ferrichrome) chelate. The inorganic ligands have a tendency to be partially associated with Mg^{2+}; inorganic complexes or ion pairs involving other cations are less significant.

At the small concentrations prevalent, citrate and salicylate do not participate in any significant way in complexing metal ions. As expected, nocardamine becomes completely tied up with Fe(III). The strong but rather nonspecific complex-former nitrilotriacetate, which is present as the most concentrated organic species (total concentration $= 10^{-6}\ M$) and is present at equilibrium predominantly as a CaNTA complex, has a significant influence on the distribution of Cu(II), Zn(II) and Ni(II).

This example epitomizes the complexity of chemical forms found in a natural water. Much work needs to be done on the problem of types of organic compounds associated with trace metals. The hypothetical system illustrates that a few trace metal chelates may plausibly be expected in natural waters, especially in habitats where organic substances become enriched. At the same time, these calculations show that free trace metal aquo ions may be

[33] F. Morel and J. J. Morgan, *A Numerical Method for Solution of Chemical Equilibria in Aqeous Systems*, California Institute of Technology, Pasadena, Calif., 1968.

present as predominant metal species even if complex forming organic matter is present at concentrations significantly higher than those of the trace metals. Even in those instances where no soluble metal chelates are formed, some metal ions may be associated with particulate organic matter of various sizes. Data [34] demonstrating marked vertical variation in kinds (particulate extractable, nondialyzable) and amounts of trace metal in the oceans seem to establish that a substantial fraction of certain trace metals is contained together with organic substances in particulate matter.

Kinetic Effects

Even in those cases where only a small fraction of a metal ion is present as a complex species, a ligand may occasionally exercise pronounced inhibitory or catalytic effects upon reactions. For example, traces of pyrophosphate or EDTA retard the oxygenation of $Mn(II)$ to higher valent manganese oxides [35]. The effect is considerably larger than that which could be accounted for by complex formation alone. The effect of pyrophosphate is much larger than that of any other ligand investigated (see Figure 6-16).

Significance of Coordination Chemistry

More knowledge on organic constituents and the influence of complex formers on physiology is necessary to treat some of the biological ramifications in the control of metal concentrations and to consider comprehensively how living cells can differentiate, more effectively than can chelating agents in solutions, between metal cations.

In dealing with natural waters the dominating influence of the solid phases with which metal ions and organic substances come into contact cannot be underestimated (Section 9-7). The composition and structure of the solid phases represents an important branch of coordination chemistry.

Perhaps it has become sufficiently clear from this discussion that the dividing line between inorganic and organic chemistry is no longer as sharp as it used to be. Coordination chemistry has become the common meeting ground for inorganic and organic chemists but also for reaction kineticists, surface chemists, geologists and biologists. The water chemist should join them because the environment he is interested in and seeks to control is really a world of coordination compounds.

[34] J. F. Slowey, L. M. Jeffrey and D. W. Hood, Report 64-27A to AEC (1964).
[35] H. Bilinski and J. J. Morgan, presented at the 157th National Meeting of the American Chemical Society, Minneapolis, Minn., April 1969.

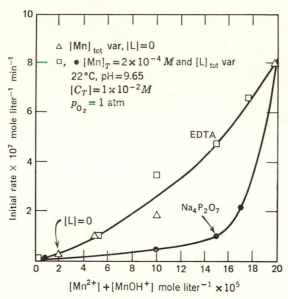

Figure 6-16 Catalytic effect of pyrophosphate on Mn(II) oxidation by oxygen. After Bilinski and Morgan [35]. Initial rate of Mn(II) oxygenation is plotted as a function of computed $[Mn^{2+}]$ and $[MnOH^+]$. The effect of EDTA upon the oxidation rate can be interpreted in terms of Mn(II) complexation (solutions containing EDTA seem to follow the same curve as systems in which the complex-former is absent). The effect of added pyrophosphate, however, is much larger than it should be if the effect were only caused by complex formation. Pyrophosphate may alter the reaction path by interacting with intermediates or products of the reaction.

PROBLEMS

6-1 Make a distribution diagram for the various Cr(VI) species as a function of pH at $Cr_T = 1$, 10^{-2} and 10^{-4} M. Neglect activity corrections.

Equilibria: $CrO_4^{2-} + H^+ = HCrO_4^-$ $\log K = 6.5$
$HCrO_4^- + H^+ = H_2CrO_4$ $\log K = -0.8$
$2HCrO_4^- = Cr_2O_7^{2-} + H_2O$ $\log K = 1.52$
$Cr_2O_7^{2-} + H^+ = HCr_2O_7^-$ $\log K = 0.07$

6-2 Find the pH at which "$Fe(OH)_3$" begins to precipitate from a 0.1 M $Fe(ClO_4)_3$ solution. Consider the dimerization of $FeOH^{2+}$.

6-3 Derive an equation that permits the computation of the charge of a hydrous oxide (more exactly, that portion of the charge caused by direct or indirect binding of H^+ or OH^-) from an alkalimetric titration curve of the hydrous oxide.

6-4 What equilibria need to be considered in assessing the fish toxicity of a galvanic waste containing Zn(II) and cyanide?

6-5 An organic compound isolated from natural waters is found to contain phenolic and carboxylic functional groups. From an alkalimetric titration curve the acidity constants $pK_1 \approx 4.1$ and $pK_2 \approx 12.5$ can be estimated. This compound forms colored complexes with Fe(III), the highest color intensity being observed in slightly acidic solutions for a mixture of equimolar concentrations of Fe(III) and the complex former. If to a 10^{-3} M Fe(III) solution at pH = 3 this complex former is added to attain a concentration of 5×10^{-2} M, a potentiometric electrode for Fe^{3+} (e.g., a ferro-ferri cell) registers a shift in $[Fe^{3+}]$ corresponding to a ΔpFe $\simeq$ 3.5. Provide a rough estimate for the stability of this complex. Constants for the hydrolysis of Fe(III) are given in Example 6-1 of Section 6-3.

6-6   Sketch a titration curve for titrating Ca^{2+} with EDTA at a pH of 10. Show the equilibrium composition of CaY^{2-}, Ca^{2+} and uncomplexed Y as a function of pCa. (H_4Y = EDTA.)

6-7   Is $BaSO_4$ appreciably soluble (S > 10^{-3} M) in a 10^{-1} M solution of citrate of pH = 8? *Information:* Acidity constants of citric acid (H_4L) pK_1 = 3.0, pK_2 = 4.4, pK_3 = 6.1, $pK_4 \simeq$ 16 BaHL is the only complex to be considered.

$$\frac{[BaHL]}{[Ba^{2+}][HL]} = 10^{2.4}$$

$$[Ba^{2+}][SO_4^{2-}] = 10^{-9}$$

ANSWERS TO PROBLEMS

6-1   The problem may be approached in a similar fashion to that given in Example 6-1. The basic equation to start with is

$$Cr_T = [CrO_4^{2-}] + [HCrO_4^{-}] + 2[Cr_2O_7^{2-}] + 2[HCr_2O_7^{-}]$$

(See plate on page 297).

6-2   pH = 1.7.

6-3 $\dfrac{C_B - C_A + [H^+] - [OH^-]}{\text{moles Me(OH)}_n}$ = mean charge. For the derivation see Example 9-2

in Section 5, Chapter 9.

6-4 $Zn^{2+} + H_2O = ZnOH^+ + H^+$; $*K_1$

$2Zn^{2+} + H_2O = Zn_2(OH)^{3+} + H^+$; $*\beta_{12}$

$Zn^{2+} + CN^- = Zn(CN)^+$; β_1

$Zn^{2+} + 2CN^- = Zn(CN)_2$; β_2

$Zn^{2+} + 3CN^- = Zn(CN)_3^-$; β_3

$Zn^{2+} + 4CN^- = Zn(CN)_4^{2-}$; β_4

$H^+ + CN^- = HCN$

Chemical species toxic to fish are primarily HCN, Zn^{2+}, $ZnOH^+$.

$$Zn_T = [Zn^{2+}] + \sum_{n=1}^{4} [Zn(CN)_n^{2-n}]_4 + [ZnOH^+] + 2[Zn_2OH^{3+}]$$

$$CN_T = [HCN] + [CN^-] + n \sum_{n=1}^{4} Zn(CN)_n^{2-n}$$

6-7   No. The solubility is increased from $10^{-4.5}$ to $10^{-3.8}$ M. The calculation is made very simple by realizing that [HL] is much larger than any other L-bound species; hence, $\alpha_3 \simeq [HL]/L_T \simeq 1.0$, and $\alpha_{Ba(HL)} = [Ba^{2+}]/Ba_T = \{1 + 10^{2.4}[HL]) \simeq 10^{-1.4}$. The conditional solubility product for $BaSO_4$ becomes $P = [Ba_T][SO_4^{2-}] = K_{s0}\alpha_{Ba(HL)}^{-1} = 10^{-7.6}$.

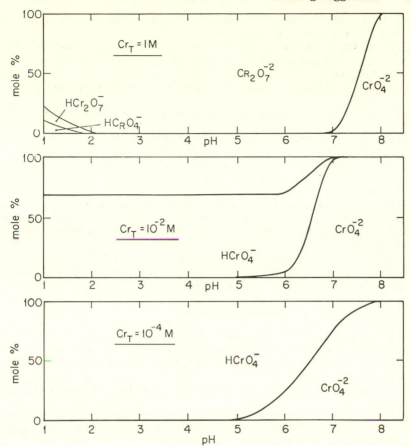

Problem 6.1

READING SUGGESTIONS

I. Introductory Texts

Basolo, F., and R. C. Johnson, *Coordination Chemistry*, Benjamin, New York, 1964.

Basolo, F., and R. G. Pearson, *Mechanism of Inorganic Reactions, A Study of Metal Complexes in Solution*, Wiley, New York, 1958.

Chaberek, S., and A. E. Martell, *Sequestering Agents*, Wiley, New York, 1959.

Graddon, D. P., *An Introduction to Coordination Chemistry*, Pergamon, New York, 1961.

Gray, H. B., *Electrons and Chemical Bonding*, Benjamin, New York, 1964.

Hunt, J. P., *Metal Ions in Solution*, Benjamin, New York, 1963.

Jones, M. M., *Elementary Coordination Chemistry*, Prentice-Hall, Englewood Cliffs, N.J., 1964.

Jørgensen, C. K., *Inorganic Complexes*, Academic Press, New York, 1963.

Martin, D. F., and B. B. Martin, *Coordination Compounds*, McGraw-Hill, New York, 1964. This is an elementary introduction.

Nancollas, G. H., *Interactions in Electrolyte Solutions*, Elsevier, New York, 1966.

Orgel, L. E., *An Introduction to Transition Metal Chemistry: Ligand Field Theory*, Wiley, New York, 1960.

Schneider, W., *Einführung in die Koordinationschemie*, Springer, Berlin, 1968.

II. Quantitative Complex Formation: Stability of Complexes and Analytical Applications

Anderegg, G., "Multidentate Ligands," in *Coordination Chemistry*, A. E. Martell, Ed., American Chemical Society Monograph, Washington, D.C., 1968.

Ringbom, A., *Complexation in Analytical Chemistry*, Interscience, New York, 1963.

Rossotti, F. J. C., and H. Rossotti, *The Determination of Stability Constants*, McGraw-Hill, New York, 1961.

Schindler, P., "Heterogeneous Equilibria Involving Oxides, Hydroxides, Carbonates and Hydroxide Carbonates," in *Equilibrium Concepts in Natural Water Systems*, Advances in Chemistry Series, No. 67, American Chemical Society, Washington, D.C., 1967.

Schwarzenbach, G., *Complexometric Titrations*, Interscience, New York, 1959.

Sillén, L. G., and A. E. Martell, *Stability Constants of Metal–Ion Complexes*, Special Publ., No. 17, The Chemical Society, London, 1964.

III. Selective and Specific Formation of Complexes

Amphlett, C. B., *Inorganic Ion Exchangers*, Elsevier, New York, 1964.

Edwards, J. O., *Inorganic Reaction Mechanisms*, Benjamin, New York, 1964.

Heller, H. J., *Some Aspects of Chelation Chemistry*, International Atomic Energy Agency, Vienna, 1963.

Schwarzenbach, G., "Die Sonderstellung des Wasserstoff Ions," *Chimia* (Aarau), **3**, 1 (1949).

Schwarzenbach, G., "The General, Selective and Specific Formation of Complexes by Metallic Cation," in *Advances in Inorganic Chemistry and Radiochemistry*, Volume 3, Academic Press, New York, 1961.

IV. Metal Ion Hydrolysis

Hem, J. D., "Aluminum Species in Water", in *Trace Inorganics in Water*, Advances in Chemistry Series, No. 73, American Chemical Society, Washington, D.C., 1968, p. 98.

Matijevic, E., et al., "Detection of Metal Ion Hydrolysis by Coagulation. III. Aluminum," *J. Phys. Chem.*, **65**, 826 (1961).

Olin, A., "Studies on the Hydrolysis of Metal Ions," *Svensk Kem. Tidskr.*, **73**, 482 (1961).

Sillén, L. G., "Quantitative Studies of Hydrolytic Equilibria," *Quart. Rev.* (London), **13**, 146 (1959).

Stumm, W., and J. J. Morgan, "Chemical Aspects of Coagulation," *J. Amer. Water Works Assoc.*, **54**, 971 (1962).

Stumm, W. and O'Melia, C. R., "Stoichiometry of Coagulation," *J. Amer. Water Works Assoc.*, **60**, 514 (1968).

Tyree, S. Y., Jr., "The Nature of Inorganic Solute Species in Water," in *Equilibrium Concepts in Natural Water Systems*, Advances in Chemistry Series, No. 67, American Chemical Society, Washington, D.C., 1967, p. 183.

V. Metal Ions in Natural Waters

Duursma, E. K., paper presented at Conference on Organic Matter in Natural Waters, University of Alaska, College, Alaska, 1968.

Goldberg, E. D., "Minor Elements in Seawater," in *Chemical Oceanography*, J. P. Riley and S. Skirrow, Eds., Academic Press, New York, 1965.

Jenne, E. A., "Controls on Mn, Fe, Co, Ni, Ci and Zn Concentrations in Soils and Water," in *Trace Inorganics in Water*, Advances in Chemistry Series, No. 73, American Chemical Society, Washington, D.C., 1968, p. 337.

Sillén, L. G., "Physical Chemistry of Seawater," in *Oceanography*, M. Sears, Ed., American Association for the Advancement of Science, Washington, D.C., 1961.

VI. Computational Aspects of Chemical Equilibrium

DeLand, E. C., *CHEMIST — The RAND Chemical Equilibrium Program*, RAND Corporation Report No. RM-5404-PR, December 1967 [Clearinghouse No. AD 664 045].

Duffin, R. J., et al., "Chemical Equilibrium Treated by Geometric Programming," in *Geometric Programming*, Wiley, New York, 1967, Appendix C.

White, W. B., et al., "Chemical Equilibrium in Complex Mixtures," *J. Chem. Phys.*, **28**, 751–755 (1958).

7 Oxidation and Reduction

7-1 Introduction

A functional description of the causal relationships effective in a natural water system must include parameters that characterize the influence of electrons on the environment.

In this chapter we stress the stability relations of pertinent redox (oxidation–reduction) components in natural water systems. However, one must be aware that concentrations of oxidizable or reducible species may be far from those predicted thermodynamically because many redox reactions are slow. In the sea or in a lake there is a marked difference in redox environment between the surface in contact with the oxygen of the atmosphere and the deepest layers of the sediments. In between are numerous localized intermediate zones, resulting from imperfections in mixing or diffusion and from varying biological activities, none of which is truly at equilibrium. The need for biological mediation of most redox processes encountered in natural waters means that approaches to equilibrium depend strongly on the activities of the biota. Moreover, quite different oxidation–reduction levels may be established within biotic microenvironments than those prevalent in the over-all environment; diffusion or dispersion of products from the microenvironment into the macroenvironment may give an erroneous view of redox conditions in the latter. Also, because many redox processes do not couple with one another readily, it is possible to have several different apparent oxidation–reduction levels in the same locale. Therefore detailed, quantitative exposition of redox conditions and processes will depend ultimately on understanding the dynamics of aquatic systems — the rates of approach to equilibrium — rather than on describing the total or partial equilibrium compositions. Elementary aspects of biological mediation and a few kinetic considerations on some redox reactions will be discussed in the latter part of this chapter.

Equilibrium considerations can greatly aid attempts to understand in a general way the redox patterns observed or anticipated in natural waters. In all circumstances equilibrium calculations provide boundary conditions toward which the systems must be proceeding. Moreover, partial equilibria (those involving some but not all redox couples) are approximated frequently, even though total equilibrium is not reached. In some instances active poising of particular redox couples allows us to predict significant oxidation–reduction levels or to estimate properties and reactions from computed redox levels. Valuable insight is gained even when differences are observed between computations and observations. The lack of equilibrium and the need for additional information or more sophisticated theory are then made clear.

Additional difficulties occur with attempts to measure oxidation–reduction potentials electrochemically in aquatic environments. Values obtained depend on the nature and rates of the reactions at the electrode surface and are seldom meaningfully interpretable. Even when suitable conditions for measurement are obtained, the results are significant only for those components whose behavior is electrochemically reversible at the electrode surface.

7-2 Redox Equilibria and the Electron Activity

There is a conceptual analogy between acid–base and reduction–oxidation reactions. In a similar way that acids and bases have been interpreted as proton donors and proton acceptors, reductants and oxidants are defined as electron donors and electron acceptors. Because there are no free electrons, every oxidation is accompanied by a reduction and vice versa; or an oxidant is a substance which causes oxidation to occur, while being reduced itself.

$$O_2 + 4H^+ + 4e = 2H_2O; \qquad \text{reduction}$$
$$4Fe^{2+} = 4Fe^{3+} + 4e; \qquad \text{oxidation}$$

$$O_2 + 4Fe^{2+} + 4H^+ = Fe^{3+} + 2H_2O; \quad \text{redox reaction}$$

THE OXIDATION STATE. As a result of the electron transfer (mechanistically the transfer may occur as a transfer of a group that carries the electron) there are changes in the oxidation states of reactants and products. Sometimes, especially in dealing with reactions involving covalent bonds, there are uncertainties in the assignment of electron loss or electron gain to a particular element. The oxidation state (or oxidation number) represents a hypothetical charge that an atom would have if the ion or molecule were to dissociate. This hypothetical dissociation, or the assignment of electrons to an atom, is carried out according to rules. The rules and a few examples are given in Table 7-1. In this book roman numerals are used to represent oxidation states and arabic numbers represent actual electronic charge. The concept of an oxidation state may often have little chemical reality, but the concept is extremely useful in discussing stoichiometry — as a tool for balancing redox reactions — and in systematic descriptive chemistry.

Sometimes the balancing of redox reactions causes difficulties. One of various approaches is illustrated in the following example.

Table 7-1 Oxidation State

Rules for Assigning Oxidation States:

(1) The oxidation state of a monoatomic substance is equal to its electronic charge.

(2) In a covalent compound, the oxidation state of each atom is the charge remaining on the atom when each shared pair of electrons is assigned completely to the more electronegative of the two atoms sharing them. An electron pair shared by two atoms of the same electronegativity is split between them.

(3) The sum of oxidation states is equal to zero for molecules, and for ions is equal to the formal charge of the ions.

Examples:

Nitrogen Compounds		Sulfur Compounds		Carbon Compounds	
Substance	Oxidation States	Substance	Oxidation States	Substance	Oxidation States
NH_4^+	$N = -III, \ H = +I$	H_2S	$S = -II, \ H = +I$	HCO_3^-	$C = +IV$
N_2	$N = 0$	$S_8(s)$	$S = 0$	$HCOOH$	$C = +II$
NO_2^-	$N = +III, \ O = -II$	SO_3^{2-}	$S = +IV, \ O = -II$	$C_6H_{12}O_6$	$C = 0$
NO_3^-	$N = +V, \ O = -II$	SO_4^{2-}	$S = +VI, \ O = -II$	CH_3OH	$C = -II$
HCN	$N = -III, \ C = +II, \ H = +I$	$S_2O_3^{2-}$	$S = +II, \ O = -II$	CH_4	$C = -IV$
SCN^-	$S = -I, \ C = +III, \ N = -III$	$S_4O_6^{2-}$	$S = +2.5, \ O = -II$	C_6H_5COOH	$C = -2/7$
		$S_2O_6^{2-}$	$S = +V, \ O = -II$		

Example 7-1 *Balancing Redox Reactions.* Balance the following redox reactions: 1. Oxidation of Mn^{2+} to MnO_4^- by PbO_2, 2. oxidation of $S_2O_3^{2-}$ to $S_4O_6^{2-}$ by O_2.

1. Reactants: Mn(II), Pb(IV); Products: Mn(VII), Pb(II)
   Oxidation: $Mn(II) = Mn(VII) + 5e$
   Reduction: $Pb(IV) + 2e = Pb(II)$

Half reactions:

$$Mn^{2+} = Mn(VII) + 5e$$
$$Mn(VII) + 4O(\text{-II}) = MnO_4^-$$
$$4H_2O = 4O(-II) + 8H^+$$

$$Mn^{2+} + 4H_2O = MnO_4^- + 8H^+ + 5e \qquad \text{(i)}$$
$$Pb(IV) + 2e = Pb^{2+}$$
$$PbO_2 = Pb(IV) + 2O(-II)$$
$$2O(-II) + 4H^+ = 2H_2O$$

$$PbO_2 + 4H^+ + 2e = Pb^{2+} + 2H_2O \qquad \text{(ii)}$$

Reactions (i) + (ii):

$$2Mn^{2+} + 5PbO_2 + 4H^+ = 2MnO_4^- + 5Pb^{2+} + 2H_2O$$

2. Reactants: S(II), O(0); Products: S(+2.5), O(−II)
   Oxidation: $2S(II) = 2S(+2.5) + e$
   Reduction: $O(0) + 2e = O(-II)$

Half reactions:

$$2S(II) = 2S(+2.5) + e$$
$$S_2O_3^{2-} = 2S(II) + 3O(-II)$$
$$2S(+2.5) + 3O(-II) = \tfrac{1}{2}S_4O_6^{2-}$$

$$S_2O_3^{2-} = \tfrac{1}{2}S_4O_6^{2-} + e \qquad \text{(iii)}$$
$$\tfrac{1}{2}O_2 + 2e = O(-II)$$
$$O(-II) + 2H^+ = H_2O$$

$$\tfrac{1}{2}O_2 + 2H^+ + 2e = H_2O \qquad \text{(iv)}$$

Reactions (iii) + (iv):

$$2S_2O_3^{2-} + \tfrac{1}{2}O_2 + 2H^+ = S_4O_6^{2-} + H_2O$$

ELECTRON ACTIVITY AND pϵ. Aqueous solutions do not contain free protons and free electrons, but it is nevertheless possible to define relative proton activity (pH $= -\log \{H^+\}$) and relative electron activity (pϵ $= -\log \{e\}$). Large positive values of pϵ (low electron activity) represent strongly oxidizing conditions while small or negative values (high electron activity) correspond to strongly reducing conditions.

In Table 7-2 the corresponding relations for pH and pϵ are derived in a stepwise manner. In order to relate pϵ to redox equilibria, we recall first the relationship derived for pH and acid–base equilibria (left-hand side of Table 7-2). An electron transfer reaction, in analogy to a proton transfer reaction, can be interpreted in terms of two reaction steps [(2) and (3) in Table 7-2].

In parallel to the convention of arbitrarily assigning $\Delta G° = 0$ for the hydration of the proton (3a), we also assign a zero free energy change for the reduction of $H_2(g)$ (3b). Equations 5a and 5b (Table 7-2) show that pH and pϵ are measures of the free energy involved in the transfer of 1 mole of protons or electrons, respectively.

Equations 9b and 10b (Table 7-2) can be expressed more generally for the reaction

$$\sum_i n_i A_i + ne = 0; \quad K \tag{1}$$

where A_i designates the participating species and n_i their numerical coefficients, positive for reactants and negative for products, in terms of the relations

$$p\epsilon = p\epsilon° + \frac{1}{n} \log \left(\prod_i \{A_i\}^{n_i} \right) \tag{2}$$

and

$$p\epsilon° = \frac{1}{n} \log K \tag{3}†$$

In accordance with (2), the quantity p$\epsilon°$ is the (relative) electron activity when all species other than the electrons are at unit activity. Relationships are even simpler and more readily comparable if all the (half) reactions are written for transfer of a single mole of electrons in the form

$$\sum \left(\frac{n_i}{n}\right) A_i + e = 0 \tag{4}$$

---

† In order to relate p$\epsilon°$ to log K, K is defined in terms of the equilibrium constant of the *reduction* (electron acceptance) reaction. (Similarly, in an acid–base reaction, pH is related to log $\bar{K}$ where $\bar{K}$ is defined in terms of the equilibrium constant for the proton acceptance reaction, i.e., log $\bar{K}$ = pK_{acidity}.)

Table 7-2 pH and pε

$pH = -\log \{H^+\}$		$p\epsilon = -\log \{e\}$	
Acid–base reaction: $HA + H_2O = H_3O^+ + A^-$; K_1	(1a)	Redox reaction: $Fe^{3+} + \tfrac{1}{2}H_2(g) = Fe^{2+} + H^+$; K_1	(1b)
Reaction (1a) is composed of two steps:		Reaction (1b) is composed of two steps:	
$HA = H^+ + A^-$; K_2	(2a)	$Fe^{3+} + e = Fe^{2+}$; K_2	(2b)
$H_2O + H^+ = H_3O^+$; K_3	(3a)	$\tfrac{1}{2}H_2(g) = H^+ + e$; K_3	(3b)
According to thermodynamic convention: $K_3 = 1$		According to thermodynamic convention: $K_3 = 1$	
Thus: $K_1 = K_2 = K_2K_3 = \{H^+\}\{A^-\}/\{HA\}$	(4a)	Thus: $K_1 = K_2 = K_2K_3 = \{Fe^{2+}\}/\{Fe^{3+}\}\{e\}$	(4b)
or $pH = pK + \log [\{A^-\}/\{HA\}]$	(5a)	or $p\epsilon = p\epsilon° + \log [\{Fe^{3+}\}/\{Fe^{2+}\}]$	(5b)
Since $pK = -\log K = \Delta G°/2.3RT$		Since $p\epsilon° = \log K = -\Delta G°/2.3RT$	
$pH = \Delta G°/2.3RT + \log [\{A^-\}/\{HA\}]$	(6a)	$p\epsilon = -\Delta G°/2.3RT + \log [\{Fe^{3+}\}/\{Fe^{2+}\}]$	(6b)
or for the transfer of 1 mole of H^+ from acid to H_2O:		or for the transfer of 1 mole of e to oxidant from H_2:	
$\Delta G/2.3RT = \Delta G°/2.3RT + \log [\{A^-\}/\{HA\}]$	(7a)	$-\Delta G/2.3RT = -\Delta G°/2.3RT + \log [\{Fe^{3+}\}/\{Fe^{2+}\}]$	(7b)
For the general case where n protons are transferred:		For the general case where n electrons are transferred:	
$H_nB + nH_2O = nH_3O^+ + B^{-n}$; β^*	(8a)	$ox + \{n/2\}H_2 = red + nH^+$; $ox + ne = red$; K^*	(8b)
$pH = \{1/n\}p\beta^* + \{1/n\}\log[\{B^{-n}\}/\{H_nB\}]$	(9a)	$p\epsilon = \{1/n\} \log K^* + \{1/n\} \log [\{ox\}/\{red\}]$;	(9b)
$pH = \Delta G/n2.3RT$		$p\epsilon = p\epsilon° + \{1/n\} \log [\{ox\}/\{red\}]$	
$= \Delta G°/n2.3RT + \{1/n\} \log [\{B^{-n}\}/\{H_nB\}]$	(10a)	$p\epsilon = -\Delta G°/n2.3RT + \{1/n\} \log [\{ox\}/\{red\}]$	(10b)
$\Delta G = -nFE$ (E = acidity potential)	(11a)	$\Delta G = -nFE_H$ (E_H = redox potential)	(11b)
$pH = -E/\{2.3RTF^{-1}\}$		$p\epsilon = E_H/\{2.3RTF^{-1}\}$	
$= -E°/\{2.3RTF^{-1}\} + \{1/n\} \log [\{B^{-n}\}/\{H_nB\}]$	(12a)†	$= E°/\{2.3RTF^{-1}\} + \{1/n\} \log [\{ox\}/\{red\}]$	(12b)†
Acidity potential:		Redox potential (Peters-Nernst equation):	
$E = E° + \{2.3RT/nF\} \log [\{H_nB\}/\{B^{-n}\}]$	(13a)	$E_H = E_H° + \{2.3RT/nF\} \log [\{ox\}/\{red\}]$	(13b)

† At 25°C, $2.3RTF^{-1} = 0.059$ (V/eq). From (10) and (12): 25°C, $p\epsilon = E/0.059$, $p\epsilon° = E_H°/0.059$.

Then

$$pe = pe^\circ + \log \left(\prod_i \{A_i\}^{(n_i/n)} \right) \qquad (5)$$

and

$$pe^\circ = \log K \qquad (6)$$

where K is given by (7)

$$\log K = \log \prod_i \{A_i\}^{-(n_i/n)} \qquad (7)$$

with the A_i at their equilibrium activities. For example, for the reduction of NO_3^- to NH_4^+:

$$K = \frac{\{NH_4^+\}^{1/8}}{\{NO_3^-\}^{1/8}\{H^+\}^{5/4}\{e\}}$$

Equilibrium Calculations

So far we have treated the electron like a ligand in complex formation reactions. Indeed in *Stability Constants of Metal–Ion Complexes* the first ligand considered in Section I (inorganic ligands) is the electron, and the equilibrium constants listed there are for redox reactions corresponding to (1).

Example 7-2   *The Formal Computation of pe-Values.*   Calculate pe values for the following equilibrium systems (25°C, $I = 0$):

(i) An acid solution 10^{-5} M in Fe^{3+} and 10^{-3} M in Fe^{2+}.
(ii) A natural water at pH $= 7.5$ in equilibrium with the atmosphere ($p_{O_2} = 0.21$ atm).
(iii) A natural water at pH $= 8$ containing 10^{-5} M Mn^{2+} in equilibrium with γ-MnO_2(s).

Stability Constants of Metal–Ion Complexes gives the following equilibrium constants:

$$Fe^{3+} + e = Fe^{2+}; \qquad K = \frac{\{Fe^{2+}\}}{\{Fe^{3+}\}\{e\}}; \qquad \log K = 12.53 \qquad (i)$$

$$\tfrac{1}{2}O_2(g) + 2H^+ + 2e = H_2O(l); \qquad K = \frac{1}{p_{O_2}^{1/2}\{H^+\}^2\{e\}^2}; \qquad \log K = 41.55 \qquad (ii)$$

$$\gamma\text{-}MnO_2(s) + 4H^+ + 2e = Mn^{2+} + 2H_2O(l); \qquad K = \frac{\{Mn^{2+}\}}{\{H^+\}^4\{e\}^2}; \qquad \log K = 40.84 \qquad (iii)$$

By using (1) and (2) the following $p\epsilon$ values are obtained for the conditions stipulated:

(i) $p\epsilon = 12.53 + \log \dfrac{\{Fe^{3+}\}}{\{Fe^{2+}\}} = 10.53$

(ii) $p\epsilon = 20.78 + \frac{1}{2}\log(p_{O_2}^{1/2}\{H^+\}^2) = 12.94$

(iii) $p\epsilon = 20.42 + \frac{1}{2}\log\dfrac{\{H^+\}^4}{\{Mn^{2+}\}} = 6.92$

Example 7-3   *Equilibrium Composition of Simple Solutions.*   Calculate the equilibrium composition of the following solutions (25°C, $I = 0$), both in equilibrium with the atmosphere ($p_{O_2} = 0.21$ atm).

(i) An acid solution (pH = 2) containing a total concentration of iron, $Fe_T = 10^{-4}$ M.

(ii) A natural water (pH = 7) containing Mn^{2+} in equilibrium with γ-MnO_2(s). The equilibrium constants were given in Example 7-2. The redox equilibria are defined by the conditions given (p_{O_2} and pH). Hence, $p\epsilon$ for both can be calculated with the equation:

$$p\epsilon = 20.78 + \frac{1}{2}\log(p_{O_2}^{1/2}\{H^+\}^2)$$

The following values are obtained:

(i) $p\epsilon = 18.43$
(ii) $p\epsilon = 13.42$

Correspondingly we find by

$$p\epsilon = 12.53 + \log\frac{\{Fe^{3+}\}}{\{Fe^{2+}\}}$$

and

$$p\epsilon = 20.42 + \frac{1}{2}\log\frac{\{H^+\}^4}{\{Mn^{2+}\}}$$

for solution (i) $\{Fe^{3+}\}/\{Fe^{2+}\} = 10^{5.9}$ or $\{Fe^{3+}\} = 10^{-4}$ and $\{Fe^{2+}\} = 10^{-9.9}$; and for solution (ii) $\{Mn^{2+}\} = 10^{-14}$ M.

There are, of course, other approaches to calculate the equilibrium composition; for example, we may first compute the equilibrium constants of the over-all redox reactions:

$\frac{1}{2}O_2(g) + 2H^+ + 2e = H_2O(l);$	$\log K = 41.55$
$2Fe^{2+} = 2Fe^{3+} + 2e;$	$\log K = -25.06$
$\frac{1}{2}O_2(g) + 2Fe^{2+} + 2H^+ = 2Fe^{3+} + H_2O(l);$	$\log K = 16.49$
$\frac{1}{2}O_2(g) + 2H^+ + 2e = H_2O(l);$	$\log K = 41.55$
$Mn^{2+} + 2H_2O(l) = \gamma\text{-}MnO_2(s) + 4H^+ + 2e;$	$\log K = -40.84$
$\frac{1}{2}O_2(g) + Mn^{2+} + H_2O(l) = \gamma\text{-}MnO_2(s) + 2H^+;$	$\log K = 0.71$

With the equilibrium constants defined for a given p_{O_2} and pH, the ratio $\{Fe^{3+}\}/\{Fe^{2+}\}$ and the equilibrium activity of Mn^{2+} can be calculated.

pϵ AS A MASTER VARIABLE. The logarithmic equilibrium expressions [e.g., (9b) in Table 7-2] lend themselves to graphical presentation in double logarithmic equilibrium diagrams. As we used pH as a master variable in acid–base equilibria, we may use pϵ as a master variable for the graphical presentation of redox equilibria.

For example in an acid solution of Fe^{2+} and Fe^{3+} (hydrolysis is neglected) the redox equilibrium

$$\frac{\{Fe^{2+}\}}{\{Fe^{3+}\}\{e\}} = K \tag{8}$$

may be combined with the concentration condition

$$[Fe^{2+}] + [Fe^{3+}] = Fe_T \tag{9}$$

to obtain the relations

$$[Fe^{3+}] = \frac{Fe_T K^{-1}}{\{e\} + K^{-1}} \tag{10}$$

and

$$[Fe^{2+}] = \frac{Fe_T\{e\}}{\{e\} + K^{-1}} \tag{11}$$

which, in logarithmic notations, can be formulated in terms of asymptotes. For $\{e\} \gg 1/K$ or pϵ $<$ pϵ°:

$$\log [Fe^{3+}] = \log Fe_T + p\epsilon - p\epsilon° \tag{12}$$

$$\log [Fe^{2+}] = \log Fe_T \tag{13}$$

Similarly for $\{e\} \ll 1/K$ or pϵ $>$ pϵ°

$$\log [Fe^{3+}] = \log Fe_T \tag{14}$$

$$\log [Fe^{2+}] = \log Fe_T + p\epsilon° - p\epsilon \tag{15}$$

These relations can be conveniently plotted (Figure 7-1). The diagram shows how pϵ changes with the ratio of $\{Fe^{3+}\}$ and $\{Fe^{2+}\}$ for log $Fe_T = -3$.

For aqueous solutions at a given pH each pϵ value is associated with a partial pressure of H_2 and of O_2:

$$2H^+ + 2e = H_2(g); \qquad \log K = 0 \tag{16}$$

or

$$2H_2O + 2e = H_2(g) + 2OH^-; \quad \log K = -28 \tag{17}$$

and

$$O_2(g) + 4H^+ + 4e = H_2O; \qquad \log K = 83.1 \tag{18}$$

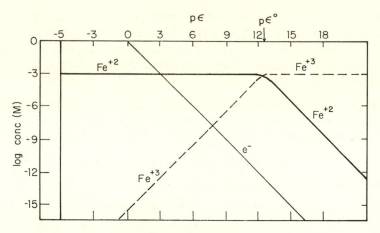

Figure 7-1   Redox equilibrium Fe^{3+}, Fe^{2+}. Equilibrium distribution of a $10^{-3} M$ solution of aqueous iron as a function of $p\epsilon$ (Example 7-2).

or

$$O_2(g) + 2H_2O + 4e = 4OH^-; \qquad \log K = 27.1 \qquad (19)$$

The equilibrium redox equations in logarithmic form are

$$\log p_{H_2} = 0 - 2pH - 2p\epsilon \qquad (20)$$

$$\log p_{O_2} = -83.1 + 4pH + 4p\epsilon \qquad (21)$$

Figure 7-2 gives a representation of (16) and (18) in logarithmic form. Thus, for example, a water of pH $= 10$ and of $p\epsilon = 8$ corresponds to $p_{O_2} = 10^{-11}$

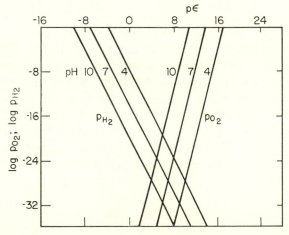

Figure 7-2   The stability of water $[H_2O(l) = H_2(g) + \frac{1}{2}O_2(g)]$. Partial pressure of H_2 and O_2 at various pH values in equilibrium with water.

atm and $p_{H_2} = 10^{-34}$ atm. Instead of using pϵ as a measure of oxidizing intensity, it is possible to characterize this intensity by specifying pH and p_{O_2} or p_{H_2}† (Figure 7.4). As Figure 7-2 also illustrates, the pϵ-range of natural waters (pH = 4–10) extends from approximately pϵ = −10 to 17; beyond these values water is reduced to H_2 or oxidized to O_2, respectively.

Example 7-4 *Redox Equilibrium* $SO_4{}^{2-}$–HS^-. Construct a diagram showing the pϵ dependence of a 10^{-4} M $SO_4{}^{2-}$–HS^- system at pH = 10 and 25°C. The reaction is

$$SO_4{}^{2-} + 9H^+ + 8e = HS^- + 4H_2O(l) \tag{i}$$

and the redox equilibrium equation is

$$p\epsilon = \frac{1}{8} \log K + \frac{1}{8} \log \frac{[SO_4{}^{2-}][H^+]^9}{[HS^-]} \tag{ii}$$

We may calculate the equilibrium constant from available data on the standard free energy of formation. The U.S. Bureau of Standards (quoted in W. M. Latimer, *Oxidation Potentials*, Prentice-Hall, New York, 1952) gives the following $\bar{G}_f^\circ$ values (kcal mole^{-1}) $SO_4{}^{2-}$: −177.34; HS^-: 3.01; $H_2O(l)$: −56.69. $\bar{G}_f^\circ$ for the aqueous proton and the electron are zero. Hence, the standard free energy change of (i) is $\Delta G^\circ = -46.41$ and the corresponding equilibrium constant ($K = 10^{-\Delta G^\circ/2.3RT}$) = 10^{34}. Hence, substituting in (ii) we obtain

$$p\epsilon = 4.25 - 1.125pH + \tfrac{1}{8} \log [SO_4{}^{2-}] - \tfrac{1}{8} \log [HS^-] \tag{iii}$$

Or, for pH = 10

$$p\epsilon = -7 + \tfrac{1}{8} \log [SO_4{}^{2-}] - \tfrac{1}{8} \log [HS^-] \tag{iv}$$

Equation (iv) has been plotted in Figure 7-3 for the condition $[SO_4{}^{2-}]$ + $[HS^-] = 10^{-4}$ M. HS^- is the predominant $S(-II)$ species at pH = 10. The lines for $[SO_4{}^{2-}]$ and $[HS^-]$ intersect at pϵ = −7; the asymptotes have slopes of 0 and ±8, respectively. Lines for the equilibrium partial pressures of O_2 and H_2 are also given in the diagram. As the diagram shows, rather high relative electron activities are necessary to reduce $SO_4{}^{2-}$. At the pH value selected, the reduction takes place at pϵ values slightly less negative than for the reduction of water. In the presence of oxygen, only sulfate can exist; its reduction is only possible under very anaerobic conditions (pϵ < −6; p_{O_2} < 10^{-68} atm).

---

† The resulting partial pressures are often mere calculation numbers; some correspond to less than one H_2 molecule in a space as large as the solar system. In 1921 Clark proposed a reduction intensity parameter defined by $r_H = -\log p_{H_2}$; fortunately r_H is no longer used and even Clark himself discouraged its use.

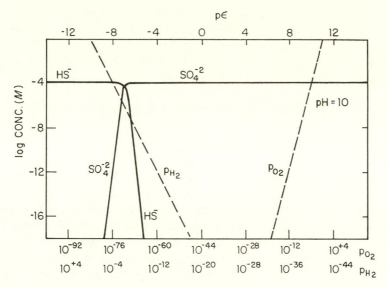

Figure 7-3 Equilibrium distribution of sulfur compounds as a function of $p\epsilon$ at pH = 10 and 25°C. Total concentration of compounds is 10^{-4} M (Example 7-4).

Example 7-5 *Activity Ratio Diagram; Equilibrium System:* SO_4^{2-}, $S^{\underline{0}}$, $S(-II)$. In the previous example, no consideration was given to solid sulfur as a possible intermediate state in the reduction of SO_4^{2-} to $S(-II)$.

1. Evaluate whether rhombic solid sulfur can be formed under the conditions specified (pH = 10, $[SO_4^{2-}] + [HS^-] = 10^{-4}$ M).

2. Establish a log conc.-$p\epsilon$ diagram for the SO_4^{2-}–$S(s)$–$H_2S(aq)$ system at pH = 4, assuming that the concentration of soluble sulfur species must not exceed 10^{-2} M.

In addition to the equilibrium constant for the reduction of SO_4^{2-} to HS^-, we need an equilibrium constant for a reaction with $S(s)$. *Stability Constants of Metal–Ion Complexes* lists

$$S(s) + 2H^+ + 2e = H_2S(aq); \quad \log K = 4.8 \qquad \text{(i)}$$

This reaction may be combined with other pertinent equilibria as follows:

$$
\begin{array}{llll}
SO_4^{2-} + 9H^+ + 8e = HS^- + 4H_2O; & \log K = 34.0 & \text{(ii)} \\
H_2S(aq) = S(s) + 2H^+ + 2e; & \log K = -4.8 & \text{(iii)} \\
H^+ + HS^- = H_2S(aq); & \log K = 7.0 & \text{(iv)} \\
\hline
SO_4^{2-} + 8H^+ + 6e = S(s) + 4H_2O; & \log K = 36.2 & \text{(v)}
\end{array}
$$

For the $S(s)$–$S(-II)$ system we can combine (iii) and (iv) to yield

$$HS^- = S(s) + H^+ + 2e; \quad \log K = 2.2 \qquad \text{(vi)}$$

312   Oxidation and Reduction

1. In order to investigate whether sulfur can be formed under the conditions given, it might be most convenient to prepare an *activity ratio diagram* (see Example 5-4), to select the activity of S(s) as a reference and describe the dependence of log ($[SO_4^{2-}]/\{S\}$) and of log ($[HS^-]/\{S\}$) as a function of pϵ.

From (v) and (vi) we obtain the relations

$$\log \frac{[SO_4^{2-}]}{\{S\}} = -36.2 + 8pH + 6p\epsilon \tag{vii}$$

$$\log \frac{[HS^-]}{\{S\}} = -2.2 - pH - 2p\epsilon \tag{viii}$$

which can be plotted conveniently for pH = 10. As Figure 7-4 illustrates, over the entire pϵ range the relative activity of solid sulfur is always smaller than that of $[SO_4^{2-}]$ and $[HS^-]$; hence elementary sulfur cannot be formed and need not be considered for the construction of the diagram in Figure 7-3.

2. For convenience, the pertinent reactions to be considered should be written in terms of the species prevailing under the given pH conditions. At pH = 5 the predominant species are SO_4^{2-} and H_2S(aq). For the construction of an activity ratio diagram, (vii) may be used together with (iii), whose redox equilibrium equation is

$$\log \frac{[H_2S]}{\{S\}} = 4.8 - 2pH - 2p\epsilon \tag{ix}$$

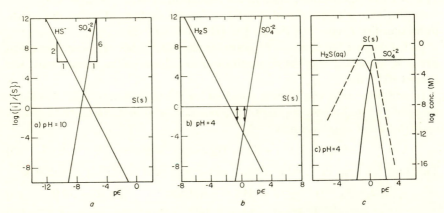

Figure 7-4   Equilibrium distribution of sulfur species (Example 7-5).

a, b: Activity ratio versus pϵ diagrams at pH = 10 and 4. Elemental sulfur is chosen as a reference. It is apparent that at pH = 10, elemental sulfur cannot exist thermodynamically, whereas at pH = 4, S(s) becomes the relatively most stable species in a narrow pϵ range.

c: Equilibrium diagram at pH = 4. Solid S is formed when its activity becomes unity. (It may be formed at lower activity if deposited as solid solution.)

The activity ratio diagram (Figure 7-4*b*) reveals that elemental sulfur can be formed in a narrow pϵ range. Only between p$\epsilon = -0.6$ and $+0.4$ are the ratios of $[H_2S]/\{S\}$ and $[SO_4{}^{2-}]/\{S\}$ smaller than 10^{-2}. For this condition the log C–pϵ diagram can be constructed (Figure 7-4*c*). Outside of the pϵ range -0.6 to $+0.4$, elemental sulfur acquires a (hypothetical) activity of less than unity (dashed line).

Combining (i), (ii) and (iv) gives

$$SO_4{}^{2-} + 3H_2S(aq) + 2H^+ = 4S(s) + 4H_2O; \quad \log K = 21.8 \qquad (x)$$

This equation indicates, as do Figures 7-4*b* and 7-4*c*, that the formation of elemental sulfur is only possible thermodynamically under acid conditions (pH < 7). Note that elemental sulfur (rhombic sulfur) may be formed as an intermediate and may persist as a metastable solid phase. Solid elemental sulfur occurs quite commonly in recent marine sediments [1].

7-3   Electron Free-Energy Levels; The Electrode Potential

In the previous section it was shown that pϵ is a measure of the free energy involved in the transfer of electrons:

$$p\epsilon = -\frac{\Delta G}{n2.3RT}; \qquad p\epsilon^\circ = -\frac{\Delta G^\circ}{n2.3RT}; \qquad p\epsilon^\circ = \frac{1}{n}\log K \qquad (22)$$

Here ΔG, ΔG° and K refer to a half-reaction written in the form of a reduction [compare (1)]. With (22) and tabulated ΔG° values (or equilibrium constants), an electron free-energy scale may be established in a similar way to the proton free-energy scale that was established earlier (Chapter 3). The concept of proton and electron free-energy levels was advanced by Gurney [2] and amplified by Reilley [3].

The Electron Free-Energy Scale

In Figure 7-5 electron free-energy levels are plotted for a few typical redox reactions. The ordinate measures the energy that is required for the transfer of electrons from one free-energy level to another. This energy can be expressed in different units, for example, kcal mole^{-1} of electrons, eV mole^{-1} and pϵ units that are all proportional to one another (see Table 7-3 for conversion factors). If we add electrons (in the form of reductants, i.e., electron complexes) to a system containing oxidants (i.e., substances of unoccupied

[1] R. A. Berner, *Geochim. Cosmochim. Acta*, **27**, 563 (1963).
[2] R. W. Gurney, *Ionic Processes in Solution*, McGraw-Hill, New York, 1953.
[3] C. N. Reilley, in *Electroanalytical Principles*, Wiley-Interscience, New York, 1963.

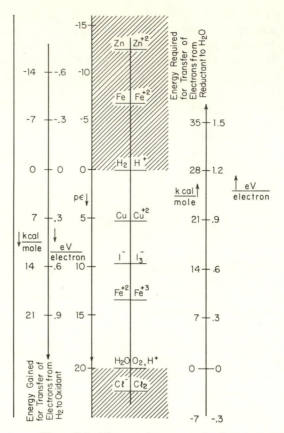

Figure 7-5   Electron free-energy levels.

electron levels), the lowest unoccupied electron level will first be filled up. With more electrons added, successive levels will be filled up. For example, a reductant such as Zn(s) will, from a thermodynamic point of view (but not necessarily from a kinetic point of view), first react with Cl_2, then successively with O_2, Fe^{3+}, Cu^{2+}, etc. Similarly, we can readily see that Fe^{3+} can oxidize I^- and Cu(s), and that Zn(s) and Fe(s) are oxidized by H^+ (metal dissolution). The energy gained in such processes per mole of electrons transferred can be read from the energy scale in terms of volts, kcal or $\Delta p\epsilon$ units. The oxidation of 1 equivalent ($\frac{1}{2}$ mole) of Fe(s) by H^+ yields an energy equivalent of 0.44 V or 10.1 kcal. But the dissolution of $\frac{1}{2}$ mole of Cu(s) by H^+ requires an energy equivalent of 0.34 V or 7.8 kcal. The electron-energy level diagram also illustrates that $\Delta G°$ of the electron transfer in the reaction $Cu^{2+} + Fe = Fe^{2+} + Cu$ equals -17.9 kcal or 0.78 eV per mole of electrons transferred. It is convenient to assign a zero reference level to a particular redox reaction such as $H^+ + e = \frac{1}{2}H_2(g)$, or $\frac{1}{4}O_2(g) + H^+ + e = \frac{1}{2}H_2O(l)$ and express differences in electron energy with regard to these reference levels.

INTENSITY AND CAPACITY. $p\epsilon$ is an intensity factor; it measures oxidizing intensity. Oxidation or reduction capacity must be expressed in terms of a quantity of system electrons which must be added or removed in order to attain a given $p\epsilon$. This is analogous to the acid or base neutralizing capacity with respect to protons; for example, alkalinity and acidity, are measured in terms of the proton condition. Thus the oxidative capacity of a system with respect to a given electron energy level will be given by the equivalent sum of all the oxidants below this energy level minus the equivalent sum of all reductants above it. For example, the oxidative capacity of a solution with respect to an electron level corresponding to a Cu^{2+} solution is (compare Figure 7-5)

$$2[Cu^{2+}] + 2[I_3^-] + [Fe^{3+}] + 4[O_2] - 2[H_2] \qquad (23)$$

THE STABILITY OF WATER. As Figure 7-5 illustrates there are reducing substances that are stronger reductants than H_2 and oxidizing substances that are stronger oxidants than O_2. Cl_2, for example, has an occupied electron level below that of O_2. Hence, thermodynamically speaking, Cl_2 is unstable in H_2O and will acquire electrons from H_2O, that is, oxidize H_2O to O_2. (This corresponds to the unstability of NH_2^- ion in water in that it acquires protons from the H_2O molecules.) The solvent exerts a leveling influence on the system and restricts the range of accessible $p\epsilon$ values (Figure 7-2).

$p\epsilon$ AND THE REDOX POTENTIAL. As (11) and (12) of Table 7-2 show, and as Figure 7-5 illustrates, the electron free-energy scale per mole of electron can be expressed in ΔG (cal), in potential E (V) or in $p\epsilon$ (dimensionless). (For various energy units and conversion factors, see Table 7-3.) The energy gained in the transfer of 1 mole of electrons from an oxidant to H_2, expressed in volts, is the redox potential E_H. (The suffix H denotes that this potential is on the hydrogen scale.) As already given in Table 7-1, $p\epsilon$ is related to E_H by

$$p\epsilon = \frac{F}{2.3RT} E_H \qquad (24)$$

E_H as given in (24) is in accord with the IUPAC (Stockholm)† Convention.

The Peters-Nernst Equation

Under certain conditions it is possible to measure E_H electrochemically. The thermodynamic relation of the potential E_H to the composition of the

† International Union of Pure and Applied Chemistry, 1953.

Table 7-3   Energy Units

$$1 \text{ erg} = 10^{-7} \text{ joule}$$

$$
\begin{aligned}
1 \text{ joule} &= 1 \text{ volt-coulomb} \\
&= 1 \text{ watt-second} \\
&= 2.8 \text{ kilowatt hours} \\
&= 9.9 \times 10^{-3} \text{ liter atmospheres} \\
&= 0.239 \text{ calorie}
\end{aligned}
$$

$$
\begin{aligned}
1 \text{ cal} &= 4.18 \text{ joule} \\
&= 4.34 \times 10^{-5} \text{ electron-volt (proton-volt)} \\
1 \text{ electron-volt} &= 23,060 \text{ calorie}
\end{aligned}
$$

$1 \text{ faraday} = 96,500 \text{ coulombs equivalents}^{-1}$ ($=$ electric charge of 1 mole of electrons)

$$
\begin{aligned}
1 \text{ volt-faraday} &= 96,500 \text{ volt-coulombs} \\
&= 23,060 \text{ cal} = 1 \text{ electron-volt}
\end{aligned}
$$

$$
\begin{aligned}
R \text{ (gas constant)} &= 8.314 \text{ joules deg}^{-1} \text{ mole}^{-1} \\
&= 1.987 \text{ cal deg}^{-1} \text{ mole}^{-1} \\
&= 0.08205 \text{ liter-atmosphere deg}^{-1}
\end{aligned}
$$

solution, as already given in Table 7-2, is generally known as the Nernst equation:

$$E_{\mathrm{H}} = E_{\mathrm{H}}^{\circ} + \frac{2.3RT}{nF} \log \frac{\{\mathrm{ox}\}}{\{\mathrm{red}\}} \tag{25}$$

$$E_{\mathrm{H}} = E_{\mathrm{H}}^{\circ} + \frac{2.3RT}{nF} \log \left(\prod_i \{A_i\}^{n_i} \right) \tag{26}$$

It is necessary to distinguish between the concept of a potential and the measurement of a potential. Redox or electrode potentials (quoted in tables in *Stability Constants of Metal–Ion Complexes* or by others [4–7]) have been derived from equilibrium data, thermal data and the chemical behavior of a redox couple with respect to known oxidizing and reducing agents, and from direct measurements of electrochemical cells. Hence there is no *a priori* reason to identify the thermodynamic redox potentials with measurable electrode potentials.

It follows from (24) that we can express (relative) electron activity in $p\epsilon$ or in volts. The use of $p\epsilon$, which is dimensionless, makes calculations simpler than the use of E_{H} because every tenfold change in the activity ratio causes a unit change in $p\epsilon$. Furthermore, because an electron can reduce a proton, the

[4] W. M. Latimer, *Oxidation Potentials*, Prentice-Hall, New York, 1952.

[5] W. M. Clark, *Oxidation Reduction Potentials of Organic Systems*, William & Wilkins, Baltimore, 1960.

[6] G. Charlot, *Selected Constants and Oxido-Reduction Potentials*, Pergamon, London, 1958.

[7] M. Pourbaix, *Atlas d'Equilibres Electrochemiques*, Gauthiers-Villars, Paris, 1963.

intensity parameter for oxidation might preferably be expressed in units equivalent to pϵ. Of course in making a direct electrochemical measurement of the oxidizing intensity an electromotive force (volts) is being measured, but the same is true in pH measurements, and a few decades ago the "acidity potential" was used to characterize the relative H^+ ion activity.

SIGN CONVENTION. There have been controversies regarding sign conventions, and the one adapted here is based on the IUPAC Convention. Tables 7-4 and 7-5 list for illustrative purposes a few representative redox potentials. According to the IUPAC Convention, all half reactions are written as reductions with a sign that corresponds to the sign of log K of the reduction reaction.

Useful reference works may use other sign conventions. For example, in Latimer's *Oxidation Potentials* the half cell equation is written in the reverse direction. For example,

$$Zn(s) = Zn^{2+} + 2e; \quad E^{\circ}_{cell} = 0.76 \text{ V} \tag{27}$$

corresponding to the over-all reaction

$$Zn(s) + 2H^+ = Zn^{2+} + H_2(g) \tag{28}$$

$E^{\circ}_{(Latimer)}$ corresponds to the emf measured in an electrochemical cell, E°_{cell}, where reaction (28) takes place. If we consider the reverse reaction

$$Zn^{2+} + H_2(g) = Zn(s) + 2H^+ \tag{29}$$

Table 7-4   Equilibrium Constants for a Few Redox Reactions

	log K, 25°C	Standard Electrode Potential, V, 25°C
$Na^+ + e = Na(s)$	-46	-2.71
$Zn^{2+} + 2e = Zn(s)$	-26	-0.76
$Fe^{2+} + 2e = Fe(s)$	-15	-0.44
$Co^{2+} + 2e = Co(s)$	-9.5	-0.28
$V^{3+} + e = V^{2+}$	-8.8	-0.26
$2H^+ + 2e = H_2(g)$	0.0	0.00
$S(s) + 2H^+ + 2e = H_2S$	$+0.47$	$+0.14$
$Cu^{2+} + e = Cu^+$	$+2.7$	$+0.16$
$AgCl(s) + e = Ag(s) + Cl^-$	$+3.7$	$+0.22$
$Cu^{2+} + 2e = Cu(s)$	$+12.0$	$+0.34$
$Cu^+ + e = Cu(s)$	$+18.0$	$+0.52$
$Fe^{3+} + e = Fe^{2+}$	$+13.2$	$+0.77$
$Ag^+ + e = Ag(s)$	$+13.5$	$+0.80$
$Fe(OH)_3(s) + 3H^+ + e = Fe^{2+} + 3H_2O$	$+18.8$	$+1.06$
$IO_3^- + 6H^+ + 5e = \frac{1}{2}I_2(s) + 3H_2O$	$+104$	$+1.23$
$MnO_2(s) + 4H^+ + 2e = Mn^{2+} + 2H_2O$	$+42$	$+1.23$
$Cl_2(g) + 2e = 2Cl^-$	$+46$	$+1.36$
$Co^{3+} + e = Co^{2+}$	$+31$	$+1.82$

Table 7-5   Equilibrium Constants of Redox Processes Pertinent in Aquatic Conditions (25°C)

Reaction	$p\epsilon°$ ($\equiv \log K$)	$p\epsilon°$ (W)
(1) $\frac{1}{4}O_2(g) + H^+(W) + e = \frac{1}{2}H_2O$	$+20.75$	$+13.75$
(2) $\frac{1}{5}NO_3^- + \frac{6}{5}H^+(W) + e = \frac{1}{10}N_2(g) + \frac{3}{5}H_2O$	$+21.05$	$+12.65$
(3) $\frac{1}{2}MnO_2(s) + \frac{1}{2}HCO_3^-(10^{-3}) + \frac{3}{2}H^+(W) + e$		
$\quad = \frac{1}{2}MnCO_3(s) + \frac{3}{8}H_2O$	—	$+8.5†$
(4) $\frac{1}{2}NO_3^- + H^+(W) + e = \frac{1}{2}NO_2^- + \frac{1}{2}H_2O$	$+14.15$	$+7.15$
(5) $\frac{1}{8}NO_3^- + \frac{5}{4}H^+(W) + e = \frac{1}{8}NH_4^+ + \frac{3}{8}H_2O$	$+14.90$	$+6.15$
(6) $\frac{1}{6}NO_2^- + \frac{4}{3}H^+(W) + e = \frac{1}{6}NH_4^+ + \frac{1}{3}H_2O$	$+15.14$	$+5.82$
(7) $\frac{1}{2}CH_3OH + H^+(W) + e = \frac{1}{2}CH_4(g) + \frac{1}{2}H_2O$	$+9.88$	$+2.88$
(8) $\frac{1}{4}CH_2O + H^+(W) + e = \frac{1}{4}CH_4(g) + \frac{1}{4}H_2O$	$+6.94$	-0.06
(9) $FeOOH(s) + HCO_3^-(10^{-3}) + 2H^+(W) + e$		
$\quad = FeCO_3(s) + 2H_2O$	—	$-1.67†$
(10) $\frac{1}{2}CH_2O + H^+(W) + e = \frac{1}{2}CH_3OH$	$+3.99$	-3.01
(11) $\frac{1}{6}SO_4^{2-} + \frac{4}{3}H^+(W) + e = \frac{1}{6}S(s) + \frac{2}{3}H_2O$	$+6.03$	-3.30
(12) $\frac{1}{8}SO_4^{2-} + \frac{5}{4}H^+(W) + e = \frac{1}{8}H_2S(g) + \frac{1}{2}H_2O$	$+5.75$	-3.50
(13) $\frac{1}{8}SO_4^{2-} + \frac{9}{8}H^+(W) + e = \frac{1}{8}HS^- + \frac{1}{2}H_2O$	$+4.13$	-3.75
(14) $\frac{1}{2}S(s) + H^+(W) + e = \frac{1}{2}H_2S(g)$	$+2.89$	-4.11
(15) $\frac{1}{8}CO_2(g) + H^+(W) + e = \frac{1}{8}CH_4(g) + \frac{1}{4}H_2O$	$+2.87$	-4.13
(16) $\frac{1}{6}N_2(g) + \frac{4}{3}H^+(W) + e = \frac{1}{3}NH_4^+$	$+4.68$	-4.68
(17) $\frac{1}{2}(NADP^+) + \frac{1}{2}H^+(W) + e = \frac{1}{2}(NADPH)$	-2.0	$-5.5‡$
(18) $H^+(W) + e = \frac{1}{2}H_2(g)$	0.0	-7.00
(19) Oxidized ferredoxin + e = reduced ferredoxin	-7.1	$-7.1§$
(20) $\frac{1}{4}CO_2(g) + H^+(W) + e = \frac{1}{24}(glucose) + \frac{1}{4}H_2O$	-0.20	$-7.20\|$
(21) $\frac{1}{2}HCOO^- + \frac{3}{2}H^+(W) + e = \frac{1}{2}CH_2O + \frac{1}{2}H_2O$	$+2.82$	-7.68
(22) $\frac{1}{4}CO_2(g) + H^+(W) + e = \frac{1}{4}CH_2O + \frac{1}{4}H_2O$	-1.20	-8.20
(23) $\frac{1}{2}CO_2(g) + \frac{1}{2}H^+(W) + e = \frac{1}{2}HCOO^-$	-4.83	-8.73

† These data correspond to $(HCO_3^-) = 10^{-3}M$ rather than unity and so are not exactly $p\epsilon°$(W); they represent typical aquatic conditions more nearly than $p\epsilon°$(W) values do.

‡ M. Calvin and J. A. Bassham, *The Photosynthesis of Carbon Compounds*, Benjamin, New York, 1962.

§ D. I. Arnon, *Science*, **149**, 1460 (1965).

∥ A. L. Lehninger, *Bioenergetics*, Benjamin, New York, 1965.

then the cell emf must have a sign opposite to that of reaction (28):

$$Zn^{2+} + 2e = Zn(s); \quad E°_{cell}(= E°_H) = -0.76 \text{ V}$$

Whereas an emf can change signs depending on the direction of the reaction, an electrode potential or $p\epsilon$ must not, according to IUPAC Convention, change signs because these parameters reflect electron activity or in an electrochemical cell reflect the electrostatic charge in the electrode. Hence we can define a standard electrode potential E_H° or a $p\epsilon^\circ$ of -0.76 volt or -12.8, respectively, for the Zn^{2+}–Zn couple independent of the direction of writing the half reaction. Note that multiplying or dividing the stoichiometric coefficients of a reaction does not affect $p\epsilon^\circ$ or E_H° but of course affects $\log K$ or ΔG°. The Nernst potential is independent of the number of moles of electrons transferred.

7-4 pɛ–pH Diagrams

An attempt has been made thus far to describe the stability relationships of the distribution of the various soluble and insoluble forms through rather simple graphical representation. Essentially two types of graphical treatments have been used: first, equilibria between chemical species in a particular oxidation state as a function of pH and solution composition; second, equilibria between chemical species at a particular pH as a function of pɛ (or E_H). Obviously these diagrams can be combined into pɛ–pH diagrams. Such pɛ–pH stability field diagrams show in a comprehensive way how protons and electrons simultaneously shift the equilibria under various conditions and can indicate which species predominate under any given condition of pɛ and pH.

The value of a pɛ–pH diagram consists primarily in providing an aid for the interpretation of equilibrium constants (free energy data) by permitting the simultaneous representation of many reactions. Of course, such a diagram, as with all other equilibrium diagrams, represents only the information used in its construction.

Natural waters are in a highly dynamic state with regard to oxidation–reduction rather than in or near equilibrium. Most oxidation–reduction reactions have a tendency to be much slower than acid–base reactions, especially in the absence of suitable biochemical catalysis. Nonetheless equilibrium diagrams can greatly aid attempts to understand the possible redox patterns in natural waters.

The Construction of pɛ–pH Diagrams

Example 7-6 *Chlorine Redox Equilibria.* Summarize in a pɛ–pH diagram the information contained in the equilibrium constants $I = 0$, $25°C$ of the following three reactions involving $Cl_2(aq)$, Cl^-, OCl^- and $HOCl$. (For convenience, in addition to the equilibrium constant, the standard redox potential, E_H° in volts, is given.)

$$HClO + H^+ + e^- = \tfrac{1}{2}Cl_2(aq) + H_2O; \quad \log K = 26.9, \; E_H^\circ = 1.59 \quad \text{(i)}$$

$$\tfrac{1}{2}Cl_2(aq) + e = Cl^-; \qquad\qquad\qquad \log K = 23.6, \; E_H^\circ = 1.40 \quad \text{(ii)}$$

$$HClO = H^+ + ClO^-; \qquad\qquad\qquad \log K = -8.3 \qquad\qquad\quad \text{(iii)}$$

Reactions (i) and (ii) correspond to the equilibrium equations

$$p\epsilon = 26.9 + \log[\text{HClO}] - \tfrac{1}{2}\log[\text{Cl}_2] - \text{pH} \qquad\qquad (iv)$$

$$p\epsilon = 23.6 + \tfrac{1}{2}\log[\text{Cl}_2] - \log[\text{Cl}^-] \qquad\qquad (v)$$

Combining (i) and (ii) gives the equilibrium expression for the reduction of HOCl to Cl⁻:

$$p\epsilon = 25.25 + \tfrac{1}{2}\log\frac{[\text{HClO}]}{[\text{Cl}^-]} - 0.5\,\text{pH} \qquad\qquad (vi)$$

and, finally, the distribution between ClO⁻ and Cl⁻ is obtained by combining (i), (ii) and (iii):

$$p\epsilon = 28.9 + \tfrac{1}{2}\log\frac{[\text{ClO}^-]}{[\text{Cl}^-]} - \text{pH} \qquad\qquad (vii)$$

The resulting four equations are plotted in Figure 7-6. The line for (v) is pH independent and thus plots as a horizontal line; it intersects with the line

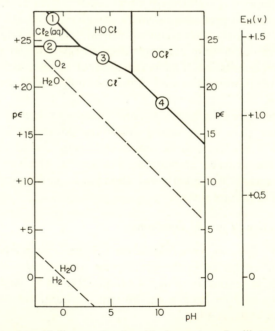

Figure 7-6 $p\epsilon$ versus pH diagram for the chlorine system. In dilute solutions Cl₂(aq) exists only at low pH. Cl₂, OCl⁻, and HOCl are all unstable or metastable in water. Numbers 1, 2, 3 and 4 refer to the equilibria described by (iv), (v), (vi) and (vii), respectively, of Example 7-6.

representing (iv) which has a $dp\epsilon/dpH$ of -1. Why do the lines discontinue at this intersection? To the right of pH $= 2$, HOCl is the more stable oxidant than $Cl_2(aq)$. (If any doubt should arise about which species predominates thermodynamically, an activity ratio diagram either at a given pє or at a given pH may immediately clarify the stability relations.) Equations (vi) and (vii) have slopes of -0.5 and -1, respectively, in the graphical representation; the lines intersect at pH $=$ pK of the hypochlorous acid.

It is convenient to introduce the equations that define the stability limit of H_2O into the same diagram.

$$O_2(g) + 4H^+ + 4e = 2H_2O \qquad \text{(viii)}$$

and

$$2H^+ + 2e = H_2(g) \qquad \text{(ix)}$$

These equations plot as

$$p\epsilon = 20.8 - pH + \tfrac{1}{4}p_{O_2} \qquad \text{(x)}$$

and

$$p\epsilon = 0 - pH - \tfrac{1}{2}\log p_{H_2} \qquad \text{(xi)}$$

Figure 7-6 reveals the following pertinent features of the equilibria involving Cl at the oxidation state $-I$, $\bar{0}$ and $+I$:

1. $Cl_2(aq)$ in dilute solutions exists only at rather low pH. Addition of Cl_2 to water is accompanied by disproportionation into HOCl and Cl^-. The disproportionation equilibrium is given by

$$Cl_2(aq) + H_2O = HClO + H^+ + Cl^- ; \quad \log K = -3.3 \qquad \text{(xii)}$$

This reaction obtains by combining (i) and (ii).

2. Cl as Cl_2 or HOCl and OCl^- is a strong oxidant, stronger than O_2; in other words, Cl_2, HOCl and OCl^- are thermodynamically unstable in water because these species oxidize water. (In the absence of suitable catalysts, however, this reaction is extremely slow.)

3. Within the entire pє–pH range of natural waters, Cl^- is the stable Cl species; it cannot be oxidized by O_2.

Example 7-7 *Aqueous Lead.* In constructing an equilibrium model for the chemistry of aqueous lead, it is necessary to identify all pertinent reactions and equilibria. We may proceed first by considering the system $Pb-H_2O$ only, that is, we ignore other ligands such as CO_3^{2-}, SO_4^{2-} or Cl^-. Then the system $Pb-H_2O-CO_2$ is treated and the additional restraints caused by the presence of CO_2 are demonstrated.

1. For the Pb–H_2O system the following reactions are pertinent:

$$PbO_2(s) + 4H^+ + 2e = 2H_2O + Pb^{2+}; \qquad \log K = 49.2, \; E_H^{\circ} = +1.455 \quad (i)$$
$$Pb^{2+} + 2e = Pb(s); \qquad\qquad \log K = -4.26, \; E_H^{\circ} = -0.126$$
$$\qquad\qquad\qquad\qquad\qquad\qquad\qquad\qquad\qquad\qquad\qquad\qquad (ii)$$

$$PbO(s) + 2H^+ = Pb^{2+} + H_2O; \qquad \log K = 12.7 \qquad\qquad (iii)$$
$$PbO(s) + 2H_2O = Pb(OH)_3^- + H^+; \quad \log K = -15.4 \qquad\qquad (iv)$$

In a $p\epsilon$ = pH diagram the lines are characterized by the equations:

$$p\epsilon = -2.13 + \tfrac{1}{2}\log[Pb^{2+}] \qquad\qquad (v)$$
$$= 24.6 - \tfrac{1}{2}\log[Pb^{2+}] - 2pH \qquad\qquad (vi)$$
$$= 10.55 - \tfrac{1}{2}\log[Pb(OH)_3^-] - \tfrac{1}{2}pH \qquad\qquad (vii)$$

[Equation (vii) obtains from combination of (i), (iii) and (iv)]

$$p\epsilon = +4.22 - pH \qquad\qquad (viii)$$

[Equation (viii) results from combinations of (ii) and (iii)]

$$p\epsilon = 11.9 + \tfrac{1}{2}\log[Pb(OH)_3^-] - \tfrac{3}{2}pH \qquad\qquad (ix)$$

[from (ii), (iii) and (iv)].

Furthermore we have the $p\epsilon$ independent relationships:

$$pH = \tfrac{1}{2}(12.7 - \log[Pb^{2+}]) \qquad\qquad (x)$$

[from (iii)]

$$pH = 15.4 + \log[Pb(OH)_3^-] \qquad\qquad (xi)$$

[(iv)].

Equations (v) through (xi) are plotted in Figure 7-7 for a concentration of soluble species of 10^{-4} M. In order to relate the $p\epsilon$ versus pH diagram to the previously discussed equilibrium diagrams, activity ratio diagrams for the pH dependence of the Pb(II) system and for the $p\epsilon$-dependence (at constant pH) of the Pb(s)–PbO(s)–PbO$_2$(s) are given in Figures 7-7b and 7-7c, respectively.

The diagram informs us that elemental lead can be readily dissolved at low pH. $PbO_2(s)$ is unstable in acid solutions. In near-neutral and slightly alkaline conditions soluble Pb^{2+} cannot exist at appreciable concentrations, the stable Pb(II) phase being solid PbO; at higher $p\epsilon$ values, oxidation to $PbO_2(s)$ should occur; in the presence of O_2, Pb^{2+} should be oxidized to $PbO_2(s)$.

2. For the Pb–H_2O–CO_2 system, in addition to (i)–(iv) we must consider

$$PbCO_3(s) = Pb^{2+} + CO_3^{2-}; \quad \log K = -12.83 \qquad\qquad (xii)$$

and

$$Pb_3(CO_3)_2(OH)_2(s) + 2H^+ = 3Pb^{2+} + 2CO_3^{2-} + 2H_2O;$$
$$\log K = -18.8 \quad (xiii)$$

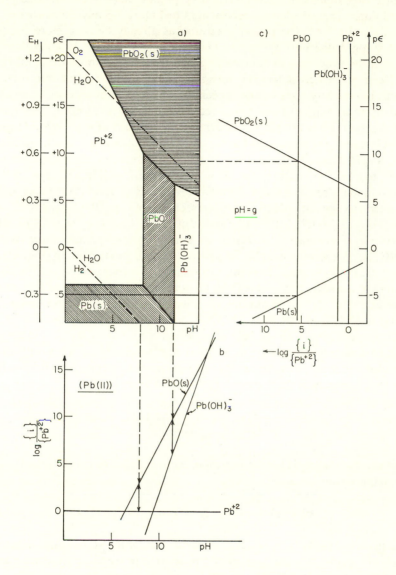

Figure 7-7 Redox equilibrium for the Pb–H_2O system ($Pb_T = 10^{-4} M$) (Example 7-7).
a: pε–pH diagram.
b: Activity ratio versus pH at constant pε (pε = 0).
c: Activity ratio diagram versus pε at constant pH (pH = 9).
The pε–pH diagram can be considered to be composed from pε-independent and pH-independent activity ratio sections.

Hence two additional solid phases may predominate under suitable pϵ–pH conditions. The calculations represented in Figure 7-8 have been carried out for $C_T = 10^{-2}\ M$. The pϵ–pH diagram has already become more complex. The straight line portions in the diagram correspond to 16 equations. Although the procedure in constructing the pϵ–pH diagram is straightforward and can be carried out in the way illustrated above, the labor involved is great. For such systems — and especially for more complex ones — auxillary activity ratio diagrams may often reduce the labor involved in constructing the diagram. For example, the activity ratio diagram of $\log \{i\}/\{Pb^{2+}\}$ versus pH (Figure 7-8b) calculated for $C_T = 10^{-2}\ M$ gives a horizontal section of the pϵ–pH diagram. For a $10^{-4}\ M$ soluble Pb-species, the transition from Pb^{2+} to $PbCO_3(s)$ occurs at a pH = 4.8 [where $\log (\{PbCO_3\}/\{Pb^{2+}\}) = 4$]. The boundary between the predominance of $PbCO_3(s)$ and $Pb_3(CO_3)_2(OH)_2(s)$ is seen to be at pH = 7.3 [intersection of $\log \{PbCO_3\}/\{Pb^{2+}\}$ with $\log \{Pb_3(CO_3)_2(OH)_2\}/\{Pb^{2+}\}$]. Finally, at pH = 12.5, $\log \{Pb_3(CO_3)_2(OH)_2\}/\{Pb(OH)_3^-\}$ becomes -4. Hence at concentrations greater than $10^{-4}\ M$, $Pb(OH)_3^-$ predominates at higher pH values; with the stipulation $[Pb(OH)_3] = 10^{-4}\ M$, $PbO(s)$ cannot be formed. Similarly, a vertical section of the pϵ–pH diagram is readily obtainable from an activity ratio diagram for a given pH. For example, for the conditions specified at pH = 5.0 the transitions from $Pb(s)$ to $PbCO_3(s)$ and from $PbCO_3(s)$ to $PbO_2(s)$ occur at pϵ = -4.5 and 15.7, respectively. With the help of such reference points the pϵ–pH diagram can be constructed readily; it is often not necessary to work out all the detailed equations because the slope of the straight line positions follows from the stoichiometry of the reaction. The slope is given by

$$\frac{\Delta p\epsilon}{\Delta pH} = -\frac{H^+ \text{ consumed}}{e \text{ consumed}} \qquad \text{(xiv)}$$

(the production of H^+ or e corresponds to a negative consumption). Thus for the reaction

$$PbO_2(s) + H^+ + 2e + H_2O = Pb(OH)_3^- \qquad \text{(xv)}$$

the slope is $\Delta p\epsilon/\Delta pH = -1/2$. In order to obtain the slopes appropriate for each pH region, the reactions should be written in terms of the species which predominate within this pH range. For example, in the pH range of pH < 6.3, carbonate occurs predominantly as $H_2CO_3^*$ reduction of $PbCO_3(s)$ to elemental $Pb(s)$ may be written as

$$PbCO_3(s) + 2H^+ + 2e = Pb(s) + H_2CO_3^* \qquad \text{(xvi)}$$

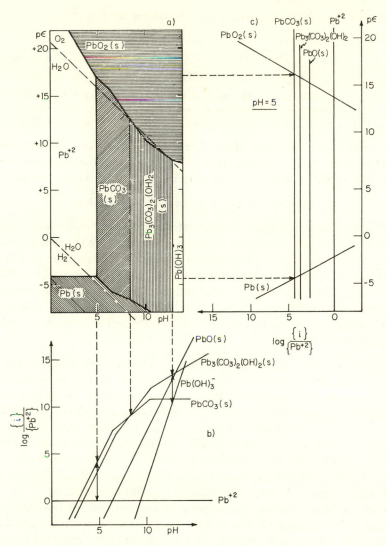

Figure 7-8 Redox equilibrium for the Pb–H$_2$O–CO$_2$ system, Pb$_T$ = 10^{-4} M, C_T = 10^{-2} M (Example 7-7).

  a: pε–pH diagram.
  b: Activity ratio versus pH at pε = 0.
  c: Activity ratio versus pε at pH = 5.

A comparison with Figure 7-7 shows that CO$_2$ has a very pronounced effect upon the solubility and the redox relations. Activity ratio diagrams facilitate the construction of the pε–pH diagram.

This reaction is obtained conveniently by adding the following reactions:

$$Pb^{2+} + 2e = Pb(s); \qquad \log K = -4.26$$
$$PbCO_3(s) = Pb^{2+} + CO_3^{2-}; \qquad \log K = -12.83$$
$$CO_3^{2-} + H^+ = HCO_3^-; \qquad \log K = +10.3$$
$$HCO_3^- + H^+ = H_2CO_3^*; \qquad \log K = +6.3$$

$$PbCO_3(s) + 2H^+ + 2e^- = Pb(s) + H_2CO_3^*; \quad \log K = -0.49$$

In this pH region the slope of the line dividing the predominance area of $PbCO_3(s)$ and $Pb(s)$ is $\Delta p\epsilon/\Delta pH = -1$, whereas in the adjacent region, $6.3 < pH < 10.3$, a slope of $-1/2$ obtains in accord with the reaction

$$PbCO_3(s) + H^+ + 2e = Pb^{2+} + HCO_3^- \qquad \text{(xvii)}$$

The diagrams in Figure 7-8 illustrate in a comprehensive way that the chemistry of aqueous lead is very strongly influenced by the reaction of the solvated Pb^{2+} with CO_3^{2-}. Similarly, as a result of the reaction of solvated metal ions with ligands other than OH^- and CO_3^{2-}, such as Cl^-, SO_4^{2-}, PO_4^{3-}, etc., additional solid phases, such as phosphates and hydroxide salts [e.g., $Pb(SO_4)_{0.25}(OH)_{1.5}(s)$, $Pb(S)_4)_{0.5}(OH)(s)$, $PbClOH(s)$, $PbCl_{0.5}(OH)_{1.5}(s)$] and soluble complexes (e.g., $PbCl_4^{2-}$) may be formed and can be considered in equilibrium diagrams.

COMPARING VARIOUS EQUILIBRIUM DIAGRAMS. The main advantage of a $p\epsilon$–pH diagram is that it provides a good survey and a clear picture of the situation, but it cannot give too much detail, especially about the concentration dependence of the predominance areas. It is possible to draw the boundary lines for various assumed activities, or to construct three-dimensional diagrams, with activity as one of its axis. There is little limit to the combinations of variables and types of phase diagrams that can be constructed, but we must not forget that the main reason for making a phase diagram is to try to understand or to solve complicated equilibrium problems.

In the authors' opinion, the activity ratio diagrams (at constant $p\epsilon$ or constant pH) consisting of only straight lines and showing quantitatively the concentration dependence provide the most useful information for the small labor involved.

7-5 Redox Conditions in Natural Waters

Only a few elements — C, N, O, S, Fe, Mn — are predominant participants in aquatic redox processes. Table 7-5 presents equilibrium constants for several couples pertinent to consideration of redox relationships in natural waters and their sediments. Data are taken principally from the second edition of *Stability Constants of Metal–Ion Complexes.* A subsidiary symbol $p\epsilon°(W)$ is convenient for considering redox situations in natural waters. $p\epsilon°(W)$ is analogous to $p\epsilon°$ except that $\{H^+\}$ and $\{OH^-\}$ in the redox equilibrium equations are assigned their activities in neutral water. Values for $p\epsilon°(W)$ for 25° thus apply to unit activities of oxidant and reductant at $pH = 7.00$. $p\epsilon°(W)$ is defined by

$$p\epsilon°(W) = p\epsilon° + \frac{n_H}{2} \log K_W \tag{30}$$

where n_H is the number of moles of protons exchanged per mole of electrons.

The listing of $p\epsilon°(W)$ values in Table 7-5 permits an immediate grading of different systems in the order of their oxidizing intensity at $pH = 7$. Any system in Table 7-5 will tend to oxidize equimolar concentrations of any other system having a lower $p\epsilon°(W)$ value. For example, we see that at $pH = 7$, NO_3^- can oxidize HS^- to SO_4^{2-}:

$$\tfrac{1}{8}NO_3^- + \tfrac{5}{4}H^+(W) + e = \tfrac{1}{8}NH_4^+ + \tfrac{3}{8}H_2O;$$
$$p\epsilon°(W) = +6.15, \log K(W) = +6.15 \tag{31}$$

$$\tfrac{1}{8}HS^- + \tfrac{1}{2}H_2O = \tfrac{1}{8}SO_4^{2-} + \tfrac{9}{8}H^+(W) + e;$$
$$p\epsilon°(W) = -3.75, \log K(W) = +3.75 \tag{32}$$

$$\tfrac{1}{8}NO_3^- + \tfrac{1}{8}HS^- + \tfrac{1}{8}H^+(W) + \tfrac{1}{8}H_2O = \tfrac{1}{8}NH_4^+ + \tfrac{1}{8}SO_4^{2-};$$
$$\log K(W) = +9.9 \tag{33}$$

or

$$NO_3^- + HS^- + H^+(W) + H_2O = NH_4^+ + SO_4^{2-}; \quad \log K(W) = 79.2 \tag{34}$$

[log $K(W)$ is the equilibrium constant for the redox reaction in neutral water, $pH = 7.00$ at 25°C; note that $p\epsilon°$, because it is a measure of oxidizing intensity, maintains the same sign independent of the direction in which the reaction is written.] Since $\Delta p\epsilon°(W)$ or log $K(W)$ is positive, the reaction is thermodynamically possible in neutral aqueous solutions at standard concentrations.

Example 7-8 *Oxidation of Organic Matter by* SO_4^{2-}. Is the oxidation of organic matter, here CH_2O, thermodynamically possible under conditions normally encountered in natural water systems?

The over-all process is obtained by combining (12) and (22) in Table 7-5.

$$\tfrac{1}{8}SO_4^{2-} + \tfrac{5}{4}H^+(W) + e = \tfrac{1}{8}H_2S(g) + \tfrac{1}{2}H_2O;$$
$$p\epsilon°(W) = -3.50, \log K(W) = -3.50 \tag{i}$$

$$\tfrac{1}{4}CH_2O + \tfrac{1}{4}H_2O = \tfrac{1}{4}CO_2(g) + H^+(W) + e;$$
$$p\epsilon°(W) = -8.20, \log K(W) = +8.20 \tag{ii}$$

$$\tfrac{1}{8}SO_4^{2-} + \tfrac{1}{4}CH_2O + \tfrac{1}{4}H^+(W) = \tfrac{1}{8}H_2S(g) + \tfrac{1}{4}CO_2(g) + \tfrac{1}{4}H_2O;$$
$$\log K(W) = +4.70 \tag{iii}$$

This may also be written

$$SO_4{}^{2-} + 2CH_2O + 2H^+(W) = H_2S(g) + 2CO_2(g) + 2H_2O;$$
$$\log K(W) = 37.6$$

Hence at standard concentrations the reaction at pH = 7 is thermodynamically possible. The same results will also hold for any equal fractions of unit activity because the numbers of molecules of sulfur-containing and carbon-containing species do not change as a result of the reaction. Furthermore, for assumed actual conditions a calculation of $\Delta G = RT \ln (Q/K)$ will show whether the oxidation can proceed thermodynamically. Thus, for a set of assumed actual conditions, such as $p_{CO_2} = 10^{-3.5}$ atm, $[CH_2O] = 10^{-6}\ M$, $[SO_4{}^{2-}] = 10^{-3}\ M$, $p_{H_2S} = 10^{-2}\ M$, a value of $10^{-31.6}$ is obtained for Q/K; hence ΔG is clearly negative, indicating that $SO_4{}^{2-}$ can oxidize CH_2O under these conditions.

Redox Intensity and the Biochemical Cycle

We have noted that only a few elements are predominant participants in aquatic redox processes. Water in solubility equilibrium with atmospheric oxygen has a well-defined $p\epsilon = 13.6$ (for $P_{O_2} = 0.21$ atm, $E_H = 800$ mV at pH 7 and 25°C). Calculations from the $p\epsilon°$ values of Table 7-5 show that at this $p\epsilon$ all of the other elements should exist virtually completely in their highest naturally occurring oxidation states: C as CO_2, HCO_3^- or $CO_3{}^{2-}$ with reduced organic forms less than $10^{-35}M$; N as NO_3^- with NO_2^- less than $10^{-7}M$; S as $SO_4{}^{2-}$ with $SO_3{}^{2-}$ or HS^- less than $10^{-20}M$; Fe as FeOOH or Fe_2O_3 with Fe^{2+} less than $10^{-18}M$; and Mn as MnO_2 with Mn^{2+} less than $10^{-10}M$. Even the N_2 of the atmosphere should be largely oxidized to NO_3^-.

Since, in fact, N_2 and organic matter are known to persist in waters containing dissolved oxygen, total redox equilibrium is not found in natural water systems, even in the surface films. At best there are partial equilibria, treatable as approximations to equilibrium either because of slowness of interaction with other redox couples or because of isolation from the total environment as a result of slowness of diffusional or mixing processes.

The maintenance of life resulting directly or indirectly from a steady impact of solar energy (*photosynthesis*) is the main cause for nonequilibrium conditions. All the predominant redox components with respect to the composition of natural waters are biologically active. Organisms, themselves a product of inorganic matter, are primarily built up from "redox elements," and their relatively constant stoichiometric composition ($C_{106}H_{263}O_{110}N_{16}P$) and the cyclic exchange between chemical elements of the water and the biomass have pronounced effects on the relative concentrations of the elements in the environment. The biologically active elements circulate in a different pattern than does water itself or inactive (conservative) solutes.

Photosynthesis, by trapping light energy and converting it to chemical energy, produces reduced states of higher free energy (high energy chemical bonds) and thus nonequilibrium concentrations of C, N and S compounds. In a simplified context, photosynthesis may be conceived as a process producing localized centers of highly negative $p\epsilon$. As shown in (20) of Table 7-5, the conversion of CO_2 to glucose at unit activities requires $p\epsilon^\circ(W) = -7.2$. Although this value may be modified somewhat for the actual intracellular activities, it does represent approximately the negative $p\epsilon$ level that must be reached during photosynthesis.

The NADP or TPN system [8], ubiquitous in living organisms and believed to play a major role in electron transport during photosynthesis, exhibits $p\epsilon^\circ(W) = -5.5$. Moreover, various ferredoxins [9], now widely considered to be the primary electron receptors from excited chlorophylls, show $p\epsilon^\circ(W)$ values in the range -7.0 to -7.5 (Table 7-5). The coincidence of this range with the $p\epsilon^\circ(W)$ value for conversion of CO_2 to glucose is suggestive.

In contrast the respiratory, fermentative and other nonphotosynthetic processes of organisms tend to restore equilibrium by catalyzing or mediating chemical reactions releasing free energy and thus increasing the mean $p\epsilon$ level.

The ecological systems of natural waters are thus more adequately represented by dynamic than by equilibrium models. The former are needed to describe the free energy flux absorbed from light and released in subsequent redox processes. Equilibrium models can only depict the thermodynamically stable state and describe the direction and extent of processes tending toward it.

When comparisons are made between calculations for an equilibrium redox state and concentrations in the dynamic aquatic environment, the implicit assumptions are that the biological mediations are operating essentially in a reversible manner at each stage of the ongoing processes or that there is a metastable steady-state that approximates the partial equilibrium state for the system under consideration.

MICROBIAL MEDIATION. As already pointed out, the nonphotosynthetic organisms tend to restore equilibrium by catalytically decomposing the unstable products of photosynthesis through energy yielding redox reactions, thereby obtaining a source of energy for their metabolic needs. The organisms use this energy both to synthesize new cells and to maintain the old cells already formed [10]. The energy exploitation is of course not 100% efficient; only a

[8] M. Calvin and J. A. Bassham, *The Photosynthesis of Carbon Compounds*, Benjamin, New York, 1962.
[9] D. I. Arnon, *Science*, **149**, 1460 (1965).
[10] P. L. McCarty, *Thermodynamics of Biological Synthesis and Growth; 2nd International Water Pollution Conference*, Tokyo, Pergamon, London, 1965.

proportion of the free energy released can become available for cell use. It is important to keep in mind that organisms cannot carry out *gross* reactions that are thermodynamically not possible. From a point of view of over-all reactions these organisms act only as redox catalysts. Therefore organisms do not "oxidize" substrates or "reduce" O_2 or SO_4^{2-}, they only mediate the reaction, or more specifically, the electron transfer, for example, of the specific oxidation of the substrate and of the reduction of O_2 or SO_4^{2-}. Since, for example, SO_4^{2-} can be reduced only at a given pϵ or redox potential, an equilibrium model characterizes the pϵ range in which reduction of sulfate is not possible and the pϵ range in which reduction of sulfate is possible. Thus, pϵ is a parameter that characterizes restrictively the ecological milieu. The pϵ range in which certain oxidation or reduction reactions are possible can be estimated by calculating equilibrium composition as a function of pϵ. This has been done for nitrogen, manganese, iron, sulfur and carbon at a pH of 7. The results are shown in Figure 7-9.

NITROGEN SYSTEM. Figure 7-9*a* shows relationships among general oxidation states of nitrogen as a function of pϵ for a total atomic concentration of nitrogen-containing species equal to 10^{-3} *M*. Maximum N_2(aq) concentration is therefore 5×10^{-4} *M*, corresponding to about p_{N_2} of 0.77 atm. For most of the aqueous range of pϵ, N_2 gas is the most stable species, but at quite negative pϵ values ammonia becomes predominant and nitrate dominates for pϵ greater than $+12$ and pH $= 7$. The fact that nitrogen gas has not been converted largely into nitrate under prevailing aerobic conditions at the land and water surfaces indicates a lack of efficient biological mediation of the reverse reaction also, for the mediating catalysis must operate equally well for reaction in both directions. It appears then that denitrification must occur by an indirect mechanism such as reduction of NO_3^- to NO_2^- followed by reaction of NO_2^- with NH_4^+ to produce N_2 and H_2O.

Because reduction of N_2 to NH_4^+ (N_2-*fixation*) at pH 7 can occur to a substantial extent when pϵ is less than about -4.5, the level of pϵ required is not as negative as for the reduction of CO_2 to CH_2O. It is not surprising then that blue-green algae are able to mediate this reduction at the negative pϵ levels produced by photosynthetic light energy. What is perhaps surprising is that nitrogen fixation does not occur more widely among photosynthetic organisms and proceed to a greater extent as compared with CO_2 reduction. Kinetic problems in breaking the strong bonding of the N_2 molecule probably are major factors here.

Because of this kinetic hindrance between "bound" nitrogen and N_2, it might also be useful to consider a system in which NO_3^-, NO_2^- and NH_4^+ are treated as components metastable with respect to gaseous N_2. A diagram for such a system, Figure 7-9*b*, shows the shifts in relative predominance of

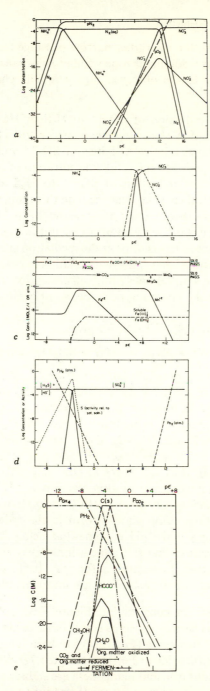

Figure 7-9 Equilibrium concentrations of pertinent redox components as a function of $p\epsilon$ at a pH of 7.0. These equilibrium diagrams have been calculated using standard equilibrium constants (Table 7-6), assuming a few concentration conditions representative of natural water systems. The assumptions made are discussed in the text.

the three species all within the rather narrow $p\epsilon$ range from 5.8 to 7.2. That each of the species has a dominant zone within this $p\epsilon$ range would seem to be a contributing factor to the observed highly mobile characteristics of the nitrogen cycle.

SULFUR SYSTEM. The reduction of SO_4^{2-} to H_2S or HS^- provides a good example of the application of equilibrium considerations to aquatic relationships. Figure 7-9d shows relative activities of SO_4^{2-} and H_2S at pH 7 and $25°C$ as a function of $p\epsilon$ when the total concentration of sulfur is 1 mM. It is apparent that significant reduction of SO_4^{2-} to H_2S at this pH requires $p\epsilon < -3$. The biological enzymes that mediate this reduction with oxidation of organic matter must then operate at or below this $p\epsilon$. Because the system is dynamic rather than static, only an upper bound can be set in this way, for the excess driving force in terms of $p\epsilon$ at the mediation site is not indicated by equilibrium computations. Since, however, many biologically mediated reactions seem to operate with relatively high efficiency for utilizing free energy, it appears likely that the operating $p\epsilon$ value is not greatly different from the equilibrium value.

Combining (11) and (14) of Table 7-5 gives

$$SO_4^{2-} + 2H^+(W) + 3H_2S = 4S(s) + 4H_2O; \quad \log K(W) = 4.86 \quad (35)$$

This equation indicates a possibility of formation of solid elemental sulfur in the reduction of sulfate at pH 7 and standard concentrations. A concentration of 1 M sulfate is unusual, however. If the concentration of SO_4^{2-} is reduced to about 0.018 M (about 1600 mg liter^{-1}) and the H_2S activity is taken correspondingly as 0.09 atm (H_2S is about half-ionized at pH 7, and the solubility of H_2S at 1 atm is about 0.1 M; thus this condition means about 0.018 M total sulfide), then $\Delta G = RT \ln Q/K$ is positive and solid sulfur cannot form thermodynamically by this reaction at pH 7.

The solubility of $CaSO_4(s)$ is about 0.016 M at $25°C$. According to the foregoing rough calculation, sulfur should form in the reduction of SO_4^{2-} in saturated $CaSO_4(s)$ only if the pH is somewhat below 7 (see Example 7-4). There are some indications that this conclusion agrees with the condition of natural sulfur formation. Elemental sulfur may be formed, however, as a kinetic intermediate or as a metastable phase under many natural conditions.

IRON AND MANGANESE. In constructing Figure 7-9c, solid $FeOOH$ ($\bar{G}_f^\circ = -98$ kcal mole^{-1}) has been assumed as stable ferric oxide. Although thermodynamically possible, magnetite ($Fe_3O_4(s)$) has been ignored as intermediate in the reduction of ferric oxide to Fe(II). As Figure 7-9 shows, in the presence of O_2, $p\epsilon > 11$, aqueous iron and manganese are stable only as solid oxidized oxides. Soluble forms are present at concentrations less than 10^{-9} M. The concentration of soluble iron and manganese, as Fe^{2+} and Mn^{2+}, increases with decreasing pϵ, the highest concentrations being controlled by the solubility of $FeCO_3(s)$ and $MnCO_3(s)$, respectively. ($[HCO_3^-] = 10^{-3} M$ has been assumed for the construction of the diagram.)

CARBON SYSTEM. With the exception of CH_4, no organic solutes encountered in natural waters are thermodynamically stable. For example, the disproportionation of acetic acid

$$CH_3COOH = 2H_2O + 2C(s); \qquad \log K = 18$$
$$CH_3COOH = CH_4(g) + CO_2(g); \quad \log K = 9$$

is thermodynamically favored, but prevented by slow kinetics. Similarly, formaldehyde is unstable with respect to its decomposition into carbon (graphite) and water

$$CH_2O(aq) = C(s) + H_2O; \quad \log K = 18.7$$

but there is no evidence that this reaction occurs.

Even though reversible equilibria cannot be attained at low temperatures, it is of considerable interest to compare the equilibrium constants of the various steps in the oxidation of organic matter. The compounds CH_4, CH_3OH, CH_2O and $HCOO^-$ given in the diagram (Figure 7-9e) represent organic material with formal oxidation states of $-IV$, $-II$, $\bar{0}$ and $+II$, respectively. The diagram has been constructed for the condition $p_{CH_4} + p_{CO_2} = 1$ atm.† (The construction of such a diagram is facilitated by first drawing an activity ratio diagram using $\{CH_2O(aq)\}$ as a reference.) The major feature of the equilibrium carbon system is simply a conversion of predominant CO_2 to predominant CH_4 with a half-way point at p$\epsilon = -4.13$. At this pϵ value, where the other oxidation states exhibit maximum relative occurrence, formation of graphite is thermodynamically possible.

Methane fermentation may be considered a reduction of CO_2 to CH_4; this reduction may be accompanied by oxidation of any one of the intermediate oxidation states.‡ Since all of the latter have p$\epsilon°$(W) values less than -6.2 (this for CH_3OH), each can provide the negative pϵ level required thermodynamically for reduction of CO_2 to CH_4 in its oxidation. Alcoholic fermentation, exemplified here by the reduction of CH_2O to CH_3OH, can occur with

---

† This condition is not fulfilled in the pϵ range where $\{C(s)\} = 1$.

‡ This statement does not imply a mechanism. Methane may be formed directly, for example, from acetic acid:

$$CH_3COOH \rightarrow CH_4 + CO_2$$

This reaction can be classified as the sum of the reactions

$$CH_3COOH + 2H_2O = 2CO_2 + 8H^+ + 8e$$
$$CO_2 + 8e + 8H^+ = CH_4 + 2H_2O$$

$p\epsilon < -3.0$. Because the concomitant oxidation of CH_2O to CO_2 has $p\epsilon°(W) = -8.2$, there is no thermodynamic problem.

MICROBIALLY MEDIATED OXIDATION AND REDUCTION REACTIONS. Although, as has been stressed, conclusions regarding chemical dynamics may not generally be drawn from thermodynamic considerations, it appears that all of the reactions discussed in the previous section, except possibly those involving $N_2(g)$ and $C(s)$, are biologically mediated in the presence of suitable and abundant biota. Table 7-6 surveys the oxidation and reduction reactions which may be combined to result in exergonic processes. The combinations listed represent the well-known reactions mediated by hetrotrophic and chemoautotrophic organisms. It appears that in most natural habitats organisms capable of mediating the pertinent redox reactions are nearly always found.

THE SEQUENCE OF REDOX REACTIONS. In a closed aqueous system containing organic material — say CH_2O — oxidation of organic matter are observed to occur first by reduction of O_2 ($p\epsilon(W) = 13.8$). This will be followed by reduction of NO_3^- and NO_2^-. As seen in Figure 7-9 the succession of these reactions follows the decreasing level. Reduction of MnO_2 if present should occur at about the same $p\epsilon$ levels as that of nitrate reduction, followed by reduction of $FeOOH(s)$ to Fe^{2+}. When sufficiently negative $p\epsilon$ levels have been reached, fermentation reactions and reduction of SO_4^{2-} and CO_2 may occur almost simultaneously.

The described sequence (Model 1, Table 7-6) would be expected if reactions tended to occur in the order of their thermodynamic possibility. In terms of the concept of electron free energy levels, the reductant (CH_2O) will supply electrons to the lowest unoccupied electron level (O_2); with more electrons available, successive levels — NO_3^-, NO_2^-, $MnO_2(s)$ etc. — will be filled up (see Figure 7-4). The described succession of reactions is mainly reflected in the vertical distribution of components in a nutrient-enriched (eutrophied) lake and in general also in the temporal succession in a closed system containing excess organic matter, such as a batch digester (anaerobic fermentation unit).

Since the reactions considered (with the possible exception of the reduction of $MnO_2(s)$ and $FeOOH)(s)$ are biologically mediated, the chemical reaction sequence is paralleled by an *ecological succession* of micro-organisms (aerobic heterotrophs, denitrifers, fermentors, sulfate reducers and methane bacteria). It is perhaps also of great interest from an evolutionary point of view that there appears to be a tendency for more energy yielding mediated reactions to take precedence over processes that are less energy yielding.

The quantity of biological growth resulting from inorganic and organic metabolism is related to the free energy released by the mediated reaction

$[-\Delta G°(\text{W})]$ and the efficiency of the micro-organisms in capturing this energy [10]. Organic carbon compounds (with exception of CH_4) are unstable over the entire pϵ range, but it is frequently assumed that anaerobic conditions are more favorable to the preservation of organic matter than aerobic conditions [11].

Model 2 of Table 7-6 illustrates the type of reaction sequence one would observe in a system of incipient low pϵ to which O_2 is added. This is the situation commonly encountered after a stream has become polluted with a variety of reducing substances. In such a case it is typically observed that aerobic respiration takes precedence over nitrification, that is, bacterially mediated nitrification is, at least partially, inhibited or repressed in the presence of organic material.

It may also be noted from Tables 7-5 and 7-6 that there is a thermodynamic possibility for N_2-reduction accompanied by CH_2O-oxidation. This is the gross mechanism mediated by nonphotosynthetic nitrogen-fixing bacteria.

The concern over so-called nonbiodegradable pollutants and the recovery from sediments of organic substances hundreds of millions of years old [12] may serve as a reminder that a state of equilibrium is not always attained, not even within geological time spans, and that microorganisms are not "infallible" in catalyzing processes toward the stable state. Equilibrium models can describe the conditions of stability of redox components in natural water systems, but more extended quantitative inferences must be made with great caution.

REDOX CAPACITY OF EARTH'S OCEAN AND ATMOSPHERE. What controls the pϵ of our present-day environment? Those parts of the environment in equilibrium with the atmosphere should have a pϵ level in accordance with an oxygen partial pressure of 0.2 atm. In the pH range of natural waters, pϵ should be between 12 and 14. It is pertinent to remember that the mass of the atmosphere is ca. 10^6 times smaller than that of the whole earth. The atmosphere is still about 10^4 times smaller in mass than the earth's crust. Hence oxidants and reductants abound in the solid phases rather than in the atmosphere or hydrosphere. Sillén [13], using Goldschmidt's [14] data and Horn's [15] modifications of a geochemical mass balance which is based on the schematic reaction

$$\text{igneous rock} + \text{volatile substances} \rightleftharpoons \text{air} + \text{seawater} + \text{sediments} \quad (36)$$

[11] P. H. Abelson, "Geochemistry of Organic Substances" in *Researches in Geochemistry*," P. H. Abelson, Ed., Wiley, New York, 1959.
[12] M. Blumer, *Science*, **149**, 722 (1965).
[13] L. G. Sillén, *Arkiv Kemi*, **24**, 431 (1965); **25**, 159 (1966).
[14] V. M. Goldschmidt, *Geochemistry*, Oxford Univ. Press, 1954.
[15] M. K. Horn, Report 64-282, Project EG-16r, Pure Oil Co., Houston, Texas, 1964.

Table 7-6 Principal Microbially Mediated Oxidation and Reduction Reactions

Oxidation	$p\epsilon°(W)$ $= -\log K(W)$	Reduction	$p\epsilon°(W)$ $= \log K$
(1) $\frac{1}{4}CH_2O + \frac{1}{4}H_2O = \frac{1}{4}CO_2(g) + H^+(W) + e$	-8.20	(A) $\frac{1}{4}O_2(g) + H^+(W) + e = \frac{1}{2}H_2O$	$+13.75$
(1a) $\frac{1}{2}HCOO^- = \frac{1}{2}CO_2(g) + \frac{1}{2}H^+(W) + e$	-8.73	(B) $\frac{1}{5}NO_3^- + \frac{6}{5}H^+(W) + e$ $= \frac{1}{10}N_2(g) + \frac{3}{5}H_2O$	$+12.65$
(1b) $\frac{1}{2}CH_2O + \frac{1}{2}H_2O$ $= \frac{1}{2}HCOO^- + \frac{3}{2}H^+(W) + e$	-7.68	(C) $\frac{1}{8}NO_3^- + \frac{5}{4}H^+(W) + e$ $= \frac{1}{8}NH_4^+ + \frac{3}{8}H_2O$	$+6.15$
(1c) $\frac{1}{2}CH_3OH = \frac{1}{2}CH_2O + H^+(W) + e$	-3.01	(D) $\frac{1}{2}CH_2O + H^+(W) + e = \frac{1}{2}CH_3OH$	-3.01
(1d) $\frac{1}{2}CH_4(g) + \frac{1}{2}H_2O$ $= \frac{1}{2}CH_3OH + H^+(W) + e$	$+2.88$	(E) $\frac{1}{8}SO_4^{2-} + \frac{9}{8}H^+(W) + e = \frac{1}{8}HS^- + \frac{1}{2}H_2O$	-3.75
(2) $\frac{1}{8}HS^- + \frac{1}{2}H_2O$ $= \frac{1}{8}SO_4^{2-} + \frac{9}{8}H^+(W) + e$	-3.75	(F) $\frac{1}{8}CO_2(g) + H^+(W) + e = \frac{1}{8}CH_4(g) + \frac{1}{4}H_2O$	-4.13
(3) $\frac{1}{8}NH_4^+ + \frac{3}{8}H_2O$ $= \frac{1}{8}NO_3^- + \frac{5}{4}H^+(W) + e$	$+6.16$	(G) $\frac{1}{6}N_2 + \frac{4}{3}H^+(W) + e = \frac{1}{3}NH_4^+$	-4.68
(4)† $FeCO_3(s) + 2H_2O$ $= FeOOH(s) + HCO_3^- \ (10^{-3})$ $+ 2H^+(W) + e$	-1.67		
(5)† $\frac{1}{2}MnCO_3(s) + \frac{3}{2}H_2O$ $= \frac{1}{2}MnO_2(s) + \frac{1}{2}HCO_3^- \ (10^{-3})$ $+ \frac{3}{2}H^+(W) + e$	-8.5		

Sequence of Microbial Mediation

Model 1: Excess organic material (water contains incipiently O_2, NO_3^-, SO_4^{2-} and HCO_3^-). Examples: Hypolimnetic layers of eutrophic lake, sediments; sewage treatment plant digester.

	Combination	log K(W)	$\Delta G°$(W),‡ kcal
Aerobic respiration	(1) + (A)	21.95	−29.9
Denitrification	(1) + (B)	20.85	−28.4
Nitrate reduction	(1) + (C)	14.36	−19.6
Fermentation§	(1b) + (D)	4.67	−6.4
Sulfate reduction	(1) + (E)	4.45	−5.9
Methane fermentation	(1) + (F)	4.07	−5.6
N-Fixation	(1) + (G)	3.52	−4.8

Model 2: Excess O_2 (water contains incipiently organic material, SH^-, NH_4^+ and possibly Fe(II) and Mn(II)). Examples: Aerobic waste treatment, self-purification in streams; epilimnetic waters.

	Combination	log K(W)	$\Delta G°$(W),‡ kcal
Aerobic respiration	(A) + (1)	21.95	−29.9
Sulfide oxidation	(A) + (2)	17.50	−23.8
Nitrification	(A) + (3)	7.59	−10.3
Ferrous oxidation†	(A) + (4)	15.42	−21.0
Mn(II) oxidation†	(A) + (5)	5.75	−7.2

† The data for pe°(W) log K(W) or $\Delta G°$(W) of these reactions correspond to $\{HCO_3^-\} = 10^{-3}$ rather than unity. The autotrophic nature of iron and manganese bacteria is in dispute.

‡ $\Delta G°(W) = -RT \ln K(W)$.

§ Fermentation is interpreted as an organic redox reaction where one organic substance is reduced by oxidizing another organic substance (e.g., alcoholic fermentation; the products in such a reaction are metastable thermodynamically with respect to CO_2 and CH_4).

has estimated the quantities of the major redox components that have participated in the formation of 1 liter of seawater and has calculated a redox titration curve for the model system. This titration curve is reproduced in Figure 7-10. It is apparent from this figure that redox couples of C, N, Fe and Mn have larger oxidation or reduction capacities than that of oxygen. Titrating the model system with electrons, that is, a strong reductant and thus reducing $p\epsilon$, corresponds to going back in time — some hundred or thousand millions of years — to a more reduced state representative of a time when the atmosphere did not contain oxygen, when carbon and hydrocarbons were buried, and when H_2 was lost to outer space. in Sillén's titration curve, the

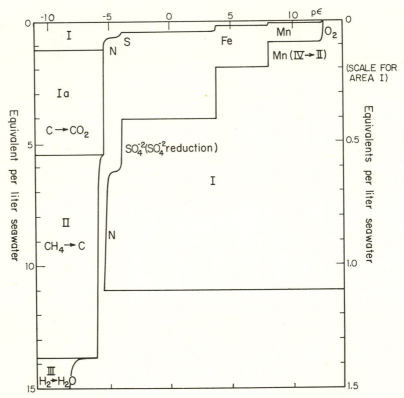

Figure 7-10   "Titration curve" showing the variation of $p\epsilon$ with the number n_e of oxidation equivalents per liter seawater added to the model system. The present state ($p\epsilon = 12.51$) is at the top of the diagram and has been chosen as the zero level, $n_e = 0$. Movement upward (n_e decreasing) means oxidation, downward (n_e increasing) is reduction. The ranges I, Ia, II and III are indicated to the left. The thin curve is $p\epsilon(n_e)$ for range I, given on a tenfold n_e scale. The main reactions are indicated. From L. G. Sillén, *Ark. Kemi*, **25**, 159 (1965). Reproduced with permission from Almqvist and Wiksells, Uppsala, Sweden.

sequence indicated in Figure 7-10, we start out with the present state ($p\epsilon$ = 12.51, corresponding to present atmosphere composition and pH = 8.1) in which 0.09 mole of electrons (equivalents) liter^{-1} of seawater is necessary to "titrate" the 0.022 mole O_2. Next we would need 0.1 equivalent of reductant per liter of seawater to reduce $MnO_2(s)$ to Mn(II). Around $p\epsilon$ = 3.7 the conversion of FeOOH(s) to Fe_3O_4 would provide the main buffering. At lower $p\epsilon$ values, SO_4^{2-} would be reduced predominantly to $FeS_2(s)$ and N_2 to NH_4^+. At $p\epsilon$ = -5.5 solid carbon would exist at equilibrium, that is, carbonates would be reduced to graphite. Ultimately $CH_4(g)$ and $H_2(s)$ would be formed, being — in the model system — somewhat representative of the early atmosphere. Such calculations are of great value in formulating consistent models for the chemical evolution of the earth and its atmosphere.

The only mechanisms for the production of free oxygen throughout geological time appear to be photosynthesis and the photo dissociation of water vapor in the upper atmosphere followed by the escape of hydrogen [16]. According to Holland [16], the former is by far the most effective. We are not really sure when photosynthesis began, but at the present time it appears that O_2 is added to the system by photosynthesis at a steady rate and is withdrawn at a steady rate from the system predominantly by decay — the photosynthesis in excess to respiration corresponds to the burial of carbon — and perhaps to a minor extent by oxidation of ferrous iron and of volcanic gases H_2, SO_2 and H_2S. According to such a hypothesis, life (i.e., photosynthesis) determines the oxidation state of the atmosphere and those parts of the hydrosphere and lithosphere in equilibrium with the atmosphere.

7-6   The Cycle of Organic Carbon

After having devoted so much attention to the discussion of mostly inorganic species, some further consideration should be given to the transformation of organic substances. The cycle of carbon in natural waters is especially important because it is interrelated with the cycles of all other elements (Figure 7-11). The organic material produced in the world oceans, furthermore, constitutes one of man's most significant marine environment resources. Discussion of the organic chemistry of natural waters is appropriate since in the context of this chapter organic compounds, because of their thermodynamic instability under aquatic conditions, may be considered as

[16] H. D. Holland, "On the Chemical Evolution of the Terrestrial and Cytherean Atmospheres" in *Origin and Evolution of Atmospheres and Oceans*, P. J. Branzatio and A. G. W. Cameron, Eds., Wiley, New York, 1964.

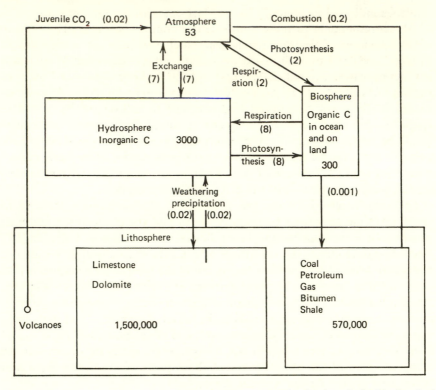

Figure 7-11   The circulation of carbon in nature is determined primarily by biochemical reactions. Amounts of carbon on the earth's surface in 10^{15} moles. Parenthesis indicate rates of turnover of carbon in 10^{15} moles year^{-1}. In 1933 Goldschmidt made quantitative estimates on the amount of carbon in the different parts of the cycle. His data have been revised by various workers. Although the estimates given for the annual turnover are quite uncertain, they clearly illustrate that most of the carbon in the earth's surface has been cycled through organisms.

intermediates in the continuous and complicated circuit of redox reactions. Despite its importance, the chemistry of aqueous organic matter is one of the least well-known sections in the chemistry of natural waters [17].

   Experimental demonstration that certain organic substances can markedly modify growth rates of microorganisms has emphasized the importance of organic constituents. Progressive concern with pollution has also stimulated research on this subject.

[17] E. K. Duursma, in *Chemical Oceanography*, J. P. Riley and G. Skirrow, Eds., Academic Press, New York, 1965.

Analytical Considerations

Organic matter in natural waters includes a great variety of organic compounds, usually present in minute concentrations that elude direct isolation and identification. Collective parameters, such as chemical oxygen demand (COD), biological oxygen demand (BOD) and total carbon by combustion, C_{org}, have therefore often been used to estimate the total quantity of organic matter present. COD is obtained by measuring the equivalent quantity of an oxidizing agent, usually permanganate or dichromate in acid solution, necessary for the oxidation of the organic constituents; the amount of oxidant consumed is customarily expressed in equivalents of oxygen. The BOD test measures the oxygen uptake in the microbiologically mediated oxidation of organic matter directly. In both tests not all the organic matter reacts with the oxidant. In determinations of total organic carbon, we measure the CO_2 produced in the oxidation or combustion of the dry residue of a water sample (from which carbonate has been removed). C_{org} and COD are capacity terms; the latter measures the reduction capacity by organic matter. If COD is expressed in moles of O_2 per liter and C_{org} in moles C per liter, the "average" oxidation state of the organic carbon present can be obtained from (37)

$$\frac{4(C_{org} - COD)}{C_{org}} = \text{oxidation state} \qquad (37)$$

The COD of a compound is very nearly proportional to its heat of formation, and a crude value of its energy of formation can often be obtained. Thus a rough estimate of the potential energy available for the metabolic needs of organisms can be made from COD data. Although measurements of C_{org}, BOD or COD are obtainable conveniently, these collective parameters lack physiological meaning. The rates of microbial growth and the over-all use of organic matter in multisubstrate media depend in a complex way on the activities of a great variety of different enzymes and on various mechanisms by which these activities are ecologically interrelated and physiologically coordinated [18].

Dissolved organic compounds occur in waters in such low concentrations that they usually have to be concentrated and separated from the inorganic salts before thay can be determined chemically. Adsorption on columns of carbon and other adsorbents, solvent extraction and coagulation with hydrolyzing metal ions are used for these purposes.

Bio-assay methods are frequently of great sensitivity and permit direct determination of very low concentrations of organic nutrients in water. Vitamin B_{12}, for example, has been determined at concentrations down to

[18] E. Stumm-Zollinger, *Appl. Microbiol.*, **14**, 654 (1966); *J. Water Pollution Control Federation*, **40**, R213 (1968).

10^{-16} liter^{-1}, with *Euglena* as an assay organism. Of special interest is the utilization of growth kinetics of microorganisms for the measurement of individual organic solutes. A combination of the exponential growth equation, $dB/dt = \mu B$, and of the equation describing the relationship between bacterial growth and substrate utilization, $-dB/dS = y$, gives an expression that describes the rate of substrate utilization as related to growth

$$-\frac{dS}{dt} = \left(\frac{\mu}{y}\right)B \tag{38}$$

where B is the concentration of organisms (dry weight cells per unit of volume) or a measure of enzymatic activity at time, t; y is the yield, that is, the fraction of the substrate converted into bacterial mass; μ is the specific growth rate constant (time^{-1}); and S is the concentration of substrate present in the solution. In accordance with a simple enzyme model, μ varies with the substrate concentration in the following way

$$\mu = \frac{\mu_0 S}{K + S} \tag{39}$$

μ_0 is the maximum specific growth rate constant and it is a characteristic for a particular substrate and microorganism ($S = K$ when $\mu = 0.5\mu_0$). For the estimation of the concentration of a substrate S, various known quantities of a substrate, usually C-14 labeled, of concentration A are added to aliquot samples containing unknown quantities of the same substrate. In the procedure adapted by Vaccaro *et al.* [19], a pure bacterial culture whose uptake constant is well established is added to aliquots of membrane filtered seawater. Combining (38) and (39), the following relation between the measurable rate of substrate utilization, $V = -dS/dt$, and the substrate originally present, S, is obtained:

$$V = \frac{\mu_0 B(S + A)}{y(K + S + A)} \tag{40}$$

Because K, μ_0, y and B are assumed to remain constant during the test period (B is constant when $B > S$ or when the uptake time is very brief), a set of measurements of V for various A values permits the evaluation of S. The method is nearly specific and quite sensitive; with bacterial isolates from natural waters K values, that is, substrate concentrations which allow half maximum uptake velocities, of the order of 10^{-8} to 10^{-5} M have been determined for various substrates. The method has been applied in some analyses

[19] R. F. Vaccaro, S. E. Hicks, H. W. Jannasch and F. C. Carey, *Limnol. Oceanog.*, **13**, 356 (1968).

in fresh water [20] and in seawater [19]. Vaccaro *et al.* [19] used the method to measure the glucose concentration along a transoceanic section. The concentrations varied from a high of 10^{-6} M to values below 10^{-8} M. The results were in reasonable agreement with those obtained using an enzymatic method.

Fate of Organic Substances in Natural Waters

Most of the carbon in the earth's crust has cycled through organisms and plants; as a result it has been incorporated into unstable but long-lived biochemical structures [21]. Table 4-1 contains a survey of carbon in its various forms in the atmosphere and biosphere. Carbon forms the link in the interaction between the inorganic environment and the living organisms. The composition of the inorganic environment thus depends on the cyclic activity of life [21]. The photosynthetic activity of the earth's plants would nearly deplete the atmosphere of the carbon dioxide within a short time period (ca. 1 year [21]) if CO_2 were not returned to the atmosphere by respiration of the heterotrophic organisms. Only a small fraction (4×10^{-4}) [21] of the solar energy supply is used for photosynthesis. The photosynthetic algae use a fraction of this energy (ca. 1/6) for their own respiration. The remaining energy is bound biochemically and becomes available to heterotrophic consumers. The principle of the food chain may be illustrated by an example given by Cole [21]. Of 1000 cal fixed by the algae of Cayuga Lake, 150 cal can be transferred to biochemical bonds in their predators, small aquatic animals. Only 30 cal can be incorporated into the next step of the food chain, for example, into protoplasm of smelt. Six of the 30 cal are further transferred to biochemical bonds in fat and muscle tissue of the man who eats the smelt. If trout is a link between smelt and man, the yield of energy to man shrinks to 1.2 cal. For each link in the food chain energy is lost to the environment, organic carbon converted into CO_2, and as by-products of metabolic reactions some organic intermediates are released to the water. Information on the catabolic pathway of organic life substances permits prediction of the type of organic substances to be found in natural waters. Table 7-7 gives a greatly condensed and simplified survey of the decomposition products of life substances and includes an abbreviated list of specific organic compounds reportedly found and identified in natural waters. For more details the reviews by Duursma [17], Strickland [22] and Degens

[20] J. E. Hobbie and R. T. Wright, *Limnol. Oceanog.*, **10**, 471 (1965).
[21] L. C. Cole, *Sci. Amer.*, 198, 83 (April 1958).
[22] J. D. H. Strickland, in *Chemical Oceanography*, J. P. Riley and G. Skirrow, Eds., Academic Press, New York, 1965.

Table 7-7 Naturally Occurring Organic Substances

Life Substances	Decomposition Intermediates	Intermediates and Products Typically Found in Nonpolluted Natural Waters
Proteins	Polypeptides $\rightarrow$ RCH(NH$_2$)COOH $\rightarrow$ $\begin{cases} \text{RCOOH} \\ \text{RCH}_2\text{OHCOOH} \\ \text{RCH}_2\text{OH} \\ \text{RCH}_3 \\ \text{RCH}_2\text{NH}_2 \end{cases}$ amino acids	NH$_4^+$, CO$_2$, HS$^-$, CH$_4$, HPO$_4^{2-}$, peptides, amino acids, urea, phenols, indole, fatty acids, mercaptans
Polynucleotides	Nucleotides $\rightarrow$ purine and pyrimidine bases	
Lipids Fats Waxes Oils	RCH$_2$CH$_2$COOH + CH$_2$OHCHOHCH$_2$OH $\rightarrow$ $\begin{cases} \text{RCH}_2\text{OH} \\ \text{RCOOH} \\ \text{shorter chain acids} \\ \text{RCH}_3 \\ \text{RH} \end{cases}$ fatty acids glycerol	CO$_2$, CH$_4$, aliphatic acids, acetic, lactic, citric, glycolic, malic, palmitic, stearic, oleic acids, carbohydrates, hydrocarbons
Carbohydrates Cellulose Starch Hemicellulose Lignin	$\left.\begin{array}{l}\text{} \\ \text{} \\ \text{} \end{array}\right\}$ C$_x$(H$_2$O)$_y$ $\rightarrow$ $\left.\begin{array}{l}\text{monosaccharides} \\ \text{oligosaccharides} \\ \text{chitin} \end{array}\right\}$ $\rightarrow$ $\left.\begin{array}{l}\text{hexoses} \\ \text{pentoses} \\ \text{glucosamine} \end{array}\right\}$ $\rightarrow$ (C$_2$H$_2$O)$_x$ $\rightarrow$ unsaturated aromatic alcohols $\rightarrow$ polyhydroxy carboxylic acids	HPO$_4^{2-}$, CO$_2$, CH$_4$, glucose, fructose, galactose, arabinose, ribose, xylose
Porphyrines and Plant Pigments Chlorophyll Hemin Carotenes and Xantophylls	$\left.\begin{array}{l}\text{} \\ \text{} \end{array}\right\}$ Chlorin $\rightarrow$ pheophytin $\rightarrow$ hydrocarbons $\rightarrow$	Pristane, carotenoids
Complex Substances Formed from Breakdown Intermediates, e.g., Phenols + quinones + amino compounds $\rightarrow$ Amino compounds + breakdown products of carbohydrates $\rightarrow$		Melanins, melanoidin, gelbshoffe Humic acids, fulvic acids, "tannic" substances

[23] should be consulted. Transformation of biogenic substances within the soil environment and a particular discussion of "humic"† acids and related complex substances are thoroughly considered in the monograph of Kononova [24].

† The term "humic" acid is rather loosely defined and is used for a number of different products; these are high-molecular-weight compounds of a complex nature. Structurally, they may be considered to be largely condensation products of phenols, quinones and amino compounds. It is probable that both humic matter and similar complex materials, soluble in water, consist of polyanions containing aromatic nuclei with functional groups (OH and COOH) linked through a kind of atomic bridge (—O—, —CH—). A typical building block may contain structures such as

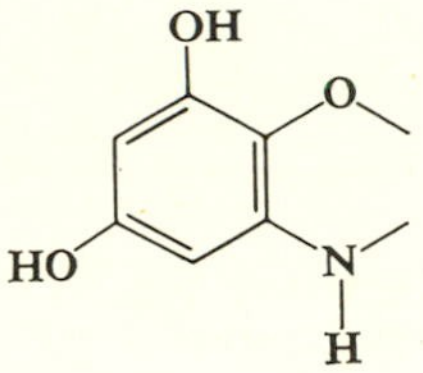

The yellow substances, referred to as Gelbstoffe, fall into the same group; usually they tend to form complexes with Fe(III).

The final products also include compounds similar to melanin which may be formed from tyrosine.

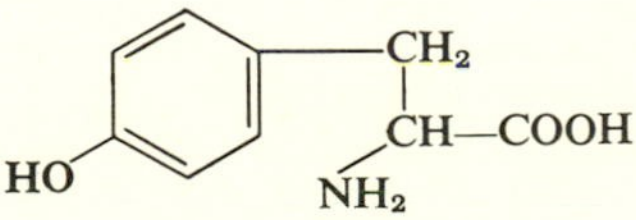

tyrosine

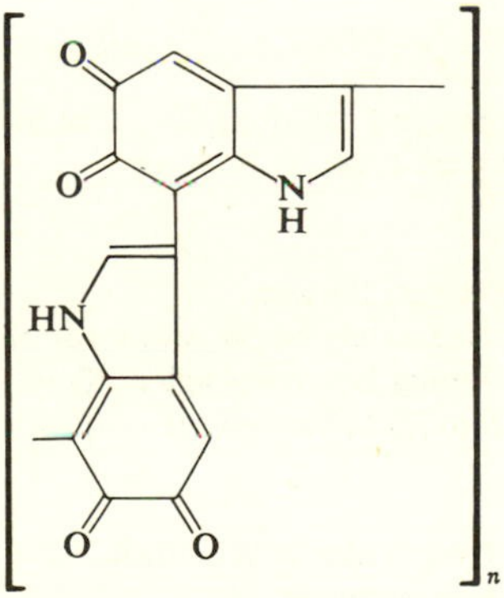

melanin

[23] E. T. Degens, *Geochemistry of Sediments*, Prentice-Hall, Englewood Cliffs, N.J., 1965.

[24] M. M. Kononova, *Soil Organic Matter* (translated from the Russian), Pergamon, New York, 1961.

The Concentration of Aqueous Organic Carbon

Odum [25] and Ryther [26] have summarized production and respiration data for several ecological systems. A few illustrative data are presented in Table 7-8. Gross primary production (productivity) is the total rate of photosynthesis. Net primary productivity is the rate of synthesis of organic matter in excess of its respiratory utilization during the period of measurement. The concentration of organic matter in a natural body of water results from an

Table 7-8   Metabolism of Some Ecosystems

	Production†	
	Gross Photosynthesis, organic matter	Total Respiration, gram m^{-2} day^{-1}
Land		
Wheat (growing season, world average)	3	0.7
Sugar cane (growing season, world average)	6.1	1.4
Forest, tropical	13	3
Forest, deciduous	6.7	1.6
Marine		
Coral reef maxima	18	—
Continental shelf	0.55	0.25
Long Island Sound (annual average)	2.1	1.2
Open ocean	0.2–0.6	0.1–0.3
Polluted Systems:		
Galveston Bay	20–58	36–87
Corpus and Houston Ship Channels	0–31	0–51
Potomac Estuary, polluted	8	12
Rivers		
White River, Indiana:		
Immediately below sewage outfall	0.15	21
30 miles below sewage outflow	16	16
60 miles below sewage outflow (zone of recovery)	41	29
Lakes		
Clear Deep Lake in Wisconsin	0.7	—
Lake Erie, summer	9.0	4.0
Lake Tahoe (Calif.-Nev.)	0.2	0.1

† The exemplifications given here are based on results by various investigators and are not precisely comparable; the data are meaningful in a semiquantitative way only.

[25] H. T. Odum, *Limnol. Oceanog.*, **1**, 102 (1956).
[26] J. H. Ryther, *Science*, **130**, 602 (1959).

interplay of net productivity and import and export (inflow, outflow, dissolution, sedimentation, etc.) of organic matter. Productive lakes have gross primary production of organic matter in the order of a few grams m^{-2} day^{-1}. Oceanic waters show gross productions of up to 1 g m^{-2} day^{-1}. The estimates of production in flowing waters are very high. Odum has suggested that streams are among the most productive biological environments. The highest primary production rates occur in the recovery zones of streams polluted with sewage; this is a consequence of the supply of fertilizing elements and of additional organic nutritive sources possibly contained in sewage. High production is not limited, however, to polluted waters. A rough calculation shows that the net production of organic matter in a productive lake (10 g organic matter or 3.6 g carbon m^{-2} day^{-1}) corresponds to a net synthesis within the euphotic zone (assumed to be ca. 10 m deep) of 0.4 mg organic carbon $liter^{-1}$ day^{-1}. At least an equivalent amount of organic matter is introduced into a lake from outside.

In view of the considerable variety of sources of organic material from outside the basin, a great diversity in the concentration of organic matter must be expected, even in the absence of human contamination. In fresh water bodies one typically encounters concentrations of a few mg C $liter^{-1}$, but occasionally (e.g., in bog or swamp waters) concentrations may be as high as 50 mg C $liter^{-1}$. In the oceans the concentrations of organic carbon ranges from 0.5–1.2 mg C $liter^{-1}$, with the higher values occurring in the surface waters. Particulate organic carbon, including planktonic organisms, generally accounts for about 10% of the total organic carbon. Over much of the ocean we have, according to Wangersky [27], a "soluble" organic carbon concentration of around 500 μg C $liter^{-1}$, and a total particulate organic carbon concentration of some 50 μg C $liter^{-1}$, all ultimately derived from a surface phytoplankton population equivalent to a concentration of some 5 μg C $liter^{-1}$. These figures are significantly higher for waters on the continental shelf and especially for regions where nutrients are returned to the surface by up-welling. Concentrations in the surface layers are generally larger than in deeper waters. Below a depth of 400–600 m the concentration variations in organic carbon become very slight. This suggests that materials present in deep waters are resistant to further metabolic decomposition. It seems then that simpler compounds released in the surface waters are quickly oxidized and utilized by organisms while the more resistant compounds sink to deeper waters. Duursma's data [17] suggest that the organic compounds occurring even in the deepest waters can still be oxidized to some extent. The mode of chemical or biological transformation is not specified, however, and little information on the chemical composition is available.

[27] P. J. Wangersky, *Amer. Sci.*, **53**, 358 (1965).

ORGANIC AGGREGATES. The nonliving particulate organic matter in waters has usually been assumed to be fecal pellets and minute plankton remains in various stages of disintegration. However, in the last decade evidence has accumulated that many of these particles are delicate platelike aggregates a few microns to several millimeters in diameter. They are amorphous in shape and contain both organic and inorganic nonliving material with inclusions of microorganisms [28–31]. It has furthermore been shown that the nuclei of such particles can be formed by adsorption of dissolved organic compounds on air bubbles and other surfaces [29]. The initial aggregates tend to increase in size by further aggregation and adsorption.

The process of particle formation on air bubbles from dissolved carbon seems to be restricted to organic molecules with molecular weights of more than 100,000 under sterile conditions [30]. In the light of this observation, the lytic products of microorganisms and polymeric excretions by the plankton deserve special consideration. The role of microorganisms in the process of particle formation is not yet clear. Barber [30] was able to form aggregates with the help of air bubbles in sterile seawater, while Sheldon, Evelyn and Parsons [31] found no aggregate formation if the bacterial concentration in the seawater was grossly reduced by filtration.

POLLUTION. Man-made pollution adds organic matter to the receiving waters and can cause impairment of water quality. Modern chemical industry has synthesized and processed a great variety of organic chemicals. These stream pollutants need to be differentiated from biogenic organic constituents. The presence of trace concentrations of hydrocarbons or surfactants in a body of water does not necessarily indicate industrial or domestic pollution; such substances may also occur "naturally." Rosen and Rubin [32] have illustrated that the method of carbon dating can be used to distinguish between petrochemical and natural sources of organic matter. Since petrochemical wastes are derived from fossil carbon sources (coal, petroleum), they do not contain $C-14$. Biogenic materials (products of photosynthesis and their decay and organic waste constituents of domestic sewage) contain essentially contemporary levels of $C-14$. Rosen and Rubin obtained the quantity of sample required for assay of $C-14$ content by adsorption of organic constituents of many thousands of liters of waters on a column of granular activated carbon.

SEDIMENTS. The quantity of organic matter carried in natural waters is small compared to that in sediments. The latter are the principal depositories of

[28] G. A. Riley, *Limnol. Oceanog.*, **8**, 372 (1963).
[29] E. R. Baylor and W. H. Sutcliffe, *Limnol. Oceanog.*, **8**, 369 (1963).
[30] R. T. Barber, *Nature*, **211**, 257 (1966).
[31] R. W. Sheldon, T. P. T. Evelyn and T. R. Parsons, *Limnol. Oceanog.*, **12**, 367 (1967).
[32] A. A. Rosen and M. Rubin, *J. Water Pollution Control Federation*, **37**, 1302 (1965).

posthumous organic debris accumulated throughout the earth's history. Most of the organic matter occurs in a finely disseminated state and is associated with fine grained sediments. In order to illustrate more vividly the ratio of inorganic to organic matter, Degens [23] has offered the following generalized picture: "From an evaluation of the total thickness of sedimentary materials that have been formed and deposited over the last 3 to 4 billion years, it can be assumed that a sediment layer of approximately 1000 m around the earth's surface has been laid down. Roughly 2% of this layer, namely ~20 m, is organic, the rest is inorganic in nature. Of this 20-m organic part, coal comprises ~5 cm and crude oil only a little more than 1 mm. The rest represents the finely disseminated organic matter in shales, limestones and sandstones." It can be inferred from the published literature that most of the chemical alternation of organic matter takes place in the environment of deposition and during early stages of diagenesis.

KINETIC CONSIDERATIONS. Much, but not all, of the organic matter produced photosynthetically is decomposed. It is generally held that anaerobic conditions are more favorable to the preservation of organic material than aerobic condition, but the reasons for this difference is not very well understood. The fact that more organic compounds are found under anaerobic conditions may be explained by the fact that an organically enriched environment becomes by necessity anaerobic. Although anaerobic microorganisms multiply more slowly than aerobic ones, their efficiency in biodegradating organic matter is not necessarily lower; because the smaller growth rate of the anaerobes is accompanied by a smaller yield, the specific rate of substrate utilization (expressed in terms of oxidation equivalents) of anaerobes is about equal to that of aerobes. However, other factors may play a significant role. Multicellular organisms cannot exist under anaerobic conditions; furthermore, conditions toxic to organisms (H_2S, high or low pH) are more likely to occur in anaerobic than in aerobic environments.

Many organic substances escape biodegradation. Because of the *intrinsic instability* of organic compounds, organic geochemistry is dominated by nonequilibrium processes. As Blumer [33] has pointed out, during diagenesis the less stable compounds are eliminated by reactions that tend to scramble the ordered structures created by organisms. Diagenesis thus leads gradually to equilibration of the sedimentary organic matter. At moderate temperatures equilibrium is, however, not reached within time spans comparable to the age of the earth.

Data by Abelson [11], Vallentyne,† Conway and Libby† on the stability of

---

† Quoted in [11].
[33] M. Blumer, in *Equilibrium Concepts in Natural Water Systems*, Advances in Chemistry Series, No. 67, American Chemical Society, Washington, D.C., 1967, p. 312.

alanine obtained by studying the (purely chemical) degradation of C−14-tagged alanine in the laboratory may serve as an illustration. Decomposition of alanine occurs at rates following first-order kinetics ($-dc/dt = kc$) where the rate constant k can be formulated as

$$k = Ae^{-E/RT} \tag{41}$$

In the temperature range of 100°C, an activation energy $E = 44$ kcal mole^{-1} and a frequency factor of 3×10^{13} are obtained. At 100°C the experimental findings indicate, for example, a half-time of ca. 10^4 years. If extrapolation to lower temperatures were allowed, half-times of more than 10^{10} years at temperatures below 50°C would result. (Extrapolation to low temperatures is not justified in this case because at low temperatures a second-order reaction involving O_2 appears.) Abelson has surveyed the stability of some of the other amino acids and the laboratory findings are generally in agreement with observations on fossil materials.

Bonds between adjacent carbon atoms in hydrocarbons and in aromatic structures are very strong. With activation energies higher than 50 kcal mole^{-1} for such bonds, half-times of more than 10^{10} years are calculated at temperatures of 100°C or lower. The publications by Abelson, Blumer, and others may be consulted for a detailed illustration on how kinetic data are of diagnostic value for the determination of fossil environmental conditions. These publications also present impressive substantiating evidence for the long duration of some important characteristics of the present-day comparative biochemical plan.

To investigate the evolution of this plan and to inquire into the chemical events of the primitive earth and the origin of life constitutes perhaps the most challenging area of biogeochemical research [11–13, 16, 33, 34].

7-7   The Electrochemical Cell

Direct measurement of pϵ values or of E_H for natural aquatic environments involves complex problems in spite of the apparent simplicity of the electrochemical technique. E_H is measured with an electrode pair consisting of an inert electrode (Au or Pt) and a reference electrode. Natural media containing large quantities of oxidizing agents certainly give high pϵ or E_H values, and those containing large quantities of reducing agents have low potentials. Even when reproducible results are obtained they often do not represent reversible Nernst potentials, and quantitative interpretation with respect to solution composition is often not justified.

[34] A. I. Oparin, *Life, Its Nature, Origin and Development*, Academic Press, paperback edition, New York, 1964.

GALVANIC AND ELECTROLYTIC CELLS. Many chemical reactions can take place in an electrochemical cell. These reactions can occur spontaneously (galvanic cell) or can be driven by an applied external potential (electrolytic cell). Electrical terminals of a cell are labeled according to the type of electrode reactions associated with them. Reduction takes place at the *cathode* and oxidation at the *anode*. Hence the electrons enter an electrochemical cell at the cathode and leave at the anode. Terminals are also called plus or minus, but this depends on whether the cell is galvanic or electrolytic (Figure 7-12).

Electrochemical cells are described by a short-hand notation. For example, the cell reaction

$$Zn(s) + Cu^{2+} = Zn^{2+} + Cu(s) \tag{42}$$

is represented as

$$Zn \mid Zn^{2+}(a_1) \parallel Cu^{2+}(a_2) \mid Cu \tag{43}$$

in which a_1 and a_2 indicate activities, a single vertical line represents a phase boundary across which there is a potential difference and a double vertical line signifies that the liquid junction is either ignored or considered to be eliminated by a salt bridge. The following convention is used for the sequence in which the diagram is written. The electrode reaction on the right is a reduction (making this electrode a cathode) and the electrode reaction on the left is an oxidation (anode).

The electromotive force of the half cell $Zn^{2+} \mid Zn$ corresponds to the electromotive force of the cell

$$Pt, H_2 \, (p_{H_2} = 1) \mid H^+ \, (a = 1) \parallel Zn^{2+} \mid Zn \tag{44}$$

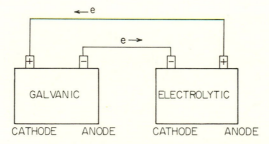

Figure 7-12   Plus and minus terminals in galvanic and electrolytic cells. A cathode is always the electrode where reduction takes place, an anode is where oxidation occurs. Electrons leave at the anode and enter at the cathode. The direction of electron flow is in the direction opposite to that of positive electricity. (After J. Waser, *Basic Chemical Thermodynamics*, Benjamin, New York, 1966.) Reproduced with permission from W. A. Benjamin, Inc.

where the electrode on the left is the standard hydrogen electrode. According to IUPAC convention such a cell emf is the electrode potential. If both Zn and Zn^{2+} are at unit activity, we speak of the standard electrode potential of the Zn electrode.

ELECTRODES. In the hydrogen electrode the platinum acts as a catalyst for the reaction between H^+ and H_2 molecules and acquires a potential characteristic of this reaction. Similarly, electrodes may be characterized by the redox couple they represent, such as Cu^{2+}/Cu, Cl_2/Cl^-, H^+/H_2, O_2/H_2O, AgCl/Ag and Hg_2Cl_2/Hg. The last two electrodes mentioned are also called electrodes of the second kind. These electrodes tend to maintain a constant electrode potential because the concentration of the cations associated with the electrode metal is kept constant by buffering through solubility product principle. Thus a AgCl/Ag electrode consists in principle of a silver electrode coated with AgCl immersed in a solution of high $[Cl^-]$. Because AgCl(s) is present, the activity of Ag^+ is given by

$$\{Ag^+\} = \frac{K_{s0(AgCl)}}{\{Cl^-\}} \qquad (45)$$

Furthermore, if some Ag^+ is reduced to Ag(s), or some Ag(s) is oxidized to Ag^+, the dissolution or precipitation of AgCl, respectively, will keep $\{Ag^+\}$ constant. Hence the electrode potential of a AgCl/Ag electrode remains constant even if some current is flowing through the half cell (nonpolarizable electrode). Of course, the current must be small enough so that the rate of the reaction

$$AgCl(s) + e = Ag(s) + Cl^- \qquad (46)$$

does not exceed that attained by the AgCl solubility product. Such a nonpolarizable electrode is a convenient reference electrode. Another important reference electrode is the calomel electrode with the half reaction

$$Hg_2Cl_2(s) + 2e = 2Hg(l) + 2Cl^- \qquad (47)$$

Its electrode potential depends on the activity of the chloride ion.

Other electrodes represent redox couples in which oxidized and reduced forms are soluble species in the solution and the electron transfer occurs at the inert electrode (Pt, Au). For example, a platinum electrode immersed in an acid solution of Fe^{2+} and Fe^{3+} (Pt/Fe^{3+}, Fe^{2+}) may under favorable conditions acquire the potential characteristics of a Fe^{3+}/Fe^{2+} couple

$$Fe^{3+} + e = Fe^{2+} \qquad (48)$$

When half cells are combined in an electrochemical cell, the emf of the cell is given by

$$E_{cell} = E_{right\ cell} - E_{left\ cell} \qquad (49)$$

or

$$E_{cell} = E_{H\,right} - E_{H\,left} \quad (50)$$

If the left electrode is a reference electrode, we may also write

$$E_{cell} = E_{H(ox\text{-}red)} - E_{ref} \quad (51)$$

Example 7-9 *Electrode Potential of* Fe^{3+}–Fe^{2+} *System.* Consider the following cell and compute its emf:

$$Hg \mid Hg_2Cl_2, KCl_{satd} \parallel HClO_4 \,(1\,M), Fe^{3+}\,(10^{-3}), Fe^{2+}\,(10^{-2}) \mid Pt$$

According to Table 7-9 the potential of the saturated calomel electrode (25°C) is 0.244 V. Hence

$$E_{cell} = E_{Fe^{3+},Fe^{2+}} - 0.244$$

and

$$E_{Fe^{3+},Fe^{2+}} = 0.771 + 0.059 \log \frac{10^{-3}}{10^{-2}} = 0.712\,V$$

$$E_{cell} = 0.468\,V$$

Example 7-10 *Solution Composition from Measured emf.* With an electrode pair consisting of an inert Pt and a saturated calomel reference electrode in a sample of a sediment–water interface of pH = 6.4 a potential difference (emf) of 0.47 V is measured. The sediment contains solid α-FeOOH and we assume that the measured potential corresponds to an oxidation–reduction potential of the aquatic environment.

(a) What is the E_H and pϵ of the sample? (b) What is the activity of Fe^{2+}? (c) Does the redox level found indicate aerobic or anaerobic conditions?

1. The emf has been measured in a cell

$$Hg, Hg_2Cl_2 \mid KCl \parallel ox, red/Pt$$

Table 7-9   Potentials of Reference Electrodes†

Temp., °C	Calomel		AgCl, Ag	
	KCl: 0.1 M	Satd.	0.1 M	Satd.
12	0.3362	0.2528	—	—
20	0.3360	0.2508	—	—
25	0.3356	0.2444	0.2900	0.1988
dE_{cell}/dT	−0.00004	−0.00058		

† The values listed (V) include the liquid junction potential. Cell: Pt, $H_2/H^+(a = 1) \parallel$ reference electrode. Cf. R. G. Bates, in *Treatise in Analytical Chemistry*, Part I, Vol. I, I. M. Kolthoff and P. J. Elving, Eds., Interscience, New York, 1959, p. 319. Reproduced with permission from Interscience Publishers.

and

$$E_{cell} = E_{H(ox-red)} - E_{ref}$$

Because $E_{ref} = 0.244$, we obtain $E_{H(ox-red)} = 0.47 + 0.244 = 0.714$ V. This is equivalent to a pϵ of $+12.1$.

2. At equilibrium FeOOH(s) is in equilibrium with Fe^{2+}, the redox reaction being $FeOOH(s) + 3H^+ + e = Fe^{2+} + 2H_2O$. We obtain $\log K$ for this reaction from values of the free energy of formation (25°C) $\Delta G°(FeOOH(s)) = -98.0$ kcal, $\Delta G°(Fe^{2+}) = -20.3$ kcal and $\Delta G°(H_2O) = -56.69$ kcal. The following values obtain for the reduction reaction: $\Delta G° = -35.68$ kcal, $\log K = 26.3 (= p\epsilon°)$. The activity of Fe^{2+} can be calculated by

$$p\epsilon = p\epsilon° + \log \frac{\{H^+\}^3}{\{Fe^{2+}\}}$$

$$12.1 = 26.3 - 3pH + pFe^{2+}$$

This gives pFe^{2+} of 5.0 for pH = 6.4.

3. In order to decide whether this corresponds to aerobic or anaerobic conditions, p_{O_2} must be computed. For

$$O_2(g) + 4H^+ + 4e = 2H_2O; \quad p\epsilon° = 20.8$$

because

$$p\epsilon = p\epsilon° + \tfrac{1}{4} \log p_{O_2}\{H^+\}^4$$
$$12.1 = 20.8 - pH + \tfrac{1}{4} \log p_{O_2}$$
$$\log p_{O_2} = 4(12.1 - 20.8 + pH)$$
$$p_{O_2} = 10^{-9.2} \text{ atm}$$

This corresponds to ca. 10^{-12} M dissolved O_2. We may properly speak of anaerobic conditions.

Example 7-11 *Standard Potential of the* Cl_2/Cl^- *Couple.* G. Faita, P. Longhi and T. Mussini [*J. Electrochem. Soc., 114,* 340 (1967)] determined the standard potential of the Cl_2/Cl^- electrode. They made emf measurements with the cell

$$\text{Pt} \mid \text{Ag} \mid \text{AgCl} \mid 1.75 \ M \ \text{HCl} \mid \text{Cl}_2 \text{ (atm)} \mid \text{Pt–Ir (45\%), Ta} \mid \text{Pt} \qquad \text{(i)}$$

The right-hand electrode consisted of a tantalum foil (attached to a Pt wire) coated with a platinum iridium alloy. This alloy, containing 45% iridium, had to be used as a Cl_2 electrode because Pt is not fully appropriate since it is subject to corrosive attack by Cl_2 in the presence of HCl.

Their results of the cell (i) are as follows:

$$25°C: E_{cell\ (i)} = 1.13596\ V$$
$$30°C: E_{cell\ (i)} = 1.13309\ V$$
$$40°C: E_{cell\ (i)} = 1.12711\ V$$
$$50°C: E_{cell\ (i)} = 1.12110\ V$$

1. What is the chemical reaction taking place in the cell?
2. What is the standard potential of the Cl_2/Cl^- electrode?
3. Determine from the experimental data $\Delta G°$, $\Delta H°$ and $\Delta S°$ of the cell reaction of the reduction of Cl_2 to Cl^-, respectively.

The standard $Ag/AgCl/Cl^-$ electrode, $E°_{AgCl/Ag}$, has the following values (volts):

$E°_{AgCl/Ag}$: 25°C, 0.22234; 30°C, 0.21904; 40°C, 0.21208; 50°C, 0.20449

Furthermore the concentration of HCl in cell (i) has been chosen such that $\{Cl^-\} = 1.0$.

1. The chemical reaction taking place in the cell (i) is composed of the half reactions

$$Ag(s) + Cl^- = AgCl(s) + e$$
$$\tfrac{1}{2}Cl_2(g) + e = Cl^-$$

$$Ag(s) + \tfrac{1}{2}Cl_2(g) = AgCl(s) \tag{ii}$$

The emf of $E_{cell\ (i)}$ is a direct measure of the free energy of the reaction $\Delta G° = -nFE°_{cell}$. Hence at 25°C, $\Delta G° = -26,198$ cal.

2. The standard potential of the Cl_2/Cl^- electrode, $\tfrac{1}{2}Cl_2(g) + e = Cl^-$, is given by

$$E_{cell\ (i)} = E_{Cl_2/Cl} - E_{AgCl/Cl} \tag{iii}$$

Thus

$$E_{Cl_2/Cl^-} = E_{cell\ (i)} + E_{AgCl/Cl} \tag{iv}$$

and at 25°C $E°_{Cl_2/Cl^-} = 1.35830\ V$. Thus $\Delta G°$ for the reaction $\tfrac{1}{2}Cl_2(g) + e = Cl^-$ (which is the same as for the reaction $\tfrac{1}{2}H_2(g) + \tfrac{1}{2}Cl_2(g) = H^+ + Cl^-$) is $-31,326$ cal mole^{-1}, corresponding to a log K value of 22.97.

3. $\Delta S°$ and $\Delta H°$ for the cell reaction $Ag(s) + \tfrac{1}{2}Cl_2(g) = AgCl(s)$ can be calculated by considering that $\Delta S = nF(dE_{cell}/dT)$ and $\Delta H = -nFE + nFT(dE_{cell}/dT)$. The data given for $E_{cell\ (i)}$ have been fitted as function of absolute temperature T (least square procedure) by $E_{cell\ (i)} = 1.28958 - (4.31562 \times 10^{-4})T - (2.7922 \times 10^{-7})T^2$. At 25°C the first derivative of this equation is 5.986×10^{-4} (V deg^{-1}). (This temperature dependence may also be obtained approximately by plotting $E_{cell\ (i)}$ versus $1/T$.) Hence $\Delta H_{(25°C)} = -30,310$ cal mole^{-1}. Similarly $\Delta S°$ and $\Delta H°$ values at 25°C (or at other

temperatures) may be obtained from the temperature dependence of $E^\circ_{Cl_2/Cl^-}$. The latter is found (25°C) to be -1.246×10^{-3} V deg^{-1}. Correspondingly $\Delta S^\circ_{(25°C)} = -28.73$ cal deg^{-1} mole^{-1} and $\Delta H_{(25°C)} = -39.91$ kcal mole^{-1}.

7-8   Evaluation of E_H

It was already pointed out that it is necessary to distinguish between the concept of a potential and the measurement of a potential. E_H measurements are of great value in those systems for which the variables are known or under control. In this section we will discuss the measurement and the indirect evaluation of redox potentials. In this context it is useful to discuss the concept of formal potentials and to illustrate the effect of complex formation upon redox potentials.

MEASURING ELECTRODE POTENTIAL; KINETIC CONSIDERATIONS. Some of the essential principles involved in the measurement of an electrode potential can be described qualitatively by a consideration of the behavior of a single electrode (platinum) immersed in an acidified Fe^{2+}–Fe^{3+} solution. To cause the passage of a finite current at this electrode, it is necessary to shift the potential from its equilibrium value. We thus obtain a curve depicting the electrode potential as a function of the applied current (polarization curve). At the equilibrium potential, that is, at the point of zero applied current, the half reaction

$$Fe^{3+} + e \rightleftharpoons Fe^{2+}$$

is at equilibrium, but the two opposing processes, the reduction of Fe^{3+} and the oxidation of Fe^{2+}, proceed at an equal and finite rate,

$$v_1 \text{ (rate of reduction)} = v_2 \text{ (rate of oxidation)} \tag{52}$$

is proportional to the rate of passage of electrons in both directions. Although the net rate of passage of electrons is equal in both directions and thus the (applied) current is zero, the passage of current in a single direction is not zero and is called the *exchange current*, i_0. (See Figure 7-12.)

The rate of electron exchange at the surface of the electrode may be expressed as moles reacting per unit time per unit surface area. Under simplifying assumptions (first-order kinetics) v_1 or v_2 will be proportional to the concentration or activity of the reactants at the surface of the electrode ($[Fe^{3+}]_e$ and $[Fe^{2+}]_e$). (For other systems we may substitute $[ox]_e$ and $[red]_e$, respectively.)

$$v_1 = k_1[Fe^{3+}]_e \tag{53a}$$
$$v_2 = k_2[Fe^{2+}]_e \tag{53b}$$

where k_1 and k_2 are velocity constants (with the units cm sec^{-1}) of oxidation and reduction, respectively, and where concentrations are conveniently expressed as moles cm^{-3}. Velocity constants are a function of activation energy, E_{act},

$$k = Z \exp\left(-E_{act}/RT\right) \tag{54}$$

In the case of electrochemical reactions, the activation energy is a function of the difference $E_H - E_H^\circ$ between the potential of the electrode, E_H, and the standard oxidation potential, E_H°, of the redox system in question. It has been shown that

$$k_1 = k_0 \exp\left(-\frac{\alpha nF}{RT(E_H - E_H^\circ)}\right) \tag{55}$$

or

$$k_1 = k_0 \times 10^{-n\alpha(p\epsilon - p\epsilon^\circ)} \tag{56}$$

and

$$k_2 = k_0 \times 10^{n(1-\alpha)(p\epsilon - p\epsilon^\circ)} \tag{57}$$

where n is the number of electrons required in the kinetic process ($n = 1$ for the Fe^{3+}–Fe^{2+} system) and α is called the transfer coefficient and is a measure of the symmetry of the energy barrier; α is the fraction of the total energy that acts to decrease the height of the energy barrier for the reduction reaction and $1 - \alpha$ is the fraction that tends to increase the height of the anodic potential barrier [35]. (α is frequently close to 0.5.)

Because current is proportional to the velocity of reduction or the velocity of oxidation

$$i_c = nFv_1A \tag{58}$$

$$i_a = nFv_2A \tag{59}$$

where A is the area of the electrode. The cathodic and anodic currents can then be given by

$$i_c = nFk_0A[Fe^{3+}]_e \times 10^{-\alpha n(p\epsilon - p\epsilon^\circ)} \tag{60}$$

$$i_a = nFk_0A[Fe^{2+}]_e \times 10^{(1-\alpha)n(p\epsilon - p\epsilon^\circ)} \tag{61}$$

The behavior of each of these currents as a function of cell emf or pϵ is shown in Figure 7-13. The net current, i, is the sum of the two opposing currents i_a and i_c. Adopting the convention $i_a > 0$ and $i_c < 0$, i is given by

$$i = i_c - i_a \tag{62}$$

[35] See, for example, C. N. Reilley, in *Treatise on Analytical Chemistry*, Part I, Volume 4, I. M. Kolthoff and P. J. Elving, Eds., Wiley-Interscience, New York, 1963, Chapter 42, p. 2109.

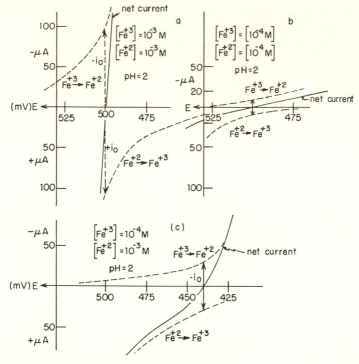

Figure 7-13 Polarization curves for various concentrations of Fe^{2+} and Fe^{3+}. Solid lines = polarization curves (electrode area = 1 cm^2); i_0 = exchange current. Dashed lines are hypothetical cathodic $(-i)$ and anodic $(+i)$ currents. Curves are schematic but based on experimental data at significant points.

Thus at any $p\epsilon$ or potential (these may now be values measured with an electrochemical cell, i.e., not equilibrium values) the net current is given by

$$i = nFk_0A([Fe^{3+}]_e \times 10^{-\alpha n(p\epsilon - p\epsilon^\circ)} - [Fe^{2+}]_e \times 10^{(1-\alpha)n(p\epsilon - p\epsilon^\circ)}) \quad (63)$$

At equilibrium where no net current flows $(i = 0, i_a = i_c)$, concentrations at the electrode surface equal bulk concentrations and $p\epsilon$ values or potentials measured correspond to equilibrium conditions, (62) and (63) yield:

$$\frac{[Fe^{3+}]}{[Fe^{2+}]} = 10^{(p\epsilon - p\epsilon^\circ)} = e^{(F/RT)(E_H - E_H^\circ)} \quad (64)$$

or in a more general sense

$$\frac{[ox]}{[red]} = 10^{n(p\epsilon - p\epsilon^\circ)} = e^{(nF/RT)(E_H - E_H^\circ)} \quad (65)$$

Equations 64 and 65 are identical with the Nernst equation.

The exchange current i_0 is given by i_c or i_a at equilibrium. By combining (60) or (61) and (64) the exchange current i_0 becomes†

$$i_0 = FAk_0[Fe^{3+}]^{1-\alpha}[Fe^{2+}]^\alpha = nFAk_0[ox]^{1-\alpha}[red]^\alpha \qquad (66)$$

and (63) can be rewritten as

$$i = i_0(10^{-\alpha n\Delta p\epsilon} - 10^{(1-\alpha)n\Delta p\epsilon}) = i_0(e^{-(\alpha nF/RT)\Delta E} - e^{(1-\alpha)(nF/RT)\Delta E}) \qquad (67)$$

where $\Delta p\epsilon$ and ΔE are the deviations in $p\epsilon$ or in E_H, respectively, from its equilibrium value that results from the passage of a net current. When ΔE is very small, the exponential terms in (67) can be expanded in the form $e^{-x} \approx 1 - x$ and (67) becomes

$$i = -i_0(nF/RT)\Delta E = -i_0 n\Delta p\epsilon(\ln 10) \qquad (68)$$

where $(F/RT)^{-1} = 0.0257$ V. Correspondingly the slope of the polarization curve, $(\Delta i/\Delta E)_{i=0}$ or $(\Delta i/\Delta p\epsilon)_{i=0}$, respectively, are given by

$$\left(\frac{\Delta i}{\Delta E}\right)_{i=0} = -\left(\frac{nF}{RT}\right)i_0 \qquad (69)$$

and

$$\left(\frac{\Delta i}{\Delta p\epsilon}\right)_{i=0} = -ni_0(\ln 10) \qquad (70)$$

The equations given above are illustrated in Figure 7-13a.

As indicated, the net current can be visualized as the summation of two opposing currents (cathodic and anodic). The rate of Fe^{3+} reduction (conventionally expressed as cathodic current) generally increases exponentially with more negative electrode potential values and is, furthermore, a function of the concentration of $[Fe^{3+}]$ and of the effective electrode area. Similar considerations apply to the rate of Fe^{2+} oxidation (anodic current), which is proportional to $[Fe^{2+}]$, electrode area and the exponential of the potential. It is obvious from the schematic representation that in the case of Fe^{2+}–Fe^{3+} an infinitesimal shift of the electrode potential from its equilibrium value will make the half reaction proceed in either of the two opposing directions, provided the concentration of these ions is sufficiently large. The measurement of the equilibrium electrode potential in such a case is feasible; the amount the potential must be shifted to obtain an indication with the measuring instrument and hence the sharpness and reproducibility of the measurement is determined by the slope of the net current in the vicinity of the balance

---

† Equation (66) can be derived by first rearranging (64) to $([Fe^{3+}]/[Fe^{2+}])^{(1-\alpha)} = 10^{(p\epsilon - p\epsilon^\circ)(1-\alpha)}$ which is inserted into (61).

point. According to (69) and (70), this slope is proportional to the exchange current, i_0, while the dependence of i_0 upon concentration of reactants and electrode area is given by (66).

Under favorable circumstances with modern instrumentation for which the current drain is quite low, reliable potential measurements can be made with systems giving i_0 greater than about 10^{-7} A. For the $Fe^{3+}-Fe^{2+}$ system k_0 at 25°C has been determined to be about 10^{-3} cm sec^{-1}. This represents a rather rapid electrode reaction. This value of k_0 gives $i_0 = 10^{-4}$ A for 10^{-3} M concentrations of Fe^{3+} and Fe^{2+} and an electrode area of 1 cm^2. Thus the exchange current and the slope of the polarization curve amply provide accurate measurement of pϵ or E_H as shown in Figure 7-13. If tenfold smaller concentrations of both ions are present, i_0 and the slopes are only one-tenth as great, but with $i_0 = 10^{-5}$ A reliable measurements can still be made, as shown in Figure 7-13b. If the concentration of only one ion is decreased, the drop in i_0 is not as great and E_H, the potential corresponding to equal cathodic and anodic currents, is shifted, as shown by Figure 7-13c. If both Fe^{3+} and Fe^{2+} are at 10^{-6} M concentrations (ca. 0.05 mg liter^{-1}), i_0 is 10^{-7} A and measurements are no longer precise. Actually, because of other effects caused by trace impurities, it becomes difficult to obtain measurements in accord with simple Nernst theory when either Fe-ion concentration is less than about 10^{-5} M.

According to (66), more precise measurements can be obtained at very low concentrations with an increase in the electrode area A. Electrodes with large areas are advantageous in this way but they also tend to magnify the effects of trace impurities or other reactions on the electrode surface itself, such as adsorption of surface-active materials leading to reduction in k_0 or in effective area A.

"SLOW" ELECTRODES. We might contrast such behavior with the conditions we would encounter in attempting the measurement of the electrode potential in water containing O_2. A schematic representation of the polarization curve for this case is given in Figure 7-14. The equilibrium electrode potential again should be located at the point where the net applied current (i.e., the sum of cathodic and anodic currents) is zero. The exact location becomes difficult to determine. Over a considerable span of electrode potentials the net current is virtually zero; similarly, the electron exchange rate, or the exchange current reflecting the opposing rates of the half reaction

$$H_2O \rightleftharpoons \tfrac{1}{2}O_2 + 2H^+ + 2e$$

is virtually zero. Operationally, a remarkable potential shift must be made to produce a finite net current, and the current drawn in the potentiometric measurement is very large compared with the exchange current. Even with

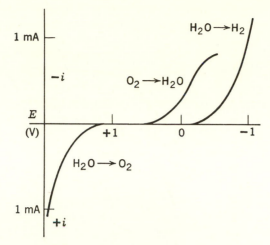

Figure 7-14   Electrode polarization curves for oxygen-containing solutions in otherwise-pure water. Curves are schematic but in accord with available data at significant points.

modern instrumentation in which the current drain can be made extremely low, the experimental location of the equilibrium potential is ambiguous. It has been estimated [36] that for O_2 (1 atm) the specific exchange current i_0/A is 10^{-10} A cm^{-2}, far less than 10^{-7}. Thus the influence of p_{O_2} on pϵ is not reflected in an experimental E_H measurement. Instead of responding to the p_{O_2}, a Pt electrode responds to spurious oxide potentials such as Pt(OH) + e^- = Pt + OH$^-$ which has a potential (in acid solutions) of 0.84 V. Other potentials corresponding to different electrode reactions may be observed.

MIXED POTENTIALS. Another difficulty arises in E_H measurements. The balancing anodic and cathodic currents at the apparent "equilibrium" potential need not correspond to the same redox process and may be a composite of two or more processes. An example of this is shown in Figure 7-15 for a Fe^{3+}–Fe^{2+} system in the presence of a trace of dissolved oxygen. The measured zero-current potential is that value where the rate of O_2 reduction at the electrode surface is equal to the rate of Fe^{2+} oxidation rather than the value of E_{eq} since at the latter point simultaneous O_2 reduction produces excess cathodic current. In addition, because the net reaction at E_m converts Fe^{2+} to Fe^{3+}, the measured potential exhibits a slow drift. Such "mixed" potentials are of little worth in determining equilibrium E_H values.

Except for the Fe^{3+}–Fe^{2+} couple at concentrations greater than about 10^{-5} M and perhaps the Mn(IV)–Mn^{2+} couple, the over-all redox systems

[36] W. Watanabe and M. A. V. Devanathan, *J. Electrochem. Soc.*, **111**, 615 (1964).

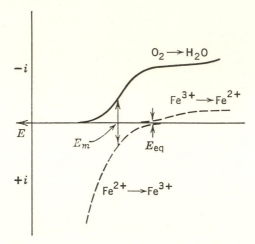

Figure 7-15 Electrode polarization curves for oxygen-containing solutions with a mixed potential resulting from oxygen reduction accompanied by Fe^{2+} oxidation. Curves are schematic but in accord with available data at significant points.

important in natural waters are not electroactive. No reversible electrode potentials are established for the $NO_3^- - NO_2^- - NH_4^+$, $SO_4^{2-} - H_2S$ or $CH_4 - CO_2$ systems. Moreover few organic redox couples yield reversible potentials. Unfortunately, many and perhaps most measurements of E_H (or pϵ) in natural waters represent mixed potentials not amenable to quantitative interpretation.

INDIRECT EVALUATION OF REDOX POTENTIALS. Of the standard redox potentials quoted in handbooks or summaries, relatively few have been determined by direct potential measurements; the others have been calculated from a combination of free energy data or from equilibrium constants. It is possible to evaluate the redox level in natural water systems by determining the relative concentrations of the members of one of the redox couples in the system and applying the electrochemical relations in reverse. As Figure 7-9 suggests, quantitative analytical information on any one of the species O_2, Fe^{2+}, Mn^{2+}, $HS^- - SO_4^{2-}$ and $CH_4 - CO_2$ gives a conceptually defined pϵ (or E_H), provided the system is in equilibrium. In practice, however, there are limitations even on this basically sound procedure. The system must be in equilibrium or must be in a sufficiently constant metastable state to make the concept of a partial equilibrium meaningful. For example, although most aqueous media are not in equilibrium with regard to processes involving N_2, the inertness of these reactions may allow us to ignore them while treating the equilibrium achieved by other species. A singular pϵ value can be ascribed to a system if equal values of E_H or $p\epsilon$ are obtained for each of the redox couples in a multi-component system; otherwise the couples are not in equilibrium and the concept of a singular redox potential becomes meaningless.

Within these limitations analytical determination of the oxidized and reduced forms of a redox couple can provide quite precise values for E_H. Depending on the number of electrons transferred in the process, determinations accurate within a factor of 2 will give E_H values within 5 to 20 mV or $p\epsilon$ values within 0.1 to 0.3 units.

Example 7-12   *Computation of* $p\epsilon$ *and* E_H *from Analytic Information.* Estimate $p\epsilon$ and E_H for the aquatic habitats that are characterized by the following analytic information.

1. Sediment containing $FeOOH(s)$ and $FeCO_3(s)$ in contact with 10^{-3} M HCO_3^- at pH = 7.0.

2. Water from the deeper layers of a lake having a dissolved O_2 concentration of 0.03 mg O_2 liter^{-1} and a pH of 7.0.

3. An anaerobic digester in which 65% CH_4 and 35% CO_2 are in contact with water of pH = 7.0.

4. A water sample of pH = 6, containing $\{SO_4^{2-}\} = 10^{-3}$ M and smelling of H_2S.

5. A sediment–water interface containing $FeCO_3(s)$ with a crust of black $FeS(s)$ at pH = 8.

1. This corresponds to the conditions of (9) in Table 7-5. $p\epsilon = -1.67$ and $E_H = -0.099$ V. This system, in equilibrium, has an infinite poising (redox-buffering) intensity. The redox level can be defined rather precisely.

2. This corresponds to (1) in Table 7-5, and $p\epsilon = 13.75 + \frac{1}{4} \log p_{O_2}$ 10^{-6} moles O_2 liter^{-1} is equivalent to a partial pressure, $p_{O_2} \approx 6 \times 10^{-4}$ and $\frac{1}{4} \log p_{O_2} = -0.8$. $p\epsilon$ for this system is 12.95 and $E_H = +0.77$ V. Note that E_H varies little with p_{O_2} or dissolved O_2 concentration. A reduction in O_2 from 10 (approximately air saturated) to 0.1 mg liter^{-1} will lower E_H by 30 mV.

3. According to (15) in Table 7-5

$$p\epsilon = -4.13 + \frac{1}{8} \log \frac{p_{CO_2}}{p_{CH_4}} = -4.16$$

$$E_H = -0.25 \text{ V}$$

4. The equilibrium is characterized by

$$\tfrac{1}{8}SO_4^{2-} + \tfrac{5}{4}H^+ + e = \tfrac{1}{8}H_2S(g) + \tfrac{1}{2}H_2O$$

$p\epsilon°$ can be calculated from the $p\epsilon°(W)$ value given in Table 7-5. $p\epsilon° = p\epsilon°(W) - (n_H/2) \log K_W$ [see (30) of Section 7-5]; $p\epsilon° = -3.5 + 8.75 = 5.25$. Hence

$$p\epsilon = 5.25 - \tfrac{5}{4}pH + \tfrac{1}{8} \log [SO_4{}^{2-}] - \tfrac{1}{8} \log pH_2S$$
$$= 2.62 - \tfrac{1}{8} \log p_{H_2S}$$

It is reasonable to set p_{H_2S} between 10^{-2} and 10^{-8} atm. This puts $p\epsilon$ between -1.6 and -2.4, corresponding to E_H of between -0.09 and -0.14 V.

5. We do not have all the information necessary but we can make a guess. A reaction defining the most likely redox poising equilibrium could be formulated as

$$SO_4{}^{2-} + FeCO_3(s) + 9H^+ + 8e = FeS(s) + HCO_3{}^- + 4H_2O$$

Using free energy values given in Latimer and assigning to FeS(s) a $\bar{G}_f^{\circ}$ of formation corresponding to α-FeS, the equilibrium constant of the reaction given above is calculated (25°C) to be $\log K = 38.0$. Correspondingly $p\epsilon^{\circ} = 4.75$ and $p\epsilon$ can be calculated from

$$p\epsilon = 4.75 - \tfrac{9}{8}pH + \tfrac{1}{8}(pHCO_3{}^- - pSO_4{}^{2-})$$

The term in parentheses has to be estimated; it is very unlikely that it is outside the range -2 and $+2$. This fixes $p\epsilon$ for this equilibrium system at $pH = 8$ between -4.0 and -4.5. Correspondingly, $-0.27 < E_H < -0.24$ V.

Example 7-13   *Redox Potential of* S(s) *Bearing Sediments.*   R. A. Berner [*Geochim. Cosmochim. Acta*, **27**, 563 (1963)] has demonstrated that a Pt electrode inserted into H_2S-containing sediments gives electrode potentials in accord with the half cell $S(s) + 2e = S^{2-}$ (S^{2-} was determined specifically with a Ag_2S/Ag electrode). Establish a relationship between electrode potential of measured E_H versus pS^{2-} (25°C).

From equilibrium constants we can make the combinations:

$$
\begin{array}{ll}
S(s) + 2H^+ + 2e = H_2S(aq); & \log K = 4.8 \\
H_2S(aq) = H^+ + HS^-; & \log K = -7.0 \\
HS^- = H^+ + S^{2-}; & \log K = -14 \\
\hline
S(s) + 2e = S^{2-}; & \log K = -16.2
\end{array}
$$

Hence we can write

$$p\epsilon = -8.1 + 0.5pS^{2-}$$

or

$$E_H = -0.48 + 0.030pS^{2-}$$

This relationship has been observed by Berner for natural and artificial H_2S-containing sediments.

There are two questions relevant in this context:

1. How does elemental sulfur participate in the electron exchange at the Pt surface? Sulfur probably does not participate in the electrode reaction;

most likely soluble species that are in equilibrium with elemental sulfur, such as polysulfides [$S(s) + S^{2-} = S_2^{2-}$; $4S(s) + S^{2-} = S_5^{2-}$] are electroactive. As shown by Peschanski and Valensi and as quoted by Berner, because of such polysulfides the Pt electrode behaves with respect to the $S(s)$–S^{2-} system as a kinetically reversible system.

2. Do such measurements indicate the redox level of the sediments? Though the response of the Pt electrode is controlled by the $S(s)$–$S(-II)$ couple, the E_H measured is representative of the redox potential of the sediment only if the $S(s)$–$S(-II)$ couple is in equilibrium with all other redox couples present in the same locale. If different apparent redox levels coexist, the Pt electrode records the couple characterized by the highest exchange current. Consider for example the case where an organic redox system which is not in equilibrium with the $S(s)$–$S(-II)$ couple has a lower apparent pϵ. Because the organic species are not electroactive, the Pt electrode is not affected by this lower partial redox level.

EFFECT OF IONIC STRENGTH AND COMPLEX FORMATION ON ELECTRODE POTENTIALS. Redox equilibria also confronts us with the problem of evaluating activity corrections or maintaining the activities under consideration as constants. The Nernst equation rigorously applies only if the activities and actual species taking part in the reaction are inserted into the equation. The activity scales discussed before, the infinite dilution scale and the ionic medium scale may be used. The standard potential or standard pϵ at the infinite dilution scale is related to the equilibrium constant for $I = 0$ of the reduction reaction by

$$\frac{F}{RT(\ln 10)}\, E_H^\circ = \frac{1}{n}\log K = \mathrm{p}\epsilon^\circ \tag{71}$$

and is usually obtained by either extrapolating the measured constant or measured potential to infinite dilution.

In a constant ionic medium the concentration quotient becomes the equilibrium constant K, and correspondingly one might define ${}^c E_H^\circ$ and $\mathrm{p}^c E^\circ$

$$\frac{F}{RT(\ln 10)}\, {}^c E_H^\circ = \frac{1}{n}\log {}^c K = \mathrm{p}^c E^\circ \tag{72}$$

COMPLEX FORMATION. How the standard potential is influenced by complex formation may be illustrated by the dissolution of zinc

$$Zn(s) + 2H^+ = Zn^{2+} + H_2(g) \tag{73}$$

which is characterized by the half reaction

$$Zn^{2+} + 2e = Zn(s); \quad \Delta G_1^\circ = -RT \ln K = -2FE_{Zn^{2+},Zn(s)}^\circ \tag{74}$$

If the dissolution of Zn occurs in a medium containing ligands that can displace coordinated H_2O from the zinc ions, that is, form complexes, such as in the reaction of Cl^- ions with Zn^{2+} to form $ZnCl_4^{2-}$

$$ZnCl_4^{2-} = Zn^{2+} + 4Cl^-; \quad \Delta G_2^\circ = RT \ln \beta_4 \tag{75}$$

where β_4 is the formation constant $[\{ZnCl_4^{2-}\}/(\{Zn^{2+}\}\{Cl^-\}^4)]$, then we may characterize the over-all half reaction by

$$ZnCl_4^{2-} + 2e = Zn(s) + 4Cl^-; \quad \Delta G_3^\circ = -RT \ln {}^+K = -2FE_{ZnCl_4^{2-},Zn(s)}^\circ \tag{76}$$

This is the sum of (69) and (70). Correspondingly, we may write any one of the following relations

$$\Delta G_3^\circ = \Delta G_1^\circ + \Delta G_2^\circ \tag{77}$$

$$\log {}^+K = \log K - \log \beta_4 \tag{78}$$

$$p\epsilon_{ZnCl_4^{2-},Zn(s)}^\circ = p\epsilon_{Zn^{2+},Zn(s)}^\circ - \tfrac{1}{2} \log \beta_4 \tag{79}$$

$$E_{ZnCl_4^{2-},Zn(s)}^\circ = E_{Zn^{2+},Zn(s)}^\circ - \frac{RT(\ln 10)}{2F} \log \beta_4 \tag{80}$$

These equations show that Cl^- stabilizes the higher oxidation state, facilitating the dissolution of Zn. The Nernst equation may now be expressed either in terms of free Zn^{2+} or in terms of $ZnCl_4^{2-}$

$$E = E_{Zn^{2+},Zn(s)}^\circ + \frac{RT(\ln 10)}{2F} \log \{Zn^{2+}\}$$

$$= E_{ZnCl_4^{2-},Zn(s)}^\circ + \frac{RT(\ln 10)}{2F} \log \{ZnCl_4^{2-}\} - \frac{2RT(\ln 10)}{F} \{Cl^-\} \tag{81}$$

FORMAL POTENTIALS. In a similar way as with conditional constants, that is, constants valid under specifically selected conditions, for example a given pH and a given ionic medium, conditional or formal potentials are of great utility.

$$\frac{F}{RT(\ln 10)} {}^FE_H^\circ = \frac{1}{n} \log P = p^F\epsilon^\circ \tag{82}$$

The measurement of a formal $p\epsilon$ or formal electrode potential consists of the measurement of the emf of an electrochemical cell in which, under the specified conditions, the analytical concentration of the two oxidation states are varied. For example, in a 0.1 M H_2SO_4 solution the formal electrode potential for Fe(III)–Fe(II) is 0.68 V in comparison to 0.77 V for the Fe^{3+}/Fe^{2+} ($I = 0$) system.

$$^FE_{(I=0.1\ M\ H_2SO_4)}^\circ = 0.68 + \frac{RT}{F} \ln \frac{Fe(III)_T}{Fe(II)_T} \tag{83}$$

In this case the formal potential includes correction factors for activity coefficients, acid–base phenomena (hydrolysis of Fe^{3+} to $FeOH^{2+}$), complex formation (sulfate complexes) and liquid junction potential used between the reference electrode and the half cell in question. Although the correction is strictly valid only at the single concentration at which the potential has been determined, formal potentials may often lead to better predictions than standard potentials because they represent quantities subject to direct experimental measurement.

Example 7-14   *Formal Potential of* Fe(III)–Fe(II) *System in Presence of* F^-. Estimate the formal potential of the Fe(III)–Fe(II) couple for solutions of $[H^+] = 10^{-2}\ M$ and $[F^-] = 10^{-2}\ M$, $I = 0.1\ M$. The following constants are available:

$$Fe^{3+} + e = Fe^{2+}; \qquad\qquad \log K\,(I = 0) = 13.0 \qquad (i)$$
$$Fe^{3+} + H_2O = FeOH^{2+} + H^+; \quad \log K_H\,(I = 0.1) = -2.7 \qquad (ii)$$
$$Fe^{3+} + F^- = FeF^{2+}; \qquad\qquad \log \beta_1\,(I = 0.1) = 5.2 \qquad (iii)$$
$$Fe^{3+} + 2F^- = FeF_2^+; \qquad\qquad \log \beta_2\,(I = 0.1) = 9.2 \qquad (iv)$$
$$Fe^{3+} + 3F^- = FeF_3; \qquad\qquad \log \beta_3\,(I = 0.1) = 11.9 \qquad (v)$$

In order to compute the formal potential we first consider the activity correction of the Fe^{3+}–Fe^{2+} electrode. Using Güntelberg approximation, $K_{(1)}$ is corrected into $^cK_{(1)}$:

$$\frac{[Fe^{2+}]}{[Fe^{3+}]\{e\}} = {}^cK = K\frac{f_{Fe^{3+}}}{f_{Fe^{2+}}} = K\frac{0.083}{0.33} = 10^{12.4}; \quad p^c\epsilon^\circ = 12.4 \qquad (vi)$$

from which $^cE_H^\circ$ is obtained as $+0.73$ V. It has been assumed that the liquid junction makes a negligible contribution. The same result is obtained if we consider that

$$^cE^\circ_{Fe^{3+},Fe^{2+}} = E^\circ_{Fe^{2+},Fe^{2+}} + \frac{RT}{F}\ln\frac{f_{Fe^{3+}}}{f_{Fe^{2+}}}$$

Next, the correction caused by hydrolysis and complex formation can be taken into account. For the conditions specified, $FeOH^{2+}$ is the only important hydrolysis species.

$$Fe(III)_T = [Fe^{3+}] + [FeOH^{2+}] + [FeF^{2+}] + [FeF_2^+] + [FeF_3] \quad (vii)$$

Under the specified conditions, ferrous iron does not form complexes with F^- and OH^-, hence

$$Fe(II)_T = [Fe^{2+}] \qquad\qquad\qquad (viii)$$

Equation vii can be rewritten as

$$Fe(III)_T = [Fe^{3+}]\left(1 + \frac{K_H}{[H^+]} + \beta_1[F^-] + \beta_2[F^-]^2 + \beta_3[F^-]^3\right) \quad (ix)$$

For the conditions specified

$$\alpha_{Fe} = \frac{Fe(III)_T}{[Fe^{3+}]} = 9.6 \times 10^5 \tag{x}$$

and (i) may be formulated as in (xi) and (xii)

$$\frac{Fe(II)_T}{Fe(III)_T\{e\}} = \frac{{}^cK}{\alpha_{Fe}} = P \tag{xi}$$

$$\log P = 6.4 = p^F\epsilon^\circ \tag{xii}$$

Correspondingly, the formal potential of the Fe(III)–Fe(II) electrode for the given conditions is ${}^FE^\circ = 0.38$ V. The potential, of course, is the same whether it is expressed in terms of actual concentrations or analytic concentrations:

$$E = 0.73 + \frac{RT}{F}\ln\frac{[Fe^{3+}]}{[Fe^{2+}]} = 0.38 + \frac{RT}{F}\ln\frac{Fe(III)_T}{Fe(II)_T} \tag{xiii}$$

or

$$p\epsilon = 12.4 + \log\frac{[Fe^{3+}]}{[Fe^{2+}]} = 6.4 + \log\frac{Fe(III)_T}{Fe(II)_T} \tag{xiv}$$

Note that from an electrode kinetic point of view the Nernst equation does not give any information as to the actual species which establishes the electrode potential. In the case of a fluoride-containing solution it is very possible that one of the fluoro-iron(III) complexes rather than the Fe^{3+} iron participates in the electron exchange reaction at the electrode. In addition to a thermodynamic effect, complexation may cause an electrode kinetic effect by changing the exchange current; the rate of electron exchange between the electrode and the complex species may be greater or smaller than the exchange with the aquo ions. Complexation usually stabilizes a system against reduction. In the example just considered, because complexation is stronger with Fe(III) than with Fe(II), the tendency for reduction of Fe(III) to Fe(II) is decreased. It is apparent that coordination with a donor group, in general, decreases the redox potential. In those relatively rare instances where the lower oxidation state is favored (e.g., complexation of aqueous iron with phenanthroline), the redox potential is increased as a result of coordination.

7-9 The Potentiometric Determination of Individual Solutes; Potentiometric Titration

Methods that involve the introduction of electrodes without contamination into the natural media are most appealing. Unfortunately, as we have seen, the platinum or gold electrode lacks specificity for measuring E_H. The glass electrode, however, has proved to be sufficiently specific and sensitive. More recently other electrodes specific for more than a dozen ions have joined the pH-type glass electrode in commercial production. Many of these electrodes are sufficiently specific and sensitive to permit measurement or monitoring of individual solution components. Progress has also been made in the development of a reliable amperometric oxygen electrode [37, 38].

An electrochemical cell can be used conveniently to study the properties of a solution quantitatively. For example, it is possible to determine $\{Ag^+\}$ with an Ag electrode, $\{H^+\}$ with a $Pt/H_2(g)$ electrode or with a glass electrode, $\{Cl^-\}$ with an AgCl/Ag electrode and $\{SO_4^{2-}\}$ with a $PbSO_4/Pb$ electrode. Table 7-10 gives a survey of most of the ion-sensitive electrodes employed in potentiometric measurements.

Table 7-10   Ion Selective Electrodes

Ion	Electrode Material	Electrode Reaction
I. Metal Electrodes		
H^+	Platinized Pt, H_2	$H^+ + e = \frac{1}{2}H_2(g)$
Ag^+, Cu^{2+}, Hg^{2+}	Ag, Cu, Hg	$Me^{n+} + ne = Me(s)$
Zn^{2+}, Cd^{2+}	Zn(Hg), Cd(Hg)	$Me^{2+} + 2e + Hg(l) = Me(Hg)$
II. Electrodes of the Second Kind		
Cl^-	AgCl/Ag	$AgCl(s) + e = Ag(s) + Cl^-$
S^{2-}	Ag_2S/Ag	$Ag_2S(s) + 2e = 2Ag(s) + S^{2-}$
SO_4^{2-}	$PbSO_4/Pb(Hg)$	$PbSO_4 + Hg + 2e = Pb(Hg) + SO_4^{2-}$
H^+	Sb_2O_3/Sb	$Sb_2O_3 + 6H^+ + 6e = 2Sb(s) + 3H_2O$
III. Complex Electrodes		
H^+	Pt, quinhydrone	$C_6H_4O_2 + 2H^+ + 2e = C_6H_4(OH)_2$
EDTA (Y^{4-})	HgY^{2-}/Hg	$HgY^{2-} + 2e = Hg(l) + Y^{4-}$
IV. Glass Electrodes		
H^+	Glass	$Na^+(aq) + H^+_{Membrane} = Na^+_{Membrane}$ $+ H^+(aq)$
Na^+, K^+, Ag^+, NH_4^+	Cation sensitive glass	Ion exchange
V. Solid-State or Precipitate Electrodes		
F^-, Cl^-, Br^-, I^-, S^{2-}, Cu^{2+}, Cd^{2+}, Na^+, Ca^{2+}, Pb^{2+}	Precipitate impregnated or solid-state electrodes	Ion exchange
VI. Liquid–Liquid Membrane Electrodes		
Ca^{2+}, Mg^{2+}, Pb^{2+}, Cu^{2+}, NO_3^-, Cl^-, ClO_4^-	Liquid ion exchange	Ion exchange

[37] D. E. Carritt and J. Kanwisher, *Anal. Chem.*, **31**, 5 (1959).
[38] K. H. Mancy and D. A. Okun, *Anal. Chem.*, **32**, 108 (1960).

In the case of metal electrodes the potential determining mechanism is a fast electron exchange, given by the indicator reaction

$$Me^{n+} + ne = Me(s) \tag{84}$$

for which the equilibrium potential is

$$E_H = E^\circ_{H_{Me^{n+},Me(s)}} + \frac{RT(\ln 10)}{nF} \log \{Me^{n+}\} \tag{85}$$

If a potential of a given indicator electrode is to be controlled by a specific redox reaction, it is clear that the essential solid phases must be present in adequate amounts and the electrode reaction must be characterized by a large enough exchange current. The magnitude of the latter determines the concentration at which the indicator solute must be contained in the solution. Obviously the Nernst equation must not be expected to hold for indefinitively decreasing activity of the potential-determining species. If the exchange current is too small the electrode response becomes too slow, and reactions with impurities or spurious oxides or oxygen at the electrode surface influence the potential and render it unstable and indefinite. Two of the highest exchange current densities are for H^+ discharge at Pt and for reduction of Hg^{2+} at a Hg surface. It may be noted, however, that metal ion electrodes often may show an unusually sensitive response because the electroactive species is actually a complex present at larger concentration than the free metal ions. For example, if a slower electrode responds properly to $\{Ag^+\}$ in a sulfide-containing medium, the electrode response is caused by the two silver complexes usually present at much higher concentrations than the free Ag^+.

For many metals the system $Me^{n+} + ne = Me(s)$ is slow and the equilibrium potential is established too slowly for using such metals as indicator electrodes.

Strongly reducing metals cannot be used as indicator electrodes because the potentials established with such metals are mixed potentials. Hence we cannot use an Fe electrode or a Zn electrode to measure $\{Fe^{2+}\}$ and $\{Zn^{2+}\}$, respectively. The current potential curve for the Zn system is given schematically in Figure 7-16a. The Zn electrode is characterized by a mixed potential because H_2O is reduced at potentials more positive than the reduction of Zn^{2+}. If a zinc amalgam electrode is used instead of a zinc electrode, the current potential curves are displaced (Figure 7-16b). At the Zn(Hg) electrode, in comparison to the Zn electrode, hydrogen discharge occurs at more negative potentials and the reduction of Zn^{2+} at more positive potentials. The reaction at the electrode is now

$$Zn^{2+} + Hg(l) + 2e = Zn(Hg)$$

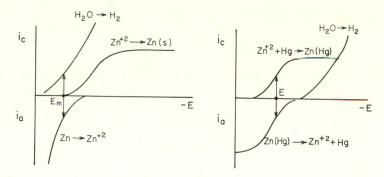

Figures 7-16a and 7-16b   The response of a zinc and of a zinc amalgam electrode to zinc ions. Because the reduction of H_2O precedes that of Zn^{2+} at the Zn electrode, a mixed potential is established. The Zn electrode cannot be used as an indicator electrode. The Zn(Hg) electrode, however, can be used to indicate $\{Zn^{2+}\}$,

and the electrode potential is given by

$$E = E_0 + \frac{RT}{nF} \ln \frac{\{Zn^{2+}\}}{\{Zn(Hg)\}} \tag{86}$$

where $\{Zn[Hg]\}$ is the activity of the Zn in the amalgam.

Glass Electrodes

When a thin membrane of glass separates two solutions an electric potential difference which depends on the ions present in the solutions is established across the glass. Glass electrodes responding chiefly to H^+ ions have become common laboratory tools and a number of treatises are concerned with their properties [39–41]. More recently modification of the glass composition has led to the development of electrodes selective for a variety of cations other than H^+.

The origin of the glass electrode potential is not discussed here, but it may be helpful to mention that the glass membrane functions as a cation ex-changer and that a Nernst potential is observed if such a membrane separates

[39] G. Eisenmann, R. Bates, G. Mattock and S. M. Friedman, *The Glass Electrode*, Wiley-Interscience, New York, 1965.
[40] G. Eisenmann, Ed., *Glass Electrodes for Hydrogen and Other Cations*, Dekker, New York, 1967.
[41] R. G. Bates, *Electrometric pH Determination, Theory and Practice*, Wiley, New York, 1964.

two solutions at two different concentrations

$$E_{\text{cell}} = \frac{RT}{F} \ln \frac{\{^1\text{H}^+\}}{\{^2\text{H}^+\}} \qquad (87)$$

Because the glass electrode contains a solution (acid) of constant $\{^2\text{H}^+\}$ inside the glass bulb, the emf measured depends only on $\{^1\text{H}^+\}$ in the external solution

$$E_{\text{cell}} = \text{const} + \frac{RT}{F} \ln \{^1\text{H}^+\} \qquad (88)$$

CALIBRATION. Typically, the cell used to measure pH (or pMe) can be represented by the scheme

glass electrode ‖ test solution | SCE (saturated calomel electrode)   (89)

If E_1 and E_2 are the values of the emf of cell (89) in test solutions of pH, with pH equal to pH_1 and pH_2, respectively, one finds

$$E_2 - E_1 = \left(\frac{RT}{F}\right)(\text{pH}_2 - \text{pH}_1) \qquad (90)$$

Thus for unit change in pH, a 59.16 mV change in cell potential should occur. This assumes that the liquid junction potential remains constant. In practice pH or pMe measurement is always made by comparing the pH or pMe of an unknown with a standard of known pH or pMe.

POTENTIOMETRIC ELECTRODES MEASURE ACTIVITIES. The standards may be chosen according to either one of the activity scales discussed before (Section 3-4). In the *infinite dilution scale*, the value of the pH of a standard buffer is determined once and for all by measurements of cells without liquid junction such as

Pt, H_2 | standard solution, AgCl | Ag   (91)

making reasonable assumptions about activity coefficients in such a way as to make pH as nearly as possible equal to

$$-\log \{\text{H}^+\} = -\log [\text{H}^+] f_{\pm} \qquad (92)$$

where $f_{\pm}$ is the mean activity coefficient of a univalent electrolyte in the standard buffer. Using the cell given above, the National Bureau of Standards has adopted a series of buffers having standard pH values. By comparing an unknown test solution with such a pH standard, or preferably with two pH standards, one on each side of the unknown, pH values that may be regarded as measures of $-\log \{\text{H}^+\}$ may be determined.

As discussed before, on the *constant ionic medium scale* the activity coefficient approaches unity as the solution approaches the pure solvent, that is, the pure ionic medium. Correspondingly, in a medium where the concentrations of the reacting species, for example, H^+, are much lower than that of the medium ions, the ionic medium scale activity coefficient, $f_{\pm}$, is close to unity and for each ionic medium a pH scale may be defined for which H^+ ion activity and H^+ ion concentration are identical. (See Figure 7-17.) The standard solution is now a solution of known $[H^+]$, that is, of known concentration of strong acid in a constant ionic medium.

Similar calibrations can be performed when any potentiometric electrode is substituted for the glass electrode in cell (89).

Other Ion-Sensitive Electrodes

In Table 7-10 other types of electrodes are enumerated. In these the glass membrane is replaced by a synthetic single crystal membrane (solid-state electrodes), by a matrix (e.g., inert silicone rubber) in which precipitate particles are imbedded (precipitate electrodes) or by a liquid ion exchange layer (liquid–liquid membrane electrodes). The selectivity of these electrodes is determined by the composition of the membrane. All these electrodes show

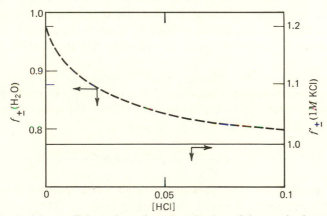

Figure 7-17   Activity coefficients depend on the selection of the standard state (modified from Schindler). On the *infinite dilution scale* the standard state is the infinitely dilute aqueous solution (i.e., a hypothetical solution of concentration 1 with properties of an infinitely dilute solution). The activity coefficient $f_{\pm \text{ HCl}}$ varies with [HCl] in accordance with a Debye-Hückel equation (dotted line, left ordinate); only at very great dilutions does f become unity ([HCl] = {HCl}). On the *ionic medium scale* the standard state is the ionic medium (i.e., infinitely diluted with respect to HCl only). In such a medium f_{HCl}' (solid line, right ordinate) is very nearly constant, that is, $f_{\text{HCl}}' = 1$ and [HCl] = {HCl}. Both activity coefficients are equally meaningful.

a response in their electrode potentials according to the Nernst equation. The so-called oxygen electrode is based on a different principle: an electrolysis cell is used and the current measured at the inert cathode is, under standardized conditions, a function of the oxygen concentration (activity). Selectivity is enhanced by covering the cathode with a membrane permeable to molecules only.

POTENTIOMETRIC TITRATION. Potentiometry may be used to follow a titration and to determine its end point. The principles have already been discussed in connection with acid–base or complex formation titrations where pH or pMe has been used as a variable. Any potentiometric electrode may serve as an indicator electrode, which either indicates a reactant or a reaction product. Usually the measured potential will vary during the course of the reaction and the end point will be characterized by a "jump" in the curve of voltage versus amount of reactant added.

Although the potentiometric measurement permits the determination of the concentration (activity) of a specific species, the potentiometric titration determines the total analytical concentration (capacity); for example, a Ca^{2+} electrode responds to free Ca^{2+} ions only and not to Ca^{2+} complexes present in the solution. Using the Ca electrode as an indicator electrode with a strong complex former (forming more stable complexes with Ca^{2+} than those already present in solution) gives the sum of the concentrations of all Ca^{2+} species. In a similar way total F^- may be determined by titrating with $La(NO_3)_3$, using a F^- electrode, or total K^+ may be measured from titration with a specific K^+ electrode using Ca tetraphenylborate, $(Ca[B(C_6H_5)_4]_2)$, as a reagent.

The result obtained by a titration is usually more precise than that obtained in a direct potentiometric measurement. It is usually not too difficult to create an end point with a precision better than 5%. In the direct potentiometric determination the relative error, F_{rel} is given by

$$F_{rel} = \frac{\Delta C}{C} = 2.3\Delta \log C \qquad (93)$$

In other words, if the measurement with a glass electrode can be reproduced within 0.04 pH units (2.5 mV), the relative error in $[H^+]$ is 10%. For a similar reproducibility with a Ca^{2+} electrode, $\Delta pCa = 0.08$ pCa units (the Nernst coefficient is half that for the pH electrode), the relative error in $[Ca^{2+}]$ is 20%.

TITRATION CURVE. The morphology of the titration curve can be calculated using the same principles discussed earlier. However, for many systems the slopes of the theoretical titration curves do not conform to the experimental values because the indicator electrode does not respond with sufficient sensitivity or selectivity. A detailed example on a redox titration, given below, will serve for illustration. As with acid–base titrations, Gran plots (see Appendix, Chapter 4) can be used with advantage for a most precise end point detection. A good example is the titration of Cl^- with $AgNO_3$ using a $AgCl/Ag$ electrode as an indicator electrode.

The stability of a redox system with respect to change in $p\epsilon$ as a result of the addition of a reductant, C_R

$$S = \frac{dC_R}{dpE} \tag{94}$$

may be called the *redox poising intensity* and is analogous to the buffer intensity in acid–base systems.

Example 7-15   *Titration of* $Fe(II)$ *with* MnO_4.   Construct a theoretical titration curve of redox potential ($E_{half\,cell}$ versus H_2 electrode) for the titration of Fe^{2+} with MnO_4^- in a dilute H_2SO_4 solution using a Pt electrode as an indicator electrode.

The stoichiometry of the reaction is given by

$$MnO_4^- + 8H^+ + 5e = Mn^{2+} + 4H_2O$$
$$5Fe^{2+} = 5Fe^{3+} + 5e$$
$$\overline{\phantom{MnO_4^- + 8H^+ + 5Fe^{2+} = 5Fe^{3+} + 4H_2O + Mn^{2+}}}$$

$$MnO_4^- + 8H^+ + 5Fe^{2+} = 5Fe^{3+} + 4H_2O + Mn^{2+} \tag{i}$$

I. The *conceptual redox potential* is given by

$$E'_H = E^{\circ\prime}_{H\,Fe^{3+},Fe^{2+}} + 0.06 \log \frac{\{Fe^{3+}\}}{\{Fe^{2+}\}} \tag{ii}$$

or

$$E'_H = E^{\circ\prime}_{H\,MnO_4^-,Mn^{2+}} + \frac{0.06}{5} \log \frac{[MnO_4^-]}{[Mn^{2+}]} \tag{iii}$$

where E'_H and $E^{\circ\prime}_H$ are formal potentials.

II. The *mass balances* are (neglecting dilution by titrant)

$$[Fe^{2+}] + [Fe^{3+}] = C_0 \tag{iv}$$
$$5[MnO_4^-] + 5[Mn^{2+}] = C_0 f \tag{v}$$

where f is the mole fraction of Fe(II) titrated (i.e., Fe(III)/[Fe(II) + Fe(III)] assuming stoichiometric completion of the reaction). (For example, at the end point $f = 1$, we have added $\frac{1}{5}$ mole of MnO_4^- per mole of Fe(II) originally present; correspondingly, at the end point we will have a concentration of Mn_T of one fifth of that of Fe_T: $[Mn_T] = \frac{1}{5}[Fe_T]$ or $5[Mn_T] = Fe_T = C_0$.)

III. The *electron balance* is

$$5[Mn^{2+}] = [Fe^{3+}] \tag{vi-a}$$

that is, apply similar considerations as in proton balance. For every Mn^{2+} formed 5 electrons are lost, and for every ferric ion formed 1 electron is gained (or every MnO_4^- can accept 5 electrons and every Fe^{2+} can donate 1 electron).

The equations derived from *equilibrium constants* [i.e., redox potential–Nernst equation, (ii) and (iii)], from *mass balances* [(iv) and (v)] and the *electron balance* (vi-a) define E' as *a function of* f (5 equations and 5 unknowns). This gives the exact solution.

Since operational E is not necessarily conceptual E, *approximate* solution may be more expedient if the assumption is made that the reaction goes virtually to completion.

(1) *Before the end point:* All MnO_4^- added is reduced and oxidizes Fe^{2+} to Fe^{3+} quantitatively

$$[Fe^{3+}] = C_0 f = 5[Mn^{2+}]; \quad [Fe^{2+}] = C_0(1 - f)$$

According to (ii)

$$E_H' \simeq 0.7 + 0.06 \log \frac{f}{1-f} \tag{vii}$$

(2) *Beyond the end point:* Essentially all the ferrous converted to ferric and $[Fe^{2+}]$ in (iv) can be neglected. Accordingly (vi-a) becomes $5[Mn^{2+}] \simeq C_0$ and the MnO_4^- added essentially stays as $[MnO_4^-]$; therefore [see also (v)]

$$5[MnO_4^-] \simeq C_0(f - 1) \tag{viii}$$

and

$$E_H' \simeq 1.5 + \frac{0.06}{5} \log (f - 1) \tag{ix}$$

At $f \simeq 2$, $E_H' = E_{H\,MnO_4^-, Mn^{2+}}^{\circ\prime}$.

(3) *At the end point:* Addition of (ii) and (iii) leads to

$$2E_H' = 0.70 + 1.5 + 0.06 \log \frac{[Fe^{3+}]}{[Fe^{2+}]} \frac{[MnO_4^-]^{1/5}}{[Mn^{2+}]^{1/5}} \tag{x}$$

At the end point $E_H' = E_H'(ep)$, and (vi) as well as the condition

$$5[MnO_4^-] = [Fe^{2+}] \tag{vi-b}$$

is fulfilled.

From an operational point of view we may consider that the Pt electrode does not appear to respond to MnO_4^- and Mn^{2+} ions. Therefore the operational potential equals the conceptual potential only under conditions where

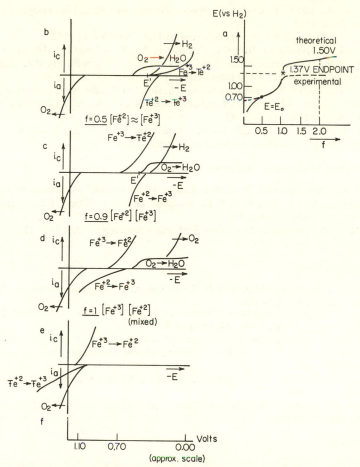

Figure 7-18   Titration of Fe(II) with MnO_4^- (Example 7-15). In *a* the experimental titration curve is compared with the calculated one. Figures *b* through *f* show schematic polarization curves for various extents of the titration.

a Pt electrode responds satisfactorily to the Fe^{2+}–Fe^{3+} couple. If $[Fe^{3+}] \gg [Fe^{2+}]$ (i.e., beyond the end point) and if $[Fe^{2+}] \gg [Fe^3]$ (i.e., in the initial portion of the titration curve), the Pt electrode indicates a mixed potential that does not change appreciably with *f*. An illustration of the titration curve together with the schematic polarization curves are given in Figure 7-18.

READING SUGGESTIONS

I. Redox Equilibria

Clark, W. M., *Oxidation–Reduction Potentials of Organic Systems*, Williams & Wilkins, Baltimore, Md., 1960.

Garrels, R. M., and C. L. Christ, *Solutions, Minerals and Equilibria*, Harper and Row, New York, 1965.

Latimer, W. M., *Oxidation Potentials*, Prentice-Hall, New York, 1952.

Sillén, L. G., in *Treatise on Analytical Chemistry*, Part I, Volume 1, I. M. Kolthoff and P. J. Elving, Eds., Interscience, New York, 1959, p. 227; in *Chemical Equilibrium in Analytical Chemistry*, Wiley-Interscience, New York, 1965.

Sillén, L. G., "Master Variables and Activity Scales," in *Equilibrium Concepts in Natural Water Systems*, Advances in Chemistry Series, No. 67, American Chemical Society, Washington, D.C., 1967, p. 45.

Sillén, L. G., and A. E. Martell, *Stability Constants of Metal–Ion Complexes*, Special Publ., No. 17, The Chemical Society, London, 1964.

II. Mechanism of Oxidation–Reduction Reactions

Duke, F. R., "Mechanism of Oxidation–Reduction Reactions," in *Treatise on Analytical Chemistry*, Part I, Volume 1, I. M. Kolthoff and P. J. Elving, Eds., Interscience, New York, 1959, Chapter 15.

Taube, H., "Mechanisms of Oxidation–Reduction Reactions," *J. Chem. Educ.*, **45**, 453 (1968).

III. Electrochemistry

Charlot, G., J. Badoz-Lambling and B. Tremillon, *Electrochemical Reactions*, Elsevier, Amsterdam, 1962, Chapter 7.

Conway, B. E., and M. Salomon, "Electrochemistry: Its Role in Teaching Physical Chemistry," *J. Chem. Educ.*, **44**, 555 (1967).

Laitinen, H. A., *Chemical Analysis*, Chapter 15. McGraw-Hill, New York, 1960.

Murray, R. W., and C. N. Reilley, *Electroanalytical Principles*, Wiley-Interscience, New York, 1963.

Tanaka, N., and R. Tamushi, "Kinetic Parameters of Electrode Reaction," *Electrochem. Acta*, **9**, 963 (1964).

Watanabe, W., and M. A. Devanathan, "Reversible Oxygen Electrodes," *J. Electrochem. Soc.*, **111**, 615 (1964).

IV. Electrodes

Durst, R. A., "Mechanism of the Glass Electrode Response," *J. Chem. Educ.*, **44**, 175 (1967).

Eisenmann, G., Ed., *Glass Electrodes for Hydrogen and Other Cations, Principles and Practice*, Dekker, New York, 1967.

Rechnitz, G. A., "Ionic Selective Electrode," *Chem. Eng. News*, **45**(24), 146 (1967).

V. Redox Reactions and Organisms

Kirschbaum, J., "Biological Oxidations and Energy Conservation," *J. Chem. Educ.*, **45**, 29 (1968).

Klotz, I. M. *Energy Changes in Biochemical Reactions*, Academic Press, New York, 1967.

Lehninger, A. L., *Bioenergetics*, Benjamin, New York, 1965.

McCarty, P. L., *Thermodynamics of Biological Synthesis and Growth: Proceedings of the Second International Water Pollution Conference, Tokyo*, Pergamon, 1965.

Redfield, A. C., B. H. Ketchum and F. A. Richards, "The Influence of Organisms on the Composition of Sea Water," in *The Sea*, Volume 2, M. N. Hill, Ed., Wiley, New York, 1963, p. 26.

VI. Redox Potential as Environmental Parameter

Abelson, P. H., "Chemical Events on the Primitive Earth," *Proc. Natl. Acad. Sci.*, **55**, 1365 (1966).

Baas-Becking, L. G. M., I. R. Kaplan and O. Moore, "Limits of the Natural Environment in Terms of pH and Oxidation–Reduction Potentials," *J. Geol.*, **68**, 243 (1960).

Bostrom, K., "Some pH Controlling Redox Reactions in Natural Waters," in *Equilibrium Concepts in Natural Water Systems*, Advances in Chemistry Series, No. 67, American Chemical Society, Washington, D.C., 1967, p. 286.

Krauskopf, K. B., *Introduction to Geochemistry*, McGraw-Hill, New York, 1967.

Kusnetzov, S. I., "Das Oxidoreduktionspotential in den Seen," *Arb. Limnol. Station Kossino*, **20**, 55 (1935).

Mortimer, C. H., "Dissolved Substances between Mud and Water," *J. Ecol.*, **29**, 280 (1941); **30**, 147 (1942).

Sillén, L. G., "Oxidation State of Earth's Ocean and Atmosphere," *Arkiv Kemi*, **24**, 431 (1965); **25**, 159 (1965).

Zobell, C. E., "Studies in Redox Potential of Marine Sediments," *Bull. Amer. Soc. Petrol. Geol.*, **30**, 477 (1946).

VII. Organic Carbon

Blumer, M., "Organic Pigments: Their Long-term Fate," *Science*, **149**, 722 (1965).

Calvin, M. *Chemical Evolution*, Oxford University Press, New York, 1969.

Cheldelin, V. H., and R. W. Newburgh, *The Chemistry of Some Life Processes*, Reinhold, New York, 1964 (paperback). The authors give an elementary treatment of what life is in chemical terms.

Duursma, E. K., "The Dissolved Organic Constituents of Sea Water," in *Chemical Oceanography*, J. P. Riley and G. Skirrow, Eds., Academic Press, New York, 1965.

Gorham, E., and D. J. Swaine, "The Influence of Oxidizing and Reducing Conditions upon the Distribution of Some Elements in Lake Sediments," *Limnol. Oceanog.*, **10**, 166 (1965).

Jannasch, H. W. "Current Concepts in Aquatic Microbiology" *Verh. Internal Verein Limnol.*, **17**, 25 (1969).

PROBLEMS

7-1 (a) Write balanced reactions for the following oxidations and reductions:

    (i) Mn^{2+} to MnO_2 by Cl_2

    (ii) Mn^{2+} to MnO_2 by OCl^-

    (iii) H_2S to SO_4^{2-} by OCl^-

    (iv) MnO_2 to Mn^{2+} by H_2S

    (b) Arrange the following in order of decreasing $p\epsilon$:

    (i) Lake sediment

    (ii) River sediment

    (iii) Seawater

    (iv) Ground water containing 0.5 mg liter^{-1} Fe^{2+}

    (v) Digester gas (CH_4, CO_2)

7-2 The II, III and IV oxidation states of Mn are related thermodynamically by these equilibria in acid solution:

$$Mn(III) + e = Mn(II), \quad \log K = 25$$
$$Mn(IV) + 2e = Mn(II), \quad \log K = 40$$

(i) What is the standard electrode potential of the couple

$$Mn(IV) + e = Mn(III)?$$

(ii) Is Mn(III) stable with respect to simultaneous oxidation and reduction to Mn(II) and Mn(IV)? Explain.

7-3 (i) Under what pH conditions can NO_3^- be reduced to NO_2^- by ferrous iron? (Use constants available in Tables 7-4 and 7-5.)

(ii) What is the ratio of the concentrations of NO_3^- to NO_2^- in equilibrium with an aqueous Fe^{2+}–$Fe(OH)_3$(s) (or FeOOH(s)) system that has a pH of 7 and a pFe^{2+} of 4?

7-4 (i) Can the oxidation of NH_4^+ to NO_3^- by SO_4^{2-} be mediated by microorganisms at pH = 7?

(ii) Is the oxidation of HS^- to SO_4^{2-} by NO_3^- thermodynamically possible at a pH of 9?

(iii) Estimate the $p\epsilon$ range in which sulfate reducing organisms can grow.

7-5 In a closed tank, water and gases have been brought into equilibrium. The water at equilibrium has the following composition: Alkalinity = 2×10^{-2} equivalents per liter; $[Fe^{2+}] = 2 \times 10^{-5} M$; pH = 6.0; Fe(III) = negligible but ferric hydroxide is precipitated.

(i) What is the partial pressure of CO_2?

(ii) What is the $p\epsilon$ of the solution?

(iii) What partial pressure of O_2 corresponds to this $p\epsilon$?

The following information is available for the appropriate temperature:

$$CO_2 + H_2O = H_2CO_3^* \qquad \log K = 1.5$$
$$H_2CO_3^* = H^+ + HCO_3^- \qquad \log K = -6.3$$
$$HCO_3^- = H^+ + CO_3^{2-} \qquad \log K = -10.3$$
$$Fe^{2+} + CO_3^{2-} = FeCO_3(s) \qquad \log K = +10.6$$
$$Fe^{3+} + 3OH^- = Fe(OH)_3(s) \qquad \log K = +38.0$$
$$O_2 + 4e + 4H^+ = 2H_2O \qquad E° = +1.23 \text{ V}$$
$$H^+ + OH^- = H_2O \qquad \log K = +14$$
$$Fe(OH)_3(s) + e = Fe^{2+} + 3OH^- \qquad E° = -1.41 \text{ V}$$

7-6 What p_{O_2} cannot be exceeded so that a reduction of SO_4^{2-} to HS^- (pH = 7) can take place? (Use Table 7-5.)

7-7 (i) Can glucose ($C_6H_{12}O_6$) be converted into CH_4 and CO_2 (e.g., in a digester)?

(ii) What is the per cent composition of the resulting gas?

7-8 Separate each of the following reactions into its half-reactions and in each case write down the schematic representation of a galvanic cell in which the reaction would take place. Wherever possible devise a cell without liquid-junction potentials.

(i) $H_2 + PbSO_4(s) = 2H^+ + SO_4^{2-} + Pb$

(ii) $AgCl = Ag^+ + Cl^-$

(iii) $3Mn^{2+} + 2MnO_4^- + 4OH^- = 5MnO_2 + 2H_2O$

(iv) $6Cl^- + IO_3^- + 6H^+ = 3Cl_2 + I^- + 3H_2O$

For each reaction above compute E_{cell} for the corresponding cell and decide the direction in which each reaction would take place spontaneously if all substances were present at unit activity.

7-9 The solubility product of mercurous sulfate is 6.2×10^{-7} at 25°, and the cell H_2(735 mm)/H_2SO_4(0.001 M), Hg_2SO_4/Hg has an emf of $+0.8630$ V at 25°. Compute the standard potential of the half-reaction $Hg_2^{2+} + 2e = 2Hg$.

7-10 0.05 mole of potassium iodide and 0.025 mole of iodine are dissolved together in a liter of water at 25°C. By appropriate analysis it is found that only 0.00126 moles of the iodine in the solution has remained in the form of I_2, the remainder having reacted to form I_3^- according to the equation:

$$I_2(aq) + I^- = I_3^-$$

Calculate the equilibrium constant. Check your result on the basis of the following information:

$$I_2(s) + 2e = 2I^-; \qquad p\epsilon^\circ = 9.1$$
$$2I^- = I_2(s) + 2e; \qquad E_H^\circ = -0.5355$$
$$I_3^- + 2e = 3I^-; \qquad p\epsilon^\circ = 9.1$$
$$3I^- = I_3^- + 2e; \qquad E_H^\circ = -0.536$$
$$\Delta G^\circ \text{ for } I_2(aq) = +3.93 \text{ kcal}$$

7-11 What is the solubility of Au(s) in seawater [cf. L. G. Sillén, *Svensk Kem. Tidskr.*, **75**, 161 (1954)]? The following equilibrium constants (18–20°C) may be used:

$$AuCl_2^- + e = Au(s) + 2Cl^-; \qquad \log K = 19.2$$
$$AuCl_4^- + 3e = Au(s) + 4Cl^-; \qquad \log K = 51.3$$
$$Au(OH)_4^- + 4H^+ + 4Cl^- = AuCl_4^- + 4H_2O; \qquad \log K = 29.64$$

7-12 Consider the equilibrium of an aqueous solution containing the solid phases $FeOOH(s)$, $FeCO_3(s)$. Cf. the reaction

$$FeOOH(s) + HCO_3^- + 2H^+ + e = FeCO_3(s) + 2H_2O$$

Does $p\epsilon$ increase, decrease or stay constant upon addition of small quantities of the following: (i) FeOOH(s); (ii) CO_2; (iii) HCl; (iv) EDTA at constant pH; (v) O_2 at constant pH; (vi) NaOH.

ANSWERS TO PROBLEMS

7-1 (a) $Mn^{2+} + Cl_2 + 2H_2O = MnO_2(s) + 2Cl^- + 4H^+$; $Mn^{2+} + ClO^- + H_2O = MnO_2(s) + 2H^+ + Cl^-$; $H_2S + 4OCl^- = SO_4^{2-} + 4Cl^- + 2H^+$; $4MnO_2(s) + H_2S + 6H^+ = SO_4^{2-} + 4Mn^{2+} + 4H_2O$. (b) (iii) ($O_2$–$H_2O$, $p\epsilon \geq 12$); (ii) (NH_4^+–NO_3^-, $p\epsilon \simeq 5$–8); (iv) (10^{-5} M Fe^{2+}–FeOOH(s), $p\epsilon \simeq -2$–0); (i) (HS^-–SO_4^{2-}, $p\epsilon \simeq -2$ to -4); (v) (CH_4–CH_2O–CO_2, $p\epsilon \simeq -3$ to -5).

7-2 (i) $E_H^\circ = +0.885$ V ($p\epsilon^\circ = 15$); (ii) No, $2Mn(III) = Mn(IV) + Mn(II)$, $\log K = 10$.

7-3 (a) For $[Fe^{2+}] \geq 10^{-4}$ M and ($[NO_2^-]/[NO_3^-]$) ≥ 1, pH ≥ 4.3 (consider the reaction $\frac{1}{2}NO_3^- + Fe^{2+} + 2\frac{1}{2}H_2O = \frac{1}{2}NO_3^- + 2H^+ + Fe(OH)_3(s)$; $\log K \simeq -4.6$). (b) $\sim 10^{-11}$.

7-4 (i) No, $\log K(W) \simeq -10$ for reaction $\frac{1}{8}NH_4^+ + \frac{1}{8}SO_4^{2-} = \frac{1}{8}NO_3^- + \frac{1}{8}H^+$ ($[H^+] = 10^{-7}$) $+ \frac{1}{8}H_2O$. (ii) Yes. (iii) For pH of natural waters (4.6 < pH < 9.4), $0 > p\epsilon > -6$.

7-5 (i) 1 atm. (ii) 4.8. (iii) 10^{-40} atm.

7-6 $p_{O_2} < 10^{-70}$ atm.

7-7 (i) Yes. (ii) 50% CO_2, 50% CH_4.

7-8 (i) $H_2(g) = 2H^+ + 2e$; $Pb(s) + SO_4^{2-} = PbSO_4(s) + 2e$;
Pb(s), $PbSO_4$(s), H_2SO_4/H_2SO_4, H_2 (Pt); $E_{cell}^\circ = +0.35$ V.
(ii) AgCl(s) $+ e = Ag(s) + Cl^-$, Ag(s) $= Ag^+ + e$
Ag(s), AgCl(s), $KCl/AgNO_3$, Ag(s), $E_{cell}^\circ = 0.58$ V.

(iii) $MnO_4^- + 2H_2O + 3e = MnO_2(s) + 4OH^-$;
$Mn^{2+} + 4OH^- = MnO_2(s) + 2H_2O + 2e$;
(Pt) $MnO_2(s)$, Mn^{2+}, $NaOH/MnO_4^-$, $MnO_2(s)$ (Pt); $E_{cell}^\circ = 1.3$ V.

(iv) $Cl_2(g) + 2e = 2Cl^-$; $IO_3^- + 6H^+ + 6e = I^- + 3H_2O$
(Pt) $Cl_2(g)$, $NaCl/NaI$, $NaIO_3$ (Pt); $E_{cell}^\circ = 0.46$ V.

7-9   For $Hg^{2+} + 2e = Hg(l)$, $E^\circ = +0.798$ V.

7-10 $K = 10^{2.88}$.

7-11 $\sim 10^{-7.2}$ M.

7-12   No change for (i), (iv) and (v); increase for (ii) and (iii); decrease for (vi).

8   The Regulation of the Chemical Composition of Natural Waters

8-1 Introduction

Natural waters acquire their chemical characteristics by dissolution and by chemical reactions with solids, liquids and gases with which they have come into contact during the various parts of the hydrological cycle. Waters vary in their chemical composition, but these variations are at least partially understandable if the environmental history of the water and the chemical reactions of the rock–water–atmosphere systems are considered. Dissolved mineral matter originates in the crustal materials of the earth; water disintegrates and dissolves mineral rocks by weathering. Gases and volatile substances participate in these processes. As a first approximation, seawater may be interpreted as the result of a gigantic acid–base titration — acid of volcanoes versus bases of rocks (oxides, carbonates, silicates) [1]. The composition of fresh water similarly may be represented as resulting from the interaction of the CO_2 of the atmosphere with mineral rocks.

Table 8-1 surveys the major solutes contained in seawater and in average river water. A recent survey of the frequency distribution of various constituents of terrestrial waters by Davies and DeWiest [2] (see Figure 8-1) makes it apparent that many of the aquatic constituents show little natural variation in their concentrations. For example, according to Figure 8-1, 80% of the water analyses for dissolved silica show concentrations between $10^{-3.8}$ and $10^{-3.2}$ M. The range of $[H^+]$ of naturally occurring bodies of water is generally $10^{-6.5}$ to $10^{-8.5}$ M. The composition of seawater is remarkably constant and the waters of different oceans differ very little in chemical analysis. Perhaps more interesting is the hypothesis supported by geological records that seawater has remained constant in its chemical composition for at least the past 100 million years.

In this chapter an attempt at the following will be made:

1. To discuss the processes involved in the acquisition of chemical substances by natural waters. In this context some of the principles of chemical weathering of rocks and of the formation of soils will be treated.

[1] L. G. Sillén, in *Oceanography*, M. Sears, Ed., American Association for the Advancement of Science, Washington, D.C., 1961.
[2] S. N. Davies and R. C. M. DeWiest, *Hydrogeology*, Wiley, New York, 1966.

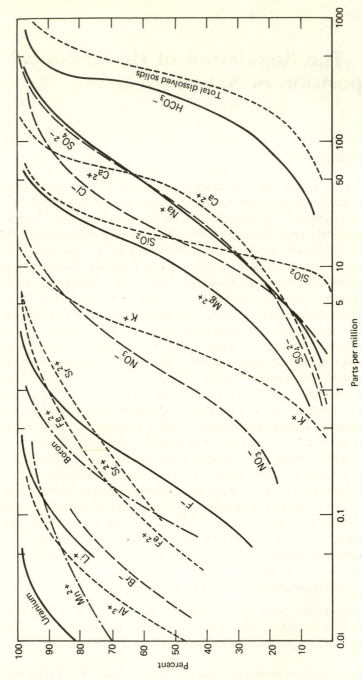

Figure 8-1 Cumulative curves showing the frequency distribution of various constituents in terrestrial water. Data are mostly from the United States from various sources [2]. Reproduced with permission from J. Wiley and Sons.

Table 8-1   Composition of Sea† and Average River Water‡

Ion	Present Ocean	Average River
Na^+, M	0.47	2.7×10^{-4}
K^+, M	1.0×10^{-2}	5.9×10^{-5}
Ca^{2+}, M	1.0×10^{-2}	3.8×10^{-4}
Mg^{2+}, M	5.4×10^{-2}	3.4×10^{-4}
F^-, M	7×10^{-5}	5.3×10^{-6}
Cl^- M	0.55	2.2×10^{-4}
SO_4^{2-}, M	3.8×10^{-2}	1.2×10^{-4}
pH	7.89	—
Alkalinity, eq liter^{-1}	2.3×10^{-3}	9.6×10^{-4}
P_{CO_2}, atm	4×10^{-4}	4×10^{-4}
Total phosphorus, M	1.5×10^{-6}	—
Ionic strength, M	0.65	—

† From H. U. Sverdrup, M. W. Johnson and R. H. Fleming, *The Oceans, Their Physics, Chemistry and General Biology*, Prentice-Hall, New York, 1942.
‡ From D. A. Livingstone, *Chemical Composition of Rivers and Lakes*, U.S. Geol. Survey Paper 440 G, 1963.

2. To give attention to the mechanisms that regulate and control the mineral composition of natural waters. Equilibrium models facilitate identification of the many variables and establish chemical boundary conditions toward which aquatic environments must proceed.

3. To consider the interrelation and interaction of the chemical environment with organisms and to illustrate how biological activity can influence the temporal and spatial distribution of aquatic constituents. In a balanced ecological system a balance between photosynthetic activity and respiratory activity seems to be maintained. In this context stream pollution is interpreted as a departure from this balance.

8-2   The Acquisition of Solutes

In order to better understand how natural waters acquire their chemical composition, the processes involved in weathering of rocks and in formation of soils must be considered. Weathering reactions — essentially caused by the interaction of water and atmosphere with the crust of the earth — take place because the original constituents of the crust, the igneous rocks, are thermodynamically unstable in the presence of water and the atmosphere. The igneous rocks have been formed under physical and chemical conditions entirely different from those presently existing at the earth's surface.

The Cycle of Rocks and the Course of Weathering

All matter on the earth's surface and in the uppermost layers of the lithosphere participate in a slow and rather complicated migration. According to Rankama and Sahama [3] and others the migration of matter may be divided into two parts: *the minor (or exogenic) cycle* and *the major (or endogenic) cycle*. The minor cycle takes place under the direct influence of water and atmospheric agents; it starts from solid crystalline rocks and ends in sedimentary rocks; reactions take place primarily in one direction only. This exogenic cycle forms but part of the major cycle of nature which takes place at deeper levels in the crust. In this endogenic cycle, sedimentary rocks that have been deposited undergo further transformations (metamorphism) into new types of rocks (schist, gneiss, lime-silicate rock, quartzite, marble, graphite) by the action of heat, pressure, and migrating fluids. Parallel with this metamorphism are changes in the level of the earth's crust (folding, faulting, thrusting, formation of mountains). Gradually a rock melt, magma, is produced. Such a magma contains volatile components. Different solid phases (igneous rocks) crystallize out sequentially from magma, producing at the same time gases, which may be released as volcanic emanations and possibly as a water solution. This endogenic cycle is not closed; it receives primary magma and heat from the interior.

Figure 8-2 shows some of the pertinent features of a part of the cycle of matter. The hydrosphere participates in the migration of matter (a) as a *conveyor* of matter in suspended and dissolved form, and (b) as a *reactant* in chemical transformations of matter. As Figure 8-2 illustrates, the atmosphere, whose composition is directly or indirectly influenced by volcanic emanations, participates in these chemical transformations by providing acids (CO_2) and oxidants (O_2).

The compositions of inland waters are related primarily to the following types of reactions of water and atmospheric gases with rock-forming minerals: (a) congruent dissolution reactions, (b) incongruent dissolution reactions and (c) redox reactions. Table 8-2 lists some of the more important rock-forming minerals and Table 8-3 gives some typical examples of weathering reactions. Highly schematic and idealized formulas are used in these tables to represent

[3] K. Rankama and Th. G. Sahama, *Geochemistry*, Univ. Chicago Press, Chicago, Ill., 1950.

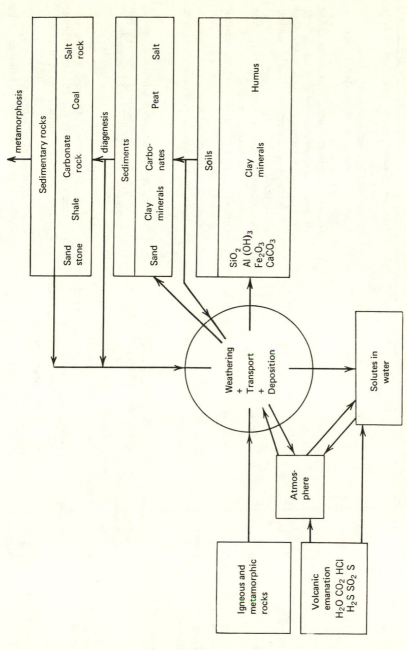

Figure 8-2 Interaction of the cycle of rocks with that of water.

Table 8-2 Important Rockforming Minerals

Silicon oxides	α-Quartz	SiO_2	Trigonal; densely packed arrayment of SiO_2 tetrahedra
	α-Tridymite	SiO_2	Orthorhombic, open structure
	α-Cristobalite	SiO_2	Tetragonal sheets of 6-membered rings of $[SiO_4]$ tetrahedra
	Opal	SiO_2	Hydrous, crypto crystalline form of cristobalite
Aluminum oxides	Corundum	$\alpha\text{-}Al_2O_3$	Oxygen in hexagonal closest packing
	Gibbsite	$Al_2O_3 \cdot 3H_2O$	Monoclinic. A layer of Al ions sandwiched between two sheets of closely packed hydroxide ions
	Boehmite	$\gamma\text{-}AlOOH$	Orthorhombic. Double sheets of octahedra with Al ions at their centers
	Diaspore	$\alpha\text{-}AlOOH$	Orthorhombic Al^{3+} in octahedrally coordinated sites
Two-layer clays† (kaolinites)	Kaolinite	$Al_4[Si_4O_{10}](OH)_8$	Sheet consisting of two layers: (1) SiO_4 tetrahedra in a hexagonal array and (2) layer of Al in 6 coordination
	Halloysite	$Al_4[Si_4O_{10}](OH)_8 \cdot 2H_2O$	
Three-layer minerals	Micas		
	Muscovite	$K_2Al_4[Si_6Al_2O_{20}](OH,F)_4$	A layer of octahedrally coordinated cations (usually Al) is sandwiched between two identical layers of $[(Si,Al)O_4]$ tetrahedra
	Biotite	$K_2(Mg,Fe)_6[Si_6Al_2O_{20}](OH,F)_4$	
	Glauconite		
	Illite	$K_xAl_4[Si_{1-x}Al_xO_{20}](OH)_4$	
Expandable three-layer clays	Montmorillonite	$(Na,K)_{x+y}(Al_{2-x}Mg_x)_2$ $[(Si_{1-y}Al_y)_8O_{20}](OH)_4 \cdot nH_2O$	Octahedral Al on Mg sheets, tetrahedral Si sheets. Al partially replaced by Mg and occasionally by Fe, Cr, Zn. In tetrahedral sheet occasional replacement of Si by Al
	Vermiculite	$(Ca,Mg)(Mg_{3-x}Fe_x)_2$ $[(Si_6Al_2)_8O_{20}](OH)_4 \cdot 8H_2O$	
	Chlorite	$(Mg,Al)_{12}[(Si,Al)_8O_{20}](OH)_{16}$	
Sulfides	Pyrite	FeS_2	Cubic, octahedral coordination of Fe by S
	Marcasite	FeS_2	Orthorhombic, octahedral coordination of Fe by S
	Pyrrhotite	FeS	Monoclinic pseudohexagonal
	Galena	PbS	Cubic

Sulfates	Baryte	$BaSO_4$	Orthorhombic
	Gypsum	$CaSO_4 \cdot 2H_2O$	Monoclinic
	Anhydrite	$CaSO_4$	Orthorhombic; more stable than gypsum above 42°C
Carbonates	Calcite	$CaCO_3$	Trigonal. Mg- and Mn-calcite
	Rhodochrosite	$MnCO_3$	Similar to calcite
	Magnesite	$MgCO_3$	Similar to calcite
	Siderite	$FeCO_3$	Similar to calcite
	Dolomite	$MgCa(CO_3)_2$	One layer calcite combined with one layer magnesite
	Huntite	$Mg_3Ca(CO_3)_4$	
	Aragonite	$CaCO_3$	Orthorhombic
	Strontianite	$SrCO_3$	Similar to aragonite
Iron oxides	Goethite	$\alpha\text{-}FeOOH$	Similar to diaspore
	Lepidocrocite	$\gamma\text{-}FeOOH$	Similar to boehmite
	Limonite	$FeOOH \cdot nH_2O$	Hydrated oxides of iron with poorly crystalline character
	Magnetite	$\alpha\text{-}Fe_2O_3$	Trigonal, occurs in sediments. Spinel type
		Fe_3O_4	8 Fe^{2+} in 4 coordination;
			16 Fe^{3+} in 6 coordination
Titanium oxide	Rutile	TiO_2	Tetragonal; band of octahedra
Magnesium hydroxide	Brucite	$Mg(OH)_2$	Trigonal, two sheets of OH parallel to basal plane with sheet of Mg ions between them
Phosphates	Apatite	$Ca_5(OH,F,Cl)(PO_4)_3$	Hexagonal
	Carbonate–apatite	$Ca_5(PO_4,OH,CO_3)_3(F,OH)$	

† An introduction to the structure of clays is given in the Appendix to this Chapter.

Table 8-3   Examples of Typical Weathering Reactions

I. Congruent Dissolution Reactions

$$SiO_2(s) + 2H_2O = H_4SiO_4 \tag{1}$$
quartz

$$CaCO_3(s) + H_2O = Ca^{2+} + HCO_3^- + OH^- \tag{2}$$
calcite

$$CaCO_3(s) + H_2CO_3^* = Ca^{2+} + 2HCO_3^-$$

$$Al_2O_3 \cdot 3H_2O(s) + 2H_2O = 2Al(OH)_4^- + 2H^+ \tag{3}$$
gibbsite

$$Al_2O_3 \cdot 3H_2O(s) + 6H_2CO_3^* = 2Al^{3+} + 6HCO_3^- + 6H_2O$$

$$Ca_5(PO_4)_3(OH)(s) + 3H_2O = 5Ca^{2+} + 3HPO_4^{2-} + 4OH^- \tag{4}$$
apatite

$$Ca_5(PO_4)_3(OH)(s) + 4H_2CO_3^* = 5Ca^{2+} + 3HPO_4^{2-} + 4HCO_3^- + H_2O$$

II. Incongruent Dissolution Reactions

$$MgCO_3(s) + 2H_2O = HCO_3^- + Mg(OH)_2(s) + H^+ \tag{5}$$
magnesite   brucite

$$Al_2Si_2O_5(OH)_4(s) + 5H_2O = 2H_4SiO_4 + Al_2O_3 \cdot 3H_2O(s) \tag{6}$$
kaolinite   gibbsite

$$NaAlSi_3O_8(s) + \tfrac{11}{2}H_2O = Na^+ + OH^- + 2H_4SiO_4 + \tfrac{1}{2}Al_2Si_2O_5(OH)_4(s)$$
albite   kaolinite   (7)

$$NaAlSi_3O_8(s) + H_2CO_3^* + \tfrac{9}{2}H_2O = Na^+ + HCO_3^- + 2H_4SiO_4 + \tfrac{1}{2}Al_2Si_2O_5(OH)_4(s)$$

$$CaAl_2Si_2O_8(s) + 3H_2O = Ca^{2+} + 2OH^- + Al_2Si_2O_5(OH)_4(s)$$
anorthite   kaolinite   (8)

$$CaAl_2Si_2O_8(s) + 2H_2CO_3^* + H_2O = Ca^{2+} + 2HCO_3^- + Al_2Si_2O_5(OH)_4(s)$$

$$4Na_{0.5}Ca_{0.5}Al_{1.5}Si_{2.5}O_8 + 6H_2CO_3^* + 11H_2O$$
plagioclase (andesine)

$$= 2Na^+ + 2Ca^{2+} + 4H_4SiO_4 + 6HCO_3^- + 3Al_2Si_2O_5(OH)_4(s) \quad (9)$$
kaolinite

$$3KAlSi_3O_8(s) + 2H_2CO_3^* + 12H_2O$$
K-feldspar (orthoclase)

$$= 2K^+ + 2HCO_3^- + 6H_4SiO_4 + KAl_3Si_3O_{10}(OH)_2(s) \quad (10)$$
mica

$$7NaAlSi_3O_8(s) + 6H^+ + 20H_2O$$
albite

$$= 6Na^+ + 10H_4SiO_4 + 3Na_{0.33}Al_{2.33}Si_{3.67}O_{10}(OH)_2(s) \quad (11)$$
Na$^+$-montmorillonite

$$KMg_3AlSi_3O_{10}(OH)_2(s) + 7H_2CO_3^* + \tfrac{1}{2}H_2O$$
biotite

$$= K^+ + 3Mg^{2+} + 7HCO_3^- + 2H_4SiO_4 + \tfrac{1}{2}Al_2Si_2O_5(OH)_4(s) \quad (12)$$
kaolinite

$$Ca_5(PO_4)_3F(s) + H_2O = Ca_5(PO_4)_3(OH)(s) + F^- + H^+ \tag{13}$$
fluoroapatite   hydroxyapatite

$$KAlSi_3O_8(s) + Na^+ = K^+ + NaAlSi_3O_8(s) \tag{14}$$
orthoclase   albite

$$CaMg(CO_3)_2(s) + Ca^{2+} = Mg^{2+} + 2CaCO_3(s) \tag{15}$$
dolomite   calcite

Table 8.3 Continued

III. Redox Reactions

$$MnS(s) + 4H_2O = Mn^{2+} + SO_4^{2-} + 8H^+ + 8e \qquad (16)$$

$$3Fe_2O_3(s) + H_2O + 2e = 2Fe_3O_4(s) + 2OH^- \qquad (17)$$
hematite   magnetite

$$FeS_2(s) + 3\tfrac{3}{4}O_2 + 3\tfrac{1}{2}H_2O = Fe(OH)_3(s) + 4H^+ + 2SO_4^{2-} \qquad (18)$$
pyrite

$$PbS(s) + 4Mn_3O_4(s) + 12H_2O = Pb^{2+} + SO_4^{2-} + 12Mn^{2+} + 24OH^- \qquad (19)$$
galena

the complex mineral phases that occur in nature; furthermore, some of the reactions listed may go through intermediates not shown in the stoichiometric equations.

Weathering of Aluminum-Silicates

Most important among the weathering reactions is the incongruent dissolution of aluminum silicates (reactions of Table 8-3) which may schematically be represented by

$$\text{cation Al-silicate} + H_2CO_3^* + H_2O$$
$$= +HCO_3^- + H_4SiO_4 + \text{cation} + \text{Al-silicate(s)} \qquad (1)$$

Essentially, a primary mineral is converted into a secondary mineral. The secondary minerals are frequently structurally ill-defined or X-ray amorphous. The structural breakdown of aluminum silicates is accompanied by a release of cations and usually of silicic acid. As a result of such reactions alkalinity is imparted to the dissolved phase from the bases of the minerals. In most silicate phases Al is conserved during the reaction, the solid residue being higher in Al than the original silicates. Because the alkalinity of the solution increases during the weathering process, the solid residue has a higher acidity than the original aluminum silicate. (Concerning the structure of clays, see the Appendix to this chapter.)

Because $H_2CO_3^*$ is usually the proton donor in the attack of water on the primary silicates, HCO_3^- is the predominant anion in most fresh waters. The gas composition in soils is usually quite different from that in the atmosphere. Because of respiration by organisms, the CO_2 composition in soils is up to a few hundred times larger than that in the atmosphere. Correspondingly, ground waters tend to contain higher concentrations of HCO_3^- and other solutes. Minerals of the kaolinite group are the main alteration products of weathering of feldspar. Besides kaolinites, montmorillonites and micas are possible products or intermediates. Mica (illite) has been identified as an intermediate in the decomposition of K-feldspar (orthoclase).

The dissolution of feldspars may go through intermediates. Virtually all bonds must be broken in the tetrahedric Al framework structure before rearrangement of the lattice with Al in 6 coordination (as in kaolinite) becomes possible. Understandably, the rate of reaction is very slow. Physical weathering processes, that is, processes which alter by physical or mechanical means the size and hence the specific surface area of the minerals, may enhance reaction rates because the weathering processes take place only at the rock–water interface.

Water composition, obviously, is influenced by these weathering reactions. While Ca^{2+}, HCO_3^-, H^+ and perhaps Mg^{2+} may be controlled by dissolution of carbonate rocks, the source of Na^+, K^+ and H_4SiO_4, and possibly also of Ca^{2+} and Mg^{2+}, is the silicate minerals that make up 70% or more of the rocks in contact with underground waters and streams [4].

Example 8-1 *Synthesizing Natural Waters Stoichiometrically.* 1. Estimate the gross composition of a water that results from the reaction of 1 mmole of CO_2 upon 1 liter of water in contact with

(i) calcite

(ii) anorthite

(iii) andesine

(iv) biotite

Considering the stoichiometry of the equations given in Table 8-3, the resulting compositions are:

(i) pCa = 3.0, pHCO$_3$ = 2.7

(ii) pCa = 3.3, pHCO$_3$ = 3.0

(iii) pNa = 3.48, pCa = 3.48, pHCO$_3$ = 3.0, pH$_4$SiO$_4$ = 3.18

(iv) pK = 3.85, pMg = 3.37, pHCO$_3$ = 3.0, pH$_4$SiO$_4$ = 3.54

Note that the weathering of anorthite leads to solutions having the same mole ratio of $[Ca^{2+}]/[HCO_3^-]$ as do waters resulting from the dissolution of calcite.

2. Estimate the gross composition of a spring water that results from the weathering of the following quantities of source minerals per liter of water: 0.2×10^{-4} moles K-feldspar, 0.15×10^{-4} moles biotite and 1.2×10^{-4} moles andesine. Further, 0.40 moles NaCl and 0.26 moles Na_2SO_4 as atmospheric salt particles from the sea have become dissolved per liter of water.

[4] O. P. Bricker and R. M. Garrels, in *Principles and Applications of Water Chemistry*, S. D. Faust and J. V. Hunter, Eds., Wiley, New York, 1967.

The composition of the resulting water is:

| | Concentration $M \times 10^4$ | | | | | | | |
Reaction	Na^+	Ca^{2+}	Mg^{2+}	K^+	HCO_3^-	SO_4^{2-}	Cl^-	H_4SiO_4
K-feldspar → kaolinite				0.20	0.20			0.40
Biotite → kaolinite			0.45	0.15	1.05			0.30
Andesine → kaolinite	0.60	0.60			1.8			1.2
NaCl, Na₂SO₄	0.92					0.26	0.40	
Spring water	1.52	0.60	0.45	0.35	3.05	0.26	0.40	1.9

Complex Formation, Ion Exchange and Redox Processes

The dissolution processes may become modified by complex formation or by changes in the oxidation state. Table 8-3, part III, illustrates that naturally occurring redox processes can have a pronounced pH controlling action. For example, the oxidation of pyrite yields ferric oxide and acid [4 moles of acidity per mole of pyrite; (18) of Table 8-3]. The acid in turn may then react with rocks. Alternatively, the reduction of metal oxides such as Mn_3O_4 and Fe_2O_3 may produce large quantities of OH^- ions.

An incongruent dissolution reaction consists essentially of a ligand exchange reaction. Thus the conversion of feldspar into kaolinite can be visualized as a hydrolysis reaction, that is, a partial exchange of silica for OH^-. Similarly in a fluoro-apatite, the F^- may become exchanged for OH^-, even some of the phosphate groups in the apatite might under suitable conditions become replaced by CO_3^{2-}. The exchange of cations, in an analogous way, reflects a change in the coordinative relations. Typically clay minerals have the ability to exchange interlayer and surface cations with cations from the solution, and these processes are also of great importance in modifying the solution composition of natural waters. Such ion exchange reactions usually proceed without modifying the gross structure of the aluminum silicates but, mechanistically, an ion exchange reaction may precede the breakdown of silicate minerals; for example, Frederickson [5] explained the effect of pH on the mineralization of water by postulating as a first step in the rock dissolution a penetration of H^+ into the lattice of the mineral and displacement of a metal ion cation by H^+. Subsequently, as the bonds become weakened, silicate groups pass into solution as silicic acid. As the crystals expand, progressively more surface area is exposed.

The dislocation of structural constituents of the minerals depends in a more general way upon the prevailing environment. Complex formers may preferentially "leach" out certain lattice constituents and thus enhance the

[5] P. F. Frederickson, *Bull. Geol. Soc. Amer.*, **62**, 221 (1951).

rock disintegration. Naturally-occurring organic matter may act as chelate complex formers. The influence of organisms and plants on the weathering reaction may be interpreted to result (in addition to pH variation) from the effect of organic chelates released by the cells. Under the influence of complex formation, for example, the formation of soluble AlY, an otherwise incongruent dissolution may become congruent

$$\underset{\text{kaolinite}}{Al_2Si_2O_5(OH)_4} + 2Y^- = \underset{\text{soluble}}{2AlY} + 2H_4SiO_4 + H_2O \qquad (2)$$

In general, increasing acidity, increasing $p\epsilon$ and increasing tendency to complex formation favor weathering intensity.

8-3 Solubility of Silica and Aluminum Silicates

The earth's surface may be considered a huge chemical laboratory in which a quantitative rock analysis is being made [3]. In this rock analysis, products are formed incipiently by congruent and incongruent dissolution of the rocks; the products then are subjected to chemical separations and redepositions (sedimentation) in new surroundings followed by aging, recrystallization and formation of new minerals (diagenesis into sedimentary rocks).

Evaluation of the solubility of minerals combined with field observations contribute to a better understanding on the functional interrelationship between minerals and their environment. Over-all weathering reactions, of course, are characterized by a decrease in free energy. The latter may be relatively small; repeated leaching with fresh water tends to decrease the reaction quotient, Q and hence $\Delta G = RT \ln (Q/K)$, that is, to "drive" the weathering reaction to the right. The thermodynamic weathering sequence could be specified if the stability relations of the various rocks could be defined reliably. Unfortunately, information on the free energy of formation of aluminum silicates is insufficient and thermochemical data for many minerals of significance are not known with any precision; conflicting values have been reported by various authors. Thermochemical data for aluminum silicate minerals at low temperature and pressure may be obtained by extrapolation from investigations on mineralogical equilibria at elevated temperatures and pressures [6, 7] or from tedious low-temperature solubility studies [8]. The uncertainties resulting from the difficulties involved together with our ignorance on the exact composition of the solid phases (e.g., extent of solid solution formation) make the establishment of a detailed thermodynamic

[6] J. J. Hemley, *Amer. J. Sci.*, **257**, 241 (1959).
[7] J. J. Hemley and W. R. Jones, *Econ. Geol.*, **59**, 538 (1964).
[8] W. L. Polzer and J. D. Hem, *J. Geophys. Res.*, **70**, 6233 (1965).

weathering sequence uncertain. By using tentative values of equilibrium constituents, the principles of chemical behavior of rocks in water can be circumscribed and stability diagrams that may help in understanding weathering transformations can be constructed. Initially, in imaginary experiments, individual minerals are added to distilled water and exposed to air containing CO_2. In the next section stability relations, that is, predominance diagrams, valid for equilibrium among general minerals will be developed.

The Chemistry of Aqueous Silica

The solubility of SiO_2 can be characterized by the following equilibria:

$$SiO_2(s, \text{quartz}) + 2H_2O = Si(OH)_4; \qquad \log K = -3.7 \qquad (3)$$

$$SiO_2(s, \text{amorphous}) + 2H_2O = Si(OH)_4; \qquad \log K = -2.7 \qquad (4)$$

$$Si(OH)_4 = SiO(OH)_3^- + H^+; \qquad \log K = -9.46 \qquad (5)$$

$$SiO(OH)_3^- = SiO_2(OH)_2^{2-} + H^+; \qquad \log K = -12.56 \quad (6)$$

$$4Si(OH)_4 = Si_4O_6(OH)_6^{2-} + 2H^+ + 4H_2O; \quad \log K = -12.57 \quad (7)$$

The equilibrium constants given are valid at 25°C. Data for (3)–(5) are those given by Lagerstrom [9] as valid for 0.5 M $NaClO_4$.

In these equations silicic acid has been written as $Si(OH)_4$ (rather than as H_4SiO_4) in order to emphasize that metalloids (Si, B, Ge) similar to metal ions have a tendency to coordinate with hydroxo and oxo ligands. Like multivalent metal ions, such metalloids tend to form multinuclear species.

The rate of crystallization of quartz is so slow in the low-temperature range that the solubility of amorphous silica represents the upper limit of dissolved aqueous silica (Siever [10]), and we first consider the solubility of amorphous silica. The equilibrium data of (2)–(5) permit computation of the solubility of amorphous SiO_2 for the entire pH range and the relative concentration of the species in equilibrium with amorphous SiO_2 (Figure 8-3). Only $Si(OH)_4$ occurs within neutral and slightly alkaline pH ranges. Under alkaline conditions the solubility of SiO_2 becomes enhanced because of the formation of monomeric and multimeric silicates. Although there is some uncertainty concerning the exact nature of the multimeric species, the experimental data of Lagerstrom [9] and Ingri [11] leave no doubt about the existence of multinuclear species. Even if multinuclear species other than $Si_4O_2(OH)_6^{2-}$ were present, the solubility characteristics of SiO_2 would not be changed markedly.

[9] G. Lagerstrom, *Acta Chem. Scand.*, **13**, 722 (1959).
[10] R. Siever, *Amer. Mineral.*, **42**, 826 (1957).
[11] N. Ingri, *Acta Chem. Scand.*, **13**, 758 (1959).

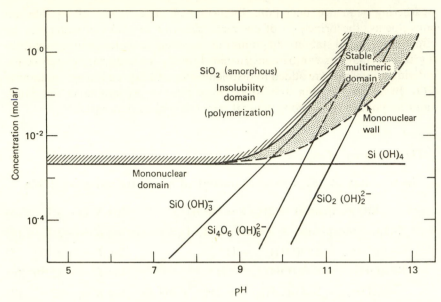

Figure 8-3 Species in equilibrium with amorphous silica. Diagram computed from equilibrium constants (25°C.) The line surrounding the shaded area gives the maximum soluble silica. The mononuclear wall represents the lower concentration limit, below which multinuclear silica species are not stable. In natural waters the dissolved silica is present as monomeric silicic acid.

Also indicated in Figure 8-3 is the "mononuclear" wall — that is, the concentration of total silica that must be exceeded to encounter multi- or polynuclear species in the solution. The mononuclear wall follows the line characterizing $[Si(OH)_4]$ up to a pH of approximately 9; the concentration of mononuclear Si then increases with increasing pH. The line has been calculated for the condition $100[Si]_{(multimeric)} = [Si]_{(monomeric)}$.

Three domains in the concentration–pH diagram are of importance: (a) the insolubility domain — that is, the range where amorphous SiO_2 precipitates —, (b) the multimeric domain where silicon polyanions are stable and (c) the monomeric domain where mononuclear Si species $[Si(OH)_4, SiO(OH)_3^-$ and $SiO_2(OH)_2^{2-}]$ prevail thermodynamically.

Natural waters are in the monomeric range, the dissolved silica being predominantly (below pH 9) present as silicic acid. Commercially available concentrated silicate solutions are in the multimeric domain. In this range silicate glass (produced in the soda ash fusion process) or silicate powders within appropriate ratios of SiO_2 to alkali dissolve when brought into contact with water. Solutions that are oversaturated with respect to amorphous silica are unstable and a precipitate will eventually be formed.

The Solubility of Aluminum Silicates

Example 8-2   *Solubility of Kaolinite.*   Calculate the solubility of kaolinite, $Al_2Si_2O_5(OH)_4(s)$, in pure water. Construct a predominance diagram in the system Al_2O_3–SiO_2–H_2O for the solid phases gibbsite ($Al_2O_3 \cdot 3H_2O$), amorphous SiO_2 and kaolinite.

The following information is available

$$Al_2Si_2O_5(OH)_4(s) + 5H_2O = Al_2O_3 \cdot 3H_2O(s)$$
$$+ 2H_4SiO_4; \quad \log K = -9.4 \tag{i}$$

$$\tfrac{1}{2}Al_2O_3 \cdot 3H_2O(s) = Al^{3+} + 3OH^-; \qquad\qquad \log K = -34.0 \tag{ii}$$

$$\tfrac{1}{2}Al_2O_3 \cdot 3H_2O(s) + OH^- = Al(OH)_4^-; \qquad \log K = -1.0 \tag{iii}$$

$$SiO_2(amorph) + 2H_2O = H_4SiO_4; \qquad\qquad \log K = -2.7 \tag{iv}$$

[The equilibrium constant of (i) is from R. M. Garrels, *Amer. Mineral.*, **42**, 789 (1957).]

Furthermore, we can combine (i) and (ii) as well as (i) and (iii) in order to obtain equilibria for (v) and (vi):

$$\tfrac{1}{2}Al_2Si_2O_5(OH)_4(s) + 2\tfrac{1}{2}H_2O = Al^{3+} + H_4SiO_4 + 3OH^-;$$
$$\log K = -38.7 \tag{v}$$

$$\tfrac{1}{2}Al_2Si_2O_5(OH)_4(s) + 2\tfrac{1}{2}H_2O + OH^- = Al(OH)_4^- + H_4SiO_4;$$
$$\log K = -5.7 \tag{vi}$$

If kaolinite is introduced into pure H_2O, it dissolves incongruently [reaction (i)]. Its solubility is very small; the electroneutrality of the solution will essentially be governed by $[H^+] \simeq [OH^-]$, or pH $\approx$ 7. While the existence of $Al(OH)_4^-$ and its formation constant have been well established, there is a great deal of uncertainty about the composition and stability of other aluminum hydrolysis species. Because these polynuclear hydroxo-complexes do not materially alter the solubility characteristics of gibbsite and kaolinite, we carry out our calculations by considering $Al(OH)_4^-$ and Al^{3+} as the predominant equilibrium species.

We may first consider congruent solubility characteristics for the three solid phases, gibbsite, $SiO_2(amorph)$ and kaolinite. The solubility of SiO_2 has been described in Figure 8-3. Figures 8-4a and b show the pH dependence of the solubility of gibbsite and kaolinite, respectively. For the calculation of the kaolinite solubility, a fixed value of $[H_4SiO_4]$ ($= 10^{-2.7}$ M) has been assumed.

With the information at hand we proceed to construct a predominance diagram (Figure 8-4c) valid for the dissolution of kaolinite in pure water (pH $\simeq$ 7). In this diagram the ordinate and abscissa represent molar concentrations of the "base" $Al(OH)_3$ and of the acid $[H_4SiO_4]$, respectively, which make up the "salt" kaolinite.

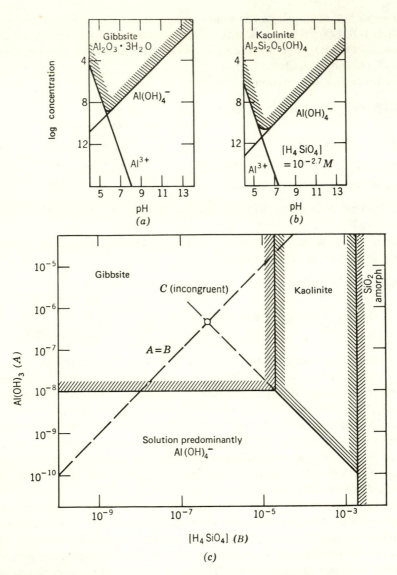

Figure 8-4   Incongruent dissolution of kaolinite (Example 8-2).
  a: Solubility of gibbsite.
  b: Hypothetical (component) solubility of kaolinite for $[H_4SiO_4] = 10^{-2.7}\,M$.
  c: Predominance diagram for the dissolution of kaolinite in pure water (pH = 7).

For our conditions of pH $\simeq 7$ we obtain the maximum solubility of gibbsite at $[Al(OH)_4{}^-] \simeq 10^{-8}\ M$ (A axis in Figure 8-4c). The maximum solubility of amorphous SiO_2 is given by (iv) as $[H_4SiO_4] = 10^{-2.7}\ M$. The transition between gibbsite and kaolinite is defined by (i) as $[H_4SiO_4] = 10^{-4.7}$. The boundary between kaolinite and the solution phase is given by (vi):

$$\log \frac{[Al(OH)_4{}^-][H_4SiO_4]}{[OH^-]} = -5.7$$

As this diagram suggests, the hypothetical maximum solubility of kaolinite at pH = 7 falls into the stability region of gibbsite at C(incongruent) $\simeq 10^{-6.5}\ M$ (this point must be at the $A = B$ axis). Hence, kaolinite dissolves incongruently according to (i) at equilibrium and coexists with gibbsite; $Al_T \simeq [H_4SiO_4] = 10^{-4.7}\ M$.

As Figure 8-4 suggests, kaolinite converts into gibbsite if, as for example, through extensive leaching, $[H_4SiO_4]$ is kept below $10^{-4.7}\ M$.

Example 8-3   *Solubility of Albite.*   Estimate the dependence of the solubility of $NaAlSi_3O_8$ on the partial pressure of CO_2. Assume the following equilibrium constant (25°C)

$$NaAlSi_3O_8(s) + H^+ + 4\tfrac{1}{2}H_2O$$
albite

$$= Na^+ + 2H_4SiO_4 + \tfrac{1}{2}Al_2Si_2O_5(OH)_4(s); \quad \log K = -1.9 \quad \text{(i)}$$
kaolinite

Reaction (i) may be combined with the protolysis reaction of $CO_2(g)$

$$CO_2(g) + H_2O = HCO_3{}^- + H^+; \quad \log K = -7.8 \qquad \text{(ii)}$$

to give

albite(s) + $CO_2(g)$ + $5\tfrac{1}{2}H_2O$
$$= Na^+ + HCO_3{}^- + 2H_4SiO_4 + \tfrac{1}{2}\ \text{kaolinite(s)}; \quad \log K = -9.7 \quad \text{(iii)}$$

If hypothetically pure albite dissolves under the influence of CO_2 according to (iii), the resulting solution is characterized by the solutes $HCO_3{}^-$, Na^+, H_4SiO_4 and H^+. Because the solutions are near neutrality, $CO_3{}^{2-}$ and silicates need not be considered. The equilibrium composition as a function of p_{CO_2} can be computed because in addition to the constants defining the equilibria of (ii) and (iii) the charge balance

$$[Na^+] \simeq [HCO_3{}^-] \qquad \text{(iv)}$$

and the stoichiometric relation

$$[Na^+] = \tfrac{1}{2}[H_4SiO_4] \qquad \text{(v)}$$

are known. Combining (iv) and (v) with the equilibrium relation of (iii), (vi) gives

$$\frac{4[HCO_3^-]^4}{p_{CO_2}} \simeq 10^{-9.7} \qquad \text{(vi)}$$

Thus, for example, for a $p_{CO_2} = 10^{-2}$, $pH_4SiO_4 = 2.78$. Then $[H^+]$ can be calculated with the equilibrium of (iii), and pH = 6.72 is obtained.

The resulting equilibrium composition is in accord with the assumptions made in (iv) and (v) and the assumed presence of kaolinite as a stable phase. In Figure 8-5 the solubility of albite, expressed as $[HCO_3^-]$, is plotted as a function of p_{CO_2}.

Weathering Sequence and Solubility

With the help of equilibrium constants given in Table 8-4, the equilibrium solubility of a few individual minerals is considered in Figure 8-5. The left

Table 8-4   Equilibrium Constants Used to Establish Stability Relations Among Minerals †

	log K, 25°C, 1 atm
(1) Na-Feldspar(s) + H$^+$ + 4½H$_2$O = ½ Kaolinite(s) + 2H$_4$SiO$_4$ + Na$^+$	-1.9
(2) 3Na-Montmorillonite(s) + 11½H$_2$O = 3½ Kaolinite(s) + 4H$_4$SiO$_4$ + Na$^+$	-9.1
(3) Ca-Feldspar(s) + 2H$^+$ + H$_2$O = Kaolinite(s) + Ca^{2+}	$+14.4$
(4) 3Ca-Montmorillonite(s) + 2H$^+$ + 23H$_2$O = 7 Kaolinite(s) + 8H$_4$SiO$_4$ + Ca^{2+}	-15.4
(5) Kaolinite(s) + 5H$_2$O = 2 Gibbsite(s) + 2H$_4$SiO$_4$	-9.4‡
(6) SiO$_2$(amorph) + 2H$_2$O = H$_4$SiO$_4$	-2.7
(7) CaCO$_3$(s)(calcite) = Ca^{2+} + CO$_3^{2-}$	-8.3
(8) HCO$_3^-$ = H$^+$ + CO$_3^{2-}$	-10.3
(9) CO$_2$(g) + H$_2$O = H$^+$ + HCO$_3^-$	-7.8
(10) CO$_2$(g) + H$_2$O = H$_2$CO$_3^*$	-1.5

† Ca-Feldspar (Anorthite) = CaAl$_2$Si$_2$O$_8$
  Na-Feldspar (Albite)   = NaAlSi$_3$O$_8$
  Na-Montmorillonite   = Na$_{0.33}$Al$_{2.33}$Si$_{3.67}$O$_{10}$(OH)$_2$
  Ca-Montmorillonite   = Ca$_{0.33}$Al$_{4.67}$Si$_{7.33}$O$_{20}$(OH)$_4$
  Kaolinite   = Al$_2$Si$_2$O$_5$(OH)$_4$
  Gibbsite   = Al$_2$O$_3 \cdot$3H$_2$O
  H$_2$CO$_3^*$ = CO$_2$(aq) + H$_2$CO$_3$

‡ Cf. R. M. Garrels, *Amer. Mineralogist*, **42**, 789 (1957).

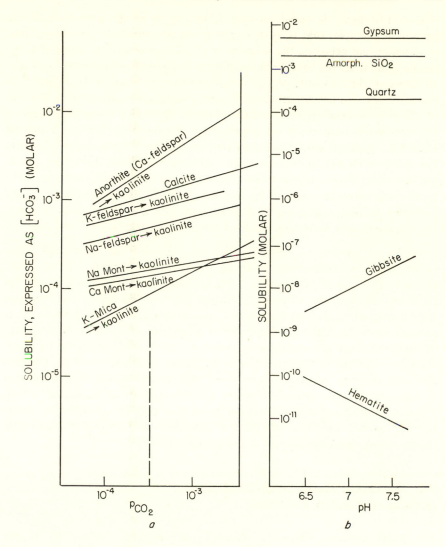

Figure 8-5   Solubility of minerals.

a: p_{CO_2}-dependent solubility of "pure" minerals. Equilibrium $[HCO_3^-]$ in reactions such as

$$\text{albite(s)} + CO_2(g) + 5\tfrac{1}{2}H_2O = Na^+ + HCO_3^- + 2H_4SiO_4 + \tfrac{1}{2}\text{kaolinite(s)}$$

or

$$\text{calcite(s)} + CO_2(g) + H_2O = Ca^{2+} + 2HCO_3^- ;$$

is used to express tendency for dissolution (see Example 8-3).

b: Congruent solubility of some minerals in the neutral pH range.

part of this figure is based on calculations such as the one given in the preceding example; that is, systems are made by introducing a mineral into pure water and exposing the solution to a selected partial pressure of CO_2; the individual curves have been calculated by assuming thermodynamic coexistence of the two solid phases specified for each reaction. In Figure 8-5*b* the solubility of a few congruently soluble minerals is indicated as a function of pH in the neutral region. The data given in Figure 8-5 suggest that following thermodynamic order of succession of minerals in the weathering sequence: gypsum, calcite, Ca-feldspar, K-feldspar, Na-feldspar, Ca- and Na-montmorillonite, quartz, K-mica, gibbsite, kaolinite, hematite. Petrographic observations appear to be consistent with this stability sequence. The weathering sequence proposed by Jackson et al. [12] is very nearly in accord with this stability sequence; in their sequence dark minerals such as hornblende, augite and biotite are grouped between calcite and the feldspars.

MOBILITY OF INDIVIDUAL ELEMENTS. Relative mobility has been used as an index of the final redistribution during the alteration of rock into soil. Most consistent in these considerations is the evidence that Ca^{2+} and Na^+ are the most mobile elements, whereas aluminum and iron belong to the group of least mobile elements, and magnesium, manganese and silicon show an intermediate mobility. Although this order of relative mobility could be predicted qualitatively from the coordinating tendency of these elements (or from the stability constants of the respective solid phases), knowledge of the thermodynamic stability relations alone does not always suffice to understand thoroughly the geochemical behavior and the manner of occurrence of an element. Kinetic and crystal chemical properties must be considered in order to interpret more quantitatively the incorporation and distribution of elements in the various solid phases and in the solution phase.

8-4   Equilibrium Models for the Establishment of Chemical Boundary Conditions

The water chemical literature abounds with detailed discussions on the solubility equilibrium of $CaCO_3$. Indeed, equilibria between calcite and other carbonate minerals play a significant role in many natural waters in controlling the solution composition. Waters in contact with carbonate minerals attain equilibrium relatively fast. However, more than 90% of the earth's crust consists of silicate minerals and, as has been illustrated, these minerals also release soluble weathering products to the waters. In order to establish

[12] M. L. Jackson, S. A. Tyler, A. C. Willis, G. A. Burbeau and R. P. Pennington, *J. Phys. Colloid Chem.*, **52**, 1237 (1948).

the boundary conditions toward which natural water systems will proceed, we will make calculations based on equilibrium of the waters with the various solid phases; conveniently the equilibrium relations can be represented in the form of a predominance diagram. The use of such diagrams as an approach to the understanding of water–rock relationships has been demonstrated by many authors, but special credit should be given to Garrels. The Gibbs phase rule is the basis for organizing and interpreting such models. The main objective of this discussion is to illustrate some of the essential regulatory factors that control the mineral composition of natural waters. It can be shown with the help of such models that, in principle, the CO_2 pressure of the atmosphere may primarily be regulated at the sea–sediment interface by reactions in which various aluminum silicates and $CaCO_3$ participate.

Stability Relations

Predominance diagrams have been calculated with the help of the equilibrium constants given in Table 8-4, and they are shown in Figure 8-6. The method of graphical representation is identical to that described by Garrels and Christ [13] and Feth, Roberson and Polzer [14]. For Figures a and b, however, equilibrium constants or solids with mineral formulas different from those used by Feth et al. and Garrels and Christ have been used. By choosing different constants and formulas, remarkably different predominance diagrams are obtained. It is important to realize that such stability diagrams represent estimates based on limited data or even tentatively assumed data; the objective is to gain qualitative or semiquantitative impressions of possible actual situations [15].

The Phase Rule for Organizing Equilibrium Models

A few simple equilibrium systems are considered in Table 8-5. They represent simple models for natural waters and, as illustrated before (Section 5-7), are constructed by incorporating the specific components into a closed system and by specifying the phases to be included. Recall that the phase rule restricts the number of independent variables F to which we can assign values according to the number of components C and phases P:

$$F = C + 2 - P \qquad (8)$$

[13] R. M. Garrels and C. L. Christ, *Solutions, Minerals, and Equilibria*, Harper and Row, New York, 1965.
[14] J. H. Feth, C. E. Roberson and W. L. Polzer, *U.S. Geol. Surv. Water Supply Papers*, **1535-I** (1964).
[15] W. Stumm and E. Stumm-Zollinger, *Chimia*, **22**, 325 (1968).

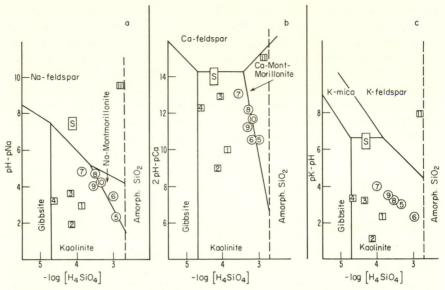

Figure 8-6 Predominance diagrams illustrating the stability relations of some mineral phases according to the equilibrium constants given in Table 8-4. In a typical weathering reaction, feldspars are converted (accompanied by increase of $[Na^+]/[H^+]$, $[Ca^{2+}]/[H^+]^2$ or $[K^+]/[H^+]$) into kaolinite. Progressive accumulation of H_4SiO_4 can lead to conversion of kaolinite into montmorillonite. For comparison, squares and circles represent analyses of Table 8-6 for surface and ground waters, respectively; S represents seawater.

For comparison Table 8-5a lists CO_2 and $CaCO_3$ equilibrium systems discussed before (Section 5-7). In the $CaCO_3$ solubility model (No. 3, Table 8-5a) a closed system containing the phases calcite, aqueous solution and gas (CO_2) can be constructed with the components H_2O, CO_2 and CaO. The system can be described with two independent variables such as temperature and pressure.

In Table 8-5b models containing the same five components — SiO_2, CaO, CO_2, H_2O, Al_2O_3 — but with different numbers of phases are compared with each other; an increase in P must be accompanied by a decrease in F. For model 4, in addition to temperature and p_{CO_2}, a concentration condition must be specified in order to define the equilibrium composition. On the other hand, the composition of model 5 can be calculated with two independent variables (e.g., temperature and p_{CO_2}).

Example 8-4 *Equilibrium Composition in the System H_2O–CO_2–CaO–Al_2O_3–SiO_2.* Compute the equilibrium composition (25°C) of a model system containing the phases Ca-montmorillonite, calcite, kaolinite, aqueous solution and a gas phase with $p_{CO_2} = 10^{-3.5}$ atm (model 5, Table 8-5).

The solutes are interrelated by the electroneutrality condition

$$2[Ca^{2+}] + [H^+] = [HCO_3^-] + 2[CO_3^{2-}] + [OH^-] \qquad (i)$$

Table 8-5 Equilibrium Models; Application of Phase Rule

	a: CO_2 and $CaCO_3$ Solubility Models			b: Aluminum Silicates and $CaCO_3$		
	1	2	3	4	5	6
Phases	Aqueous Solution $CO_2(g)$	Aqueous Solution Calcite(s)†	Aqueous Solution $CO_2(g)$ Calcite(s)	Aqueous Solution $CO_2(g)$ Kaolinite Ca-montmorillonite	Aqueous Solution $CO_2(g)$ Kaolinite Ca-montmorillonite Calcite	Aqueous Solution $CO_2(g)$ Kaolinite Ca-montmorillonite Calcite Ca-feldspar
P	2	2	3	4	5	6
Components	H_2O, CO_2	H_2O, CO_2, CaO	H_2O, CO_2, CaO	H_2O, CO_2, CaO, Al_2O_3, SiO_2		
C	2	3	3	5	5	5
F	2	3	2	3	2	1
Variables‡	$t = 25°C$ $-\log p_{CO_2} = 3.5$	$t = 25°C$ $-\log p = 0$† $[Ca^{2+}] = C_T$§	$t = 25°C$ $-\log p_{CO_2} = 3.5$	$t = 25°C$ $-\log p_{CO_2} = 3.5$ $8[Ca^{2+}] = [H_4SiO_4]$§	$t = 25°C$ $-\log p_{CO_2} = 3.5$	$t = 25°C$
Composition						
pH	5.7	9.9∥	8.3	7.4	8.3	9.0
$pHCO_3$	5.7	4.1	3.0	3.9	3.0	3.4
pCa		3.9	3.3	4.2	3.3	3.7
pH_4SiO_4				3.2	3.6	3.7
						$-\log p_{CO_2} = 4.5$

† $H_2CO_3^*$ is treated as a nonvolatile acid. The system is under a total pressure of 1 atm.

‡ By specifying p_{CO_2}, the total pressure P is determined ($P = p_{CO_2} + p_{H_2O}$). For the calculation, constants valid at $P = 1$ atm were used.

§ This additional constraint is necessary for defining the system; other conditions could be specified.

∥ $pCO_3 = 4.4$.

we are justified in neglecting the protolysis of H_4SiO_4 in formulating (i). In (i), $[Ca^{2+}]$ as well as $[HCO_3^-]$ and $[CO_3^{2-}]$ can be expressed as a function of $[H^+]$ and p_{CO_2}. $[Ca^{2+}]$ can be formulated through the solubility product of calcite, K_{s0}

$$[Ca^{2+}] = \frac{K_{s0}}{[CO_3^{2-}]} = \frac{K_{s0}}{C_T \alpha_2} \tag{ii}$$

and C_T can be expressed with the help of Henry's law, K_H

$$C_T = \frac{[HCO_3^-]}{\alpha_1} = \frac{[CO_3^{2-}]}{\alpha_2} = \frac{K_H p_{CO_2}}{\alpha_0} \tag{iii}$$

After substitution of (ii) and (iii) into (i), (i) can be solved for $[H^+]$, for example, by trial and error; subsequently C_T, $[HCO_3^-]$, $[CO_3^{2-}]$ and $[Ca^{2+}]$ are obtained. Thus far the computation is the same as that for the $CaCO_3$–CO_2–H_2O system (model 3 in Table 8-5). $[H_4SiO_4]$ can now be computed from the equilibrium constant of the Ca-montmorillonite–kaolinite equilibrium (K in Table 8-4).

$$[H_4SiO_4]^8 = \frac{K[H^+]^2}{[Ca^{2+}]} \tag{iv}$$

The result is given in Table 8-5 (model 5).

Note that $[H_4SiO_4]$ varies with $p_{CO_2}^{1/8}$. Hence, waters of this type have nearly constant $[H_4SiO_4]$.

INFINITE BUFFER INTENSITY. The activities in a system such as model 3 or 5 (Table 8-5) remain constant and independent of the concentration of the components as long as the phases coexist in equilibrium. The composition of the solution does not change if water (isothermal dilution) or the base $Ca(OH)_2$ is added. Such coexistence in equilibrium of the appropriate number of phases constitutes a chemostat or pH-stat [16].

CO_2-MANOSTAT. We now add anorthite (model 6) to the system already considered. After specifying the temperature, no other degree of freedom remains for the given number of components and phases; then p_{CO_2} in the gas phase of the model will be determined by the equilibria and cannot be varied. This model illustrates the possibility that the CO_2 content of the atmosphere is regulated at the sea–sediment interface by equilibria of reactions in which various aluminum silicates and $CaCO_3$ participate.

THE OCEAN MODEL OF SILLÉN [16]. The models given in Table 8-5 can be enlarged with the addition of each additional component to an equilibrium

[16] L. G. Sillén, in *Equilibrium Concepts in Natural Water Systems*, Advances in Chemistry Series, No. 67, American Chemical Society, Washington, D.C., 1967.

system resulting in either a new phase or an additional degree of freedom. Sillén has proposed equilibrium systems of different complexity as models for the oceans. His *nine-component system* contains HCl, H_2O, CO_2, SiO_2, Al_2O_3, NaOH, KOH, MgO and CaO. The first three components correspond to the volatile substances originating in the interior of the earth which, together with the other components, are contained in igneous rocks and participate in forming the sea. The nine components are distributed in nine phases, aqueous solution, quartz (Si), kaolinite (Al), chlorite (Mg), mica-illite (K), montmorillonite (Na), phillipsite or some other zeolite (Ca), calcite (CO_2) and a gas phase. (The items in parentheses are the components whose activities Sillén believes may become fixed by that phase.) By fixing the temperature and [Cl^-], the composition of the sea and p_{CO_2} of the atmosphere will be determined.

COMPARISON OF EQUILIBRIUM MODELS WITH NATURAL FRESH WATERS. Calculations of simple equilibrium models give concentrations of solutes quite representative of those encountered in natural waters. For example, model 5 of Table 8-5, calculated for $p_{CO_2} = 10^{-2}$ atm and representative of a gas composition in contact with ground water, gives an equilibrium composition (pH = 7.5, pCa = 3.0, $pHCO_3$ = 2.7, pH_4SiO_4 = 3.3) very similar to that given for the median composition of terrestrial waters (Figure 8-1).

A few examples of *analyses of waters* of various types are compared in Table 8-6.

It is tempting to try to relate the composition of the waters to the different geologic environments. Because such relations depend upon so many interdependent variables, and because there is substantial variance in each genetic type of water, no reliable generalizations are possible with such a small number of examples. The publication by White, Hem and Waring [17] should be consulted for a discussion of the composition of a large number of waters in relation to their genesis. A few qualitative characteristics are apparent from Table 8-6. As expected, ground waters have a higher ionic strength than surface waters. This results from the difference in CO_2 partial pressure to which these two types of waters are exposed. In comparing waters Nos. 5, 6 and 7, the influence of calcium plagioclase, which is less stable than the silicic rocks, becomes apparent. Sandstone beds are important aquifers; sandstones — as sedimentary rocks — may contain admixtures of carbonate besides grains of quartz. Thus the pH of sandstone waters tends to be higher than that of igneous rocks. Shales are laminated sediments whose constituent particles are of the clay grade. Most of these fine-grained sediments were

[17] D. E. White, J. D. Hem and G. A. Waring, "Chemical Composition of Subsurface Waters," *U.S. Geol. Surv. Profess. Papers,* **440-F** (1963).

Table 8-6 Examples of Natural Waters

	1	2	3	4	5	6	7	8	9	10	11
Type†	Stream	Stream	Stream	Lake Erie	Ground Water	Ground Water	Ground Water	Ground Water	Ground Water	Ground Water	Closed Basin Lake
Type of Rocks Being Drained	Granite	Quartzite	Sandstone‡		Granite	Gabbro plagioclase	Sandstone	Shale	Limestone	Dolomite	Soda Lake
pH	7.0	6.6	8.0	7.7	7.0	6.8	8.0	7.3	7.0	7.9	9.6
pNa	4.0	4.6	4.3	3.4	3.4	3.0	3.3	2.6	3.0	3.5	0.0
pK	4.7	5.1	4.8	4.3	4.0	4.5	4.0	4.2	3.7	—	1.7
pCa	4.0	4.3	3.1	3.0	3.5	3.1	3.0	2.5	2.7	2.8	4.5
pMg	4.6	5.1	4.0	3.4	3.8	3.2	3.5	2.5	3.4	2.8	4.6
pH_4SiO_4	3.8	4.2	4.1	4.7	3.2	3.0	3.9	3.5	3.7	3.4	2.8
$pHCO_3$	3.6	4.0	2.9	2.7	2.9	2.5	2.6	2.1	2.3	2.2	0.4
pCl	5.3	5.8	5.3	3.6	4.0	3.5	3.7	4.0	3.2	3.3	0.3
pSO_4	4.5	4.7	3.7	3.6	4.2	4.0	3.2	2.2	3.4	4.7	2.0
$-\log$ (ionic strength)	3.5	3.8	2.7	2.5	2.8	2.4	2.4	1.7	2.2	2.2	0.0

† 1–3: "Small Streams in New Mexico," *U.S. Geol. Surv. Bull.*, **1535F** (1961). 4: J. Kramer, *Geochim. Cosmochim. Acta*, **29**, 921 (1965). Types 5–10 are from *U.S. Geol. Surv. Bull.*, **440F** (1963). 5: Granite McCormick Co. (Table 1). 6: Harrisburg (Table 2). 7: Home Wood (Table 3). 8: Cuyahoga (Table 5). 9: Edwards limestone (Table 6). 10: Precambrian dolomite (Table 7). 11: "Albert Summer Lake Basin, Oregon," *North. Ohio Geol. Soc.*, J. L. Rau, Ed., 1966, p. 181.

‡ With slate and limestone beds.

deposited in saline environments. The relatively high $[HCO_3^-]$ of slate ground waters may be due to high CO_2 concentrations resulting from oxidation of organic carbon. Waters from limestone and dolomite contain substantial quantities of Ca^{2+}, Mg^{2+} and HCO_3^-; the concentrations of these solutes are usually controlled by carbonate equilibria.

COMPOSITION AND SILICATE EQUILIBRIA. In Figure 8-6 values of $[Na^+]/[H^+]$, $[K^+]/[H^+]$ and $[Ca^{2+}]/[H^+]^2$ of the waters in Table 8-6 are plotted on the silicate diagrams as functions of $[H_4SiO_4]$. Most of the resulting points fall into the stability field of kaolinite. These waters, if they were in equilibrium, would be in equilibrium with kaolinite. They are not in equilibrium with feldspars; or, in other words, feldspars at the CO_2 pressures encountered are unstable and are degraded. Points characterizing surface waters tend to be located closer to the kaolinite–gibbsite boundary. Ground waters — and this applies to most ground waters, not only to those for which the analyses are given in Table 8-6 — are characterized in these diagrams by points that fall close to the montmorillonite–kaolinite boundary.

Bricker and Garrels [4] and Garrels and Mackenzie [18] have given the following tentative interpretation on the role of silicate minerals in influencing water composition: When "aggressive" waters, high in CO_2 and low in dissolved solids, encounter silicates high in cations and silica, such as feldspars, they leach silica and cations and leave an aluminosilicate residue with an increased Al–Si ratio (usually kaolinite). Initial water attack yields a gibbsite residue but reaction is so rapid that it is under exceptional conditions that the silica in solution can be kept low enough to prevent the gibbsite from being converted to kaolinite, or to prevent kaolinite from forming in addition to the initial small amount of gibbsite. As the waters continue to attack feldspar, the pH rises, cations and silica increase in concentration and kaolinite forms until the cations and silica content rise high enough so that montmorillonite begins to form. At that stage, kaolinite apparently tends to be converted to montmorillonite, accounting for the limitation of silica content to about 60 ppm at the two-phase boundary kaolinite–montmorillonite.

If $[Ca^{2+}]$ and $[HCO_3^-]$ become sufficiently large, calcite precipitates; solubility equilibrium represents an upper limit for soluble carbonic constituents and calcium.

The upper limits of soluble silica content of natural waters are far less than saturation with amorphous silica ($\sim 2 \times 10^{-3}$ M); $[H_4SiO_4]$ appears to be controlled by equilibrium between the waters and various silicate phases.

[18] R. M. Garrels and F. T. Mackenzie, in *Equilibrium Concepts in Natural Water Systems*, Advances in Chemistry Series, No. 67, American Chemical Society, Washington, D.C., 1967.

EQUILIBRIUM WITH CALCITE. For the waters of Table 8-6, $[Ca^{2+}]/[H^+]$ is plotted as a function of $[HCO_3^-]$ in Figure 8-7a where the line defined by the equilibrium

$$CaCO_3(s) + H^+ = Ca^{2+} + HCO_3^-; \quad K = 10^2 \ (25°C) \quad (9)$$

defines $CaCO_3$ saturation for 25°C. No exact analysis is possible because

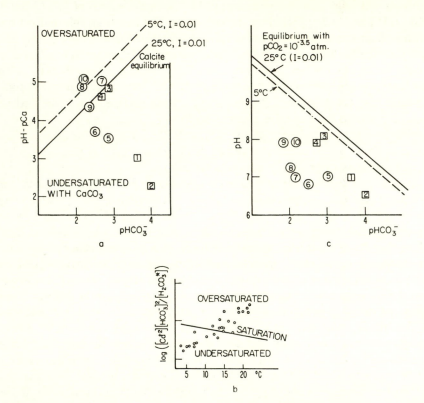

Figure 8-7 Comparison of actual composition with equilibrium systems.

  a: Comparison of actual composition of waters in Table 8-6 with the equilibrium composition for the reaction

$$CaCO_3(s) + H^+ = Ca^{2+} + HCO_3^-$$

  b: In a lake the water layers at the surface tend to be oversaturated and the bottom waters tend to be undersaturated with respect to $CaCO_3$. The comparison is made with the help of the reaction

$$CaCO_3(s) + H_2CO_3^* = Ca^{2+} + 2HCO_3^-$$

Data are for Lake Zürich.

  c: Most natural waters, especially ground waters, are oversaturated with respect to the CO_2 content of the atmosphere. Actual data of waters in Table 8-6 are compared with the equilibrium

$$CO_2(g) + H_2O = H^+ + HCO_3^-$$

temperature and ionic strength effects would have to be considered. Further-more, the pH values quoted in Table 8-6 are somewhat uncertain. Neverthe-less, it appears that those waters in contact with $CaCO_3$ (sandstone, shale, limestone, dolomite) are close to $CaCO_3$ saturation equilibrium or are over-saturated with respect to calcite. Lake Erie water falls into the same category.

In many *lake waters* precipitation and dissolution of calcite play a major role in buffering pH and water composition. In Figure 8-7*b* saturation of the water of Lake Zürich (Switzerland) with respect to calcite is represented as a function of temperature [19], that is, concentration quotients of analytical data for the reaction

$$CaCO_3(s) + H_2CO_3^* = Ca^{2+} + 2HCO_3^- + H_2O \qquad (10)$$

are compared with the equilibrium constant of (10) (corrected for ionic strength) as a function of temperature. Figure 8-7*b* shows that the lake water is oversaturated with $CaCO_3$ at high temperature and undersaturated at low temperature. Correspondingly, in the upper layers calcite is precipitated while in the lower layers it is partly dissolved. The discrepancy from equilibrium is caused in large part by photosynthesis (elevation of pH in upper layers) and by respiration (decrease in pH) in the deeper layers (Figure 8-8).

EQUILIBRIUM WITH CO_2 OF THE ATMOSPHERE. As illustrated by Figure 8-7*c*, all the waters of Table 8-6 contain more CO_2 than that corresponding to equilib-rium with the atmosphere. Understandably, ground waters influenced by a soil atmosphere enriched in CO_2 are further away from equilibrium with the atmosphere than are surface waters. The equilibrium line in Figure 8-7*c* represents the equilibrium reaction

$$CO_2(g) + H_2O = H^+ + HCO_3^- \qquad (11)$$

that is, $[H^+]$ for a water in CO_2 equilibrium with the atmosphere is defined solely by $[HCO_3^-]$ (or by the alkalinity of the water) and p_{CO_2} (Section 4-4). Most fresh waters have pH values close to neutrality (for alkalinities between $10^{-2.5}$ and $10^{-3.5}$ eq liter^{-1}), and hence are oversaturated in CO_2 with respect to an equilibrium with the atmosphere. Accordingly aeration of fresh waters leads frequently to an increase in pH when equilibrium conditions are approached. The conclusions to be drawn from this are (a) equilibration with atmosphere CO_2 is rather slow and (b) reactions that tend to depress the pH of natural waters, such as ion exchange, $CaCO_3$ deposition and respiration, tend to be more rapid than the CO_2 exchange process between these waters and the atmosphere.

[19] A similar analysis has been made for the water of the Great Lakes by J. R. Kramer (in *Equilibrium Concepts in Natural Water Systems*, Advances in Chemistry Series, No. 67, American Chemical Society, Washington, D.C., 1967).

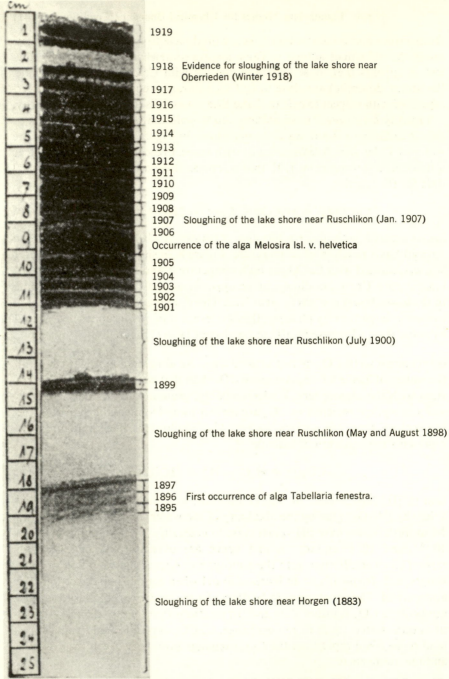

Figure 8-8 Sediment layers from Lake Zürich (1895–1919) [After F. Nipkow, *Z. Hydrol.*, **1**, 101 (1920)]. The alternative sequence of layers result from deposition of $CaCO_3$ during the summer (oversaturation with respect to $CaCO_3$ because of pH increase due to photosynthesis) and accumulation of black iron(II) sulfide containing sludge during winter (anaerobic conditions). Reproduced with permission from Birkhauser Verlag, Basel, Switzerland.

SEAWATER EQUILIBRIUM MODELS. Many fresh water systems represent highly dynamic arrangements and are considerably more transitory than the sea. Hence it is attractive to compare the real ocean with an equilibrium model. We know that the real system is not completely at equilibrium, but Sillén [1] and others have shown that the correspondence between equilibrium conditions and the real oceans is sufficiently close to justify accepting the equilibrium model as a useful first approximation.

SILLÉN'S OCEANIC MODEL. The classical geochemical material balance assumes that all sediments were ultimately derived from igneous rocks: primary rocks + volatile substances ⇌ sediments + seawater + air. Goldschmidt's [20] balance suggested that for each liter of present seawater, 600 g of igneous rocks had reacted with about 1 kg of volatile substances (H_2O, HCl, CO_2, etc.); during the process around 600 g of sediment and 3 liters of air were also formed. This balance became the basis of Sillén's [1] equilibrium model. He recalculated Goldschmidt's figures using the units moles liter^{-1} seawater (Table 8-7). In an imaginary experiment (Figure 8-9) the components of the real system (as given by Goldschmidt's values) have reached equilibrium.

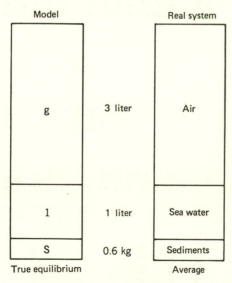

Figure 8-9   Sillén's Simplified Ocean Model. The true system (right) of 1 liter of sea water with the corresponding 0.6 kg of sediments and 3 liters of air. On the left the equilibrium model with the same amounts of the various components at true equilibrium (with the exception of N_2 is shown. After Sillén [16]. Reproduced with permission from American Chemical Society.

[20] V. M. Goldschmidt, *Fortschr. Mineral. Krist. Petr.*, **17**, 112 (1933).

Table 8-7 Balance of Materials for the Formation of 1 Liter of Seawater (unit = moles liter^{-1} seawater)†

Component	From‡			Now in			
	Primary Rock	Volatile	Air	Sediments	Solid Phase	Seawater	Main Species
H$_2$O		54.90				54.9	H_2O
Si (SiO$_2$)	6.06			6.06	SiO_2	$10^{-4.0}$ to $10^{-4.5}$	H_4SiO_4
Al (AlO$_{1.5}$, Al(OH)$_3$)	1.85			1.85	$Al_2Si_2O_5(OH)_4$	10^{-7} to $10^{-7.7}$	$Al(OH)_3$?
Cl (HCl)	0.01	0.54		0.29		0.55	Cl^-
Na (NaO$_{0.5}$, NaOH)	0.76					0.77	Na^+
Ca (CaO,Ca(OH)$_2$)	0.56			0.55	Silicates	0.01	Ca^{2+}
Mg (MgO,Mg(OH)$_2$)	0.53			0.48	(pH = 8.1)	0.05	Mg^{2+}
K (KO$_{0.5}$,KOH)	0.41			0.40		0.002	K^+
C		0.60		0.62	$CaCO_3$, $CaMg(CO_3)_2$	0.002	HCO_3^-
(CO$_2$)	0.02	0.53		0.55			
(C(s))	0.02	0.07		0.07			
O$_2$		0.027	0.027			10^{-6} to 10^{-8}	O_2
Fe	0.55			0.55	FeOOH	10^{-6} to 10^{-8}	$Fe(OH)_3$?
(FeO,Fe(OH)$_2$)	0.32			0.18			
(FeO$_{1.5}$,FeOOH)	0.23			0.37			
Ti (TiO$_2$)	0.06			0.06	TiO_2	$10^{-6.7}$ to $10^{-8.6}$	$Ti(OH)_4$?
S	0.01	0.06		0.04		0.03	SO_4^{2-}
F (HF)	0.03			0.03		$10^{-4.15}$	F^-, MgF^+
P (PO$_{2.5}$,H$_3$PO$_4$)	0.02			0.02	$Ca_5(PO_4)_3OH$	$10^{-5.5}$ to $10^{-6.0}$	HPO_4^{2-}
Mn (MnO$_{1\ to\ 2}$)	0.01			0.01	MnO_2	$10^{-6.7}$ to $10^{-7.9}$	$Mn(OH)_{3,4}$
N$_2$		0.101	0.101				N_2

† After Sillén [1, 16].
‡ Calculated from the estimates of Goldschmidt [20].

Sillén's formal approach in constructing his model consisted of adding the components in sequence starting with the more important ones — Si and Al. After each addition the conditions of equilibrium are computed. Almost correct values result for the concentrations of many major and of some minor constituents of seawater. Further progress in equilibrium chemistry and further elaborations by Sillén [16, 21] and by others have confirmed the usefulness of this model.

KRAMER'S MODEL. As an outgrowth of Sillén's calculations, Kramer [22] similarly derived an inorganic ocean from the solution equilibrium of nine solid phases (alumino silicates, carbonates, phosphates and sulfates), Table 8-8. Because chloride does not appear to have a solubility regulation in water — most of the chloride was probably derived from "de-gassing" the crust — its concentration is assumed to be that of the present ocean. His ocean model fits the present ocean values very closely. The relatively small discrepancies in $[Ca^{2+}]$, $[Mg^{2+}]$, $[Alk]$ and p_{CO_2} could, according to Kramer, be due to poor estimates of equilibrium constants. By arbitrarily decreasing the assumed value of $[Cl^-]$ a model composition quite similar to "fresh" water obtains. Kramer suggests that the presence or absence of Cl^- seems to be the most important factor in determining "fresh" or saline water concentrations. In the formal calculations this is due to the effect of Cl^- upon the electroneutrality; this is another way of saying that "saline" water results from the

Table 8-8   Kramer's Ocean Model [22]

Equilibrium Concentration, M	Liquid Phase	Solids Controlling Concentrations
0.45	Na^+	Na-mont (E site)†
9.7×10^{-3}	K^+	K-illite (E site)†
0.55 (defined)	Cl^-	0.55 (assumed)
3.4×10^{-2}	SO_4^{2-}	$SrCO_3$; $SrSO_4$
6.1×10^{-3}	Ca^{2+}	Phillipsite
6.7×10^{-2}	Mg^{2+}	Chlorite
2.7×10^{-6}	PO_4^{3-}	OH-apatite
$(1.7 \times 10^{-3}$ atm$)$	CO_2	Calcite
2.4×10^{-5}	F^-	F-CO_3-apatite
4.7×10^{-9}	H^+	Given by electroneutrality

† In interpreting mass action equilibrium, for some clays two discrete exchange sites, "C" sites and "E" sites, perhaps corresponding to interlayer and edge sites, may be considered.
Reproduced with permission from Pergamon Press.

[21] L. G. Sillén, *Arkiv Kemi*, **25**, 159 (1966).
[22] J. R. Kramer, *Geochim. Cosmochim. Acta*, **29**, 921 (1965).

interaction of HCl upon minerals, where "fresh" waters are formed by the interaction of weaker acids (CO_2) and by hydrolysis reactions. By assuming that very ancient oceans had a lower [Cl^-], Kramer implies that their composition may have approximated that of fresh water.

HOLLAND'S HISTORY OF OCEAN WATER. Holland's [23] paper, "The History of Ocean Water and Its Effect on the Chemistry of the Atmosphere," is a further extension of Sillén's approach to the factors that control the chemistry of ocean water. Holland believes that we are sufficiently close to an understanding of these controls to employ them together with stratigraphic data to set some limits on the variations of the chemical composition of ocean water in the past. These limits in turn set restraints on the chemical composition of our atmosphere in the past. Holland, in defining these limits, recognizes three classes of effects that control the composition of the ocean-atmosphere system: (a) the concentration of a first group of constituents (Cl^-, N_2, rare gases) is controlled largely by the *degassing* of the earth; (b) the concentrations of the major cations, of H^+ and p_{CO_2}, of the atmosphere are controlled by *mineral equilibria*; and (c) *biological reactions* control the oxygen content of the atmosphere and the SO_4^{2-} content of seawater. It has already been mentioned in Section 7-5 that the O_2 content of the atmosphere at any time is a product of the balance of O_2 production by photosynthesis and its consumption in organic decay, oxidation of volcanic gases and oxidation of surface rocks.

Holland suggests that the CO_2 pressure in the atmosphere is being and has been crudely buffered by the coexistence of chlorite, dolomite, calcite and quartz throughout much of geological time. The crudeness of the buffering is due in large part to the sluggishness of dolomite nucleation and crystal growth. Since present-day seawater is somewhat supersaturated in respect to dolomite, its precipitation is apparently achieved today only in areas of high temperatures or abnormally high supersaturation.

Example 8-5 p_{CO_2} *Control during Recent Geologic Past.*   Holland [23] shows that the CO_2 pressure in the atmosphere, at least in the recent geologic past, has been between two boundaries. The lower boundary in the CO_2 pressure of seawater is given by

$$CaCO_3(s) + SO_4^{2-} + 2H^+ + H_2O = CaSO_4 \cdot 2H_2O(s) + CO_2(g) \quad (i)$$

Because gypsum is not a typical constituent of normal marine sediments, it seems unlikely that p_{CO_2} has been less than that given by (i). The upper boundary in p_{CO_2} is given by the conversion of calcite into dolomite.

$$CaMg(CO_3)_2(s) + 2H^+ = CaCO_3(s) + Mg^{2+} + CO_2(g) + H_2O \quad (ii)$$

[23] H. D. Holland, *Proc. Nat. Acad. Sci. U.S.*, **53**, 1173 (1965).

Since in areas of active dolomite precipitation $[Mg^{2+}]/[Ca^{2+}] = 20$, the CO_2 pressure concordant with this condition seems to be near the maximum p_{CO_2} to be expected.

Estimate the range of CO_2 pressure in the atmosphere. The pH is buffered by clays and constant (pH = 8.1).

The equilibrium constant of (i) and (ii) can be computed from the following information valid at 25°C.

$pK_{so}(CaCO_3) = 8.1$; $pK_{so}(CaMg(CO_3)_2) = 16.7$; $pK_{so}(CaSO_4 \cdot 2H_2O) = 4.6$; $pK(CO_2(g) + H_2O = CO_3^{2-} + 2H^+) = 15.9$, for example, for (i):

$$CaCO_3(s) = Ca^{2+} + CO_3^{2-}; \qquad \log K_{so} = -8.1$$
$$Ca^{2+} + SO_4^{2-} + 2H_2O = CaSO_4 \cdot 2H_2O; \quad \log(1/K_{so}) = 4.6$$
$$CO_3^{2-} + 2H^+ = CO_2(g) + H_2O; \qquad \log K = 18.3$$

$$CaCO_3(s) + SO_4^{2-} + 2H^+ + H_2O = CaSO_4 \cdot 2H_2O(s) + CO_2(g);$$
$$\log K_{(i)} = 14.8$$

Correcting for ionic strength:

$$K'_{(i)} = p_{CO_2}/[SO_4^{2-}]\{H^+\}^2 = K_{(i)}f_{SO_4^{2-}}$$

with $f_{SO_4^{2-}}$ in the water = 0.12 (Table 6-6, part III) $\log K'_{(i)}$ becomes 13.88. For $[SO_4^{2-}] = 3.8 \times 10^{-2}$ a CO_2 pressure (atm) of $\log p_{CO_2} = -3.75$ obtains. Reaction ii has an equilibrium constant of $\log K_{(ii)} = 9.7$. Corrected for activity ($K'_{(ii)} = p_{CO_2}[Mg^{2+}]/\{H^+\}^2$), $\log K'_{(ii)} = 10.14$ is obtained. Without oversaturation for $[Mg^{2+}] = 5.4 \times 10^{-2}$ M and pH = 8.1, an equilibrium $p_{CO_2} = 10^{-3.7}$ atm is obtained for (ii). Because of the sluggishness of dolomite precipitation a considerable oversaturation in the reaction

$$2CaCO_3(s) + Mg^{2+} = CaMg(CO_3)_2 + Ca^{2+} \qquad \text{(iii)}$$

is necessary to form dolomite. The equilibrium constant of (iii) is $K_{(iii)} = 3.1$. Hence the stipulated condition $[Ca^{2+}]/[Mg^{2+}] = 0.05$ represents a 60-fold oversaturation. The CO_2 pressure related to this oversaturation is obtained by considering that $[Ca^{2+}]$ is one-twentieth of seawater concentration of magnesium ($[Mg^{2+}] = 5.4 \times 10^{-2}$ M). p_{CO_2} for this $[Ca^{2+}]$ is then given by the equilibrium

$$CaCO_3(s) + 2H^+ = Ca^{2+} + CO_2(g) + H_2O; \quad \log K = 10.2 \qquad \text{(iv)}$$

with $[Ca^{2+}] = 2.7 \times 10^{-3}$ M and $f_{Ca^{2+}} = 0.28$, the CO_2 pressure (atm) becomes $\log p_{CO_2} = -2.9$. CO_2 pressures of this magnitude or larger would lead to the precipitation of essentially all of the Mg^{2+} brought into the sea by rivers as dolomite. These two boundary conditions suggest that p_{CO_2} has been less than $10^{-2.9}$ and more than $10^{-3.7}$ atm.

8-5   Buffering

Sillén [1] has doubted the commonly held view that pH buffering in the oceans is caused by the CO_2–HCO_3^--CO_3^{2-} equilibrium. Sillén's analysis suggests that heterogeneous equilibria of silicate minerals comprise the principal pH buffer systems in oceanic waters. Garrels [24], Holland [23] and Pytkowicz [25] have expanded upon the quantitative aspects of Sillén's proposal.

In the few examples presented here, general pH buffering systems are compared. As before (Sections 3-9 and 4-5) pH buffer intensity, $\beta_{C_j}^{C_i}$, is defined for the incremental addition of C_i to a closed system of constant C_j at equilibrium

$$\beta_{C_j}^{C_i} = \frac{dC_i}{d\mathrm{pH}} \tag{12}$$

We can now compare, for example, the buffer intensities, with respect to strong acid, of the following systems (Figure 8-10): (a) a carbonate solution of constant C_T, β_{C_T}; (b) an aqueous solution in equilibrium with calcite, β_{CaCO_3}; (c) a carbonate solution in equilibrium with a gas phase of constant p_{CO_2}, $\beta_{p_{CO_2}}$; and (d) a solution in equilibrium with both kaolinite and anorthite, $\beta_{an-kaol}$. As explained earlier (Section 3-9), the buffer intensity is found analytically by differentiating the appropriate function of C_i for the system with respect to pH.

The buffer intensity of a homogeneous carbonate system has been derived before (see Table 4-2); for heterogeneous systems, the buffer intensities are derived in a similar way. Example 8-6 serves as an illustration.

Example 8-6 *Buffer Intensity of a Heterogeneous System.* Calculate the buffer intensity with respect to strong acid (or strong base), $\beta_{CaCO_3(s)}$, of a solution in equilibrium with calcite.

It is convenient to start out with the charge condition of a closed aqueous solution of $CaCO_3(s)$:

$$2[Ca^{2+}] + [H^+] = [HCO_3^-] + 2[CO_3^{2-}] + [OH^-] \tag{i}$$

After addition of a strong acid, say HCl (i.e., $[Cl^-] = C_A$), the charge balance becomes

$$C_A = 2[Ca^{2+}] + [H^+] - [HCO_3^-] - 2[CO_3^{2-}] - [OH^-] \tag{ii}$$

The buffer intensity is obtained after differentiation of (ii) with respect to

[24] R. M. Garrels, *Science*, **148**, 69 (1965).
[25] R. M. Pytkowicz, *Geochim. Cosmochim. Acta*, **31**, 63 (1967).

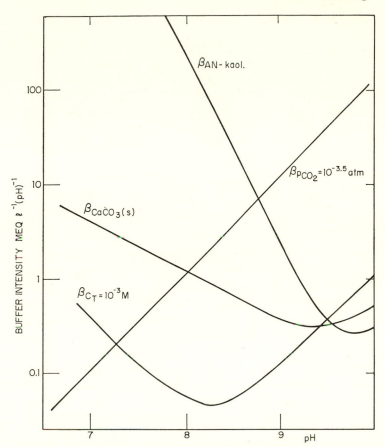

Figure 8-10   Buffer intensity versus pH for some heterogeneous systems and for the homogeneous dissolved carbonate system. Buffer intensities: β_{C_T} (dissolved carbonate, $C_T = 10^{-3}$ M), β_{CaCO_3} (carbonate solution in equilibrium with calcite), $\beta_{p_{CO_2}}$ (carbonate solution in equilibrium with $p_{CO_2} = 10^{-3}$ atm) and $\beta_{an\text{-}kaol}$ (solution in equilibrium with anorthite and kaolinite).

$-$pH. Because

$$[Ca^{2+}] = C_T = \left(\frac{K_s}{\alpha_2}\right)^{0.5} \tag{iii}$$

(compare Section 5-2), the titration curve of (ii) can be calculated as a function of $[H^+]$ or pH. Instead of a cumbersome numerical differentiation, it may be more convenient to consider the pH dependence of the individual terms on the right-hand side of (ii). The variation of $\log [Ca^{2+}]$, $\log [HCO_3^-]$ and $\log [CO_3^{2-}]$ with pH has been illustrated in Figure 5-4. In the pertinent pH region, $pK_1 > pH < pK_2$:

$$\frac{d \log [Ca^{2+}]}{-dpH} = 0.5 \tag{iv}$$

$$\frac{d \log [HCO_3^-]}{-dpH} = 0.5 \tag{v}$$

$$\frac{d \log [CO_3^{2-}]}{-dpH} = -0.5 \tag{vi}$$

From these relations we readily obtain

$$\frac{d[Ca^{2+}]}{-dpH} = \frac{d \log [Ca^{2+}]}{-dpH} \frac{d[Ca^{2+}]}{d \log [Ca^{2+}]} = (0.5)2.3[Ca^{2+}] \tag{vii}$$

Similarly

$$\frac{d[HCO_3^-]}{-dpH} = 1.15[HCO_3^-] \tag{viii}$$

$$\frac{d[CO_3^{2-}]}{-dpH} = -1.15[CO_3^{2-}] \tag{ix}$$

$$\frac{d[OH^-]}{-dpH} = -2.3[OH^-] \tag{x}$$

Thus with the help of Figure 5-4 the individual terms in (ii) are obtained for every pH value (Figure 8-10). The following relationship holds in good approximation for the pH range stipulated:

$$\beta_{CaCO_3(s)}^{C_A} \simeq \left(\frac{1}{-dpH}\right)(2d[Ca^{2+}] - d[HCO_3^-])$$

$$= \frac{d[Ca^{2+}]}{-dpH}$$

$$= 1.15(K_s/\alpha_2)^{0.5} \tag{xi}$$

In the case of the heterogeneous equilibria with silicates, highly simplified models are considered where, for example, an acid (or base) is added to an aqueous system of kaolinite and anorthite. The electroneutrality would be

$$[H^+] + 2[Ca^{2+}] = [OH^-] + C_A \tag{13}$$

where C_A is the concentration of strong acid added. $[Ca^{2+}]$ is given by $K[H^+]^2$ where K is the equilibrium constant for the anorthite–kaolinite system [(3) of Table 8-4], and hence $\beta_{an-kaol}$ can be evaluated. As Figure 8-10 shows, the homogeneous buffer intensity β_{C_T} is relatively small. For fresh water systems β_{CaCO_3} and $\beta_{p_{CO_2}}$ are of considerable practical interest.

$CaCO_3(s)$ is an efficient buffer in the neutral and acid pH range. If, for example, a large quantity of acid is discharged into a natural water system containing $CaCO_3(s)$, an initially large decrease in the pH of that system occurs. The extent of this decrease depends largely upon the magnitude of the fraction of total buffer intensity attributable to dissolved buffer components. Ultimately, however, the decrease in pH resulting from addition of the acid leads to dissolution of solid calcium carbonate and the establishment of a new position of equilibrium. Thus the final change in pH is much less than the initial decrease which is resisted only by the intensity contribution of dissolved buffer components. The addition of a strong base in large quantity, conversely, leads to a deposition of calcium carbonate, thus reducing the pH shift that would occur in the absence of dissolved calcium.

The aluminum silicates provide considerably more resistance toward pH changes. The equilibrium system anorthite–kaolinite at pH = 8 has a buffer intensity a thousand times higher than that of a 10^{-3} M carbonate solution. As has been shown, equilibrium systems consisting of a sufficient number of coexisting phases attain, in principle, an infinite buffer intensity.

What actually protects water from pH changes is also dependent upon the kinetics of the heterogeneous reactions. Investigations on the rate of the buffering reactions are very much in need [24, 25]. While reactions with solid carbonates and ion exchange processes are faster than alteration reactions of solid silicates, the latter probably regulate the pH of seawater over long periods of time.

From a kinetic point of view we must also consider that biochemical processes effect pH regulation and buffer action in natural water systems. Photosynthetic activities decrease CO_2, whereas respiratory activities contribute CO_2.

For fresh waters there is a further restraint to pH rise: the CO_2 reservoir of the atmosphere. For a given p_{CO_2} the pH is a function of alkalinity. In order to lift the pH of a water in equilibrium with the atmosphere from 8 to 9, alkalinity must increase by nearly 5 meq liter^{-1} (either by base addition or by evaporation) (see Figure 4-3). Hence only soda lakes, that is, lakes containing substantial amounts of soluble carbonates and bicarbonates can attain high pH values (Example 11 of Table 8-6); for example, Sierra Nevada spring waters discharged to the east of the Sierra and evaporated in a plaza of the California desert. As the example given below illustrates, generally a nearly neutral Na–Ca–HCO_3^- water is converted by evaporation into an alkaline Na–HCO_3^-–CO_3^{2-} water (soda lake). Buffering of pH (see Figure 8-10) occurs during the precipitation of solid phases.

Example 8-7 *Isothermal Evaporation of a Natural Water.* In order to illustrate the principles of the procedures and calculations a highly simplified situation is considered. A solution containing 1×10^{-4} M Ca^{2+}, 1×10^{-4} M Na^+, 3×10^{-4} eq liter^{-1} alkalinity and 1×10^{-4} M H_4SiO_4 is evaporated isothermally (25°C) and remains in equilibrium with atmospheric CO_2 (log $p_{CO_2} = -3.5$). Compute pH and solution composition as a function of the concentration factor.

During evaporation the ionic strength continuously changes, and for exact calculations corrections of ionic strength should be made. An iterative procedure is necessary and we first obtain approximate pH values and molarities by using constants valid at $I = 0$. With these tentative values the ionic strength can be estimated as a function of the concentration factor with sufficient precision in order to correct the equilibrium constants used. Here we make our calculations for $I = 0$. The values calculated without further correction must be regarded as approximations that may contain uncertainties of up to 0.3 logarithmic units for a concentration factor of 100.

The solution initially has a pH given by

$$\frac{[H^+][HCO_3^-]}{p_{CO_2}} = 10^{-7.8} \tag{i}$$

Accordingly, pH = 7.8 assuming $[Alk] = [HCO_3^-]$. At this pH the solution is undersaturated with respect to $CaCO_3(s)$ and amorphous silica. These two solid phases may reasonably be assumed to be the only ones that will precipitate upon evaporation. Virtually all the alkalinity is present as HCO_3^- initially. At higher concentration factors CO_3^{2-} becomes an important species and, instead of (i), we may write [compare (11) of Section 4-3]

$$[Alk] = \left(\frac{K_H p_{CO_2}}{\alpha_0}\right)(\alpha_1 + 2\alpha_2) + \frac{K_W}{[H^+]} - [H^+] \tag{ii}$$

The relationship between $[Alk]$ and pH has been calculated before for Figure 4-3.

We may now proceed in steps and concentrate the solution by a factor of two ($R = 2$), that is, the composition is now $[Alk] = 6 \times 10^{-4}$ eq liter^{-1}, $[Ca^{2+}] = 2 \times 10^{-4}$ M, $[Na^+] = 2 \times 10^{-4}$ M, $[H_4SiO_4] = 2 \times 10^{-4}$ M and the pH from (ii) or Figure 4-3 is 8.1. We now have to determine whether the water, as a result of evaporation and pH change, has become oversaturated. At saturation with calcite and the atmosphere

$$[Ca^{2+}] = \frac{K_{s0}\alpha_0}{K_H \alpha_2 p_{CO_2}} \tag{iii}$$

$[Ca^{2+}]$ can be calculated from (iii) as a function of pH; this has already been done in Figure 5-6.

At a concentration factor of four the point is reached where $CaCO_3$ starts to precipitate: $[Alk] = 1.2 \times 10^{-3}$ eq liter^{-1}, $[Ca^{2+}] = 4 \times 10^{-4}$ M,

$[Na^+] = 4 \times 10^{-4} M$, $[H_4SiO_4] = 4 \times 10^{-4} M$ and pH = 8.4. Further evaporation is accompanied by $CaCO_3$ precipitation. In the electroneutrality equation

$$[Na^+] = [Alk] - 2[Ca^{2+}] \qquad (iv)$$

the left-hand side is given by $R[Na_0^+]$; the terms on the right-hand side can be expressed as a function of pH [(ii) and (iii) or Figures 4-3 and 5-6]. Most conveniently, for a given pH, [Alk] and $[Ca^{2+}]$ are first calculated, then R is obtained from $[Na^+]$ in (iv). $[H_4SiO_4]$ increases in concentration until the saturation with respect to amorphous SiO_2 is attained.

Figure 8-11 plots the equilibrium composition as a function of the concentration factor. Note that there is buffering of $[H^+]$ and of $[Ca^{2+}]$ and [Alk]

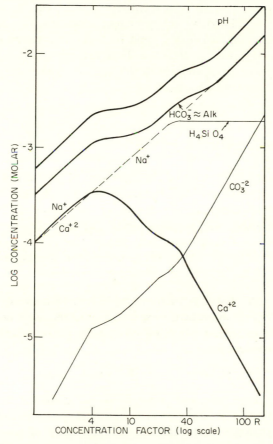

Figure 8-11 Isothermal evaporation of a water in the Na^+–Ca^{2+}–SiO_2–CO_2–H_2O system. The equilibrium composition is plotted as a function of the concentration factor. Isothermal evaporation of natural waters leads to waters high in pH and carbonate.

during the bulk precipitation of calcite. Garrels and Mackenzie [18] have calculated in a similar way (including, however, corrections for activity) the change in composition resulting from the evaporation of typical Sierra Nevada spring water. In this case, which corresponds to a natural situation, another solid phase, sepiolite $MgSi_3O_6(OH)_2(s)$, had to be considered. As these authors point out, the concentrated waters resemble the composition of some natural soda lakes. However, the waters resulting from such an isothermic evaporation of natural waters, although of high ionic strength, have little resemblance with seawater; in other words, simple concentration of fresh water does not produce ocean water.

HETEROGENEOUS ION BUFFERS. In the same way as $[H^+]$ is kept constant by heterogeneous equilibria, the concentration of other cations and anions in natural waters is buffered by heterogeneous reactions. A water that is in equilibrium with solid $CaCO_3$ will tend to maintain a rather constant pCa even if Ca^{2+} is introduced to the water from external sources. Furthermore, ion exchange phenomena and the formation of solid solutions represent special cases of heterogeneous ion buffers.

8-6   Steady-State Models

It might be argued that much of the chemistry of the oceans and of fresh water systems depends on the kinetics of various physical and chemical processes and on biochemical reactions rather than on the equilibrium conditions described above. The simplest model that describes systems open to their environment is the time-invariant steady-state model. Because the sea has remained constant for the recent geologic past, it may be well justified to interpret the ocean in terms of a steady-state model.

THE RESIDENCE TIME. Input is balanced by output in a steady-state system. The concentration of an element in seawater remains constant if it is added to the sea at the same rate as it is removed from the ocean water by sedimentation. For each element E

$$(d[E]/dt)_{\text{input}} = (d[E]/dt)_{\text{sedimentation}} \tag{14}$$

Input into the oceans consists primarily of (a) dissolved and particulate matter carried by streams, and (b) volcanic and hot spring material introduced directly. Note that the model considered is an ocean of constant volume, constant temperature and pressure, and uniform composition. The latter assumption is justified if the rate of ocean stirring is large compared to $d[E]/dt$.

Table 8-9   Residence Times of a Few Elements in Seawater

Element, E	Residence Time, years	$-\text{Log }[E]$, M	Log (Relative Oversaturation),[†] $\log([E]/[E_{eq}])$
Fe	1.4×10^2	6.8	4 to 6
Al	1.0×10^2	6.4	1
Mn	1.4×10^3	7.0[‡]	
Si	8×10^3	4.0	
Cu	5×10^4	7.3	0 to 2
Zn	2×10^5	6.8	-3 to 0
Sr	1.9×10^7	4.0	0.0 to -0.9
Ca	8×10^6	2.0	-0.4 to $+0.5$
Na	2.6×10^8	0.35	

† Cf. P. Schindler in *Equilibrium Concepts in Natural Water Systems*, Advances in Chemistry Series, No. 67, American Chemical Society, Washington, D.C., 1967, p. 196. Schindler assumed the following solid phases and solutes in equilibrium with them. Fe: amorphous $FeOOH(s)$ and $\alpha\text{-}FeOOH(s)$, $Fe(OH)_2{}^+$ and $Fe(OH)_4{}^-$. Al: $\alpha\text{-}Al(OH)_3(s)$, $Al(OH)_4{}^-$. Cu: $CuO(s)$ and $Cu_{0.003}Me(II)O_{1.997}(s)$, $CuCO_3(aq)$. Zn: $ZnO(s)$ and $Zn_{0.003}Me(II)O_{1.997}(s)$, $ZnSO_4(aq)$. Sr: $SrCO_3(s)$ and $(Ca_{0.999}Sr_{0.001})CO_3(s)$, Sr^{2+}. Ca: $CaCO_3$ (calcite, s), Ca^{2+}.
‡ Present as Mn^{2+}; the stable phase is $MnO_2(s)$; relative oversaturation $= 10^9$.
Reproduced with permission from American Chemical Society.

The rate of sedimentation is controlled largely by the rate at which an element is converted (precipitation, coprecipitation, ion exchange, biological activity) into an insoluble and settleable form. Hence the reactivity of the elements influences the time during which the elements spend on the average as constituents in the seawater. The residence time, τ, of an element is defined as

$$\tau = \frac{[E]}{d[E]/dt} \tag{15}$$

For most elements residence times have been determined on the basis of estimates on the input by runoff from the lands [26] or from calculations on sedimentation times [27]. Remarkably similar results are obtained by these two methods. A few examples of residence times are given in Table 8-9. Elements that are highly oversaturated (e.g., Al, Fe) have short residence times, that is, times that correspond to the period necessary for ocean mixing.† On the other hand, elements with low reactivity such as Na and Li have

† Estimates on the rate of mixing obtained from radioactive dates of bottom water vary from 10 to 10^3 years for one revolution [G. G. Rocco and W. S. Broecker, *J. Geophys. Res.*, **68**, 4501 (1963)].
[26] J. F. Barth, *Theoretical Petrology*, Wiley, New York, 1952.
[27] E. D. Goldberg and G. O. S. Arrhenius, *Geochim. Cosmochim. Acta*, **13**, 153 (1958).

very long residence times; times that are perhaps within one or two orders of magnitude of the age of the ocean. Elements that are neither strongly over-saturated nor undersaturated in seawater — for example, Ni, Cu, Zn, Ba — exhibit intermediate residence times (10^4 to 10^6 years).

That Ca spends so much time in the ocean may be explained by the fact that the upper layers of the sea are oversaturated and the deeper layers under-saturated with calcite. Hence, as in many lakes, Ca is precipitated and redis-solved many times before it becomes incorporated in sediments.

The residence time of Mn is related to its removal from solution by oxida-tion of Mn^{2+} to $MnO_x(s)$. As shown in Chapter 10, laboratory results on the rates of Mn(II) oxidation are compatible with the residence time of Mn. The residence time of Sr is four times larger than that of Ca. The fate of the two elements may be interrelated because part of Sr may precipitate in $CaCO_3$ as solid solution. Because of long residence time the contamination of the ocean with Sr^{90} is a long-range hazard.

THE SYNTHESIS OF CLAYS IN SEAWATER. Silicon, soluble in water as monomeric orthosilicic acid H_4SiO_4 up to a concentration of 2×10^{-3} M (25°C), is one of the few elements that occurs in seawater at a lower concentration than in fresh waters. Hence, most of the silicon that enters the ocean is removed from solution. The relatively short residence time of Si indicates that H_4SiO_4 participates readily in reactions and thus disappears from solution. Possible reactions are (a) incorporation into organisms, (b) precipitation of mag-nesium silicates and (c) synthesis of clays. Mackenzie and Garrels [28] have elaborated on the synthesis of clays.

Mackenzie and Garrels point out that simple evaporative processes do not change stream waters into oceans. As has been illustrated in Example 8-7 in the previous section, isothermal evaporation of a nearly neutral Na^+–Ca^{2+}–HCO_3^- water leads to a highly alkaline Na^+–HCO_3^-–CO_3^{2-} water; Ca^{2+} and other multivalent ions are removed by precipitation and the concentration of H_4SiO_4 increases until silica gel precipitates (2×10^{-3} M).

Reactions must take place in the ocean that obviously prevent it from becoming a soda lake; alkalinity, cations and H_4SiO_4 that would tend to become excessive upon concentration have to be removed. Mackenzie and Garrels suggest that H_4SiO_4 and HCO_3^- react with cations and the small but significant X-ray amorphous Al-silicate fraction of the suspended load of streams according to reactions of the type

amorphous Al-silicate $+ HCO_3^- + H_4SiO_4 +$ cations
$$= \text{cation Al-silicate} + CO_2 + H_2O \quad (16)$$

[28] F. T. Mackenzie and R. M. Garrels, *J. Sediment Petrol.*, **36**, 1075 (1966); *Amer. J. Sci.*, **264**, 507 (1966). R. M. Garrels, *Science*, **148**, 69 (1965).

Such reactions are essentially "reverse weathering" reactions and take place before deposition and burial.

Mackenzie and Garrels have provided additional support for such a reaction type by evaluating the mass balance of the present stream discharge and by considering the volume of sediment in the geological column of the ocean. Recent work by the same authors has shown that the ocean must be considered as a chemical system with a rapid compositional response to added detrital silicates. Such observation is in accord with the short residence time of Si.

Fresh Water Systems

Steady-state models can also aid in understanding fresh water systems (ground waters, lakes, rivers). For example, the steady state is often presupposed in the application of natural or artificial isotopes (C^{14}, Si^{32}, tritium) in order to investigate exchange processes between atmosphere and ground waters, to study mixing in lakes or to evaluate local hydrological cycles.

The concept of a residence time can also be applied to fresh water systems. For a lake, (14) must be changed into

$$(d[E_I]/dt)_{\text{input}} = (d[E_S]/dt)_{\text{sedimentation}} + (d[E_0]/dt)_{\text{outflow}} \qquad (17)$$

It is frequently convenient to define a relative residence time, τ_{rel}, that is, a time relative to the residence time of water, $\tau_{\text{H}_2\text{O}}$:

$$\tau_{\text{rel}} = \frac{\tau_E}{\tau_{\text{H}_2\text{O}}} = \frac{[E]}{(d[E_I]/dt)_{\text{inflow}}\tau_{\text{H}_2\text{O}}} \qquad (18)$$

$\tau_{\text{rel}} \simeq 1$ for substances that have no tendency to undergo reactions. Reactive constituents, however, have residence times different from those of water; τ_{rel} can be smaller or larger than unity. Al, for example, is incorporated rather readily into sediments of a lake, hence $\tau_{\text{rel}} < 1$ for Al. Fe, on the other hand, precipitates in the upper layers of a lake as $Fe(OH)_3$ but becomes redissolved as Fe^{2+} in the usually anaerobic bottom layers; because of this cycle τ_{rel} for Fe becomes larger than unity. Elements that participate in the biochemical cycle, for example, fertilizing substances like phosphate, ammonium and nitrate, have residence times usually much larger than those of water. Algae that incorporate fertilizing substances in the top layers sink into the bottom layers where they become mineralized. After spring and fall circulation these algal nutrients become available again for the synthesis of algae. Hence, such algal nutrients may have very high values of τ_{rel}; they become progressively trapped in a lake, and their concentration becomes much larger than that obtained by calculation from input and hydrographic data.

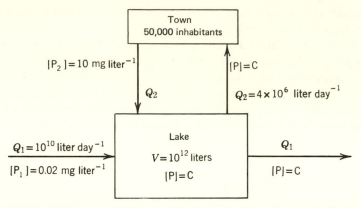

Figure 8-12   Simple model situation for evaluating steady-state concentration (Example 8-8).

Example 8-8   *Effect of Recycling on Steady-State Concentration of Algal Nutrients in a Lake.*   Assume the following simple model: A lake with a volume of 10^{12} liters and an inflow (runoff) balanced by outflow of 10^{10} liters day^{-1}. A town of 50,000 inhabitants uses lake water (4×10^6 liters day^{-1}) for water supply purposes and discharges its waste into the lake. The phosphorus content of the waste water and of the inflow to the lake is 10 and 0.02 mg P liter^{-1}, respectively. Estimate the steady-state concentration of P in this lake, assuming that P is retained because it recycles in the lake 10 times as long as water ($\tau_{rel} = 10$), but otherwise treating the lake as a completely mixed system. Incorporation of P into the sediments and exchange of P between lake water and sediment are ignored in this simple model.

Figure 8-12 summarizes the problem. The following steady-state mass balance between inflow (runoff and waste) and outflow (water supply and effluent) holds

$$\frac{d[P]}{dt} = 0 = \frac{Q_1[P_1] + Q_2[P_2]}{V} - \frac{(Q_1 + Q_2)C}{V} \tag{i}$$

If P was nonreactive ($\tau_{rel} = 1$), its steady-state concentration would be $C = 0.024$ mg P liter^{-1}. With a relative residence time of 10, the steady-state concentration by (18) would be 10 times larger, that is, $C' = 0.24$ mg P liter^{-1}.

8-7   Interaction between Organisms and Abiotic Environment

Aquatic organisms influence the concentration of many substances directly by metabolic uptake, transformation, storage and release. In order to understand the chemistry of an aquatic habitat, the causal and reciprocal relationship between organisms and their environment must be taken into consideration. Steady-state models may be applied to evaluate the interaction between biotic and abiotic variables in a highly idealized way.

Photosynthesis and Respiration

Energy-rich bonds are produced as a result of photosynthesis, thus distorting the thermodynamic equilibrium. Bacteria and other respiring organisms catalyze the redox processes that tend to restore chemical equilibrium. In a simplified way we may consider a stationary state between photosynthetic production, $P = dp/dt$ (rate of production of organic material, p = algal biomass) and heterotrophic respiration, R (rate of destruction of organic material) [29] (Figure 8-13) and chemically characterize this steady state by a simple stoichiometry

$$106CO_2 + 16NO_3^- + HPO_4^{2-} + 122H_2O + 18H^+ \ (+\text{trace elements;}$$

$$\text{energy)}$$

$$P \ \big\updownarrow \ R$$

$$\{C_{106}H_{263}O_{110}N_{16}P_1\} + 138O_2 \qquad (19)$$

$$\text{algal protoplasm}$$

The balance between P and R, as shown earlier (Section 7-5), is responsible for regulating the concentration of O_2 in the atmosphere. Thus biological processes in the sea — approximately 70% of the total photosynthesis takes place in the sea [30] — regulate p_{O_2} and in turn the redox intensity of the interface between atmosphere and hydrosphere.

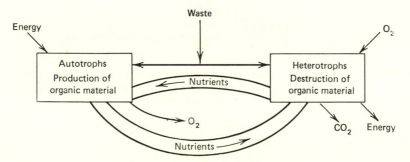

Figure 8-13   Biochemical relationships in streams.

[29] E. P. Odum, *Fundamentals of Ecology*, Saunders, Philadelphia, Pa., 1961.
[30] L. C. Cole, *Sci. Amer.*, **198**, 83 (April 1958).

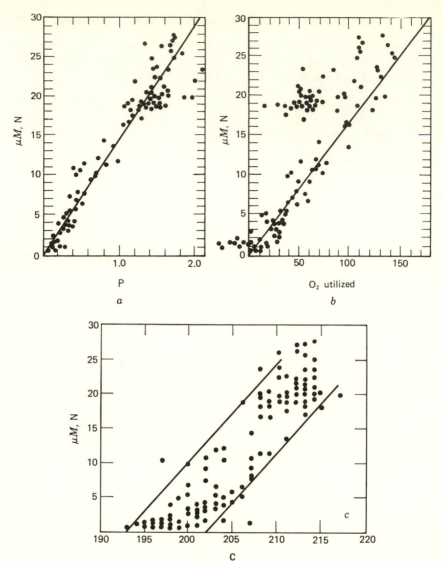

Figure 8-14 Stoichiometric correlations between nitrate, phosphate, oxygen, sulfide and carbon. The correlations can be explained by the stoichiometry of reactions such as (19).

a: Correlation between nitrate nitrogen and phosphate phosphorus in waters of western Atlantic.

b: Correlation between nitrate nitrogen and apparent oxygen utilization in same samples. Open circles represent samples from above 1000 m.

For *a* and *b* the concentrations are in μM. Phosphorus corrected for salt error. (After Redfield, 1934.)

c: Correlation between nitrate nitrogen and carbonate carbon in waters of western Atlantic. Concentrations in μM. (After Redfield, 1934.)

a and *b* reproduced with permission from Interscience Publishers.

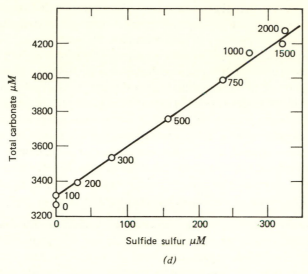

(d)

d: Relation of sulfide sulfur and total carbonate carbon in waters of the Black Sea. Numbers indicate depth of samples. Slope of line corresponds to $\Delta S^{2-}/\Delta C = 0.36$. (From data of Skopintsev et al., 1958, as quoted in [32].)

DEPARTURE FROM STEADY STATE. Figure 8-14 shows a correlation between dissolved nitrate and dissolved phosphate for the waters of the western Atlantic [31]. The molar ratio is $\Delta N/\Delta P = 16$. The fact that in other seas, too, the same constant correlation is found is a further example of how organisms can influence the composition of natural waters. The slope of the correlation curve in Figures 8-14*a* is readily understood if we consider that the differences in the concentration of phosphate and nitrate result from deviations in the steady state of *P* and *R* [32]. As suggested by the stoichiometry of (19) in the photosynthetic zone, phosphate and nitrate are eliminated from the water in a 1:16 ratio and used up for algae growth. Settled algae become mineralized in the deeper water layers and, as a result, phosphate and nitrate are released into solution again with a ratio of 1:16. Because O_2 participates in the photosynthesis and respiration reactions, a corresponding correlation between phosphate and O_2 ($\Delta O_2/\Delta P \simeq 138$) and between nitrate and O_2 ($\Delta O_2/\Delta N \simeq 9$) (Figure 8-14*b*) obtains. Figure 8-14*c* shows the correlation between nitrate–nitrogen and carbonate–carbon ($\Delta C/\Delta N = 106:15$). That the correlations found are in accord with the stoichiometric composition of plankton is obvious from Table 8-10.

[31] A. C. Redfield, *James Johnstone Memorial Volume*, 1934, p. 177, Liverpool.
[32] A. C. Redfield, B. H. Ketchum and F. A. Richards, in *The Sea*, Volume 2, M. N. Hill, Ed., Wiley-Interscience, New York, 1963.

It is remarkable that the summation of the complicated processes of the
P–R dynamics, in which so many different organisms participate, results in
such simple stoichiometry. The stoichiometric formulation of (19) reflects in
a simple way *Liebig's law of the minimum*. It follows from Figure 8-14*a* and *b*
that seawater becomes exhausted simultaneously in dissolved phosphorus and
nitrogen as a result of photosynthetic assimilation. We infer that nitrogen and
phosphorus together determine the extent of organic production if temporary
and local deviations are not considered. We might consider the possibility that
originally phosphorus (e.g., from apatite) was the sole minimum nutrient, but
that the concentration of nitrogen has been adjusted in the course of evolution
as a result of nitrogen fixation and denitrification to the ratio presently found.
Alternatively, we could also argue that the stoichiometric composition of the
organisms as a result of evolution has become the same as that in the sea.

ANOXIC CONDITIONS. The usual sequence of various redox reaction with
organic matter is observed where the accumulation of organic matter is great.
As shown in Table 7-6, oxygen dissolved in water first becomes exhausted,
then the oxidation of organic matter continues with nitrate serving as the
oxidant; subsequently organic fermentation reactions and redox reactions
with SO_4^{2-} and CO_2 as electron acceptors occur. The reduced products of the
oxidation of organic matter accumulate in the water in addition to the products of the oxidation of organic matter. The formal equations given in Table
7-6 can be modified to develop stoichiometric models that predict how the
components of the anoxic water will change as a result of the mineralization
of settled plankton. Such stoichiometric models have been developed by
Richards [33], who corroborated the validity of these models for a large
number of anoxic basins and fjords. For example, under conditions of
sufficiently low pϵ, planktonic material becomes oxidized by SO_4^{2-} which is

**Table 8-10   Ratios of the Elements Involved in the Oxidation of Organic
Matter in Seawater at Depth and Those Present in Plankton of
Average Composition, by Atoms (after Richards and Vaccaro,
1956)**

Seawater Analyses	ΔO	ΔC	ΔN	ΔP	Ref.
Northwest Atlantic	−180	105	15	1†	Redfield (1934)
Cariaco Trench, upper layers	−235	—	15	1	Richards and Vaccaro (1956)
Plankton analyses	−276‡	106	16	1	Fleming (1940)

† Corrected for salt error after Cooper (1938).
‡ Estimated assuming 2 atoms 0 ⇌ 1 atom C and 4 atoms 0 ⇌ 1 atom N; as quoted
in [32].

[33] F. A. Richards, in *Chemical Oceanography*, J. P. Riley and G. Skirrow, Eds.,
Academic Press, New York, 1965, Chapter 13.

reduced to S($-$II). For the oxidation of one $\{C_{106}H_{263}O_{110}N_{16}P_1\}$, approximately 424 electrons are necessary. (Note that the carbon in the plankton formula has a formal oxidation state of approximately 0.) In the reduction of SO_4^{2-} to S($-$II), 53 SO_4^{2-} ions can provide the 424 electrons necessary; hence, one would expect a ratio $\Delta S(-II)/\Delta C = 0.5$. As Figure 8-14$d$ shows, this ratio is somewhat less for the waters of the Black Sea.

Pollution of Natural Waters

Photosynthesis and respiration play an important role in the self-purification of natural waters. Temporally or spatially localized disturbance of the stationary state between photosynthesis and respiration leads to chemical and biological changes reflecting pollution. The condition $P > R$ is characterized by a progressive accumulation of algae which ultimately leads to an organic overloading of the receiving waters. On the other hand, when $R > P$ the dissolved oxygen may be used up (biochemical oxygen demand) and ultimately NO_3^-, SO_4^{2-} and CO_2 may become reduced to $N_2(g)$, NH_4^+, HS^- and $CH_4(g)$. The balance between P and R is necessary to maintain a water in nonpolluted and aesthetically pleasing condition. When $P \simeq R$ the organic material is decomposed by respiratory (heterotrophic) activity as fast as it is produced photosynthetically; O_2 produced by photosynthesis can be used for the respiration (see Figure 8-13). Pacific coral reefs [34] — ecosystems with very high rates of assimilation and metabolism — are an example that high value of P and R, as long as they are balanced with respect to each other, do not produce disorders and do not reflect water pollution [35, 36].

DISTURBANCE OF THE P–R BALANCE. A steady state between P and R is a prerequisite for the maintenance of a constant chemical composition of the water (chemostasis). Chemostasis, on the other hand, is a prerequisite for the maintenance of a relative constant structure of the population of organisms (homeostasis). As chemostasis depends on the coexistence of a sufficient number of phases, homeostasis depends — though for other reasons — on a sufficiently large heterogeneity of the population structure.

Stream pollution may be interpreted as a departure from the P–R balance. This balance is disturbed if a natural water receives an excess of organic bacterial nutrients or an excess of inorganic algae nutrients. The P–R balance may also be disturbed by a physical separation of the P- and R-organisms. In a stratified lake a vertical separation results from the fact that algae remain

[34] H. T. Odum, *Limnol. Oceanog.*, **1**, 102 (1956).
[35] W. Stumm and J. J. Morgan, *Transactions of the 12th Annual Conference on Sanitary Engineering*, Univ. of Kansas Press, Lawrence, Kans., 1962, p. 16.
[36] W. Stumm and M. W. Tenney, *Proceedings of the 12th Municipal and Industrial Waste Conference*, North Carolina State Univ. Press, Raleigh, N.C., 1963.

photosynthetically active only in the euphotic upper layers; algae that have settled under the influence of gravitation serve as food for the R-organisms in the deeper layers of the lake (Figure 8-13). Organic algal material that has been synthesized with excessive CO_2 or HCO_3^- in the upper layers of the lake becomes biochemically oxidized in the deeper layers; most of the photosynthetic oxygen escapes to the atmosphere and does not become available in the deeper water layers. An excessive production on the surface of the lake ($P \gg R$) is paralleled by anaerobic conditions at the bottom of the lake ($R \gg P$).

Eutrophication of Lakes

A vertical separation of P and R activities exists in every lake. The continuous sequence of nutrient assimilation and mineralization of organic matter accompanied by the physical cycle of circulation and stagnation leads to a retention of the fertilizing constituents (increase in residence time) and a progressive accumulation of organic matter in the water. This enrichment in nutrients and the enhancement of productivity and respiration is called eutrophication. The eutrophication of a lake is a natural aging process that leads ultimately to the filling up of the lake with sediments. Though uninfluenced by the activity of man, many lakes have disappeared [37] since the ice period. By altering the natural patterns of our environment, civilization may accelerate or retard the advancement of eutrophication. The discharge of domestic wastes and agricultural drainage have autocatalytically accelerated the nutrient enrichment of many lakes.

PROMOTION OF ALGAL GROWTH BY NITROGEN AND PHOSPHORUS. An estimate on the extent of the possible effect of algal nutrients (potential fertility) is obtained from the schematic stoichiometric equation (19) on production and respiration [35]. If phosphorus is the limiting factor, 1 mg of P is able to synthesize, according to (19), approximately 0.1 g of algal biomass (dry weight) in one single cycle of the limnological transformation. After settling to the deeper layers, this biomass exerts a biochemical oxygen demand of approximately 140 mg for its mineralization.

From this simple calculation it is obvious that the organic material that may be introduced into the lake by domestic wastes (20–100 mg organic matter liter^{-1}) is small in comparison to the organic material that can be biosynthesized from fertilizing constituents (3–8 mg P liter^{-1}). Aerobic biological waste treatment with a heterotrophic enrichment culture mineralizes substantial fractions of bacterially oxidizable organic substances but is not capable of eliminating more than 20–50% of nitrogen and phosphate con-

[37] A. D. Hasler, *Ecology*, **28**, 383 (1947).

stituents. This inefficiency of biological treatment with respect to removal of algal nutrients can be understood by comparing the elemental composition of domestic sewage with the stoichiometric relation between C, N and P in bacterial sludges. Most municipal wastes are nutritionally unbalanced for a heterotrophic enrichment process in the sense that they are deficient in organic carbon (see Figure 8-15) [36, 38].

One may ask the question, however: "To what extent is it possible to represent and interpret complicated natural processes in terms of simplified and schematic stoichiometric equations?" We know that the physiological conditions of growth differ from organism to organism, that there are algae which are capable of storing excess (luxury) phosphate, and that special growth factors, for example, biotin and niacin, can influence algal growth. With analytical data of individual lakes [39] we can see, however, that as in the oceans, simple stoichiometric relations hold. As shown in Figure 8-16, simple correlations between ΔP, ΔN and ΔO_2 exist in these lakes. The observed molar ratios may differ somewhat from the ratio postulated in (19), but systematic analytical errors, which are very difficult to avoid in the analysis of soluble phosphorus, may influence the slope of the curve. Especially pronounced is the evidence in the case of Lake Zürich (Figure 8-16) (intersection of ordinate by regression line) that nearly all phosphorus is eliminated by photosynthetic assimilation when the water still contains substantial concentrations of nitrogen compounds. This lends support to the hypothesis that in Lake Zürich, besides localized or temporary deviations, phosphorus must be indicted as the major production controlling factor. In other lakes, however, nitrogen may become the key element in fertilization; but we must also be aware that any incipient deficiency in nitrogen can, at least partially, be relieved by nitrogen fixation.

TRACE ELEMENTS. Of course a deficiency in any trace element (Mo, Mn, Si) can limit productivity. However, in a lake the total supply usually exceeds by far the need for trace constituents for a potential maximum production; thus only temporary and localized limitations of productivity seem possible.

BIOMASS AND PRODUCTIVITY. Although maximum possible production can be estimated quite reliably by stoichiometry, no simple correlation exists between biomass and productivity. Even if two lakes have a similar productivity, mass development of algae may occur in one ($P > R$) while the other remains unenriched ($P \leq R$). Riley [40], on the basis of experimental observations,

[38] W. Stumm, *Proceedings of the International Conference on Water Pollution Control*, Volume II, Pergamon, New York, 1963, p. 216.
[39] P. Zimmermann, *Schweiz. Z. Hydrol.*, **23**, 342 (1962). I. Ahlgren, *Schweiz. Z. Hydrol.*, **29**, 53 (1967).
[40] G. A. Riley, in *The Sea*, Volume 2, M. N. Hill, Ed., Wiley-Interscience, New York, 1963.

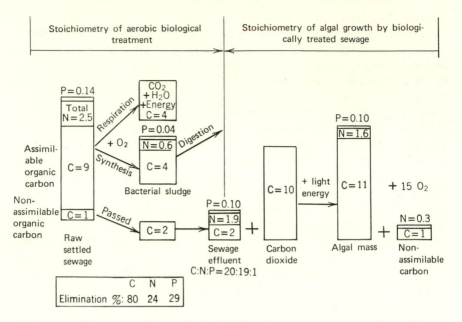

Figure 8-15 Stoichiometry of aerobic biological waste treatment. Comparing the relative composition of "average" domestic sewage with the mean stoichiometric relations between C, N and P in bacterial organisms shows in a schematic way that only a fraction of the phosphate and nitrogeneous material can become incorporated into the sludge. The inorganic nutrients as released by biologically-treated sewage effluents can be converted into algal cell material. CO_2 (or HCO_3^-) and other essential elements are usually available in sufficient quantities relative to nitrogen and phosphorus.

formulated the relation between biomass and productivity, $P = d[p]/dt$, as follows

$$P = \frac{d[p]}{dt} = (p_h - r - h)[p] \tag{20}$$

where p_h, r and h are empirically determined coefficients for photosynthesis, phytoplankton respiration and respiration by other organisms, respectively. Because the proportionality factor $(p_h - r - h)$ is time and location dependent, no simple proportionality exists between $[p]$ and P.

Differential Distribution of Nutrients in Estuaries, Fjords and the Mediterranean

We have illustrated how the characteristics of the circulation in lakes can markedly influence the distribution of the biochemically important elements.

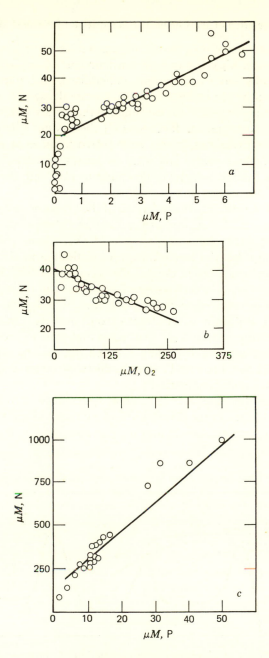

Figure 8-16 Correlation between concentrations of nitrate, phosphate and oxygen in lakes. For *a* and *b* (Lake Zürich), data collected at various depths during the summer stagnation period have been plotted. For *b* only results from the deeper water layers were considered. *c* (Lake Norrvicken) plots analytical results obtained during winter stagnation. (The lake was covered with ice.)

In *estuaries* and most fjords the circulation creates a trap in which nutrients tend to accumulate. Algae grown from nutrients carried seaward in the surface outflow eventually settle and become mineralized. The nutrients are then carried landward by the counter current of more dense seawater moving in from the outer sea to replace that entrained in the surface outflow. As pointed out by Redfield, Ketchum and Richards [32], the amount of circulation varies greatly in different estuaries. It may be expected to increase with the rate of production of organic matter and with the length of the basin, to decrease with turbulence and the velocity of flow, and to vary with the relative depth and velocity of the surface and deep layers.

A reversal of the currents characteristic of the estuarine circulation, that is, *an antiestuarine circulation*, leads to an impoverishment in nutrients of a natural body of water. An example is the *Mediterranean* which is the most impoverished large body of water known. Its antiestuarine circulation results from the fact that evaporation exceeds the accession of fresh water by about 4% [32]. Hence a much larger volume of water flows in through the Strait of Gibraltar than is required to replace the loss by evaporation. The excess escapes through the Strait as a counter current moving toward the Atlantic below the inflowing surface layer. Nutrients that tend to accumulate in the deeper layers are continuously being pumped out toward the Atlantic and toward the Black Sea. Figure 8-17 from Redfield, Ketchum and Richards [32] shows the situation diagrammatically.

As these cases illustrate, counter-current systems are particularly effective in producing changes in the distribution of nutrients along the direction of flow and lead to accumulation in the direction from which the surface current is flowing.

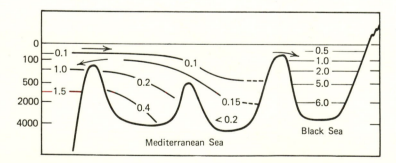

Figure 8-17 Influence of circulation upon distribution of phosphorus (from Redfield, Ketchum, and Richards [32]). Distribution of phosphorus in Black Sea, Mediterranean, and off-lying Atlantic Ocean (diagrammatic). Arrows indicate direction of currents in the Bosphorus and Strait of Gibraltar. Contours, phosphate phosphorus in μ moles l^{-1}. Depths in meters. Reproduced with permission from Interscience Publishers.

Water Pollution Control

An essential objective of water pollution control consists in the maintenance of the balance between photosynthesis and respiration. Keeping wastes away from waters or purifying water is frequently necessary in order to avoid gross pollutional effects (anaerobic conditions, mass development of algae, turbidity, etc.). Where prevention of fertilization is a more important consideration than the direct effects of biochemical oxygen demand, as is usually the case in lakes, treatment for removal of algal nutrients (especially phosphate) has more beneficial effects than biological treatment.

Stream pollution control, however, consists not only of waste treatment. Among the other possibilities for control are various biological and physical means of stream management directed toward reducing the relative residence time of fertilizing elements. Manipulations that interfere with or retard the circulation of nutrients might be considered. For example, it appears possible to increase direct consumption of algae by ecological and fish cultivation, that is, by an increase in ecological niches and in the number of herbivores (fish and zooplankton). It may also be possible to influence the physical cycle of water circulation, for example, by modifying hydraulically the influents and effluents of a lake through the installation of arrangements that permit periodic discharges of the bottom water layers into the effluent.

Example 8-9 *Vertical Exchange under Steady-State Condition.* Estimate the gradient in the phosphorus concentration, $\delta[P]/\delta x$ (Mm^{-1}) in the deeper portion of a lake during a stagnation period that results from the decomposition of sinking algae. Net photosynthesis amounts to 0.06 g carbon m^{-2} day^{-1}. The vertical eddy diffusivity D_E is found to be approximately 2×10^{-2} cm^2 sec^{-1}. About 50% of the algae grown in the euphotic zone sink into the deeper water layers.

In an oversimplified model we assume that the phosphorus produced from the decomposing algae is in stoichiometric relation to the carbon assimilated as given by (19). Furthermore, we assume a steady state in which the phosphorus produced by regeneration per unit area and per unit time, $\delta Q_p/\delta t$, equals the loss of phosphorus from sinking by eddy diffusion:

$$\frac{\delta Q_p}{\delta t} = D_E \frac{\delta[P]}{\delta x} \tag{i}$$

0.06 g carbon m^{-2} day^{-1} corresponds to an ultimate regeneration of 5.45×10^{-14} moles P cm^{-2} sec^{-1}, of which 50% sinks to the deeper layer; hence, $\delta Q_p/\delta t = 2.72 \times 10^{-14}$ moles P cm^{-2} sec^{-1}. From (i), a gradient in phosphorus concentration

$$\frac{\delta[P]}{\delta x} = 1.36 \times 10^{-7}\ Mm^{-1}$$

is obtained.

PROBLEMS

8-1 P. C. Hess [*Amer. J. Sci.*, **264**, 289 (1966)], by extrapolating hydrothermal results to 25°C and 1 atm and by utilizing mineral assemblages existing in the field, has evaluated the following phase equilibria in the Na_2O, Al_2O_3, SiO_2, H_2O system:

analcite-albite
$$NaAlSi_2O_6 \cdot H_2O(s) + H_4SiO_4 = NaAlSi_3O_8(s) + 3H_2O; \quad \log K = -2.7 \qquad (i)$$

montmorillonite-kaolinite
$$3Na_{0.33}Al_{2.33}Si_{3.67}O_{10}(OH)_2(s) + H^+ + 11\tfrac{1}{2}H_2O$$
$$= 3\tfrac{1}{2}Al_2Si_2O_5(OH)_4(s) + 4H_4SiO_4 + Na^+; \quad \log K = -7.3 \qquad (ii)$$

analcite-kaolinite
$$NaAlSi_2O_6 \cdot H_2O(s) + H^+ + \tfrac{3}{2}H_2O$$
$$= \tfrac{1}{2}Al_2Si_2O_5(OH)_4(s) + Na^+ + H_4SiO_4; \quad \log K = 5.1 \qquad (iii)$$

gibbsite-kaolinite
$$\tfrac{1}{2}Al_2O_3 \cdot 3H_2O(s) + H_4SiO_4 = \tfrac{1}{2}Al_2Si_2O_5(OH)_4(s) + 3H_2O; \quad \log K = 4.7 \qquad (iv)$$

Construct a phase diagram of $\log [Na^+]/[H^+]$ versus $\log [H_4SiO_4]$.

8-2 In many natural waters Cl^- and SO_4^{2-} are not regulated by equilibria. As a possible regulator for SO_4^{2-} in seawater, Kramer [*Geochim. Cosmochim. Acta*, **29**, 921 (1965)] has proposed a $SrCO_3(s)$–SO_4^{2-} equilibrium. Sr^{2+} concentration could be stated as a function of the carbonate concentration, and $[SO_4^{2-}]$ in turn as a function of $[Sr^{2+}]$. Sr and S–SO_4^{2-} are of approximately the same abundance in crystal rocks, hence neither of the constituents would likely become exhausted.

(i) To what extent could this equilibrium control the maximum sulfate level in fresh waters? Assume a model (25°C, $I = 0$) consisting of the phases calcite, $SrCO_3(s)$, $SrSO_4(s)$ aqueous solution and a gas phase containing CO_2. Assume $p_{CO_2} = 10^{-3.5}$ atm. The following constants are available: $pK_{s0}(SrCO_3) = 9.15$; $pK_{s0}(SrSO_4) = 6.55$.

(ii) Show that your answer is in accord with the restraints of the phase rule.

8-3 Compute semiquantitative plots for species concentration and pH versus concentration or dilution factor of the following systems:

(i) Isothermal *dilution* of 10^{-2} *M* $NaHCO_3$ in equilibrium with $p_{CO_2} = 10^{-3.5}$ atm.

(ii) Isothermal *evaporation* of a 10^{-4} *M* $Ca(HCO_3)_2$ solution in equilibrium with $p_{CO_2} = 10^{-3.5}$ atm.

8-4 F. A. Richards, (in *Chemical Oceanography*, J. P. Riley and G. Skirrow, Eds., Academic Press, New York, 1965) gives experimental data on the correlation between phosphate concentration and sulfide concentration (given in micromoles per liter) obtained for samples taken at various depths in an anoxic basin. The basin is permanently stratified. Why is the slope of the curve approximately 50?

8-5 In a well-eutrophied and stratified reservoir, the reservoir outlet can be set at different depths. How would you modify the outlet depth over the seasons in order to minimize the various nuisances of overfertilization?

ANSWERS TO PROBLEMS

8-2 (i) pH = 8.3, pHCO$_3$ = 3.0, pCa = 3.3, pSr = 4.15, pSO$_4$ = 2.4. Most fresh waters have smaller sulfate concentrations. (ii) P = 5, C = 5 (SrO, CaO, CO$_2$, H$_2$O, SO$_3$), $F = 2$ (temperature and p_{CO_2}).

Appendix † Clay Minerals

I. Layer Structure: Clay minerals are primarily crystalline aluminum or magnesium silicates with stacked-layer structures.‡ Each unit layer is in turn a sandwich of silica and gibbsite or brucite sheets. In the *silica* or *tetrahedral* (*T*) *sheet*, silicon atoms are each surrounded by four oxygen atoms in a tetrahedral arrangement; these tetrahedra are connected in an open hexagonal pattern in a continuous two-dimensional array. The *gibbsite* or *brucite* layer, or *octahedral* (*O*) *sheet*, consists of two layers of oxygen atoms (or hydroxyl groups) in a hexagonal closest packed arrangement with aluminum or magnesium atoms, respectively, in the octahedral sites.

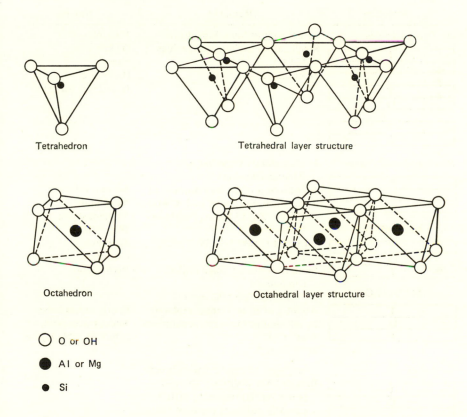

Tetrahedron   Tetrahedral layer structure

Octahedron   Octahedral layer structure

○ O or OH

● Al or Mg

• Si

† This draws on a report prepared by Joel Gordon (Harvard University).
‡ R. E. Grim, *Clay Mineralogy*, 2nd ed., McGraw-Hill, New York, 1968, and H. van Olphen, *An Introduction to Clay Colloid Chemistry*, Wiley-Interscience, New York, 1963.

These sheets are stacked into two- or three-layer units (T-O or T-O-T) in which the oxygen atoms at the vertices of the tetrahedra in the T-sheet also form the basic hexagonal pattern of the O-sheet (so that a T-O layer is 5 atoms thick and a T-O-T layer, 7 atoms). The unshared oxygens in the O-sheet are hydroxyl groups. Isomorphous substitution of Al(III) for Si(IV) in the T-sheet (or Mg(Fe,Zn)(II) for Al(III) in the O-sheet) may lead to negatively charged layers. In clay minerals, the T-O or T-O-T units are stacked one on another, with layers of water and/or interlayer or surface cations (to compensate the negative charge) beyond the units. These cations may be exchangeable, and interlayer water may be absorbed by a dry clay, causing it to swell.

II. Types and Examples

Structure	Remarks	Names
A. Two-Layer Clays	Little isomorphous substitution	Kaolinite
	Small cation exchange capacities (CEC)	Dickite
		Nacrite
	Nonexpanding	Halloysite
		(Interlayer water)
B. Three-Layer Clays	*1. Expanding (Smectites or Montmorillonites)*	
	Substitution of a small amount of Al for Si in T-sheet and of Mg, Fe, Cr, Zn, Li for Al or Mg in O-sheet	Montmorillonite
		Nontronite
		Volkhonskyite
		Hectorite
	Large CEC (M^{m+} = Na$^+$, K$^+$, Li$^+$, Ca^{2+}, . . .)	Saponite
		Sauconite
	Swell in water or polar organic compounds	Vermiculite
	2. Nonexpanding (Illites)	
	About $\frac{1}{4}$ of Si in T-sheet replaced by Al, similar O-sheet substitutions	Poorly crystallized
	Small CEC	Micas (muscovite,
	M^{m+} = K$^+$	biotite, phlogopite)
C. Chlorites	Three-layer alternating with brucite	
	Brucite layer positively charged [some Al(III) replacing M(II)], partially balances negative charge on T-O-T (mica) layer	
	Low CEC, nonswelling	

In the structure column:

A. Two-Layer Clays: stacked units of T, O; T, O; T, O

B. Three-Layer Clays: T, O, T / M^{+m}, nH_2O / T, O, T / M^{+m}, nH_2O / T, O, T

C. Chlorites: T, O, T / O (Brucite) / T, O, T

D. Fibrous Clays	Different type of structural units consisting of double silica chains (tetrahedral) joined to one-dimensional O-layers and containing interstitial water	Attapulgite Palygorskyite Sepiolite

In addition, there are *mixed-layer clays* in which unit layers of different clay minerals are stacked more or less regularly, and poorly crystallized or amorphous clays known as *allophanes*.

READING SUGGESTIONS

I. Rocks, Minerals and Sediments

Arrhenius, G., "Pelagic Sediments," in *The Sea*, Volume III, M. N. Hill, Ed., Wiley-Interscience, New York, 1966.

Deer, W. A., R. A. Howie and J. Zussman, *An Introduction to the Rock Forming Minerals*, Wiley, New York, 1966.

Degens, E. T., *Geochemistry of Sediments*, Prentice-Hall, Englewood Cliffs, N.J., 1965.

Grim, R. E., *Clay Mineralogy*, McGraw-Hill, New York, 1953.

II. Weathering

Anderson, D. H., and H. Hawkes, "Relative Mobility of the Common Elements in Weathering of Some Schist and Granite Areas," *Geochim. Cosmochim. Acta*, **14**, 204–210 (1958).

Carroll, D., "Rainwater as a Chemical Agent of Geologic Processes — A Review," *U.S. Geol. Surv. Water Supply Papers*, **1535-G** (1962), 18 pp.

Feth, J. H., C. Roberson and W. Polzer, "Sources of Mineral Constituents in Water from Granitic Rocks, Sierra Nevada California and Nevada," *U.S. Geol. Surv. Water Supply Papers*, **1535-I** (1964), 70 pp.

Jackson, M. L., "Weathering Sequence of Clay-sized Minerals in Soils and Sediments: II. Chemical Weathering of Layer Silicates," *Soil Sci.*, **16**, 3–6 (1948).

Muller, G., "Diagenesis of Argillaceous Sediments," in *Diagenesis in Sediments*, G. Larsen and G. V. Chilingar, Eds., Elsevier, Amsterdam, 1967.

Yaalon, D. H., "Weathering Reactions," *J. Chem. Educ.*, **36**, 73 (1952).

III. Equilibrium Models and Evolution of Ocean and Atmosphere

Bricker, O., and R. M. Garrels, "Mineralogical Factors in Natural Water Equilibria," in *Principles and Applications of Water Chemistry*, S. D. Faust and J. V. Hunter, Eds., Wiley, New York, 1967.

Goldberg, E. D., "Chemistry in the Oceans," in *Oceanography*, M. Sears, Ed., American Association for the Advancement of Science, Washington, D.C., 1961.

Holland, H. D., "The History of the Ocean Water and its Effect on the Chemistry of the Atmosphere," *Proc. Nat. Acad. Sci. U.S.*, **53**, 1173 (1965).

Kramer, J. R., "History of Sea Water. Constant Temperature–Pressure Equilibrium Models Compared to Liquid Inclusion Analyses," *Geochim. Cosmochim. Acta*, **29**, 921 (1965).

Kramer, J. R., "Equilibrium Models and Composition of the Great Lakes," in *Equilibrium Concepts in Natural Water Systems*, Advances in Chemistry Series, No. 67, American Chemical Society, Washington, D.C., 1967, p. 243.

Li, Y. H., T. Takahashi and W. S. Broecker, "Degree of Saturation of $CaCO_3$ in the Oceans," *J. Geophys. Res.*, **74**, 5507 (1969).

Mackenzie, F. T., and R. M. Garrels, "Chemical Mass Balance between Rivers and Oceans," *Amer. J. Sci.*, **764**, 507 (1966).

Morgan, J. J., "Applications and Limitations of Chemical Thermodynamics in Natural Water Systems," in *Equilibrium Concepts in Natural Water Systems*, Advances in Chemistry Series, No. 67, American Chemical Society, Washington, D.C., 1967, p. 1.

Rubey, W. W., "Geological History of Sea Water," *Bull. Geol. Soc. Amer.*, **62**, 1111 (1951). (Reprinted in *The Origin and Evolution of Atmospheres and Oceans*, P. J. Brancazio and A. G. W. Cameron, Eds., Wiley, New York, 1964.)

Siever, R. "Sedimentological consequences of a Steady-State Ocean-Atmosphere." *Sedimentology*, **11**, 5 (1968).

Sillén, L. G., "The Physical Chemistry of Sea Water," in *Oceanography*, M. Sears, Ed., American Association for the Advancement of Science, Washington, D.C., 1961.

IV. Influence of Biota upon Composition of Natural Waters

Odum, E. P., *Fundamentals of Ecology*, Saunders, Philadelphia, Pa., 1961.

Odum, H. T., "Primary Production in Flowing Waters," *Limnol. Oceanog.*, **2**, 102 (1957).

Redfield, A. C., B. H. Ketchum and F. A. Richards, "The Influence of Organisms on the Composition of Sea Water," in *The Sea*, Volume II, M. N. Hill, Ed., Wiley-Interscience, New York, 1966.

Stumm, W., and J. J. Morgan, "Stream Pollution by Algal Nutrients," in *Transactions of the 12th Annual Conference on Sanitary Engineering*, Univ. Kansas Press, Lawrence, Kansas, 1962.

9   The Solid–Solution Interface

9-1   Introduction

Most chemical reactions that occur in natural waters take place at phase discontinuities, that is, at the atmosphere–hydrosphere and at the lithosphere–hydrosphere interfaces. In this chapter the discussion is restricted to the solid–solution interface. This is still a very broad topic; even in single and reasonably well-defined systems the interfacial and colloidal properties involve so many variables and represent such complicated physical and chemical interactions that it is difficult to describe such systems as a unified subject.

The significance of the solid–solution interface in natural waters becomes apparent when one considers the state of subdivision of the solids typically present in natural waters. The dispersed phase in a natural body of water consists predominantly of inorganic colloids, such as clays, metal oxides, metal hydroxides and metal carbonates, and of organic colloidal matter of detrital origin, as well as living microorganisms (algae and bacteria). Most of the clay minerals have physical dimensions smaller than 1–$2\,\mu$ ($1\,\mu = 10^{-6}$ m). Montmorillonite, for example, has plate diameters of 0.002 to $0.2\,\mu$ and plate thicknesses in the range of 2 to 20 mμ. Many solid phases have specific *surface energies* of the order of a few million ergs m^{-2}; hence, for substances that expose specific surface areas of a few hundred m^2 g^{-1}, the total surface energy is of the order of 10^9 ergs g^{-1} (24 cal g^{-1}).

OBJECTIVES. We will consider a few pertinent interfacial phenomena in simple systems and to indicate ways in which the ideas that account for the properties of the single systems may be used for a qualitative understanding of some of the processes occurring in nature. Since a comprehensive survey of solid–solution phenomena would essentially cover all the divisions of colloid and surface chemistry, only a few areas have been selected for treatment.

Forces acting at interfaces are composed of extensions of forces acting within the two phases; phenomena particular to interfaces result from an unbalance of such forces. Solids in natural waters have electrically charged surfaces. One side of the interface assumes a net electrostatic charge, either positive or negative, and an equivalent number of counter ions of opposite charge form a diffuse layer in the aqueous phase. Such an electric double layer

445

exists at all interfaces in natural waters. Because of electrostatic repulsion, particles tend to remain dispersed, that is, in a colloidal state.

Historically, two broad theories have been advanced to explain the basic mechanism of colloid stability and characteristic properties of solid–solution phenomena.

1. The older, *chemical theory* [1] assumes that colloids are aggregates of defined chemical structure, that the primary charge of surfaces arises from the ionization of complex ionogenic groups present on the surface of the particles, and that the destabilization of colloids is due to such chemical interactions as complex formation and proton transfer.

2. The newer, *physical theory* emphasizes the concept of the electrical double layer and the significance of predominantly physical factors, such as counter ion adsorption, reduction of zeta potential and ion pair formation in the destabilization of colloids.

The physical or double-layer theory has been developed in great detail and has, in its various forms of simplification, found wide acceptance. It has become a most effective tool in the interpretation of interfacial phenomena. This theory has virtually replaced and superceded the older chemical theory.

These two theories are not, however, as mutually exclusive as they might appear to be on first sight. We want to call attention to the fact that purely chemical factors must be considered in addition to the theory of the double layer in order to explain, in a more quantitative way, many of phenomena pertinent to solid–solution interfaces in natural water systems.

A description of the causal relationships effective in natural water systems must include surface chemical reactions since these, in addition to those of biological activity, profoundly influence the temporal and spatial distribution of constituents. Because of the large extent of interfaces available, metal oxides and clays are of relevance in the regulation of the water composition.

9-2   Forces at Interfaces

Atoms, molecules and ions exert forces upon each other. At an interface there will inevitably be an unbalance of such forces; the properties of interfaces can be related to this unbalance. Ideally, the energy of interaction at an interface can be interpreted as a composite function resulting from the sum of attraction and repulsion forces, but our insight into intramolecular and interatomic forces is far from satisfactory. At a qualitative — and necessarily introductory — level we briefly enumerate the principal types of forces involved.

[1] W. Pauli and E. Valko, *Elektrochemie de Kolloide*, Julius Springer, Berlin, 1929.

Generally speaking, *chemical forces* extend over very short distances; a covalent bond can be formed only by merging of electron clouds. *Electric forces* extend over longer distances. Basically, these forces are defined by Coulomb's law. The electrostatic force of attraction or repulsion between two point charges separated by distance x is given by

$$F = -e_1 e_2 / x^2 \tag{1}$$

The potential energy $E(x)$ referred to infinite separation is then given by

$$E(x) = -\int_{\infty}^{x} F dx = (z_+)(z_-)e^2/x \tag{2}$$

There are other forces, principally of electrical nature, present in molecular arrays whose constituents possess a permanent dipole (H_2O, NH_3). The energy resulting from such dipole–dipole interaction is also called the *orientation energy*. Similarly, a charged species can induce a dipole (induction energy).

London (1930) showed that there is an additional type of force between atoms and molecules which is always attractive. Its origin can be qualitatively explained by considering that neutral molecules or atoms constitute systems of oscillating charges producing synchronized dipoles that attract each other; hence, this force is also, basically, an electric force. It is known as the *dispersion force* or the *London–van der Waals force*. These forces are, for example, primarily responsible for the departure of real gases from ideal behavior or for the liquefaction of gases. The energy of interaction resulting from these dispersion forces (2–10 kcal mole^{-1}) is small compared to that of a covalent bond or electrostatic (ion pair) bond ($\gg 10$ kcal mole^{-1}), but large compared to that of orientation or induction energy (< 2 kcal mole^{-1}). The van der Waals attraction energy between two atoms is inversely proportional to the sixth power of distance over small distances. The total interaction between two semi-infinite flat plates is obtained by summing the pairwise interactions of all the constituent atoms (the summation is equivalent to a triple integration). The resulting attractive energy per unit area between two plates is inversely proportional to the second power of the distance

$$V_A \simeq \frac{A}{12 \pi x^2} \tag{3}$$

where A is the Hamaker constant. Its value depends on the density and polarizability of the material and is typically of the order of 10^{-12} to 10^{-13} ergs. This equation is valid for distances up to approximately 200 Å; for larger distances corrections must be made that make the attractive energy decay as $1/x^3$.

Because of the very small size of the hydrogen ion, its largest coordination number is 2. In a *hydrogen bond*, H^+ accommodates 2 electron pair clouds in order to bind two polar molecules. While van der Waals interactions are principally spherically symmetric, hydrogen bonding occurs at preferred molecular orientation. Hydrogen bonding, however, is in the same energy range (2–10 kcal mole^{-1}) as van der Waals interaction. The unusually high boiling point of water, for example, in comparison to $H_2S(l)$, presents evidence for hydrogen bonding in addition to van der Waals association.

Surface Tension

Because of the unbalance of forces at an interface, molecules in the surface or interfacial region are subject to attractive forces from adjacent molecules which result in a net attraction into the bulk phase. The attraction tends to reduce the number of molecules in the surface region (increase in inter-molecular distance). Hence, work must be done to bring molecules from the interior to the interface and, in turn, to increase the area of the interface. The minimum work required to create surface $d\bar{A}$ is $\gamma d\bar{A}$, where $\bar{A}$ is the interfacial area and γ is the surface tension or interfacial tension. γ is thus the interfacial Gibbs free energy for the condition of constant temperature, pressure and composition:

$$\gamma = \left(\frac{\delta G}{\delta \bar{A}}\right)_{T,p,n} \tag{4}$$

γ has the units dynes cm^{-1} or ergs cm^{-2}.

In water the intermolecular forces producing surface tension are essentially composed of (a) London–van der Waals dispersion forces, $\gamma_{H_2O(L)}$ and (b) hydrogen bonds, $\gamma_{H_2O(H)}$

$$\gamma = \gamma_{H_2O(L)} + \gamma_{H_2O(H)} \tag{5}$$

In water about one-third of the interfacial tension is due to van der Waals attraction and the remainder is due to hydrogen bonding (Table 9-1). Similarly, for other liquids, the surface tension can be divided into two parts: the part due to dispersion forces and the part due to more specific chemical forces (e.g., hydrogen bonding, pi electron interactions in aromatic solvents, metallic bonds in metals) [2]. For liquid hydrocarbons the intermolecular

[2] F. M. Fowkes, "Attractive Forces at Interfaces," in *Chemistry and Physics at Interfaces*, American Chemical Society, Washington, D.C., 1965. This comprehensive presentation unifies several fields of surface chemistry. It gives a background useful in deciding what molecules or chemical groups should bond most strongly to a given substrate and also what contaminants are most likely to be strongly attached to surfaces or are capable of penetrating into interfaces. A more recent detailed discussion on work of adhesion by the same author is in *J. Colloid Interf. Sci.*, **28**, 493 (1968).

Table 9-1   Attractive Forces at Interfaces.† Surface Energy, γ, and London–van der Waals Dispersion Force Component of Surface Energy, $\gamma_{(L)}$

	γ, dynes cm^{-1} or ergs cm^{-2}, 20°C	$\gamma_{(L)}$,
Liquids ‡		
Water	72.8	21.8
Mercury	484	200
n-Hexane	18.4	18.4
n-Decane	23.9	23.9
Carbon tetrachloride	26.9	—§
Benzene	28.9	—
Nitrobenzene	43.9	—
Glycerol	63.4	37.0
Solids ‖		
Paraffin wax	—§	25.5
Polyethylene	—	35.0
Polystyrene	—	44.0
Silver	—	74
Lead	—	99
Anatase (TiO_2)	—	91
Rutile (TiO_2)	—	143
Ferric oxide	—	107
Silica	—	123 (78)
Graphite	—	110

† Based on information provided by Fowkes [2].
‡ $\gamma_{(L)}$ values for water and mercury have been determined (Fowkes) by measuring the interfacial tension of these liquids with a number of liquid saturated hydrocarbons. The intermolecular attraction in these liquids is entirely due to London–van der Waals dispersion forces for all practical purposes. $\gamma_{(L)}$ was derived (Fowkes) from contact angle measurements.
§ A dash indicates that no value is available.
‖ $\gamma_{(L)}$ of solids were derived (Fowkes) from contact angle measurements or from measurements of equilibrium film pressures of adsorbed vapor on the solid surface.

attraction is almost entirely due to London dispersion forces; for Hg(l), on the other hand, the contribution due to metallic bonds is significant.

As shown by Fowkes [2], the interfacial tension between two phases (whose surface tensions — with respect to vacuum — are γ_1 and γ_2) is subject to the resultant force field made up of components arising from attractive forces in the bulk of each phase and the forces, usually the London dispersion forces, operating across the interface itself. Then the interfacial tension between two phases γ_{12} is given by

$$\gamma_{12} = \gamma_1 + \gamma_2 - 2(\gamma_{1_{(L)}}\gamma_{2_{(L)}})^{1/2} \qquad (6)$$

The geometric mean of the dispersion force components $\gamma_{1_{(L)}}$ and $\gamma_{2_{(L)}}$ may be interpreted as a measure of the interfacial attraction resulting from dispersion forces between adjacent dissimilar phases. $\gamma_{(L)}$ values of solid phases (Table 9-1) may be used to compare adsorbent properties of various materials since $\gamma_{(L)}$ of the solid phase is a quantitative measure of the available energy of the solid surface for interaction with adjacent media [2]. Of the materials listed, graphite and oxides (rutile, ferric oxide) are strong adsorbents for hydrocarbons. Low-energy organic solids are weak adsorbents. The much weaker intermolecular forces in soft, organic solids of low molecular weight are reflected in their lower surface free energies; hence the dispersion force contribution to surface free energy can also be used to estimate the long-range attraction forces between solid bodies. Values for the Hamaker constant have been estimated this way [2]. Adsorption of organic films onto a solid converts the surface into a low-energy surface. Similarly, adsorbed water greatly decreases the surface energy of glasses, silica, alumina and metals [3]. As more than a monolayer of H_2O is adsorbed, the surface energy of the solid may approach the surface tension of a bulk water surface (72 ergs cm^{-2} at 20°C). Similarly, as pointed out by Baier et al. [3], a subtle change in the surface composition of a biological membrane can greatly decrease or increase its wetability and also its adhesiveness. An outermost substituent which decreases the interfacial tension, such as $-CH_3$ or $-CH_2-$, would decrease adhesiveness, whereas one which increases γ (such as $-C_6H_5$, $-OH$, $-SH$, $-COOH$ or $-NH_2$) would increase adhesiveness [3].

ADSORPTION FROM SOLUTION. Adsorption of a solute molecule on the surface of a solid can involve removing the solute molecule from the solution, removing solvent from the solid surface and attaching the solute to the surface of the solid. The net energy of interaction of the surface for the adsorbate may result from chemical interactions, such as covalent bonding, from electrostatic interactions (e.g., ion exchange reactions), from van der Waals attraction, from hydrogen bonding or from orientation energies. For some solutes, solid affinity for the solute can play a subordinate role in comparison to the affinity of the aqueous solvent. Organic dipoles and large organic ions are preferentially accumulated at the solid–solution interface primarily because their hydrocarbon parts have low affinity for the aqueous phase. Simple inorganic ions (e.g., Na^+, Ca^{2+}, Cl^-), even if they are specifically attracted

[3] R. E. Baier, E. G. Shaffrin and W. A. Zisman, *Science*, **162**, 1360 (1968).

to the surface of the colloid, may remain in solution because they are readily hydrated. Less hydrated ions (for example, Cs^+, $CuOH^+$ and many anions) seek positions at the interface to a larger extent than easily hydrated ions.

Such considerations are also contained in the qualitative rules that a polar adsorbent adsorbs the more polar component of nonpolar solutions preferentially whereas a nonpolar surface prefers to adsorb a nonpolar component from a polar solution. In accord with these generalizations is Traube's rule, according to which the tendency to adsorb organic substances from aqueous solutions increases systematically with increasing molecular weight for a homologous series of solutes. Thus the interaction energy due to adsorption increases rather uniformly for each additional CH_2 group.

ADSORPTION ISOTHERMS. Adsorption is most often described in terms of isotherms which show the relationship between the bulk activity temperature. Adsorption isotherms and two of the simpler equations commonly used to characterize adsorption equilibria are given in Figure 9-1 and Table 9-2.

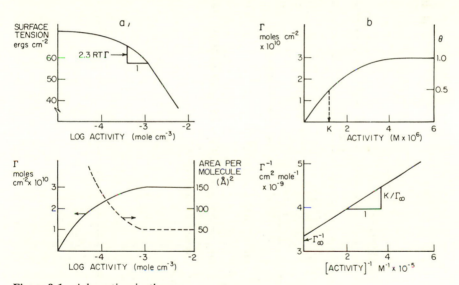

Figure 9-1 Adsorption isotherms.

 a: Gibbs adsorption. The surface excess, Γ, and hence the area occupied per molecule or ion of adsorbed substance, can be computed from the slope of a semilogarithmic plot of surface (or interfacial) tension versus log activity of sorbate [(4) of Table 9-2]. The application of the Gibbs equation to adsorption with electrostatic interaction is illustrated in Table 9-4.

 b: Langmuir adsorption isotherm. From the adsorption isotherm [plotted in accordance with (5) and (6) of Table 9-2], the equilibrium constant K and the adsorption capacity, Γ_∞, obtain by plotting Γ^{-1} versus the reciprocal activity of the sorbate [(7) of Table 9.2].

Table 9-2   Adsorption Equilibria†

	Gibbs‡§	Langmuir‡§
Adsorbed substance per unit area $\Gamma = n_i/A$ $= \Gamma_\infty \theta$ (for monolayer adsorption)	$\Gamma_i = -\left(\dfrac{\delta\gamma}{\delta\mu_i}\right)_{T,P,\text{ all }\mu\text{'s except }\mu_i\text{ and }\mu_{H_2O}}$ (1) $\Gamma_i = -\dfrac{1}{RT}\left(\dfrac{\delta\gamma}{\delta\ln a_i}\right)_{T,P,\text{ all }\mu\text{'s except }\mu_i\text{ and }\mu_{H_2O}}$ (2)	$\dfrac{\theta}{1-\theta} = K^{-1}a_i$ (5) $\Gamma_i = \Gamma_{i\infty}\dfrac{a_i}{K + a_i}$ (6)
Example: Adsorption of a hydrocarbon (HC)	$d\gamma = -RT\Gamma_{HC}d\ln\{HC\}$ (3)	$\dfrac{1}{\Gamma_i} = \dfrac{1}{\Gamma_{i\infty}} + \dfrac{K}{\Gamma_{i\infty}}\dfrac{1}{\{HC\}}$ (7)
Example: Adsorption of a hydrocarbon (HC) and of an alcohol (ROH)	$d\gamma = -RT\Gamma_{HC}d\ln\{HC\} - RT\Gamma_{ROH}d\ln\{ROH\}$ (4)	$\Gamma_{ROH} = \dfrac{\Gamma_{ROH\infty}K_{ROH}^{-1}\{ROH\}}{1 + K_{ROH}^{-1}\{ROH\} + K_{HC}^{-1}\{HC\}}$ (8)
Basis for application: $\mu_{i\,(\text{interface})} = \mu_{i\,(\text{bulk phase})}$	Reversible equilibrium	(1) Reversible equilibrium up to monolayer $(\theta = 1)$.‖ K is related to the free energy of adsorption: $K = \exp[\Delta G^\circ/RT](G')$ (2) Fixed site adsorption (immobility of adsorbate)# (3) Homogeneity in surface (energy of adsorption independent of θ)

Table 9.2 continued

† μ = chemical potential

γ = interfacial tension, erg cm^{-2}

a_i = activity of species i, M

θ = fraction of surface covered with adsorbate

R = gas constant

K = adsorption equilibrium constant, M (if $K = a_i$, $\theta = 0.5$)

Γ_i = adsorption density, moles cm^{-2}

‡ In most cases, concentrations may be used instead of activities; however, the bulk concentration can no longer be defined in terms of quantities added if the solutes have associated to form micelles. Similar difficulties arise when the adsorbates in the interface undergo irreversible changes, such as denaturation reactions.

§ Γ_i is defined in such a way that in dividing the interface, the adsorption density of water is zero (Gibbs convention).

‖ A similar adsorption model for multilayer adsorption has been set forth in the *Brunauer, Emmett and Teller (BET)* equation which has the form:

$$X = \frac{X_\infty B a_i}{(a_{is} - a_i)[1 + (B - 1)a_i/a_{is}]}$$

where X and X_∞ are the observed and ultimate quantity of substance adsorbed per unit area, B, an energy related constant, and a_{is} is the saturation activity of the solute.

Varying energies of adsorption can arise because of heterogeneity of the surface, that is, the free energy decreased with coverage. The Tempkin isotherm is based on such a model. In the *Tempkin isotherm*, θ is a function of log a_i at intermediate coverages ($0.2 < \theta < 0.8$).

By means of the *Gibbs equation* a quantitative relation between adsorption and interfacial tension is established. In a qualitative sense the equation illustrates that dissolved material which lowers the surface or interfacial tension of a solution tends to accumulate at the interface because less work is then required to increase the area of the interface. While simple electrolytes in water raise the interfacial tension, that is, increase the cohesive forces of the H_2O molecules, the interfacial tension is reduced by many nonelectrolytes and especially by polar–nonpolar substances which work against the adhesive forces of the water molecules. Almost all organic substances found in natural waters fall into the latter category and are thus concentrated or adsorbed at the surface or at the solid–liquid interface. In the absence of specific chemical or electrostatic attraction forces of the sorbent, ions cannot be adsorbed at interfaces because they tend to increase the interfacial tension.

Molecules or ions that possess a hydrophobic and hydrophilic part tend to orient at interfaces. Detergents are notable examples of such substances with a dual constitution; the polar group is immersed in the water and the hydrophobic group exposed to air, oil or the solid phase.

Many experimental adsorption isotherms are satisfactorily described by the *Langmuir equation* (Figure 9-1). However, a satisfactory fit of the experimental points to the equation does not necessarily imply that the conditions that form the basis of the theoretical Langmuir model are fulfilled.

9-3 The Electric Double Layer

Figure 9-2 shows that near neutral pH values, clays, most insoluble oxides, organic pollutants, bacteria and algae are characterized by a negative surface potential as indicated by negative values of the electrophoretic mobility.

In this section we will illustrate how charge affects some of the properties of solid surfaces, how it influences sorption and ion exchange behavior, and how it determines the state of dispersion of the solids.

Origin of Surface Charge

There are three principal ways in which the surface charge may originate.

1. The charge may arise from *chemical reactions* at the surface. Many solid surfaces contain ionizable functional groups: —OH, —COOH, —OPO$_3$H$_2$. The charge of these particles becomes dependent on the degree of ionization (proton transfer) and consequently on the pH of the medium. For example, the electric charge of a silica surface in water can be explained by the acid–base behavior of the silanol (—Si—OH) groups found on the surface of hydrated silica

$$\text{—Si—OH}_2{}^+ \overset{K_1}{\rightleftharpoons} \text{—Si—OH} \overset{K_2}{\rightleftharpoons} \text{—Si—O}^- \qquad (7a)$$

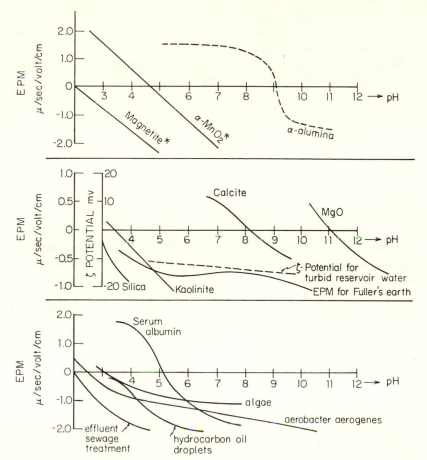

Figure 9-2 pH dependence of charge. In the neutral pH range the suspended solids typically encountered in natural waters are negatively charged. These simplified curves are based on results by different investigators whose experimental procedures are not comparable and may depend upon solution variables other than pH. The curves should exemplify trends and are meaningful in a semiquantitative way only.

Most oxides and hydroxides exhibit such amphoteric behavior; thus the charge is strongly pH dependent, being positive at low pH values. Similarly, for an organic surface, for example, that of a bacterium, one may visualize the charge as resulting from protolysis of functional amino and carboxyl groups, for example,

$$
\begin{array}{ccccc}
\text{COOH} & & \text{COO}^- & & \text{COO}^- \\
/ & & / & & / \\
\text{R} & \underset{}{\overset{K_1}{\rightleftharpoons}} & \text{R} & \underset{}{\overset{K_2}{\rightleftharpoons}} & \text{R} \\
\backslash & & \backslash & & \backslash \\
\text{NH}_3{}^+ & & \text{NH}_3{}^+ & & \text{NH}_2
\end{array}
\qquad (7b)
$$

At low pH a positively charged surface prevails; at high pH, a negatively charged surface. At some intermediate pH (the isoelectric point or the zero point of charge, pH_{ZPC}) the charge will be zero.

Charge can also originate by processes in which solutes become coordinatively bound to solid surfaces, for example,

$$S(s) + S^{2-} = S_3^{2-}(surface) \tag{8}$$

$$Cu(s) + 2H_2S = Cu(SH)_2^{2-}(surface) + 2H^+ \tag{9}$$

$$AgBr(s) + Br^- = AgBr_2^-(surface) \tag{10}$$

$$FeOOH(s) + HPO_4^{2-} = FeOHPO_4^-(surface) + OH^- \tag{11}$$

$$R(COOH)_n + mCa^{2+} = R - [(COO)_nCa_m]^{-n+2m}(surface) + nH^+ \tag{12}$$

$$MnO_2 \cdot H_2O(s) + Zn^{2+} = MnOOHOZn^+(surface) + H^+ \tag{13}$$

The alteration in surface charge results from chemical reactions; these phenomena are frequently referred to in the literature as adsorption or specific adsorption.

2. Surface charge at the phase boundary may be caused by lattice imperfections at the solid surface and by *isomorphous replacements* within the lattice. For example, if in any array of solid SiO_2 tetrahedra an Si atom is replaced by an Al atom (Al has one electron less than Si), a negatively charged framework is established

$$\left[\begin{array}{ccccc} HO & O & O & O & OH \\ \diagdown \diagup & \diagdown \diagup & \diagdown \diagup & \diagdown \diagup \\ Si & Al & Si & Si \\ \diagup \diagdown & \diagup \diagdown & \diagup \diagdown & \diagup \diagdown \\ HO & O & O & O & OH \end{array} \right]^-$$

Similarly, isomorphous replacement of the Al atom by Mg atoms in networks of aluminum oxide octahedra leads to a negatively charged lattice. Clays are representative examples where such atomic substitution causes the charge at the phase boundary. Sparingly soluble salts also carry a surface charge because of lattice imperfections. An introduction to the structure of clays is given in the Appendix of Chapter 8.

3. A surface charge may also be established by *ion adsorption*. Preferential adsorption of one type of ion on the surface can arise from London–van der Waals interactions and from hydrogen bonding.† Typical examples are adsorption of a surfactant ion on a clay surface, a humic acid or natural color

---

† Specific adsorption caused by these forces should be distinguished from chemical reactions as outlined above under (2).

anion on a silica surface and an isopolycation (such as a multinuclear hydroxo–metal complex) on a calcite surface.

Constant Potential and Constant Charge Surfaces

Electrical charge on the surface and electrical potential difference between the surface and the bulk of the solution are related. Ideally, we can distinguish two models: (a) the charge on the surface is fixed and remains independent of solution composition and (b) the surface potential remains constant and its magnitude is not affected by the presence of indifferent electrolytes.

A constant charge is established, for example, on a mineral containing isomorphous substitution in its lattice. Thus the plate of clay surfaces is characterized by a constant charge. Similarly, a solid surface covered with a monolayer of fatty acids has a fixed number of charges at constant pH.

Alternatively, the potential of a solid surface may be controlled externally. We may keep the potential of a metal electrode constant by means of an external potentiostatic arrangement, and the potential of a surface can be fixed by keeping constant the bulk concentration of certain solutes whose specific chemical reaction with the surface gives rise to the potential. The potential of a solid surface is determined by a solute (potential-determining species) whenever we can formulate a chemical reaction between the surface of the solid and constituents of the solution. Typical examples are

$$Ag(s) = Ag^+ + e \tag{14}$$

$$AgBr(s) + e = Ag(s) + Br^- \tag{15}$$

$$AgBr(s) + Ag^+ = Ag_2Br^+(surface) \tag{16}$$

$$CuS(s) + H_2S = Cu(SH)_2(surface) \tag{17}$$

$$FePO_4(s) + H_2PO_4^- = Fe(HPO_4)_2^-(surface) \tag{18}$$

$$FeOOH(s) + H^+ = Fe(OH)_2^+(surface) \tag{19}$$

$$FeOOH(s) + OH^- + H_2O = Fe(OH)_4^-(surface) \tag{20}$$

$$SiO_2(s) + OH^- = SiO_2(OH)^-(surface) \tag{21}$$

(Potential determining species are italicized.)

Constant potential surfaces are common in electrochemistry and to many of the colloids used in colloid chemistry; gold sols, silver halogenide sols and the iron oxide sols are included in this category. Because H^+ and/or OH^- are potential-determining ions, metal oxides may be interpreted to have constant potential surfaces at constant pH of the solution.

The electric potential drop Ψ_0 across the double layer (i.e., the difference in potential between the surface and the bulk of the solution) is given, in principle, by the free energy change $-\Delta G$ involved in changing the solution composition from that under consideration into that of a reference state in which $\Psi_0 = 0$ (zero point of change).† In the case of

$$-\text{Me}-\text{OH(s)} + \text{OH}^- = -\text{Me}-\text{O}^-(\text{surface}) + \text{H}_2\text{O}$$

$$-\Delta G = F\Psi_0 = RT \ln \frac{\{\text{OH}_0^-\}}{\{\text{OH}^-\}} \qquad (22)$$

Because H^+ and OH^- are interrelated, we can generalize

$$\Psi_0 = \frac{RT}{F} \ln \frac{\{\text{H}^+\}}{\{\text{H}_0^+\}} = -\frac{RT}{F} \ln \frac{\{\text{OH}^-\}}{\{\text{OH}_0^-\}} \qquad (23)$$

where F is the Faraday and $\{\text{OH}_0^-\}$ is the OH^- ion activity at the zero point of charge. *The pH of zero point of charge*, pH_0 or pH_{ZPC}, is different for different metal oxides and is a measure of the acidity and basicity of the hydrated surface oxide groups. Ideally, for a unit change in pH, Ψ_0 varies 58 mV at 20°C.

For many systems more than one potential-determining species may be present in the solution. For example, the surface potential of elemental sulfur can be interpreted as resulting from the reaction

$$\text{S(s)} + \text{H}_2\text{S(aq)} = \text{S}_2^{2-}(\text{surface}) + 2\text{H}^+ \qquad (24)$$

for which

$$-\Delta G = F\Psi_0 = RT \ln \frac{\{\text{H}_0^+\}^2}{\{\text{H}^+\}^2} \frac{\{\text{H}_2\text{S}\}}{\{\text{H}_2\text{S}_0\}} \qquad (25)$$

At constant pH the surface potential of the sulfur sols varies with the activity of the potential-determining species:

$$\Psi_0(\text{pH}) = \frac{RT}{F} \ln \frac{\{\text{H}_2\text{S}\}}{\{\text{H}_2\text{S}_0\}} = \frac{RT}{F} \ln \frac{\{\text{HS}^-\}}{\{\text{HS}_0^-\}} = \frac{RT}{F} \ln \frac{\{\text{S}^{2-}\}}{\{\text{S}_0^{2-}\}} \qquad (26)$$

As far as the solution phase is concerned, it is immaterial whether H_2S, HS^-, S^{2-} or a polysulfide is considered as the potential-determining species.

There are cases where both constant charge and constant potential surfaces are present simultaneously or where both mechanisms act together. In clays the double layer at the plate is produced by a fixed charge resulting from isomorphous substitution in the lattice, whereas the double layer at the edges where $\text{Si}-\text{O}$ or $\text{Al}-\text{O}$ bonds are broken (perpendicular to the sheets) is fixed by H^+ and OH^- as potential-determining species.

Potential and Charge Distribution in Double Layer

The simplest theory for the distribution of charge and potential in a double layer has been given by Gouy and later by Chapman. They considered that ions at a charged solid–solution interface are subject to coulombic attraction

† For a more rigorous derivation see reference [5].

and repulsion forces, but because of thermal motion they tend to be more diffusively distributed in the solution (diffuse double layer). The theory shows that the charge density in the solution decreases rapidly with the distance from the surface and that the potential Ψ declines (exponentially for small values of Ψ) as a function of distance. The theory treats ions as point charges which do not interact with each other.

QUALITATIVE TREATMENT.† Table 9-3 gives a brief derivation of the equations. An example treating these equations numerically will be given in Example 9-1. This discussion attempts to provide a qualitative understanding of the most relevant features of this model (Figure 9-3).

The theory shows that the charge density resulting from the distribution of ions in the solution decreases rapidly with distance from the surface. The net double-layer charge in the solution is reflected in the disturbance of the electroneutrality of the ions close to the surface; it is represented by the total area between the two concentration curves in Figures 9-3b and e which respectively represent the charge resulting from the surplus of cations, σ_+, and deficiency of anions, σ_-, in the double layer. The net charge in the solution is of equal magnitude and opposite sign to the charge on the solid side of the interface (shaded areas in Figures 9-3a, d and g). Figures 9-3c and f illustrate the approximately exponential decline of the potential Ψ as a function of the distance. In a medium of low ionic strength the disturbance in the electro-neutrality extends further into the solution than in the case of a solution with high ionic strength. The center of charge falls at a plane of distance $1/\kappa$, referred to as the double-layer thickness, and which depends on the electrolyte concentration. For water at 20°C, (8) of Table 9-3 gives

$$\kappa^{-1} \simeq 3.0 \times 10^{-8} I^{-0.5} \text{ (cm)} \tag{27}$$

where I is the ionic strength (M).

To a first approximation the double layer may be visualized as a parallel plate condenser of variable distance κ^{-1} between the two plates and with its capacitance

$$C = \frac{\epsilon}{4\pi\kappa^{-1}} \tag{28}$$

† For a more detailed discussion see [4–6].
[4] A. W. Adamson, *Physical Chemistry of Surfaces*, 2nd ed., Wiley-Interscience, New York, 1967.
[5] P. L. de Bruyn and G. E. Agar, in *Froth Flotation*, D. W. Fuerstenau, Ed., American Institute of Mining Metallurgical and Petroleum Engineers, New York, 1962.
[6] H. van Olphen, *An Introduction to Clay Colloid Chemistry*, Wiley-Interscience, New York, 1963.

Table 9-3   Gouy–Chapman Theory of Single Flat Double Layer†

I. VARIATION OF CHARGE DENSITY IN SOLUTION

Equality of electrochemical potential, $\bar{\mu}$ $(= \mu + zF\Psi)$ of every ion, regardless of position.

Electrochemical Potential

$$\bar{\mu}_{+(x)} = \bar{\mu}_{+(x = \infty)}; \qquad \bar{\mu}_{-(x)} = \bar{\mu}_{-(x = \infty)} \tag{1}$$

$$zF(\Psi_{(x)} - \Psi_{(x = \infty)}) = RT \ln \frac{n_{+(x)}}{n_{+(x = \infty)}} = -RT \ln \frac{n_{-(x)}}{n_{-(x = \infty)}} \tag{2}$$

Cations $\qquad\qquad n_+ = n_{+(x = \infty)} \exp\left(-zF\Psi_{(x)}/RT\right)$ \hfill (3)

Anions $\qquad\qquad n_- = n_{-(x = \infty)} \exp\left(zF\Psi_{(x)}/RT\right)$ \hfill (4)

Charge Density $\qquad q = zF(n_+ - n_-)$ \hfill (5)

II. LOCAL CHARGE DENSITY AND LOCAL POTENTIAL

Ψ and q are related by Poisson's equation

Poisson's Equation $\qquad\qquad d^2\Psi/dx^2 = -(4\pi/\epsilon)q$ \hfill (6)

Combining (3), (4), and (5) with (6) and considering that $\sinh x = (e^x - e^{-x})/2$ gives the *Double-Layer Equation*

$$d^2\Psi/dx^2 = \kappa^2/(zF/RT) \sinh (zF\Psi/RT) \tag{7}$$

where

$$\kappa = \left(\frac{8\pi \bar{n} e^2 z^2}{\epsilon kT}\right)^{1/2} = \left[\frac{8\pi e^2}{\epsilon kT}\left(\frac{NI}{10^3}\right)\right]^{1/2} \tag{8}$$

For convenience the following substitutions can be made:

$$y = zF\Psi/RT; \qquad \bar{z} = zF\Psi_0/RT; \qquad \xi = \kappa x \tag{9}$$

Considering (9), (8) becomes

Substituted Double-Layer Equation

$$d^2y/d\xi^2 = \sinh y \tag{10}$$

For boundary conditions, if $\xi = \infty$, $dy/d\xi = 0$ and $y = 0$

first integration: $\qquad\qquad dy/d\xi = 2 \sinh (y/2)$, or \hfill (11)

$$d\Psi/dx = \frac{RT}{zF} 2\kappa \sinh (y/2) \tag{11a}$$

and for boundary conditions, if $\xi = 0$, $\Psi = \Psi_0$ or $y = \bar{z}$

second integration: $\qquad e^{y/2} = \dfrac{e^{\bar{z}/2} + 1 + (e^{\bar{z}/2} - 1)e^{-\xi}}{e^{\bar{z}/2} + 1 - (e^{\bar{z}/2} - 1)e^{-}}$ \hfill (12)

Simplified Equations for $\Psi_0 < 25$ mV

Instead of (7) $\qquad\qquad d^2\Psi/dx^2 = \kappa^2\Psi$ \hfill (7a)

Instead of (12) $\qquad\qquad \Psi = \Psi_0 \exp (-\kappa x)$ \hfill (12a)

III. TOTAL DOUBLE-LAYER CHARGE AND Ψ_0

Total Charge: $\qquad \sigma = -\displaystyle\int_0^\infty q\,dx = \epsilon/4\pi \int_0^\infty (d^2\Psi/dx^2)\,dx$

$$= -\epsilon/4\pi [d\Psi/dx]_{x=0} \tag{13}$$

Table 9.3 continued

Inserting (11a):

$$\sigma = [(2/\pi)\epsilon kT\tilde{n}]^{\frac{1}{2}} \sinh (zF\Psi_0/2RT) \tag{14}$$

Simplified equation for small Ψ_0:

$$\sigma \simeq (\epsilon\kappa/4\pi)\Psi_0 \tag{14a}$$

† μ chemical potential
$\bar{\mu}$ electrochemical potential
Ψ local potential, V
Ψ_0 surface potential, V
q (volumetric) charge density, coulombs cm^{-3}
σ total surface charge, coulombs cm^{-2}
x distance from surface, cm
n_+ local cation concentration, moles cm^{-3}
n_- local anion concentration, moles cm^{-3}
$n_{(x=\infty)}$ bulk ion concentration, moles cm^{-3}
$\tilde{n}$ number of ion pairs, cm^{-3} ($=Nn_{x=\infty}$)
N Avogadro's number, 6.03×10^{23} mole^{-1}
I ionic strength, mole liter^{-1}
z valence of ion
κ reciprocal thickness of double layer, cm^{-1}
e charge of electron, 1.6×10^{-19} coulombs
kT Boltzmann constant times absolute temperature, 0.41×10^{-20} V coulomb at 20°C
$RT = N \times kT = 2.46 \times 10^3$, V coulomb mole^{-1}
ϵ dielectric constant, 89×10^{-12} coulombs V^{-1} cm^{-1}
F Faraday $= 6 \times 10^{23} \times e$, coulombs equiv^{-1} $= 96,500$ coulombs equiv^{-1}
$F\Psi/RT = e\Psi/kT = 1$ for $\Psi = 25$ mV at 20°C

Accordingly, an increase in ionic strength results in a decrease in double-layer thickness. This is equivalent to a increase in capacitance and, hence, an increase in the σ/Ψ_0 ratio. The response of the two types of double layers (constant charge and constant potential) to an increase in electrolyte concentration will be different:

$$\text{Increasing } I: [\Psi_0] \text{ decreases if } \sigma \text{ is constant} \tag{29}$$
$$[\sigma] \text{ increases if } \Psi_0 \text{ is constant}$$

These relationships are evident from Figure 9-3. Equations 28 or 14a of Table 9-3 are applicable only to small double-layer potentials ($\Psi_0 < 25$ mV), otherwise (13) or (14) of Table 9-3 have to be used. Equation 13 shows that the total surface charge is reflected in the initial slope of the potential distance curve (Figures 9-3c and f).

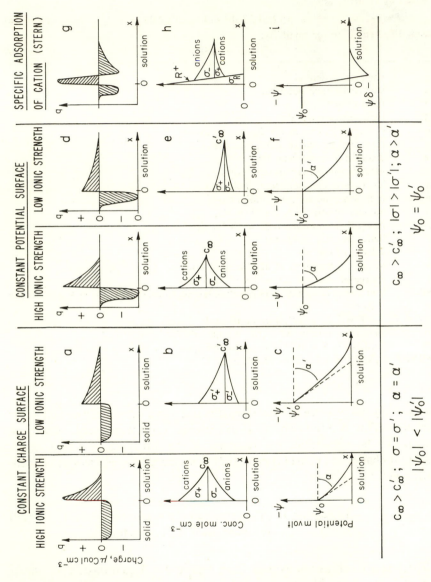

Figure 9-3 Distribution of charge, ions and potential at an interface.

Simple Applications

Despite the simplified assumptions upon which it is based, the Gouy–Chapman model gives an invaluable, though not always quantitatively applicable, description of the configuration of the double layer. A few implications readily derivable from the theory are given below. Example 9-1 illustrates how to handle the equations numerically. One of the pertinent applications of double-layer theory — the stability of colloids — will be taken up in Section 9-7.

pH AT SURFACES. Obviously the concentrations of ions in the immediate proximity of charged surfaces is different from that of the bulk of the solution. The Gouy theory [(3) and (4) of Table 9-3] predicts that

$$2.3 \log \frac{[H^+]_{x=0}}{[H^+]_{x=\infty}} = -\frac{F\Psi_0}{RT}$$

Hence

$$pH_{surface} - pH_{bulk} = \frac{1}{2.3} \frac{F\Psi_0}{RT} \tag{30}$$

or

$$pH_{surface} - pH_{bulk} = 16.9\Psi_0 \text{ at } 20°C \tag{31}$$

Obviously the pH on the surface of a clay, an oxide surface or a protein is remarkably different from that in the bulk of the solution (see part 2 of Example 9-1). Surface catalytic effects can sometimes be accounted for by such a pH shift. It can be readily shown that for a given electrolyte concentration ΔpH between surface and bulk is independent of the type of buffer (e.g., NH_4^+, NH_3, or HCO_3^-, CO_3^{2-}) and of the buffer intensity of the solution. For a constant charge surface ΔpH decreases as the concentration of inert electrolyte increases and, in turn, Ψ_0 decreases. As illustrated in Example 9-1, ΔpH in seawater is about 4–8 times smaller than in fresh water. Equations similar to (30) can be written to estimate the accumulation or depletion of surfaces with respect to ions. Because the Gouy theory assumes point charges in solution, calculations tend to overestimate the change in concentration. Because of their finite size ions cannot approach nearer to the surface than a few angstroms. At this distance of closest approach Ψ is somewhat smaller than Ψ_0.

Example 9-1   *Surface Potential of Kaolinite.*   Consider a kaolinite mineral whose negative face charge is attributable to atomic substitutions in 0.10% of all Si atoms. The surface area determination by BET (Brunauer, Emett and Teller method) has yielded a specific surface area of 25 $m^2 \ g^{-1}$. Assume that the Gouy–Chapman theory is valid.

1. Estimate the surface potential for suspension in both (i) fresh water (ionic strength $= 10^{-2} M$) and (ii) seawater (ionic strength $= 0.65 M$). For lack of any simple theory applicable to mixed electrolytes, it may be assumed that a water of ionic strength I is comparable to a solution of a monovalent symmetric electrolyte of equal ionic strength. [See, for example, K. M. Joshi and R. Parsons, *Electrochim. Acta*, **4**, 129 (1961).]

2. Estimate, for either case, how much the pH in the immediate proximity of the surface differs from that in the bulk of the solution.

1. We first compute the total surface charge density. One mole of kaolinite $[Al_2Si_2O_5(OH)_4]$ equals ca. 200 g and contains 2 moles of silicon atoms. For an atomic substitution of 1 Si in 1000, 1 g kaolinite contains 10^{-5} moles of substitutions. The surface charge is readily computed by dividing the amount of substitutions by the surface area

$$\sigma = -\frac{10^{-5} \text{ eq g}^{-1} \times 9.65 \times 10^4 \text{ Coul eq}^{-1}}{25 \times 10^4 \text{ cm}^2 \text{ g}^{-1}}$$

$$= -3.9 \times 10^{-6} \text{ Coul cm}^{-2}$$

we then estimate κ^{-1} for the two conditions specified. We may use (27)

$$\text{(i)} \ \kappa^{-1} = 2.8 \times 10^{-7} \text{ cm}$$
$$\text{(ii)} \ \kappa^{-1} = 3.5 \times 10^{-8} \text{ cm}$$

The potential Ψ_0 can now be calculated. Trying (14*a*) of Table 9-3 first, we obtain

$$\Psi_0 = \frac{4\pi\sigma\kappa^{-1}}{\epsilon}$$

$$= -\frac{1.25 \times 10 \times 3.9 \times 10^{-6} \text{ Coul cm}^{-2} \times 2.8 \times 10^{-7} \text{ cm}}{73 \times 10^{-12} \text{ Coul V}^{-1} \text{ cm}^{-1}}$$

$$\Psi_0 \simeq -0.19 \text{ V } (I = 0.01)$$

Similarly

$$\Psi_0 \simeq -0.023 \text{ V } (I = 0.65)$$

However, the simplified (14*a*) of Table 9-3 is applicable only for cases where $\Psi_0 < 25$ mV. In order to get a better estimate, (14) is applied. Here we need values for $\bar{n}$ (the number of ion pairs cm^{-3}). Hence

$$\bar{n}_{(I=0.01)} = 10^{-2} \text{ mole liter}^{-1} \times 6 \times 10^{23} \text{ mole}^{-1} \times 10^{-3} \text{ liter cm}^{-3}$$
$$= 6 \times 10^{18} \text{ cm}^{-3}$$
$$\bar{n}_{(I=0.65)} = 4 \times 10^{20} \text{ cm}^{-3}$$

Using these values in (14) we obtain

$$\sinh \frac{F\Psi_0}{2RT} = \frac{\sigma}{[(2/\pi)\epsilon kT\bar{n}]^{0.5}}$$

$$= \frac{-3.9 \times 10^{-6} \text{ Coul cm}^{-2}}{(0.64 \times 6 \times 10^{18} \text{ cm}^{-3} \times 7.3 \times 10^{-11} \text{ Coul volt}^{-1}\text{cm}^{-1}} \\ \times 0.41 \times 10^{-20} \text{ V Coul})^{0.5}}$$

$$= -3.7$$

From tables of hyperbolic functions

$$\frac{F\Psi_0}{2RT} = -2.02$$

and

$$\Psi_{0 \text{ (fresh water)}} = -0.10 \text{ V}$$

For $I = 0.65$, one obtains

$$\sinh \frac{F\Psi_0}{2RT} = -9.45$$

$$\frac{F\Psi_0}{2RT} = -0.44$$

$$\Psi_{0 \text{ (sea water)}} = -0.022 \text{ V}$$

[Note that (14) gives about the same result as (14a) for the concentrated electrolyte.]

2. The accumulation of H^+ by the negatively charged surface may be estimated by applying (30)

"Fresh water": $pH_{(surface)} = pH_{bulk} - 1.69$

"Seawater": $pH_{(surface)} = pH_{bulk} - 0.37$

This calculation overestimates ΔpH because ions are treated as point charges. More realistically, the calculation should be made for Ψ of the plane of closest approach of the hydrated hydrogen ion because these ions cannot come closer to the surface than this distance.

SURFACE CHARGE DENSITY AND ION EXCHANGE CAPACITY. The extent of isomorphous substitution determines the total surface charge. It is frequently inferred that the surface charge density, related to the number of atomic substitutions, is equal to the ion exchange capacity. This is not exactly correct. The cation exchange capacity of a negative double layer has been defined [6] as the excess of counter ions in the double layer which can be exchanged for other cations. This ion exchange capacity corresponds to the area marked σ_+

in Figure 9-3. But the surface charge density is given by the sum of σ_+ (charge due to surplus of cations) and σ_- (charge due to deficiency of anions). With the help of (3) and (4) of Table 9-3, σ_+ and σ_- can be computed from the change of local potentials with distance

$$\sigma_+ = \int_{x=0}^{x=\infty} zF(n_+ - n_{x=\infty})dx; \qquad \sigma_- = \int_{x=0}^{x=\infty} zF(n_- - n_{x=\infty})dx \quad (32)$$

It can be shown that σ_+/σ_- remains independent of electrolyte concentrations only for constant potential surfaces. However, for constant charge surfaces, such as for double layers on clays, σ_-/σ_+ increases with increasing electrolyte concentration, the deficiency of anions and the surplus of cations decreasing with increasing electrolyte concentration. Hence, the cation exchange capacity (as defined by σ_+) increases with dilution and becomes equal to the total surface charge at great dilutions. Ramifications concerning the experimental determination of ion exchange capacity have been discussed by van Olphen [6].

PREFERENTIAL CONCENTRATION OF MULTIVALENT IONS. The Gouy theory predicts that multivalent ions are concentrated in the double layer to a much larger extent than monovalent ions. Thus the ratio $[Ca^{2+}]/[Na^+]$ is much higher in the proximity of a negatively charged surface than in the medium. The fact that selectivity for greater concentration of bivalent cations decreases with increasing ionic strength can also be derived from the Gouy model. Experimental data are in accord with these generalized predictions.

The Zeta Potential

The potential difference between a charged surface or its plane of closest approach and the bulk of the solution (Ψ_0 or Ψ_δ) cannot be measured directly, but useful information concerning this potential difference can be obtained from *electrokinetic measurements. Electrophoresis* refers to the movement of charged particles relative to a stationary solution in an applied potential gradient, whereas in *electroosmosis* the migration of solvent with respect to a stationary charged surface is caused by an imposed electric field. The *streaming potential* is the opposite of electroosmosis and arises from an imposed movement of solvent through capillaries; conversely a *sedimentation potential* arises from an imposed movement of charged particles through a solution.

All these measurements concern only the double layer under moving conditions and may therefore be interpreted only in terms of the *zeta potential* or charge at a plane of shear (the slipping plane at the envelope of water adhering to the charged surface). Operationally the zeta potential can be computed from electrophoretic mobility and other electrokinetic measurements. For example, for nonconducting particles whose radii are large when compared with their double-layer thicknesses, the zeta potential ξ is related to the electrophoretic mobility m_e by

$$m_e = \frac{\xi \epsilon}{4 \pi \eta} \tag{33}$$

where η is the viscosity of the solution. Frequently many corrections that are difficult to evaluate must be considered in the computation of ξ. The position of the shearing plane is not known, but generally ξ is smaller in magnitude than Ψ_0 or Ψ_δ [7, 9]. The measurement of electrophoretic mobility, especially as applied in water treatment, has been discussed extensively by Black [7] and by Packham [8]. The concept of the zeta potential has been reviewed recently by Bier and Cooper [9] and by Sennet and Olivier [10].

9-4   Adsorption and Electrostatic Interaction

After having discussed some of the main features of the structure of the electrical double layer, we may concern ourselves with the influence of electrostatic interaction on the adsorption of solutes at interfaces. These phenomena have been investigated extensively by using the surfaces of metallic electrodes. Particular emphasis has been placed on mercury surfaces where the sign and magnitude of the surface potential can be varied widely without causing electrochemical reactions. Furthermore, the mercury–solution interface is readily amenable to precise quantitative measurements (interfacial tension, differential capacitance, current, etc.). Much of the quantitative information available today is based on investigations with this interface [11]. The Hg–solution interface serves as an excellent model for interpreting electrochemical changes at interfaces. The observations made on this system have provided experimental corroboration for the validity of those thermodynamic properties that characterize interfaces.

Applicability of the Gibbs Adsorption Equation

We have already illustrated the application of the Gibbs adsorption equation to the solid–liquid interface (Table 9-2). Here we extend the treatment to

[7] A. P. Black, *J. Amer. Water Works Assoc.*, **52**, 492 (1960).
[8] R. F. Packham, *Proc. Soc. Water Treat. Exam.*, **13**, 316 (1964).
[9] M. Bier and F. C. Cooper, in *Principles and Applications of Water Chemistry*, S. D. Faust and J. V. Hunter, Eds., Wiley, New York, 1967.
[10] P. Sennett and J. P. Olivier, in *Chemistry and Physics of Interfaces*, American Chemical Society, Washington, D.C., 1965.
[11] D. C. Grahame, *Chem. Rev.*, **41**, 441 (1947).

ions and charged surfaces. As before [(2) of Table 9-2], the surface coverage Γ is inversely related to the concentration (activity) dependence of interfacial tension. For solid–liquid interfaces the interfacial tension cannot be measured directly, but is indirectly obtainable (although not always with the desired precision) from contact angle and adsorption density measurements. By interpreting measurements of the interfacial tension at the mercury–solution interface with the help of the Gibbs equation, we can obtain quantitative information about the charge distribution in the double layer, that is, the adsorption or desorption of ions that can be utilized to predict adsorption and desorption of ions at other interfaces. Some of these generalizations have been discussed by de Bruyn and Agar [5] and a few pertinent examples are given in Figure 9-4 and Table 9-4.

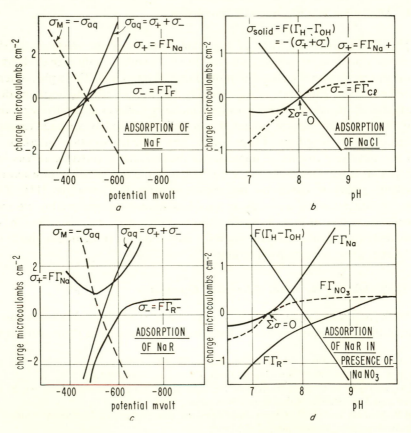

Figure 9-4   Distribution of charge components of the electric double layer.

Table 9-4   Application of Gibbs Adsorption Equation†

I. SIMPLE ELECTROLYTES AND SPECIFICALLY SORBED IONS
Adsorption (desorption) of NaF

$$-d\gamma = \Gamma_F d\mu_F + \Gamma_{Na} d\mu_{Na} \tag{1}$$

Substitutions: ‡ $\Gamma_F \simeq \Gamma_{Na}; \quad d\mu_{NaF} = d\mu_F + d\mu_{Na}$

$$-d\gamma = \Gamma_F d\mu_{NaF} = \Gamma_{Na} d\mu_{NaF} \tag{2}$$

$$-d\gamma = 2RT\Gamma_F d \ln \{NaF\} \tag{3}$$
$$-d\gamma = 2RT\Gamma_{Na} d \ln \{NaF\}$$

Adsorption of Salt with Polar Ion, NaR

$$-d\gamma = \Gamma_R d\mu_R + \Gamma_{Na} d\mu_{Na} \tag{4}$$

Substitution: ‡ $\Gamma_R \simeq \Gamma_{Na}$

$$-d\gamma = 2RT\Gamma_R d \ln \{NaR\} \tag{5}$$

Adsorption of NaR *in Presence of* NaNO$_3$

$$-d\gamma = \Gamma_R d\mu_R + \Gamma_{Na} d\mu_{Na} + \Gamma_{NO3} d\mu_{NO3} \tag{6}$$

Substitution: ‡ $\Gamma_{Na} \simeq \Gamma_{NO3} + \Gamma_R$

$$-d\gamma = \Gamma_R d\mu_{NaR} + \Gamma_{NO3} d\mu_{NaNO3} \tag{7}$$

Conditions: (1) If $\{NaNO_3\} \gg \{NaR\}$ and I = constant:

$$-d\gamma = RT\Gamma_R d \ln \{R^-\} \tag{8}$$

(2) If $\{NaR\}$ = constant and assuming Γ_{NO3} negligible:

$$-d\gamma = RT\Gamma_R d \ln \{Na^+\} \tag{9}$$

II. POTENTIAL DETERMINING IONS
Adsorption of H^+ *and/or* OH^- *in Presence of Electrolyte* (NaCl, NaOH, HCl)

$$-d\gamma = \Gamma_{OH} d\mu_{OH} + \Gamma_H d\mu_H + \Gamma_{Na} d\mu_{Na} + \Gamma_{Cl} d\mu_{Cl} \tag{10}$$

Substitutions $\Gamma_{OH} = -\Gamma_H$ and $d\mu_{OH} = -d\mu_H$ (11)

$$-d\gamma = (\Gamma_{OH} - \Gamma_H) d\mu_{OH} + \Gamma_{Na} d\mu_{Na} + \Gamma_{Cl} d\mu_{Cl} \tag{12}$$

Solid surface charge density $\sigma_s = F(\Gamma_H - \Gamma_{OH})$ (13)

Solution surface charge $\sigma_{aq} = F(\Gamma_{Na} + \Gamma_{Cl})$ (14)

   density $\sigma_s = -\sigma_{aq}$ (15)

Because H^+ and OH^- are potential determining ions§

$$d\mu_{OH} = RT d \ln \{OH\} = RT dpH \tag{16}$$

$$d\mu_{OH} = -F d\Psi_0 \tag{17}$$

Rewriting (12) $-d\gamma = F(\Gamma_H - \Gamma_{OH}) d\Psi_0 + \Gamma_{Na} d\mu_{Na} + \Gamma_{Cl} d\mu_{Cl}$ (18)

At constant I $-d\gamma = \sigma_s d\Psi_0 = -RT\sigma_s dpH$ (19)

Adsorption of Aluminum Species ‖

$$-d\gamma = (\Gamma_{OH}^* - \Gamma_H^*) d\mu_{OH} + \Gamma_K d\mu_K + \Gamma_{NO3} d\mu_{NO3} \tag{20}$$

$[Al_n(OH)_x^{3n-x}]$ on $Al(OH)_3(s)$ or $Al_2O_3(s)$ in presence of acid (HNO_3) or base (KOH)

$$\sigma_s = F(\Gamma_{OH}^* - \Gamma_H^*) \tag{21}$$

At constant I $-d\gamma = \sigma_s d\Psi_0 = -RT\sigma_s dpH$ (22)

---

† See Figure 9-4; for a more comprehensive discussion consult de Bruyn and Agar [5].
‡ H^+ and OH^- are neglected as solutes and in the charge balance of adsorbed species.
§ See (23) of Section 9-3.
‖ See Example 9-2.

Charge attributable to cations and anions in the double layer are plotted as a function of potential (Figure 9-4a, c) or pH (Figure 9-4b, d). The curves given are schematic but in accord at significant points with results obtained for the mercury–solution interface by D. C. Grahame and B. A. Soderberg [$J.$ $Chem.$ $Phys.$, **22**, 449 (1954)]. These results are based on interpretation of interfacial tension measurements with the Gibbs adsorption equation (Table 9-4).

The potential scale is relative to a reference electrode. Relative values reflect changes in Ψ_0 (at the point of zero charge, $\Psi_0 = 0$). In Figures 9-4b and d the potential axis has been replaced by a pH axis [(19) of Table 9-4]. Γ_+ and Γ_- are the excesses of cations and anions present above what would be present in a column of the solution of unit cross section extending from the surface into the interior of the solution. σ_+ and σ_- are the charge (positive and negative, respectively, for positive adsorption) attributable to cations and anions. Figures 9-4c and d illustrate that a specifically sorbable ion affects the potential of zero charge and pH_{ZPC}. They also show the possibility of adsorption of specifically sorbable species against electrostatic attraction.

SIMPLE ELECTROLYTES. In the absence of specific forces of adsorption, the double layer at a solid–solution interface is characterized by a rather simple structure. The surface charge on the electrode or on the solid is equal but opposite in sign to the charge on the solution side. The latter results from the relative accumulation of ions in the double layer

$$\sigma_{aq} = F(\Gamma_+ + \Gamma_-) \tag{34}$$

The adsorption density may be positive or negative. For positive adsorption of cations and anions, σ_+ and σ_- are positive and negative, respectively (and vice versa for negative adsorption). Obviously σ_+ is positive for a negatively charged surface ($\Psi < \Psi_{(zero\ charge)}$ or pH > pH_{ZPC}) and σ_- is negative for a positively charged surface. As Figures 9-4a and b illustrate, cations are desorbed (negative values of σ_+) at positively charged surfaces; similarly, negative adsorption (positive values for σ_-) occurs at negatively charged surfaces. At an uncharged surface (at the potential of zero charge) simple cations and anions are not adsorbed; actually there is a slight negative adsorption for both cations and anions because simple electrolytes usually tend to increase the interfacial tension [(3) of Table 9-4].

SPECIFICALLY SORBABLE ION. Interfacial tension will be depressed, however, upon addition of a salt, NaR, containing an ion, R^-, that is specifically adsorbed at the interface [(5) of Table 9-4]. The adsorbed anion profoundly affects the structure of the double layer. As Figures 9-4c and d illustrate, the potential of zero charge and the pH are shifted markedly. Because of the adsorption of R^-, cations (Na^+) become sorbed even at positive surface charges of the electrode or the solid. The addition of a simple salt (i.e., NaCl or NaNO$_3$) to a solution containing a constant concentration of NaR decreases the positive sorption tendency of R$^-$ [(9) of Table 9-4].

POTENTIAL-DETERMINING IONS. Formal application of the Gibbs equation can be extended to the adsorption of potential-determining ions. Charge on an oxide or hydroxide surface develops from "adsorption" of OH$^-$ and/or H$^+$ ions. Note that dissociation of a H$^+$ from a surface hydroxide group is equivalent to, and usually cannot be distinguished experimentally from, the binding of an OH$^-$ to the surface; hence, mathematically, $\Gamma_{OH} = -\Gamma_H$ (11). The "adsorbed" potential determining ions become part of the solid surface (13).

pH-DEPENDENT SURFACE CHARGE. If H$^+$ and OH$^-$ are potential-determining ions, the potential as a variable can be replaced by pH as a variable [(16) and (17) of Table 9-4]. Conveniently, the variation in interfacial composition can be shown as a function of pH (Figures 9-4b and d). Equation 23 can be rewritten (20°C) in the form

$$\Psi_0 = \frac{RT}{F} \ln \frac{\{H^+\}}{\{H^+\}_0} = -0.058[pH - pH_{ZPC}) \tag{35}$$

For small Ψ_0 values ($\Psi_0 < 25$ mV) the simplified equation valid for a plate condenser [(14a) of Table 9-3] can be used in combination with (35)

$$F\Gamma = \sigma = -(\epsilon\kappa/4\pi)0.058[pH - pH_{ZPC}) \tag{36}$$

If pH is not much different from pH$_{ZPC}$, the surface charge is linearly dependent upon pH as indicated in Figure 9-4b and d. As these figures illustrate; capacity for cation sorption increases with increasing pH. Anion adsorption capacity increases with decreasing pH [12].

Free Energy of Adsorption

THE STERN THEORY. The Gouy–Chapman double-layer model can be refined and made more realistic. In his theory Stern (1924) has divided the region near the surface into two parts, the first consisting of a layer of ions adsorbed at the surface forming a compact double layer (the Stern layer) and the second consisting of the diffuse double layer (the Gouy layer). He has thus accounted for the finite size of ions and the possibility of their specific adsorbability. Stern assumed that the fraction of available sites occupied by ions is related to the unoccupied fraction by Boltzmann expressions [viz. (3) of Table 9-3]. If n_c is the number of counter ions of valence z adsorbed cm^{-2} adsorbed from

[12] K. A. Kraus et al., *Proceedings of the Second International Conference on the Peaceful Uses of Atomic Energy, Geneva, 1958*, Volume 28, United Nations, 1958, p. 3.

a bulk concentration of mole fraction x_0 of the solute and n_s is the number of possible adsorption sites in the surface, the Stern relation can be expressed as

$$\frac{n_c}{n_s - n_c} = x_0 e^{(zF\Psi_\delta + \Phi)/RT} \tag{37}$$

where Ψ_δ is the potential at the solution side of the boundary of the Stern layer and Φ is the energy (per mole) of specific adsorption due to non-electrostatic forces. The left-hand side of (37) denotes the number of occupiable sites. For a dilute solution, the corresponding ratio is the mole fraction of the solute which, for water as a solvent, can be replaced (dilute solution, 25°C) by $x_0 = c/55.4$, where c is the molar concentration. Hence for a single ionic species, (37) can be rewritten in the form of a Langmuir equation [see (5) of Table 9-2]

$$\frac{\theta}{1 - \theta} = \frac{c}{55.4} e^{(zF\Psi_\delta + \Phi)/RT} \tag{38}$$

where $-(zF\Psi_\delta + \Phi)$ has the significance of a standard free energy of adsorption. If Φ is larger than $zF\Psi_\delta$, a reversal of the total surface charge occurs (Figure 9-4d) because $zFn_c = \sigma_c$ exceeds the original charge density σ_s on the surface.

The compact Stern layer can be treated as a flat plate condenser of thickness δ in which the potential drops linearly with distance from Ψ_0 to Ψ_δ:

$$\sigma_s = \frac{\epsilon}{4\pi\delta} (\Psi_0 - \Psi_\delta) \tag{39}$$

where $\sigma_s = \sigma_c + \sigma_{\text{diffuse}}$.

The dielectric constant ϵ in this Stern layer must be assumed to be considerably lower than ϵ of bulk water. The diffuse layer outside of the Stern layer is treated the same way as in the Gouy theory, simply substituting Ψ_δ for Ψ_0.

Adsorption Against Electrostatic Repulsion

Following the assumptions of Stern, the over-all standard free energy of adsorption equals the sum of the total specific ("chemical") adsorption energy and the electrochemical work involved in the adsorption

$$\Delta \bar{G}^\circ = \Phi + zF\Psi_\delta \tag{40}$$

Experimentally it is quite difficult to separate the energy of adsorption into its chemical and coulombic components. Despite this operational restriction, it is instructive to formulate a numerical example. For the adsorption of an ion on a surface of similar charge, (40) indicates that the chemical contribution to the over-all standard free energy of adsorption opposes the electrostatic contribution. For the adsorption of a monovalent organic ion to a surface of similar charge and against a potential drop (Ψ_δ) of 100 mV, the electrostatic term in (40), $zF\Psi_\delta$ is 2.3 kcal mole^{-1}. The standard chemical adsorption energy for typical sorbable monovalent organic ions is of the order of -2 to -8 kcal mole^{-1}, thus indicating that the electrostatic contribution to adsorption easily can be smaller than the chemical contribution. It is therefore at least qualitatively understandable that the addition of a suitable counter ion decreases the surface potential and may produce a reversal of the surface charge at higher concentrations.

Somasundaran, Healy and Fuerstenau [13] have investigated the effects of alkyl chain length on the adsorption of alkylammonium ions on silica surfaces at pH 6.5–6.9. These authors consider that, in addition to electrostatic effects, surfactant adsorption is influenced by the van der Waals energy of interaction between CH_2 groups on adjacent adsorbed surfactant molecules. The van der Waals cohesive energy per CH_2 group is found to be approximately 1 RT (at 25°C, $RT = 595$ cal mole^{-1}). For a 12-carbon alkylamine, the van der Waals cohesive energy is therefore of the order of 7 kcal mole^{-1} and usually exceeds the electrostatic contribution.

Davies and Rideal [14] have reviewed the results of several investigators on the specific interaction energy arising from polarization (complex formation) and van der Waals forces for a number of surfaces. Davies and Rideal, for example, give for the specific energy of interaction on a carboxylate surface for Na^+, Mg^{2+}, Ca^{2+} and Cu^{2+} values of ~ -0.6, ~ -1.3, ~ -2.6 and -5.8 kcal mole^{-1}, respectively.

POLYMERS. That van der Waals forces contribute significantly to the specific interaction energy is apparent from the observation that macromolecules have a strong tendency to accumulate at interfaces. For example, on an incipiently negatively charged silica surface, polystyrene sulfonate

$$[CH_2—CH—C_6H_4—SO_3]_n{}^{n-}$$

is adsorbed readily, while p-toluene sulfonate ($CH_3—C_6H_4—SO_3^-$) does not adsorb from $10^{-4}\ M$ solutions.

Each polymer molecule or polymeric ion can have many groups or segments that potentially can be adsorbed; these groups are often relatively free of mutual interaction. Usually the extent of adsorption, but not necessarily its kinetics, increases with increasing molecular weight and is affected by the number and type of functional groups in the polymer molecule. Hydrosyl,

[13] P. Somasundaran, T. W. Healy and T. W. Fuerstenau, *J. Phys. Chem.*, **68**, 3562 (1964).
[14] J. T. Davies and E. K. Rideal, *Interfacial Phenomena*, Academic Press, New York. 1961.

phosphoryl and carboxyl groups can be particularly effective in causing adsorption; the significance of such functional groups directs attention to the specificity of the chemical interactions involved in the adsorption process. Adsorption of anionic polymers on negative surfaces is common.

HYDROLYZED METAL IONS. As has been pointed out in Chapter 6, the hydrolysis products of multivalent ions are adsorbed more readily at particle–water interfaces than nonhydrolyzed metal ions. This tendency to be adsorbed, even against electrostatic repulsion, is especially pronounced for polynuclear polyhydroxo species. No adequate theory for this enhanced adsorption by hydrolysis is available, but a few qualitative reasons can be given [15]. First, hydrolyzed species are larger and less hydrated than nonhydrolyzed species. Second, the enhancement of adsorption is apparently due to the presence of a coordinated hydroxide group. Simple hydroxide ions are bound strongly at many solid surfaces and are frequently potential-determining ions: hydroxo-metal complexes may similarly or to an even larger extent be adsorbed to the solid surface. Alternatively, the replacement of an aquo group by a hydroxo group in the coordination sheath of a metal atom may render the complex more hydrophobic by reducing the effective change of the central ion or ions and hence decrease the interaction between the central metal atom and the remaining aquo groups. This might then, in turn, enhance the formation of covalent bonds between the metal atom and specific sites on the solid surface by reducing the energy necessary to displace water molecules from the coordination sheath. Finally, adsorption becomes especially pronounced for polyhydroxo–polymetal species because more than one hydroxide group per "molecule" can become attached at the interface. Most species containing hydroxide groups in the ionic structure, cationic as well as anionic hydroxo complexes, have been observed to adsorb at solid–liquid interfaces. In addition to metal ion hydrolysis products, we may cite polysilicates, polyphosphates and heteropoly ions (for example, phosphotungstate and silicotungstate).

9-5   The Surface Chemistry of Oxides, Hydroxides and Oxide Minerals

Most of the solid phases in natural waters contain oxides or hydroxides. Structural oxides at the surface hydrate and convert into surface hydroxide groups. Because of the significance of these interfaces in interacting with constituents of natural waters, a few pertinent surface chemical properties are outlined here.

[15] W. Stumm and C. R. O'Melia, *J. Amer. Water Works Assoc.*, **60**, 514 (1968).

The Zero Point of Charge

The pH of zero point of charge is a convenient reference for predicting how the charge of minerals and other colloids depends on pH. Parks [16] has written a comprehensive and lucid presentation of this subject. Our discussion is largely influenced by Park's work; the reader is encouraged to peruse his paper.

Alkalimetric or acidimetric titration curves for hydrous metal oxides or other ampholytes provide a quantitative explanation for the manner in which the change of the solid surface depends on the pH of the medium. This may be illustrated by considering the acidimetric and alkalimetric titration curve of X-ray amorphous $Al(OH)_3$.

Example 9-2 *Evaluation of* pH-*Dependent Change Characteristics from Alkalimetric and Acidimetric Titration Curves.* We will demonstrate how pH_{ZPC} and the charge density of an $Al(OH)_3$ suspension can be calculated from an experimental titration curve (Figure 9-5).

In titrating a suspension of $Al(OH)_3$ in an inert electrolyte (say $10^{-2}\ M$ KNO_3) with KOH (C_B) or with HNO_3 (C_A), we can write the proton condition

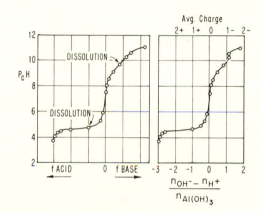

Figure 9-5 Alkalimetric and acidimetric titration of a suspension of $Al(OH)_3$. Below pH 5 and above pH 9.5 $Al(OH)_3(s)$ dissolves. In the diagram on the right the net number of moles of OH^- bound per mole of $Al(OH)_3$ has been calculated from the titration curve. The break at pH = 10.5 is evidence for the presence in solution of the $Al(OH)_4^-$ ion with one OH^- bound per $Al(OH)_3(s)$ [from Stumm and Morgan, *J. Amer. Water Works Assoc.*, **54** 1971 (1962)]. Reproduced with permission from American Water Works Association.

[16] G. A. Parks, "Aqueous Surface Chemistry of Oxides and Complex Oxide Minerals," in *Equilibrium Concepts in Natural Water Systems*, Advances in Chemistry Series, No. 67, American Chemical Society, Washington, D.C., 1967, p. 121.

(see Sections 3-8 and 6-3) assuming a variety of Al species to be formed on the surface or in the solution

$$C_B - C_A + [H^+] - [OH^-] = [Al(OH)_4{}^-] + 2[Al(OH)_5{}^{2-}] - [Al(OH)_2{}^+]$$
$$- 2[Al(OH)^{2+}] - \tfrac{1}{2}[Al_8(OH)_{20}{}^{4+}] - 3[Al^{3+}]$$
$$= \sum \frac{x - 3n}{n} [Al_n(OH)_x^{3n-x}] \qquad (i)$$

On the right-hand side of this equation aluminum species that have bound or lost hydroxide are enumerated. It does not matter whether these species actually exist and whether they are present on the surface of the solid phase or in solution. The right-hand side of (i) gives the net number of moles of OH^- ions, $n_{OH^-} - n_{H^+}$, bound to $Al(OH)_3$

$$C_B - C_A + [H^+] - [OH^-] = n_{OH^-} - n_{H^+} \qquad (ii)$$

The mean charge δ (or more precisely that portion of the charge that is due to OH^- or H^+ per mole of $Al(OH)_3$, $n_{Al(OH)_3}$) can be calculated as a function of pH from the difference between total added base or acid and the equilibrium OH^- and H^+ ion concentration in solution (Figure 9-5)

$$\delta = \frac{n_{OH^-} - n_{H^+}}{n_{Al(OH)_3}} = \frac{C_B - C_A + [H^+] - [OH^-]}{n_{Al(OH)_3}} \qquad (iii)$$

Within the pH range where the $Al(OH)_3$ does not dissolve appreciably, δ gives the mean surface charge per mole of $Al(OH)_3$. If the surface area S is known, (ii) can be rewritten as

$$\frac{\sigma_s}{F} = \Gamma^*_{OH^-} - \Gamma^*_{H^+} = \frac{n_{OH^-} - n_{H^+}}{S} = \frac{C_B - C_A + [H^+] - [OH^-]}{S} \qquad (iv)$$

The asterisk is used to indicate that all possible hydroxo complexes are included.

The zero point of charge† is given by the pH at which the surface charge σ_s is zero

$$pH = pH_{ZPC} \qquad (v)$$

if $\Gamma^*_{H^+} = \Gamma^*_{OH^-}$.

Many other methods in addition to alkalimetric titration have been used to determine pH_{ZPC}. Electrokinetic methods such as electrophoretic mobility measurements are commonly used, but ZPC may also be inferred, at least qualitatively, from a great variety of other measurements (pH of optimum coagulation, optimum sedimentation rate, heat of immersion data, etc.).

---

† Parks [16] distinguishes between ZPC and isoelectric point (IEP); ZPC is the pH at which the solid surface charge from all sources is zero, whereas IEP is a ZPC arising from interaction of H^+, OH^-, the solid and water alone; however, many authors do not distinguish between ZPC and IEP.

It is evident that oxides formed by metal ions that are strong acids (see Figures 6-3 and 6-6) have low pH_{ZPC} values, and that less acidic metal ions have high pH_{ZPC} values. We can predict the approximate acidity of aquo metal ions by using a simple electrostatic model, for example,

$$Zr^{4+} > Sc^{3+} > Be^{2+} > Mg^{2+}$$
$$Fe^{3+} > Cr^{3+} > Cu^{2+} > (Co^{2+}, Ni^{2+}, Fe^{2+}, Mn^{2+}) \qquad (41)$$
$$Tl^{3+} > Ga^{3+} > Pb^{2+} > (Cd^{2+}, Zn^{2+}) > Ag^{+}$$

We can likewise predict the pH_{ZPC} of metal oxides from electrostatic considerations. Parks [17] has shown that the ZPC of a simple oxide is related to the appropriate cationic charge and radius of the central ion.

SOLUTES AFFECTING ZPC. As already indicated for controlled potential surfaces, specifically sorbable ionic species shift pH_{ZPC} from that observed in the absence of such species. Specific adsorption of anions would be expected to produce a negative surface charge under otherwise identical conditions. On an oxide surface, the ZPC would be shifted to a lower pH because more H^+ has to be bound in order to neutralize the anion's negative charge. Figure 9-4 suggests how a specifically sorbable anion may affect the surface charge. (In constructing this figure, the simplifying assumption has been made that the extent of H^+ binding is not affected by the adsorbed anion.) Adsorption of cationic impurities would tend to shift pH_{ZPC} to more alkaline values.

As shown by Parks and illustrated by a few examples of Table 9-5, pH_{ZPC} of a composite oxide is approximately the weighted average of the pH_{ZPC} values of its components. Predictable shifts in ZPC occur in response to state of hydration, cleavage habit and crystallinity [16, 17]. Effects of surface complex formation will be discussed below, and the influence of structural charge will be illustrated in the next section.

pH_{ZPC} OF MINERALS OTHER THAN OXIDES OR HYDROXIDES. pH_{ZPC} of salt-type minerals depends, sometimes in a complicated way, upon pH and on the concentration (activities) of all potential-determining ions. Thus in the case of calcite, possible potential-determining species, in addition to H^+ and OH^-, are $H_2CO_3^-$, HCO_3^-, CO_3^{2-}, Ca^{2+} and $CaOH^+$; various mechanisms of charge development are possible. It is not meaningful to refer to a pH_{ZPC} of such non-oxides unless the solution composition is also specified.

As will be illustrated here for calcite, pH_{ZPC} usually corresponds to the pH of minimum solubility. In the absence of complications such as those caused by structural or adsorbed impurities, the zero point of charge of the solid should correspond to the pH of charge balance (electroneutrality) of potential-determining ions. In the case of calcite in equilibrium with its solution, this

[17] G. A. Parks, *Chem. Rev.*, **65**, 177 (1965).

Table 9-5   Zero Point of Charge†

Material	pH$_{ZPC}$
α-Al$_2$O$_3$	9.1
α-Al(OH)$_3$	5.0
CuO	9.5
Fe$_3$O$_4$	6.5
γ-Fe$_2$O$_3$	6.7
"Fe(OH)$_3$"(amorph)	8.5
MgO	12.4
MnO$_2$	2–4.5
SiO$_2$	2.0
ZrSiO$_4$	5
Kaolinite	4.6
Montmorillonite	2.5
Albite	2.0

† The values are from different investigators who have used different methods and are not necessarily comparable. They are given here for illustration. For references and a more detailed list of pH$_{ZPC}$ values and methods of measurements, the publications by G. A. Parks should be consulted [*Chem. Rev.*, **68**, 177 (1965) and "Aqueous Surface Chemistry of Oxides and Complex Oxide Minerals; Isoelectric Point and Zero Point of Charge," in *Equilibrium Concepts in Natural Water Systems*, Advances in Chemistry Series, No. 67, American Chemical Society, Washington, D.C., 1967].

pH corresponds to $2[Ca^{2+}] + [H^+] = [HCO_3^-] + 2[CO_3^{2-}] + [OH^-]$. For a calcite suspension in equilibrium with a constant partial pressure of CO_2 (see Figure 5-6) the potential-determining cationic and anionic species at highest concentration are Ca^{2+} and HCO_3^-, respectively; hence the condition $2[Ca^{2+}] = [HCO_3^-]$ gives the pH of charge balance and, thus, pH$_{ZPC}$. At $p_{CO_2} = 10^{-3.5}$ atm, Figure 5-6 shows that this point is at pH $\simeq 8.4$ (0.3 pH units to the right of the intersection of the $[Ca^{2+}]$ and $[HCO_3^-]$ line). Note that this point also corresponds to the pH of a suspension of $CaCO_3$ to which no acid or base has been added. Adding acid to such a suspension will increase $[Ca^{2+}]$; adding base will increase total carbonate. Hence, pH$_{ZPC}$ calculated from the isoelectric point at which the charges of the potential-determining species are equivalent electrically corresponds — under the conditions specified — to the pH of minimum solubility. Somasundaran and Agar [18] have investigated the ZPC of calcite by measuring streaming

[18] P. Somasundaran and G. E. Agar, *J. Colloid Interfac. Sci.*, **24**, 433 (1967).

potential and flotation response. For the conditions specified before, the ZPC was shown to lie within the range of pH = 8 to 9.5 in accordance with the prediction.

The concept that pH_{ZPC} corresponds to a pH of minimum solubility (in the absence of complications) permits us to predict at least semiquantitatively how certain variables affect the ZPC. For example, Figure 5-6 suggests that for solutions in equilibrium with the atmosphere, pH_{ZPC} will decrease with decreased solubility of metal carbonates and will be approximately pH = 7.5 for siderite and pH = 6.5 for $ZnCO_3(s)$. Similarly, we could postulate pH_{ZPC} values for $FePO_4$ and $AlPO_4$ according to the pH of minimum solubility between pH values of 5 and 6.

Surface Chemical Interactions of Hydrous Metal Oxides

The amphoteric behavior of suspended hydrous metal oxides or hydroxides can be compared, at least operationally, with amphoteric polyelectrolytes. Because of the pH-dependent charge characteristics, these oxides can exhibit either cation or anion exchange characteristics, depending upon pH. But in addition to their purely electrostatic effects on the ionic solutes, hydrous oxides show a strong tendency to interact chemically with anions as well as cations.

COMPLEX FORMATION. These chemical interactions may be explained most satisfactorily in terms of complex formation. If we represent a surface group of a hydrous metal oxide by —Me—OH, we can interpret the interactions with anions as a ligand exchange, for example,

$$\overset{|}{\underset{|}{-\text{Me}}}-\text{OH} + \text{HPO}_4{}^{2-} = \overset{|}{\underset{|}{-\text{Me}}}-\text{OPO}_3\text{H}^- + \text{OH}^- \qquad (42)$$

and the interaction with cations as a coordination with electron acceptors, for example,

$$\overset{|}{\underset{|}{-\text{Me}}}-\text{OH} + \text{Zn}^{2+} = \overset{|}{\underset{|}{-\text{Me}}}-\text{OZn}^+ + \text{H}^+ \qquad (43)$$

It is difficult in some instances to distinguish between exclusively coulombic interactions and complex formation interactions; we must realize that there is a continuous transition between purely electrostatic ion exchange reactions and coordination by covalent bond formation. Almost purely ionic, for example, is the interaction of $NO_3{}^-$, $ClO_4{}^-$ or Na^+, K^+ on a hydrous metal oxide surface (Boltzmann distribution of hydrated ions in double layer).

Covalent bonds, on the other hand, would be formed, for example, between a solid metal sulfide surface and Hg^{2+} or Ag^+ ions, or between a copper oxide surface and HS^- ions. In instances intermediate between these extremes it is still appropriate to speak of complex formation because these interactions are accompanied by displacement of H_2O molecules and by dislocation of electrons. Qualitatively the tendency for coordination increases with decreasing difference of the electronegativity values of the surface ligand atom (e.g., oxygen) and the metal ions in solution (Section 6-5).

In *ligand exchange* the structural metal ion can exchange its coordinative partner and can replace its ligand OH^- by another base. In order to evaluate semiquantitatively the coordinating tendency of the structural metal ion, the affinity of its aqueous counterpart toward OH^- ions and other ligands should be compared. For example, in the case of a hydrous ferric oxide that may interact with phosphates, we may compare the reactions:

$$Fe^{3+}(aq) + OH^- = Fe(OH)^{2+}; \qquad \log K = 11.8 \quad (44)$$

$$Fe^{3+}(aq) + HPO_4^{2-} = FeHPO_4^+; \qquad \log K = 8.4 \quad (45)$$

$$Fe^{3+}(aq) + 3OH^- + (x - 1.5)H_2O = \tfrac{1}{2}Fe_2O_3 + xH_2O; \quad \log K \simeq 38 \quad (46)$$

$$Fe^{3+}(aq) + PO_4^{3-} = FePO_4; \qquad \log K \simeq 22 \quad (47)$$

These equilibrium constants show that OH^- has a stronger affinity for Fe^{3+} than does HPO_4^{2-} or PO_4^{3-}. However, the extent of complex formation of Fe^{3+} with HPO_4^{2-} or PO_4^{3-} depends on the ratio $[HPO_4^{2-}]/[OH^-]$ or $[PO_4^{3-}]/[OH^-]^3$, respectively. Thus with decreasing pH phosphate will have an increasing tendency to enter into the coordination sheath of the surface Fe(III) ion. The structural incorporation of another base into the surface of the metal oxide reduces the pH_{ZPC}; this may be interpreted to result from the smaller proportion of OH^- ions that is now (in the presence of another base) necessary to produce electrical neutrality at the surface.

Complex Formation with Cations

In the chemical interaction of a hydrous oxide with cations, the surface groups —Me—OH may be treated as complex-forming species; complex formation involves (a) the separation of a proton from the covalent bond —Me—OH = —Me—O$^-$ + H$^+$, and (b) the coordination with a solute cation. As with complex formation in solution, the tendency to complex formation of cations increases with increasing pH and with increasing basicity of the —Me—O group. Qualitatively, this basicity of the surface OH-group increases with decreasing pH_{ZPC}. Metal ions that tend to form complexes with oxo-groups will be coordinated preferably. It is, therefore, understandable that the charge characteristics of hydrous oxides are strongly dependent not only upon pH, but also upon the type and concentration of metal ions in solution. The addition of salts of metal ions to a suspension of a hydrous metal oxide at pH values above pH_{ZPC} increases the acidity of the dispersion medium but decreases that of the solid phase.

The Silica Surface

Silica is one of the allegedly simpler oxides [16]. Because of its significance as the most important interface in natural water systems we will discuss some of the surface chemical properties of amorphous silica in more detail. The bulk solid phase consists of a network of oxygen bridges between the silicon atoms. At the surface oxygen atoms are free to react with water; surface —Si—OH groups are formed on immersion. The acidity of the —Si—OH groups to a large extent determines how the charge of the silica surface depends upon pH, that is, in which pH range silica behaves as a cation exchanger. Si^{4+} as a central ion has a very strong electron affinity; hence, the oxygen atoms bound to the silicon have low basicity and the silica surface is a weak acid, that is, the proton on a —Si—OH group can be readily displaced in part even at neutral and slightly acid pH values or replaced by other metal ions [19–21].

As with all complex formation equilibria, the position of the equilibrium

$$(HG)_n + M^{+n} = MG_n + nH^+ \tag{48}$$

depends not only upon the basicity of the silica framework G $[(SiO^-)_n]$ but also on its affinity for the metal ion in question. Those metal ions which have a strong tendency to form complexes with oxygen-containing anions, that is, those containing a high charge with a small ionic radius, can compete with H^+ ions even in slightly acid solutions.

THE ACIDITY OF THE SILICA FRAMEWORK. What is the acidity of the surface silanol group? There is no simple answer to this question. At first we might consider the acidity of orthosilicic acid, $Si(OH)_4$ [$pK_1 \simeq 9.5$ (25°C, $I = 0.5$)] and might ask whether a surface silanol group is more or less acidic than the same group in orthosilicic acid. Obviously the oxygen bridges on the silicon central ion will tend to distort the electron cloud away from the silanol group, thus rendering the surface —Si—OH more acidic. The acidity of the silica framework depends on its surface charge. After some of the surface silanol groups have deprotonated, the remaining silanol groups will tend, because of electrostatic influence, to hold the protons more strongly; hence, the acidity

[19] J. Rydberg, *Acta Chem. Scand.*, **9**, 1252 (1955).
[20] S. Ahrland, I. Grenthe and B. Noren, *Acta Chem. Scand.*, **14**, 1059 (1960).
[21] L. R. Snyder and J. W. Ward, *J. Phys. Chem.*, **70**, 3941 (1966).

constant will decrease with increasing degree of neutralization, $\bar{f} = [G]/([HG] + [G])$. The situation is very similar to that encountered with polyelectrolytic acids, where we can no longer specify the individual microscopic acidity constants and where the loss of protons becomes increasingly more difficult as more functional groups become negatively charged.

Formally, the acidity constant K' of a polynuclear acid with identical acid groups can be derived [22] as

$$K' = [H^+]\frac{\bar{f}}{1 - \bar{f}} \tag{49}$$

or in logarithmic form

$$pK' = pH - \log\frac{\bar{f}}{1 - \bar{f}} \tag{50}$$

where $\bar{f}$ is the fraction of acid groups that has lost protons. In Figure 9-6 titration curves for a polyacrylic acid are compared with those of SiO_2, MnO_2 and a clay (kaolinite in H form). Obviously in all cases, because of the electrostatic interactions, pK' increases with increasing $\bar{f}$. Conveniently the acidity constant can be interpreted to be composed of two factors [22, 23], namely, an intrinsic acidity constant K'_{int} which is independent of the degree of neutralization (for a neutral polynuclear acid, $K' = K'_{int}$ if $\bar{f} = 0$) and a factor that represents the electrostatic interaction, $e^{F\Psi/RT}$. Hence

$$pK' = pK'_{int} - \frac{F\Psi}{2.3RT} \tag{51}$$

with increasing $\bar{f}$, Ψ becomes more negative and pK' increases. For a given $\bar{f}$ the charge density is fixed and the absolute value of Ψ decreases with increasing ionic strength [(14) of Table 9-3]. Accordingly, $pK'_{f=0.5}$ (i.e., the "average" pK' value) decreases with increasing ionic strength† (Figure 9-6).

Schindler and Kamber [24] have determined the intrinsic acidity constant of surface silanol groups of silica gel at 25°C in solutions of constant ionic strength 0.1 M ($NaClO_4$). Their value, $pK_{int} = 6.8 \pm 0.2$, indicates that the acidity of a surface silanol group is considerably larger than that of orthosilicic acid ($pK = 9.5$).

† An "average" equilibrium constant, K_{av}, quite representative of the free energy of the protolysis reaction per mole is obtained from

$$\ln K_{av} = \int_0^1 \ln K' d\bar{f}$$

[22] E. J. King, *Acid–Base Equilibria*, Macmillan, New York, 1965. (Chapter 9 gives a lucid discussion of polyprotic acids and of the interpretation of their alkalimetric titration curves.)
[23] C. Tanford, *Physical Chemistry of Macromolecules*, Wiley, New York, 1961, p. 572.
[24] P. Schindler and H. R. Kamber, personal communication, 1968.

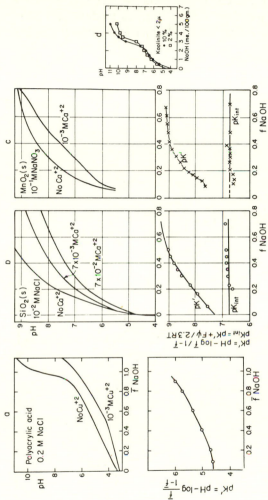

Figure 9-6 Alkalimetric titration of the "polymeric acids," polyacrylic acid, SiO_2, MnO_2 and H-kaolinite. In a, b and c: pH is plotted in the presence and absence of a complex-forming bivalent cation (Ca^{2+} or Cu^{2+}) as a function of the degree of neutralization f. Acidity constants and complex-formation constants can be calculated from the titration curves. The extent of complex formation is related to the extent of the shift in the titration curve caused by the complex forming cation.

In the lower figures, the acidity constant K' calculated from the titration curves is plotted as a function of $\bar{f}$. K' decreases with increasing $\bar{f}$ because with increasing negative surface charge it becomes more difficult for the surface OH group to dissociate a H^+. This electrostatic effect can be considered [Eq. (51)] and an intrinsic acidity constant K_{int}, independent of $\bar{f}$, can be calculated. Amorphous SiO_2 and amorphous MnO_2 (closely related to δ-MnO_2) have similar values of pK_{int}, and they also have similar pH_{ZPC}.

d: (cf. C. E. Marshall, *The Physical Chemistry and Mineralogy of Soils*, Wiley, New York, 1964) shows pH titration curves for two concentrations of kaolinite incipiently in the H^+ form.

Complex Formation Tendency

Figure 9-6 shows that the alkalimetric titration curves of polyacrylic acid and of colloidal suspensions of SiO_2 and MnO_2 are affected by the presence of metal ions. The displacement caused by the metal ions results from the displacement of bound H^+ by metal ions (48). Various models have been developed to express the complex-formation equilibria quantitatively [25–27],

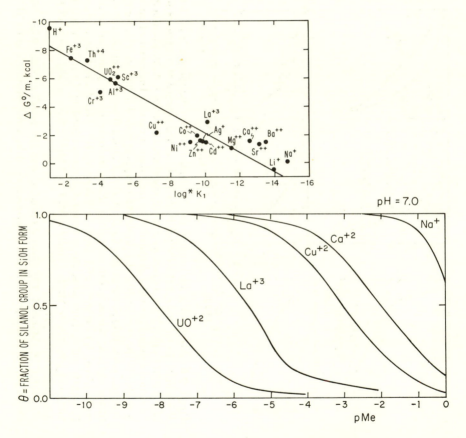

Figure 9-7 Distribution of metal ions on amorphous silica. The curves have been computed from equilibrium constants [D. L. Dugger et al., *J. Phys. Chem.*, **68**, 757 (1964)]. In order to show the similarity between surface SiO^- groups and OH^- ions as ligands, the free energy of the reaction $Me^{n+} + n(SiO^-) = Me(OSi)_n$ is compared with the equilibrium constant of the reaction $Me^{n+} + OH^- = MeOH^{(n-1)+}$. Reproduced with permission from American Chemical Society.

and equilibria of silica with various metal ions have been determined by different investigators [20, 25, 28].

A few of the apparent equilibrium constants for the reaction

$$Me^{n+} + n(-SiOH) = Me(OSi-)_n + nH^+ \qquad (52)$$

reported by Maatman's group [25, 28] are represented in terms of a distribution diagram (Figure 9-7). This figure illustrates that silica surfaces may exhibit considerable affinity toward metal ions within pH range of natural waters and may play a regulatory role as a H^+ ion buffer and in the distribution of certain trace metals between solid and solution phases.

In Figure 9-7 the free energy of the complex formation reaction

$$Me^{n+} + n(-SiO^-) = Me(OSi)_n \qquad (53)$$

for various metal ions is plotted as a function of their first hydrolysis constants (log $*K_1$) [25]. As pointed out by Dugger et al. [25], the environment of the oxygen atoms of $-SiO^-$ and aqueous OH^- are similar enough so that the reactions of each with metal ions are comparable.

9-6   Ion Exchange

Mixed oxides can be prepared in which a second cation of different charge than the parent cation is introduced into the structure, often resulting in remarkable alterations of the surface chemical behavior of these oxides. As we have seen, atomic substitutions and other lattice defects produce a fixed number of charges (anionic sites in the solid phase). Substitutions also strongly affect the acidity of structural OH groups. For example, the acid strength in the sequence $Si(OH)_4$, $PO(OH)_3$, $SO_2(OH)_2$ and $ClO_3(OH)$ increases from very weak (silicic acid) to very strong (perchloric acid) as can be predicted qualitatively from electrostatic and structural considerations. As Pauling [29] has pointed out, an aluminum tetrahedron with corners shared with silicon tetrahedra is similar to the perchlorate ion and, hence, we should expect to obtain a strong acid by replacing the K^+ ion of mica by hydrogen ion. Most clays behave as weak acids (Figure 9-6).

[25] D. L. Dugger et al., *J. Phys. Chem.*, **68**, 757 (1964).
[26] M. H. Kurbatov, G. B. Wood and J. D. Kurbatov, *J. Phys. Chem.*, **55**, 258 (1951).
[27] H. P. Gregor, L. B. Luttinger and E. M. Loebl, *J. Phys. Chem.*, **59**, 34 (1955).
[28] J. Stanton and R. W. Maatman, *J. Colloid Sci.*, **13**, 132 (1963).
[29] L. Pauling, *The Nature of the Chemical Bond*, 3rd Ed. Cornell Univ. Press, Ithaca, N.Y., 1960.

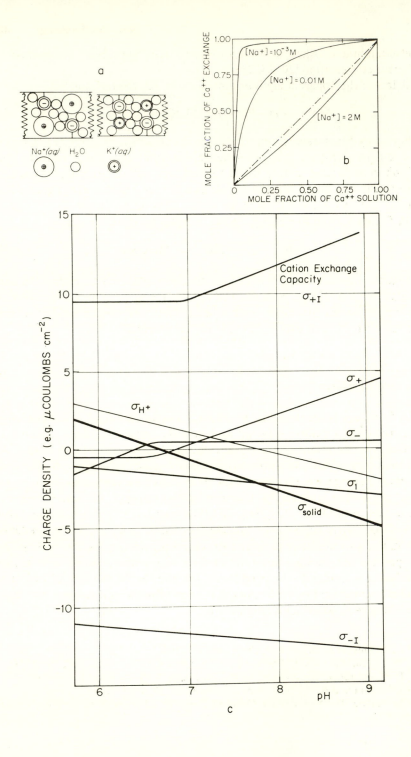

a

Na⁺(aq)   H₂O   K⁺(aq)

b

MOLE FRACTION OF Ca⁺⁺ EXCHANGE

$[Na^+] = 10^{-3}M$

$[Na^+] = 0.01 M$

$[Na^+] = 2 M$

MOLE FRACTION OF Ca⁺⁺ SOLUTION

Cation Exchange Capacity

σ_{+I}

σ_+

σ_{H^+}

σ_-

σ_1

σ_{solid}

σ_{-I}

CHARGE DENSITY (e.g. μCOULOMBS·cm⁻²)

pH

c

Simple Models

Some of the pertinent characteristics of the ion exchange process can be visualized from the model depicted in Figure 9-8a [30]. Here the fixed charges are indicated as anionic sites. The ion exchange framework is held together by lattice forces or interlayer attraction forces. These forces are represented in the model by mechanic springs. These forces may be very elastic in organic exchangers and quite nonelastic in zeolites. In clays such as montmorillonites, because of the difference in osmotic pressure between solution and interlayer space, water penetrates into the interlayer space. Depending upon the hydration tendency of the counter ions and the elasticity of the interlayer forces, different interlayer spacings may be observed. A composite balance between electrostatic, van der Waals forces and osmotic forces influences the swelling pressure in the ion exchange phase and, in turn, also the equilibrium position of the ion exchange equilibrium. Buser's [30] model illustrates that the extent of swelling increases with the extent of hydration of the counter ion. Furthermore, the swelling is much smaller for a bivalent cation than for a monovalent counter ion. Because only half as many Me^{2+} are needed as Me^+ to neutralize the charge of the ion exchanger, the osmotic pressure difference between solution and ion exchange framework becomes smaller. Thus if strongly hydrated Na^+ replaces less hydrated Ca^{2+} and Mg^{2+} in soils, the resulting swelling adversely affects the permeability of soils. Similarly, in waters of high relative $[Na^+]$ the bottom sediments are less permeable to water. Complex formation (ion pair formation) between the counter ions and the anionic sites (oxo groups) also reduces the swelling pressure.

From a geometrical point of view, clays could be packed rather closely. Muds containing clays, however, have a higher porosity than sand. The higher porosity of the clays is caused, in part, by the higher water content (swelling) which in turn is related to the ion exchange properties.

[30] W. Buser, *Chimia* (Aarau), **9**, 73 (1955).

Figure 9-8 Ion exchange.

a: Buser's model [*Chimia*, **9**, 73 (1955)]. Incorporation of well-hydrated Na^+ ions leads to larger volume and higher swelling pressure than incorporation of less-hydrated K^+ ions. Reproduced with permission from Schweiz Chemikerverband, Zürich, Switzerland.

b: Typical exchange isotherms for the reaction $Ca^{2+} + 2\{Na^+R^-\} \rightleftharpoons \{Ca^{2+}R_2\} + 2Na^+$. In dilute solutions the exchanger shows a strong preference for Ca^{2+} over $Na^{+\cdot}$ This selectivity decreases with increasing ion concentration. 45° line represents the isotherm with no selectivity.

c: Schematic representation of distribution of charge components as a function of pH on the surface of a one-component exchanger.

THE DISTRIBUTION OF CHARGE. Figure 9-8c illustrates schematically the charge densities of ions on and surrounding a one-component ion exchanger as a function of pH. Such an exchanger has a finite number of "built-in" anionic sites, σ_{-I}. For the interpretation of the distribution of exchangeable cations in the proximity of the solid surface, it is convenient to distinguish between cations within the ion exchange framework (interlayer cations and complex-bound cations) and cations in the diffuse double layer. A large part of the negative fixed charge density σ_{-I} is compensated by cations within the ion exchange framework σ_{+I}. The extent of this compensation and, hence, the magnitude of σ_{+I} depend on the type and concentration of cation and on the pH of the solution. There is a slight pH dependence of σ_{-I} because progressively less H^+ becomes dissociated from structural OH groups the lower the pH. A *net* intrinsic structural charge σ_i

$$\sigma_i = \sigma_{+I} + \sigma_{-I} \tag{54}$$

together with the charge imposed by potential-determining H^+ and OH^-, $\sigma_H = F(\Gamma_H - \Gamma_{OH})$, make up the total charge density σ_{solid} of the ion exchange structure. This is now balanced by a charge in the diffuse part of the double layer, $\sigma_{solution} = \sigma_+ + \sigma_-$. Combining the charge density of the structural cations with that of the cations in the diffuse double layer gives the total cation exchange capacity (CEC)

$$(CEC) = \sigma_{+I} + \sigma_+ \tag{55}$$

As Figure 9-8c illustrates, the CEC increases with increasing pH above ZPC but stays nearly constant at lower pH values. This is in accord with experimental observations. Some anion exchange capacity exists as well on the acid side of pH_{ZPC}. Because of the net intrinsic (negative) charge σ_i, pH_{ZPC} is shifted to a lower value than that observed in the absence of a structural charge.

Ion Exchange Equilibria

The double-layer theory predicts correctly (see Section 9-3) that the affinity of the exchanger for bivalent ions is much larger than that for monovalent ions, and that the selectivity for higher-valent ions decreases with increasing ionic strength of the solution. According to the Gouy–Chapman theory (in which ions are treated as point charges) there should be no ionic selectivity of the exchanger between different equally charged ions.

Relative affinity may be defined quantitatively by *formally* applying the mass law to exchange reactions:

$$\{Na^+R^-\} + K^+ = \{K^+R^-\} + Na^+ \tag{56}$$

$$2\{Na^+R^-\} + Ca^{2+} = \{Ca^{2+}R_2^{2-}\} + 2Na^+ \tag{57}$$

where R^- symbolizes the negatively charged network of the cation exchanger. A selectivity coefficient Q can then be defined by

$$Q_{(NaR \rightarrow KR)} = \frac{X_{KR}}{X_{NaR}} \frac{[Na^+]}{[K^+]} \tag{58}$$

$$Q_{(NaR \rightarrow CaR)} = \frac{X_{CaR}}{X_{NaR}^2} \frac{[Na^+]^2}{[Ca^{2+}]} \tag{59}$$

where X represents the equivalent fraction of the counter ion on the exchanger (e.g., $X_{CaR} = 2[Ca_R^{2+}]/(2[Ca_R^{2+}] + [Na_R^+])$. The selectivity coefficients may be treated as mass law constants for describing, at least in a semiquantitative way, equilibria for the interchange of ions, but these coefficients are neither constants nor are they thermodynamically well defined. Because the activities of the ions within the lattice structure are not known and vary depending on the composition of the ion exchanger phase, the coefficients tend to deviate from constancy. Nevertheless, it is expedient to use (59) for illustrating the concentration dependence of the selectivity for higher charged ions. In Figure 9-8b the equivalent fraction of Ca^{2+} on the exchanger is plotted as a function of the equivalent fraction of Ca^{2+} in the solution. For a hypothetical exchange with no selectivity, the exchange isotherm is represented by the dashed line. In such a case the ratio of the counter ions is the same for the exchanger phase as in the solution. The selectivity of the exchanger for Ca^{2+} increases markedly with increased dilution of the solution, but in solutions of high concentration the exchanger loses its selectivity. The representation in this figure makes it understandable why a given exchanger may contain predominantly Ca^{2+} if in equilibrium with a fresh water; in seawater, however, the counter ions on the exchanger would be predominantly Na^+. Figure 9-8b also illustrates why, in the technological application of synthetic ion exchangers such as water softeners, Ca^{2+} can be selectively removed from dilute water solutions whereas an exhausted exchanger in the Ca^{2+} form can be reconverted into a {NaR} exchanger with a concentrated brine solution or with undiluted seawater.

Wiklander [31] has determined the distribution of Ca^{2+} and K^+ for three clay minerals at various equivalent concentrations of Ca^{2+} and K^+ (Table 9-6). His results demonstrate the concentration dependence of the selectivity; they also show that marked differences exist between various clays.

[31] L. Wiklander, in *Chemistry of the Soil*, F. E. Bear, Ed., 2nd Ed. Reinhold, New York, 1964.

Table 9-6   Ion Exchange of Clays with Solutions of $CaCl_2$ and KCl of Equal Equivalent Concentration†

		Ca^{2+}/K^+ Ratios on Clay			
Clay	Exchange Capacity, meq g^{-1}	Concentration of Solution, $2[Ca^{2+}] + [K^+]$, meq liter^{-1}			
		100	10	1	0.1
Kaolinite	0.023	—	1.8	5.0	11.1
Illite	0.162	1.1	3.4	8.1	12.3
Montmorillonite	0.810	1.5	—	22.1	38.8

† From L. Wiklander, *Chemistry of the Soil*, F. E. Bear, Ed., 2nd ed. Reinhold, New York, 1964. Reproduced with permission from Reinhold Publishing Corp.

Thermodynamic Relations

The equilibrium relations for ion exchange reactions have been discussed by many writers and various approaches have been used [32, 33]. For example, equations for ion exchange equilibria have been derived from the concept of Donnan equilibria but the resulting formulations [essentially the same as those given in (58, 59)] contain terms for the activities in the ion exchanger phase (but the activity coefficients for these species cannot be evaluated).

H. C. Thomas and his coworkers [34–36] have recognized that the ion exchange equilibria between clay minerals and solutions of salts cannot be described, even roughly, by a mass law expression in terms of stoichiometric concentrations. Considering the clay phase as a separate electrolyte phase, a wet montmorillonite is at the simplest a 3-M solution in which the anions are fixed in position and in which the motion of the cations is otherwise highly inhibited [35].

Typical results from Thomas' laboratory [35] are reproduced in Figure 9-9. Experimental results on the cesium–potassium exchange in montmorillonite are plotted as an adsorption isotherm (Figure 9-9a). Obviously Cs^+ is being sorbed preferentially over K^+. Despite the use of various experimental approaches (batch and chromatographic methods) and approaching equilibrium from both sides, good experimental agreement is demonstrated. In order

[32] F. Helfferich, *Ion Exchange*, McGraw-Hill, New York, 1962.
[33] R. M. Garrels and C. L. Christ, *Solutions, Minerals and Equilibria*, Harper and Row, New York, 1965.
[34] G. L. Gaines and H. C. Thomas, *J. Chem. Phys.*, **21**, 258 (1954).
[35] C. A. Faucher and H. C. Thomas, *J. Chem. Phys.*, **22**, 258 (1954).
[36] G. R. Frysinger and H. C. Thomas, *J. Phys. Chem.*, **64**, 224 (1960).

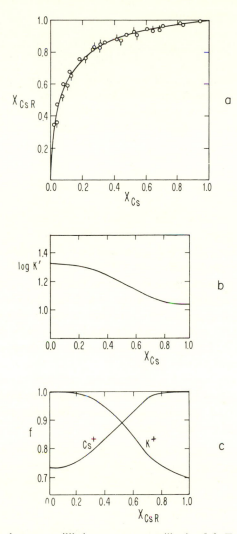

Figure 9-9 Ion exchange equilibria on montmorillonite [cf. Faucher and Thomas, *J. Chem. Phys.*, **22**, 258 (1954)]. Exchange between K^+ and Cs^+ from chloride solutions.

a: The exchange adsorption isotherm. The fraction of exchange sites occupied by Cs^+ is plotted as a function of the mole fraction of Cs^+ in the solution (circles without tails, chromatographic measurements; circles with tails up, batch experiments on Cs^+-clay; circles with tails down, batch experiments on K^+-clay).

b: The stoichiometric equilibrium constant is plotted as a function of X_{Cs}.

c: Deduced activity coefficients in mixed clay phase as a function of X_{CsR}.

Reproduced with permission from American Institute of Physics.

to quantize the ion exchange, the results can be converted first into stoichiometric equilibrium "constants"

$$K' = \frac{C_{\text{CsR}}}{X_{\text{KR}}} \frac{[\text{K}^+]}{[\text{Cs}^+]} \tag{60}$$

These are plotted as log K' as a function of X_{CsR}, the fraction of exchange sites occupied by cesium (Figure 9-9b). K' values vary from ca. 20 ($X_{\text{CsR}} = 0$) to ca. 11 for $X_{\text{CsR}} = 1$. A thermodynamically meaningful equilibrium constant, $K = 13.1$, can be obtained from these data through the use of the expression

$$\ln K = \int_0^1 \ln K' dX_{\text{CsR}} \tag{61}$$

This approach is similar to that outlined in the preceding section for computing the acidity constant K_{av} of the amorphous silica. The simplifications and assumptions inherent in this approach have been outlined by Gaines and Thomas [34]. The above expression for $\ln K$ is exact only if the activity of water in the ion exchange phase is equal to that in the solution phase. With $K = 13.1$ the activity coefficients of the mixed K^+ and Cs^+ clays have been evaluated. Although considerable deviations from ideality are observed, the data suggest that Cs^+ in Cs^+-rich clay and K^+ in a K^+-rich clay behave in nearly ideal fashion (Figure 9-9c). When present in small amounts, the ion behaves in a nonideal manner [35].

An interesting approach for determining cation exchange constants has been made by Hanshaw [37]. He used compacted clay samples as membrane electrodes and derived ion exchange constants from the theory of the membrane response of glass electrodes.

AFFINITY SERIES. With the help of ion exchange constants, a general order of affinity can be given. Eisenmann [38] has worked out the consequences of the assumption that the cation exchange equilibria are dominated by (a) Coulombic interactions between the counter ions (in various states of hydration) and the fixed groups of the exchanger, and (b) ion-dipole and ion-induced dipole interactions between the counter ions and water molecules (ionic hydration). Where (a) is weak compared to (b) the "normal" affinity sequence (Hofmeister series)

$$\begin{gather} \text{Cs}^+ > \text{K}^+ > \text{Na}^+ > \text{Li}^+ \\ \text{Ba}^{2+} > \text{Sr}^{2+} > \text{Ca}^{2+} > \text{Mg}^{2+} \end{gather} \tag{62}$$

[37] B. B. Hanshaw, *Clays and Clay Minerals, Proceedings of the 12th National Conference*, Pergamon, London, 1964, p. 397.
[38] G. Eisenmann, *Biophys. J.*, **2** [2], 259 (1962).

is observed, that is, the ion with the larger hydrated radius tends to become displaced by the ion of smaller hydrated radius. As interactions of the first kind predominate over those of the second, the selectivity may become reversed, that is, the affinity increases with decreasing crystal radii ($Li^+ >$ $Na^+ > K^+ > Rb^+ > Cs^+$). With most clays the normal series prevails, while with some zeolites and glasses the reversed selectivity may be observed. For a more refined interpretation the reader is referred to Eisenmann [38]. The buffering effect of clays on $[H^+]$ is evident from alkalimetric titration curves (Figure 9-6).

The preference of clays and most other natural ion exchangers for K^+ over Na^+ provides a ready explanation of why Na^+/K^+ ratios in natural waters and especially in seawater are much larger than unity although K^+ is only slightly less abundant than Na^+ in igneous rocks. Ion exchange processes continuously remove K^+ from solution and return it to the solid phase. Some of the K^+, however, is also removed from water by the conversion of clays, for example, montmorillonite into illite and chlorite.

9-7   Aggregation of Colloids

In dealing with colloids, the term "stability" has an entirely different meaning than in thermodynamics. A solution containing colloids is said to be stable if during the period of observation it is slow in changing its state of dispersion. The times for which sols are stable may be years or fractions of a second. The large interface present in these systems represents a large amount of free energy which by recrystallization or agglomeration tends to reach a lower value [39]; hence, thermodynamically, the lowest energy state is attained when the sol particles have united to large crystals.

Historically, two classes of colloidal systems have been recognized: hydrophobic and hydrophilic colloids. In colloids of the second kind there is a strong affinity between the particles and water; in colloids of the first kind this affinity is absent.

There exists a continuous transition between hydrophobic and hydrophilic colloids. Gold sols, silver halogenides and nonhydrated metal oxides are typical hydrophobic colloid systems. Gelatin, starch, gums, proteins, etc., and all biocolloids (viruses, bacteria) are hydrophilic. Many colloid surfaces relevant in water systems contain bound H_2O molecules at their surfaces. Amorphous silica and metal oxide surfaces are at least partially solvated and are thus intermediates between hydrophobic and hydrophilic colloids.

[39] J. Th. G. Overbeek, in *Colloid Science*, H. R. Kruyt, Ed., Elsevier, New York, 1952.

Hydrophobic and hydrophilic colloids have a different stability in the same electrolyte solution. According to Kruyt, the stability of hydrophobic colloids is governed by only two factors — charge and hydration.

Kinetics of Particle Agglomeration

The rate of particle agglomeration depends on the frequency of collisions and on the efficiency of particle contacts (as measured experimentally, for example, by the fraction of collisions leading to permanent agglomeration). We address ourselves first to a discussion of the frequency of particle collision.
FREQUENCY OF COLLISIONS BETWEEN PARTICLES. Particles in suspension collide with each other as a consequence of two mechanisms of particle transport:

1. Particles move because of their thermal energy (Brownian motion). Coagulation resulting from this mode of transport is referred to as perikinetic.
2. If colloids are sufficiently large or the fluid shear rate high, the relative motion from velocity gradients can exceed that caused by Brownian (thermal) effects (orthokinetic coagulation).

The time-dependent decrease in the concentration of particles (n = number of particles cm^{-3}) in a monodisperse suspension due to collisions by Brownian motion can be represented by a second-order rate law

$$-\frac{dn}{dt} = k_p n^2 \tag{63}$$

or

$$\frac{1}{n} - \frac{1}{n_0} = k_p t$$

As given by von Smoluchowski (1917), k_p can be expressed as

$$k_p = \alpha_p 8 D \pi a \tag{64}$$

where D is the Brownian diffusion coefficient and a the radius of the particle. α_p is the fraction of collisions leading to permanent agglomeration and is an operational parameter for the stability ratio. The diffusion coefficient in (64) can be expressed as (Einstein-Stokes) $D = kT/6\pi\eta a$, where η is the absolute viscosity. With this substitution we obtain

$$-\frac{dn}{dt} = \alpha_p \frac{4kT}{3\eta} n^2 \tag{65}$$

The rate constant k_p is of the order $2 \times 10^{-12} \, cm^3 \, sec^{-1}$ for water of 20°C and for $\alpha_p = 1$. Thus, for example, a turbid water containing 10^6 particles ml^{-1} would reduce its particle concentration by half within a period of ca. 6 days (5×10^5 sec), provided that all particles were completely destabilized and that the particles were sufficiently small so that collisions resulted from Brownian motion only.

The rate of decrease in the concentration of particles due to agglomeration under the influence of a velocity gradient, du/dz, can be described by first-order kinetics

$$-\frac{dn}{dt} = k_0 n \qquad (66)$$

or

$$\ln \frac{n}{n_0} = -k_0 t$$

with

$$k_0 = \alpha_0 \frac{4}{\pi} \mathbf{V}_m \frac{du}{dz} \qquad (67)$$

where $\mathbf{V}_m$ = volume of total solid mass suspended per volume of medium and α_0 = fraction of collisions leading to permanent agglomeration. A numerical example might again illustrate approximately the meaning of this rate law. For 10^6 particles of diameter $d \simeq 1\,\mu\,\mathrm{cm}^{-3}$, $\mathbf{V}_m$ becomes approximately $5 \times 10^{-7}\,\mathrm{cm}^3\,\mathrm{cm}^{-3}$. For $\alpha_0 = 1$ and for an agitation characterized by a velocity gradient of 5 sec^{-1} (this corresponds to a slow stirring in a beaker — about one revolution sec^{-1}), k_0 is of the order of $3 \times 10^{-6}\,\mathrm{sec}^{-1}$. Hence a period of ca. 3.7 days would lapse until the concentration of particles has been halved as a result of orthokinetic coagulation. Increasing the agitation within certain limits would increase the rate of agglomeration proportionally.

The over-all rate of decrease in concentrations of particles of any size is given by (65) and (66) by assuming additivity of the separate mechanisms [41]

$$-\frac{dn}{dt} = \alpha_p \frac{4kT}{3\eta} n^2 + \alpha_0 \frac{4(du/dz)\mathbf{V}_m}{\pi} n \qquad (68)$$

It can be shown that the second term becomes negligible for particles with a diameter of $d \ll 1\,\mu$, whereas for representative velocity gradients of $(du/dz) > 5\,\mathrm{sec}^{-1}$ only the second term is significant for particles of size $d \geq 1\,\mu$. It may be assumed that usually $\alpha_0 = \alpha_p$ [42].

The rate laws given in (65) and (66) have been corroborated experimentally by many researchers even though some of the assumptions used for their derivation (linear velocity gradients, monodispersity, etc.) are not fulfilled. Levich [40] has given a survey on the influence of more complex velocity patterns on the kinetics of particle agglomeration. By comparing actual agglomeration rates with those theoretical collision frequencies, collision efficiency factors α can be determined; in this manner a relative measure of colloid stability is obtained [41, 42].

A Physical Model for Colloid Stability

To what extent can theory predict the collision efficiency factor? Two groups of researchers, Verwey and Overbeek and Derjagin and Landau, independently of each other, have developed such a theory (VODL theory) [43] by quantitatively evaluating the balance of repulsive and attractive forces which interact when particles approach each other. Their model has undoubtedly become a most effective tool in the interpretation of many empirical facts in colloid chemistry.

It is not possible to derive this theory quantitatively here, but in order to discuss some of the relevant implications and to illustrate some of the quantitative dependencies we will refer to a few formulations without detailing derivations and conditions of application. The reader is referred to Overbeek [39] and Verwey and Overbeek [43] for a detailed quantitative development of one version of this theory and to van Olphen [6] for a comprehensive discussion on the stability of clays.

By using the Gouy description of the double layer it is in principle possible to calculate the energy of repulsion between two particles, that is, the work that must be performed to bring a particle from infinity to a given distance of another particle [43]. This energy of repulsion is a function of the charge or potential of the colloid and the ionic makeup of the solution; it increases with decreasing distance between the two particles.

For small energy interaction between two plates separated by a distance of $2d$ the following approximate result for the repulsive energy per unit area is obtained

$$V_R = 64nkT\kappa^{-1}\gamma^2 e^{-2\kappa d} \tag{69}$$

where

$$\gamma = \frac{e^{\bar{z}/2} - 1}{e^{\bar{z}/2} + 1}$$

and the other symbols are those used in Table 9-3. To apply the Stern theory it is necessary to replace Ψ_0 by Ψ_δ in the expression $\bar{z} = (zF\Psi_0/RT)$. In Figure 9-10a the energy of repulsion V_R (the repulsive potential) is plotted

[40] V. G. Levich, *Physicochemical Hydrodynamics*, Prentice-Hall, Englewood Cliffs, N.J., 1962.

[41] D. L. Swift and S. K. Friedlander, *J. Colloid Sci.*, **19**, 621 (1964).

[42] H. H. Hahn and W. Stumm, *J. Colloid Interfac. Sci.*, **28**, 134 (1968).

[43] E. J. W. Verwey and J. Th. G. Overbeek, *Theory of the Stability of Lyophobic Colloids*, Elsevier, Amsterdam, 1948.

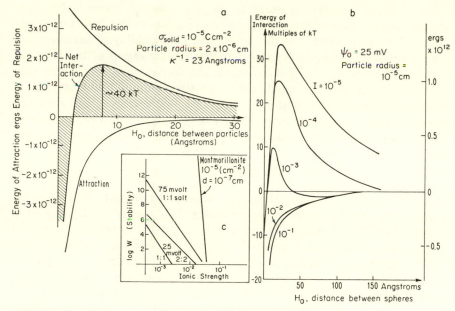

Figure 9-10 Physical model for colloid stability (VODL theory).

a: The net energy of interaction (energy of repulsion minus energy of attraction) of two constant charge surfaces (e.g., clays) is plotted as a function of particle distance for a given ionic strength $(1.8 \times 10^{-2}\ M$ 1:1 electrolyte). In addition to the conditions specified, the following assumptions were made in the calculations: the surface potential calculated for the given κ remains independent of particle distance; conditions of flat double layer are fulfilled $(\kappa a > 1)$; $A = 10^{-12}$ ergs.

b: Net energy of interaction for spheres of constant potential surface for various ionic strengths (1:1 electrolyte) (cf. Verwey and Overbeek).

c: Dependence of stability upon ionic strength. From curves such as those given in *a*, stabilities (related to height of energy barrier) for a few types of colloids have been calculated.

as a function of particle separation. As (69) shows, V_R decreases in an exponential fashion with increasing separation. The distance characterizing the repulsive interaction is similar in magnitude to the thickness of a single double layer (κ^{-1}). Because κ^{-1} is proportional to $I^{1/2}$ (27), the range of the repulsion depends primarily on the ionic strength of the electrolyte.

The energy of attraction due to London-van der Waals attractive forces as given earlier in (3) for two thick plates separated by a distance $2d$ is approximately

$$V_A = \frac{A}{48\pi d^2} \qquad (70)$$

where A is the Hamaker constant. In the lower part of Figure 9-10a, V_A is plotted as a function of particle separation. This curve varies little for a given value of A which depends on the density and polarizability of the dispersion phase but is essentially independent of the ionic makeup of the solution. Because of experimental inaccuracies, the absolute value of A is known only to within a range of magnitude (10^{-12} to 10^{-13} ergs). Thus A becomes a somewhat "adjustable constant." Summing up repulsive and attractive energies gives the total energy of interaction (Figure 9-10a). Conventionally, the repulsive potential is considered positive and the attractive potential negative

$$V = V_R - V_A \tag{71}$$

At very small separations, attraction outweighs repulsion, and at intermediate separations repulsion predominates. This energy barrier is usually characterized by the maximum (net repulsion energy) in the total potential energy curve, $V_{\max}$ (Figure 9-10b). The potential energy curve shows, under certain conditions, a secondary minimum at larger interparticle distances ($d \approx 10^{-6}$ cm). This secondary minimum depends on the ratio of the constants of (69) and (70) and on the dimensions of the particles involved; it is seldom deep enough to cause instability but might help to explain certain loose forms of adhesion or agglomeration.

THE STABILITY RATIO. A stable colloidal dispersion is characterized by a high energy barrier, that is, by a net interaction energy. Fuchs has defined a stability ratio W that is related to the area which is enclosed by the resultant curve of the energy of interaction. W is a factor by which agglomeration is slower than in the absence of an energy barrier. The conceptually defined W should correspond to the operationally determined α of (64), ($W = \alpha^{-1}$). In first approximation the stability ratio W is related to the height of the potential energy barrier $V_{\max}$

$$W = \frac{\kappa^{-1}}{2a} \, e^{V_{\max}/kT} \tag{72}$$

This equation indicates that the stability ratio increases with increasing ratio of the double-layer thickness to the particle radius κ^{-1}/a, and depends exponentially on the relative height of the energy barrier $V_{\max}$; the latter is conveniently expressed in units of kT. If $V_{\max}$ exceeds the value of a few kT, relatively stable colloids will be found. For example, if $V_{\max} \simeq 15kT$, only 1 out of 10^6 collisions will be successful. Figure 9-10b shows how $V_{\max}$ typically decreases with an increase in electrolyte concentration. For the same example, log W versus the ionic strength curve is plotted in Figure 9-10c.

Some of the pertinent interactions that affect colloid stability are readily apparent from these figures. In addition to the factors that determine the surface potential of the colloids, the most predominant influence upon stability is exerted by charge and concentration of the ions of the solution. Compacting the double layer (reducing κ^{-1}) by increasing the concentration of the electrolytes and their charge causes destabilization. The physical theory, based on a combination of van der Waals attraction with the repulsion of the Gouy–Chapman double layers has provided an analytical basis for the *valency rule of Schulze and Hardy*. Under simplifying conditions it is possible to derive the relation that the critical coagulation concentration CCC of mono-, di- and trivalent ions are in the ratio $(1/z)^6$ or 100:1.6:0.13. Experimental evidence on the coagulation of hydrophobic colloids by nonadsorbable ions is often in reasonable agreement with this rule.

The influence of the *aggregate size* on stability is complex. As a rule, increasing stability is observed with increasing aggregate particle radius. During coagulation the colloidal aggregates become larger, their stability increases and the agglomeration process slows down and may eventually come to a halt [43]. In order to predict the effects of stability upon agglomeration rate and resulting aggregate size (settling) we need more quantitative information on the finite aggregate size. Theoretically we would expect that the final aggregate size increases with increasing electrolyte concentration.

As indicated in Section 9-3, clays may carry different double-layer structures on the same particle: a negatively charged plate and a positively charged edge.† Therefore when a suspension of platelike particles aggregates, different modes of particle association may occur: face to face (FF), edge to face (EF) and edge to edge (EE) [6]. Hence the electrical energies of interaction for the three types of associations are governed by three different combinations of the double layers. Furthermore, because of geometric factors the van der Waals interaction energy will be different for the three types of association. Especially important is the fact that, owing to the opposite sign of the charge of the edge and face double layers, edge to face association (self-coagulation) can already take place in dilute electrolytes. Obviously, theoretical stability calculations for clays, simply based on considerations of the plate charge, thus have very little meaning concerning the state of aggregation. The different forms of particle association prevalent at different electrolyte concentrations cause very striking and technologically important variations in the rheological properties of clay–water systems. For a comprehensive discussion the reader is referred to van Olphen [6].

† According to van Olphen [6], at the edges of the plates the silica and alumina sheets are disrupted and primary bonds are broken. On such surfaces, similar to the surfaces of silica and alumina particles, the electric double layer is created by potential-determining ions, predominantly by OH^- or H^+.

Chemical Factors Affecting Colloid Stability

Purely chemical factors must be considered in addition to the theory of the double layer in order to explain the stability of most colloids in natural waters. Table 9–7 summarizes some of the more important general characteristics of colloid destabilization.

HYDROPHILIC COLLOIDS. The influence of electrolytes upon the stability of hydrophilic colloids is in a phenomenological and qualitative way similar to that in hydrophobic systems. By their screening effect the added indifferent salts lessen the mutual electrostatic interaction of the charged groups. Polyvalent cations will effectively screen the negatively charged groups of the hydrophilic particles but other factors besides those of valency and concentration play a role. It has been shown that the stability of hydrophilic colloids is strongly influenced by the pH of the medium and the composition of the ionized groups. Furthermore, remarkable differences in coagulative behavior toward multivalent cations are observed for colloids having different functional groups. This is caused by the fixation of multivalent cations onto the ionized group, accompanied by a reduction of the effective charge of the particles and by alteration of their surface solvation and in turn of their solubility. Complex formation with cations may further lead to a reversal of the charge of the particles. The charge-neutralizing or charge-reversing reaction may be visualized by the following schematic example

$$[R(COO^-)_n]^{n-} + mMe^{2+} = [R(COO_nMe_m)]^{(2m-n)+} \tag{73}$$

Equation 53 illustrates a potential charge neutralization for silica surfaces. Multivalent metal ions are able to form soluble or insoluble complexes with substances containing carboxyl, hydroxyl, sulfato and phosphato groups. In natural surface waters, for example, high concentrations of organic material are frequently associated with high concentrations of iron [44]. Many organic hydrophilic colloids form highly colored complexes with ferric iron.

The marked difference in the coagulative response of colloids with carboxyl, sulfate and phosphate functional groups toward a given multivalent cation, as well as the different sequence of cations with regard to the relative position of their critical concentration for each type of a colloid, are explainable if we consider the great variation in the affinity of inorganic cations toward the different functional groups [45]; for example, 10–100 times smaller concentrations of Pb, Cd, Cu and Zn are needed for coagulation of phosphate colloids than for coagulation of carboxyl colloids of similar charge density [3].

[44] J. Shapiro, *Limnol. Oceanog.*, **2**, 161 (1957).
[45] W. Stumm and J. J. Morgan, *J. Amer. Water Works Assoc.*, **54**, 971 (1962).

Table 9-7 Modes of Destabilization and Their Characteristics†

Phenomena	Physical Double-Layer Theory (Coagulation)	Chemical Bridging Model (Flocculation)	Aggregation by Adsorbable Species (Adsorption Coagulation)‡
Electrostatic interaction	Predominant	Subordinate	Important
Chemical interactions, and adsorption	Absent	Predominant	Important
Zeta potential for optimum aggregation	Near zero	Usually not zero	Not necessarily zero
Addition of an excess of destabilizing species	No effect	Restabilization due to complete surface coverage	Restabilization usually accompanied by charge reversal
Relationship between optimum dosage of destabilizing species and the concentration of colloid (or concentration of colloidal surface)	CCC virtually independent of colloid concentration	Stoichiometry, a linear relationship between flocculant dose and surface area	Stoichiometry possible but does not always occur
Physical properties of the aggregates which are produced	Dense, great shear strength but poor filtrability in cake filtration	Flocs of 3-dimensional structure; low shear strength but excellent filtrability in cake filtration	Flocs of widely varying shear strength and density

† From W. Stumm and C. R. O'Melia, *J. Amer. Water Works Assoc.*, **60**, 514 (1968).
‡ Aggregation by hydrolyzed metal ions fits into this category.
Reproduced with permission from American Water Works Association.

For the coagulation of sulfate colloids, 5–10 times larger concentrations of the same bivalent metals are needed than for the coagulation of carboxyl colloids [3]. Such a sequence finds its analogy in the relative solubilities of the metal phosphates, acetates and sulfates. In the case of amorphous silica and hydrous metal oxide sols, the Schulze–Hardy rule is not obeyed. In the neutral pH range concentrations of $[Ca^{2+}]$ as large as 1 M are necessary for coagulation.

The effect of anions that can form complexes with colloidal surface groups is similar to the cases discussed; for example, phosphates or organic substances containing functional hydroxyl or carboxyl groups interact chemically with the aluminum edge of a clay particle. The resultant alteration of the edge charge (conversion from positive to negative double layer) modifies the stability of the clay.

Adsorption and Colloid Stability

Sorbable species are observed to destabilize colloids at much lower concentrations than nonsorbable ions. The extent to which a species is sorbable is reflected in the concentration necessary to produce aggregation. Typical destabilization regions (the range of concentrations over which colloid destabilization is accomplished) are presented in Figure 9-11. It is apparent that there are dramatic differences in the coagulating abilities of simple ions (Na^+, Ca^{2+}, Al^{3+}), hydrolyzed metal ions (highly charged multimeric hydroxo metal species) and species of large ionic or molecular size (alkylamines, polyelectrolytes, polymeric molecules). It is useful to consider here a few general features which may be derived from the data presented in Figure 9-11. For species having equal counter ion charges it is apparent that the CCC decreases with increasing ion sorbability (e.g., NO_3^- and p-toluene sulfonate, CH_3—C_6H_4—SO_3^-), and the extent of adsorption of a species increases with the size of the species, for example, $C_{12}H_{25}NH_3^+$ and $C_{16}H_{33}NH_3^+$.

Sorbable species that coagulate colloids at low concentrations may restabilize these dispersions at higher concentrations. When the destabilization agent and the colloid are of opposite charge, this restabilization is accompanied by a reversal of the charge of the colloidal particles. Purely coulombic attraction will not permit an attraction of counter ions in excess of the original surface charge of the colloid and, according to the Gouy–Chapman model for the double layer, charge reversal cannot occur. That adsorption can occur against electrostatic repulsion has already been discussed in Section 9-4. Figure 9-12 documents that insiduous traces of hydrolyzed metal ions or surfactants alter the colloid stability. Simple adsorption models have been developed that depict colloid stability as a

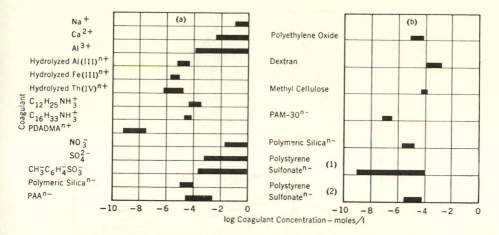

Figure 9-11 Typical destabilization regions for a few representative coagulants [from Stumm and O'Melia, *J. Amer. Water Works Assoc.*, **60**, 514 (1968)].

a: Coagulation regions observed when the colloid and the coagulant are of opposite charge.

b: Data obtained for the coagulation of colloids by uncharged molecules or by ions of like charge.

These simplified figures are based on experimental results by various investigators that have used different experimental procedures and systems; the coagulation regions are therefore not always precisely comparable. The location of the coagulation regions are exemplifications, that is, they attempt to show essential features and are meaningful in a semiquantitative way only. With sorbable species, the exact location of the coagulation region on the concentration axis depends upon the concentration and surface potential of the colloids, the pH, the presence of other cations and anions, and the temperature. Interactions with polymers are dependent upon molecular weight and steric factors. $PDADMA^{n+}$ = polydiallyldimethylammonium^{n+}. PAA^{n-} = poly-acrylic acid^{n-}. $PAM-30^{n-}$ = polyacrylamide^{n-} 30% hydrolyzed.

Reproduced with permission from American Water Works Association.

function of concentration of destabilizing species and of the surface concentration of the dispersed phase [46–48].

POLYMERS. Natural and synthetic macromolecules have been used successfully as aggregating agents in water and waste treatment and for sludge conditioning. Natural anionic polymers are of great importance in sorptive and colloidal destabilization reactions of negatively charged particles in

[46] P. Somasundaran and D. W. Fuerstenau, *J. Phys. Chem.*, **20**, 90 (1966).
[47] R. H. Ottewill, M. C. Rastogi and A. Watanabe, *Trans. Faraday Soc.*, **56**, 854 (1960).
[48] C. R. O'Melia and W. Stumm, *J. Colloid Interfac. Sci.*, **23**, 437 (1967).

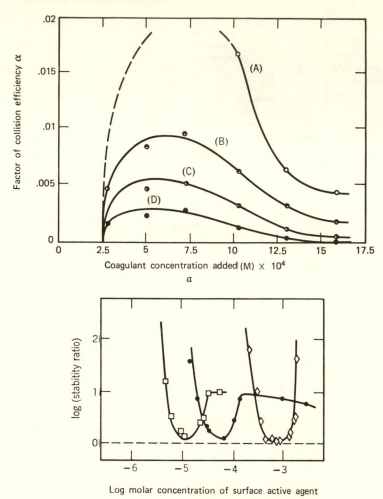

Figure 9-12   Surfactants and hydrolyzed metal ions affect colloid stability.

a: Effects of pH and Al(III) concentration upon relative colloid stability of amorphous silica expressed as collision efficiency factor. Colloidal silica: Ludox LS 0.3 g/liter^{-1}; (*A*) pH = 5.75; (*B*) pH = 5.5; (*C*) pH = 5.25; (*D*) pH = 5.0 [cf. Hahn and Stumm, *J. Colloid Interfac. Sci.*, **28**, 134 (1968)]. Reproduced with permission from Academic Press,

b: Effect of anionic surface active agents upon the stability of positive silver iodide sols [cf. Ottewill and Watanabe, *Kolloid Z.*, **170**, 38 (1959)]. The curves are, from left to right, for Manoxol OT, sodium dodecylbenzene sulfonate and sodium dodecyl sulfonate. Reproduced with permission from Steinkopff Verlag.

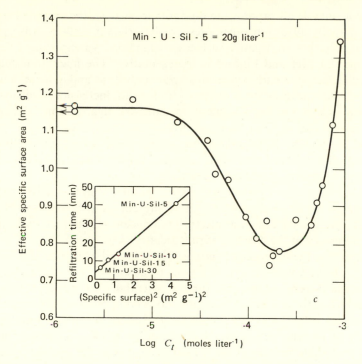

c: Effective specific surface area of silica (Min-U-Sil) suspensions (20 g/liter⁻¹) as a function of the concentration of hydrolyzed iron(III) applied. Effective surface areas are calculated from refiltration time measurements using the calibration curve presented in the insert. The linear relationship observed in the insert conforms to the predictions, based on the Kozeny equation [after O'Melia and Stumm, *J. Colloid Interfac. Sci.*, **23**, 437 (1967)]. Reproduced with permission from Academic Press.

biological systems; for example, the cohesion of tissue cells [49] and the aggregation of microorganisms [50] have been interpreted in terms of colloid–polymer interactions. It is essential to recognize that colloid destabilization by these materials cannot be characterized by the double-layer model. La Mer and coworkers and others have developed a chemical bridging theory that provides a more acceptable model for understanding the ability of polymers to destabilize colloidal suspensions. In its simplest form the chemical bridging theory proposes that a polymer molecule can attach itself to the surface of a colloidal particle at one or more adsorption sites with the remainder of the molecule extending into the solution. These extended segments can then interact with vacant sites on another colloidal particle. Failing

[49] B. A. Pethica, *Exp. Cell Res.*, *Suppl.*, **8**, 123 (1961).
[50] P. J. Mill, *J. Gen. Microbiol.*, **35**, 58 (1964).

to find a suitable adsorption site on another particle, the extended segments can eventually adsorb at other sites on the original surface. Adsorption of anionic polymers on negative surfaces is common. An experiment described by Kane, La Mer and Linford is representative. The macromolecule poly-acrylamide acquires carboxyl groups upon hydrolysis and assumes a negative charge. These investigators observed that with increasing hydrolysis and negative charge on the flocculant, progressive aggregation of negatively charged particles of silica occurred. Mechanisms that have been considered for the attachment of polyanions on anionic solid suspension are hydrogen bonding, anion interchange with adsorbed anions (such as OH^-) or inter-action with cations on the colloid surface.

Flocculation of a polymer–sol system of like charge results only if an appropriate concentration of a salt is present in the solution. However, flocculation usually occurs at electrolyte concentrations much smaller than those necessary in the absence of polymers.

Concerning the influence of cations, Nemeth and Matijevic [51] have con-cluded that the polyelectrolyte functions by adsorbing onto the surface of the colloid and thereby reducing the surface potential, thus making it possible for the electrolyte to be effective by double-layer compression. Sommerauer et al. [52], on the other hand, accounted for the effect of counter ions upon the stability of an anionic polymer–sol system by invoking complex formation in the immediate vicinity of the sol surface between the counter ion and the functional groups of the adsorbed polyelectrolyte.

Many bacteria and algae have a tendency to adhere to interfaces and to each other (*bioflocculation*). Because of the hydrophilic surface, the stability of a microbial dispersion does not depend primarily on the electrostatic repulsive forces between the cells. Reduction of surface potential is not a pre-requisite for bioflocculation; some bacterial suspensions can form stable dispersions at the pH_{ZPC}. It has been proposed [53, 54] that flocculation of microorganisms is effected by an interaction (bridging mechanism) of polymers excreted by the microorganisms or exposed at the microbial surface under suitable physiological conditions.

Naturally occurring aggregates that contain much of the nonliving organic matter in seawater commonly consist of pale yellowish or brownish amor-phous matrices with inclusions of bacteria, silt particles, and sometimes

[51] R. Nemeth and E. Matijevic, *Kolloid-Z. Z. Polym.*, **225**, 155 (1968).
[52] A. Sommerauer, D. L. Sussman and W. Stumm, *Kolloid-Z. Z. Polym.*, **225**, 147 (1968).
[53] M. W. Tenney and W. Stumm, *J. Water Pollution Control Federation*, **37**, 1370 (1965).
[54] P. L. Busch and W. Stumm, *Environ. Sci. Technol.*, **2**, 49 (1968).

phytoplankton [55] (*organic aggregates*). Recent evidence [56, 57] suggests that the first stage in the formation of these aggregates is a physical-chemical process; aggregates appear to be formed by adsorption of dissolved organic matter on bubbles and other naturally occurring surfaces in the sea. It is quite likely that polymers of microbial origin participate in the formation and further agglutination of these aggregates.

9-8   The Role of Coagulation in Natural Waters†

The aggregation of colloids is of great importance in the transport and distribution of matter in natural waters. Although dissolved substances tend to be distributed by convective mass transfer, the distribution of suspended matter is also influenced by the forces of gravity. Whether a particle will settle depends on its density, its size and the water movement. Aggregation processes not only affect the distribution of suspended matter, but they also play a role in the transformation of solutes because many dissolved substances, especially hydrolyzed metal ions and organic material, adsorb onto colloids or react chemically with colloid surface groups. It is estimated that all the rivers on the earth bring ca. 12 km^3 of sediment into the oceans per year [58]. The majority of run-off water first collects in rivers and lakes, and only very fine-grained material, predominantly clays and fine silt, are carried into the sea. Some rivers are very turbid.

Obviously colloid chemical reactions can profoundly influence the temporal and spatial distribution of dissolved and suspended constituents. In order to describe the various transportation processes of colloids and their effect on the ecological interdependence in natural water systems and on the ultimate distribution of suspended matter, one needs answers to the following types of questions:

1. What is the rate of coagulation of various naturally occurring colloids under specified conditions of pH and ionic strength?

2. What are the sizes of agglomerates found within the time period of interest under the given solution conditions?

3. How fast does suspended matter settle when the solution parameters are changed in such a way as to effect aggregation of the colloidal phase, and

† This section draws on a paper of the same title prepared by H. H. Hahn and W. Stumm. and presented at 156th Meeting, American Chemical Society, Atlantic City, Sept. 1968.
[55] G. A. Riley, *Limnol. Oceanog.*, **8**, 372 (1963).
[56] E. R. Baylor and W. H. Sutcliffe, *Limnol. Oceanog.*, **8**, 369 (1963).
[57] R. T. Barber, *Nature*, **211**, 257 (1966).
[58] P. H. Kuenen, *Marine Geology*, Wiley, New York, 1950.

what is the rate of decrease in the turbidity (naturally occurring colloids or pollutants) in a body of water at given flow velocities or detention times?

4. How far are discharged colloidal pollutants transported by the movement of natural waters before they are removed through sedimentation, and what is the variation of composition of colloidal aggregates found in the sediments as it depends on the distance from the point of discharge?

5. In what sequence are various colloids of different stability removed from a water of changing composition, such as is the case in the transition from a river water to the open sea; what are the effects upon the composition of the sediments in that transition zone?

Colloid chemistry can help to answer these questions but we do not have adequate theories or sufficient empirical information concerning these complicated processes to provide quantitative and generally valid answers.

COLLOID STABILITY IN NATURAL WATERS. It is important to emphasize that coagulation phenomena in natural waters are quite specific. Because of the great variety of possible colloid–solute interactions the computation of a stability ratio on the basis of a simplified double-layer model might give unrealistic estimates of colloid stability. For example, for clays in fresh waters stability ratios in the order of 10^{100} or higher may be computed for a given surface charge density (e.g., $10 \ \mu C \ cm^{-2}$ in a $10^{-3} \ M$ solution) (see Figure 9-10c). According to such stability ratios, clays would remain in colloidal dispersion for geological time spans, but such predictions do not incorporate the effects of face-to-edge aggregation or destabilization by adsorption of traces of a hydrolyzed metal ion. Experimentally determined α values for various colloids encountered in natural water systems should be useful for predicting agglomeration rates.

The double layer alone can also not explain the colloid chemical behavior of most other inorganic colloids in natural water. The Schulze-Hardy rule is not fulfilled for sols of amorphous silica. These do not, for example, coagulate readily if suspended in seawater.

The VODL theory predicts that the colloid stability decreases with increasing ionic strength; hence, stability ratios for seawater calculated are close to 1. Most colloids are less stable in *seawater* than in fresh water. Nevertheless the collision efficiency will be quite small for most hydrophilic colloids (organic aggregates, biocolloids and inorganic colloids well hydrated or coated with organic "colloid-protective" material). Arrhenius [59] infers from measurements of the amount of particulate aluminum (Al is a useful indicator reflecting the quantity of aluminosilicates that is believed to form a major part of the inorganic mineral suspension in seawater) that there are in North

[59] G. Arrhenius, in *The Sea*, Vol. 3, M. N. Hill, Ed., Wiley, New York, 1963.

Pacific deep water about 3×10^7 particles ml^{-1} of particle diameter 0.01–0.5 μ. From the rate of fallout on the ocean floor and from the concentration of the suspended colloids, Barth (1952) obtained estimates of an average passage time of more than 200 but considerably less than 600 years for these fine particles. The passage time is defined as the time during which the mass of particles originally present in the entire water column has been reduced to $1/e$ of the original value. In two hundred years (6×10^9 sec) the original concentration of the colloids would be reduced by perikinetic coagulation to 100 particles ml^{-1} (65) if all collisions were successful.† A collision efficiency factor of $\alpha \simeq 10^{-5}$ is necessary to adjust the rate of coagulation to the same order of magnitude as the rate of removal actually observed in the open ocean. In view of the fact that the organic constituents are present in concentrations much larger than those of the suspensoid [59], α-values in this order of magnitude do not seem unreasonable. Numerous other factors are of importance, for example, as pointed out by Arrhenius [59], aggregation in the gut of filter feeding animals possibly contributes significantly to the removal of suspended matter.

Example 9-3 *Effect of Detention Time on Residual Turbidity.* Estimate the decrease in turbidity resulting from coagulation in a reservoir as a function of detention time. Assume "completely mixed" conditions (negligible settling), spacially uniform composition and turbidity to be proportional to the number concentration of aggregates.

In a steady-state model inflow into the reservoir will be balanced by outflow plus the decrease in concentration of aggregates resulting from coagulation. Using (63) and (66) for perikinetic and orthokinetic coagulation, respectively, (i) and (ii) for the steady-state balance are obtained:

$$\frac{Q}{V} n_0 - \frac{Q}{V} n - k_p n^2 = 0 \qquad \text{(i)}$$

$$\frac{Q}{V} n_0 - \frac{Q}{V} n - k_0 n = 0 \qquad \text{(ii)}$$

where Q is the inflow or outflow rate ($m^3 \ sec^{-1}$), V is the volume of the reservoir (m^3) and V/Q is the theoretical detention time τ (sec). By solving (i) and (ii) for n, the steady-state concentration of aggregates can be estimated:

$$\text{perikinetic:} \quad n = \frac{\tau^{-1} \pm \sqrt{\tau^{-2} + 4k_p n_0 \tau^{-1}}}{2k_p} \qquad \text{(iii)}$$

$$\text{orthokinetic:} \quad n = \frac{\tau^{-1} n_0}{\tau^{-1} + k_0} \qquad \text{(iv)}$$

† There appears to be a printing error regarding the magnitude of the perikinetic rate constant in [59].

AGGREGATE SIZE. It was pointed out that there is a lack of a quantitative theory and of empirical data that would allow prediction of the size to which particles will ultimately coagulate. According to earlier discussion the limiting aggregate size should increase with decreasing colloid stability, and this is supported by experimental evidence. Whitehouse, Jeffrey and Debrecht [60] have determined sedimentation rates of colloidal aggregates in seawater of different dilution. The equivalent diameter (as calculated from the Stokes settling equation) depends directly upon the electrolyte concentration. The aggregate size in sediments is related to the electrolyte concentration. Usually the aggregate size increases with increasing salinity.

EFFECT OF HORIZONTAL MOVEMENT. In a typical *estuarine circulation* [61], particles carried down with the river water will, in the stratified area, sink from the upper into the lower water layer — the gradient in electrolyte concentration probably enhances this settling because of agglomeration — and be carried back upstream. This process may repeat itself. Thus concentrations of suspended matter in an estuary may be many times higher than those found in the river or the open sea. The mechanism by which fresh water flows freely seaward while suspended particles are retained is operative even if colloid stability were not affected by electrolyte concentration. The destabilization by electrolytes can make the retention process more efficient because it influences the settling velocity of these materials. One size of material may be trapped more efficiently than another.

GRAVITY FORCES. Can colloid chemistry aid in explaining the variation in sediment composition that parallels the increase in electrolyte content near a river inflow into the sea? A good example of such variation in abundance of different clays has been described by Grim and Johns [62] (Figure 9-13). We might argue that different clays become sequentially destabilized in meeting progressively increasing salinities. Although such effects may plausibly influence the spatial distribution of colloids in sediments, insufficient theoretical and experimental justification is available to establish this effect as the principal explanation for the observed variation in clay distribution. Actually, chemical transformation of montmorillonite into illite and chlorite under the influence of the increasing salinity appears to be a more reliable explanation,

[60] U. G. Whitehouse, L. M. Jeffrey and J. B. Debrecht, *Clays and Clay Minerals, Proceedings of the 7th National Conference*, National Academy of Sciences–National Research Council, Washington, D.C., 1960, p. 1.
[61] H. Postma, in *Pollution and Marine Ecology*, T. A. Olson and F. J. Burgess, Eds., Wiley-Interscience, New York, 1967.
[62] R. E. Grim and W. D. Johns, *Clays and Clay Minerals, Proceedings of the 2nd National Conference*, National Academy of Sciences–National Research Council, Washington, D.C., 1953, p. 81.

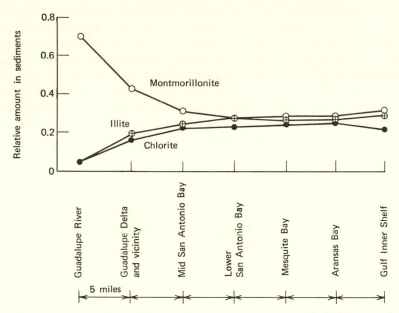

Figure 9-13   Variation in sediment composition as function of distance from point of river inflow. From data reported for Northern Gulf of Mexico by Grim and Johns [62]. Reproduced with permission from National Academy of Sciences.

but numerous other factors such as sorting by current circulation must be considered.

Infiltration and permeation through soils is affected by aggregate size. According to the Kozeny equation for flow through porous media, the filtration rate is inversely related to the square of the effective particle surface area and thus depends on the extent of coagulation. Trace constituents may affect the state of aggregation. Organic matter, especially polysaccharides excreted by microorganisms, and amorphous precipitates such as iron(II) sulfide tend to disperse the particles in the medium. On the other hand, electrolytes (Ca^{2+}), polyelectrolytes and certain metal ion hydrolysis products may, under suitable conditions, increase the permeability because of particle agglomeration and a decrease in swelling. A simple laboratory experiment may illustrate the effect of minute quantities of hydrolyzed Fe(III) upon the filtration rate through a porous medium of silica particles (Figure 9-12c).

WASTE DISPOSAL. Postma [61] has pointed out that suspended material is distributed in a different fashion than dissolved material. This has not been fully appreciated in publications on waste disposal.

Most of the organic pollutants discharged from an effluent of a sewage treatment plant are in the form of colloidal solids. While dissolved matter is usually transported away from the outfall, suspended wastes may be deposited in places not far removed from their point of origin. The spatial distribution will depend not only on water movements but also on the state of dispersion of the colloidal material, which in turn affects its sedimentation and flotation. Accumulation of trace metals, radio nuclides, organic matter and toxic material on colloidal interfaces is reversible; thus adsorbed materials may be released into solution as the particles are carried from one environment to the other. Uptake or loss is accompanied by coagulation or redispersion and occurs as a result of changes in adsorption and ion exchange equilibria [61].

READING SUGGESTIONS

I. Forces at Interfaces

Baier, R. E., E. G. Shafrin and W. A. Zisman, "Adhesion: Mechanisms That Assist or Impede It," *Science*, **162**, 1360 (1968).

Fowkes, F. M., "Attractive Forces at Interfaces," *Ind. Eng. Chem.*, **56**, 40 (1964).

II. Electric Double Layer and Surface Chemistry

Bruyn, P. L. de, and G. E. Agar, "Surface Chemistry of Flotation," in *Froth Flotation*, D. W. Fuerstenau, Ed., American Institute of Mining, Metallurgical and Petroleum Engineers, New York, 1962. This is a lucid discussion concerned with the thermo-dynamics and electrochemistry of surfaces and the electric double layer model of the interface.

Olphen, H. van, *An Introduction to Clay Colloid Chemistry*, Wiley-Interscience, New York, 1963. A readily understandable treatment of modern concepts in colloid and surface chemistry and their application to clay systems. Contains a detailed discussion on the crystal structure of clay systems.

Parks, G. A., "Aqueous Surface Chemistry of Oxides and Complex Oxide Minerals; Isoelectric Point and Zero Point of Charge," in *Equilibrium Concepts in Natural Water Systems*, Advances in Chemistry Series, No. 67, American Chemical Society, Washington, D.C., 1967.

Somasundaran, P., T. W. Healy and D. W. Fuerstenau, "Surfactant Adsorption at the Liquid–Solid Interface — Dependence of Mechanism on Chain Length," *J. Phys. Chem.*, **68**, 3562 (1964).

Vold, M. J., and R. D. Vold, *Colloid Chemistry; the Science of Large Molecules, Small Particles, and Surfaces*, Reinhold, New York, 1964. Available as a paperback edition, this exposition gives an introductory, essentially nonmathematical but nevertheless fairly rigorous account of the more important principles of colloid chemistry.

III. Ion Exchange

Amphlett, C. B., *Inorganic Ion Exchangers*, Elsevier, Amsterdam, 1964.

Garrels, R. M., and C. L. Christ, "Ion Exchange and Ion Sensitive Electrodes," in *Solutions, Minerals and Equilibria*, Harper and Row, New York, 1965, Chapter 8. The phenomena of ion exchange and of membrane electrodes are shown to be aspects of the behavior predicted for rigid, negatively charged frameworks that permit gain, loss or migration of cations without disturbance of the framework.

IV. Coagulation and Flocculation

(a) Polyelectrolytes

LaMer, V. K., and T. W. Healy, "Adsorption Flocculation Reactions of Macromolecules at the Solid–Liquid Interface," *Rev. Pure Appl. Chem.*, **13**, 112 (1963).

(b) Hydrolyzed Metal Ions, and Application in Water Treatment

Black, A. P., "Electrokinetic Characteristics of Hydrous Oxides of Aluminum and Iron," in *Principles and Application of Water Chemistry*, S. D. Faust and J. V. Hunter, Eds., Wiley, New York, 1967.

Stumm, W., and J. J. Morgan, "Chemical Aspects of Coagulation," *J. Amer. Water Works Assoc.*, **54**, 971 (1962).

Stumm, W., and C. R. O'Melia, "Stoichiometry of Coagulation," *J. Amer. Water Works Assoc.*, **60**, 514 (1968).

IO   Case Studies: Phosphorus, Iron and Manganese

10-1   Introduction

This chapter illustrates how information based on equilibria and kinetic results obtained in the laboratory can be used in arriving at interpretations of the mode of heterogeneous distribution, transportation and accumulation of a few elements in water systems. The literature on observed chemical facts in natural waters is usually rather rich in phenomena but lacking in details. Models are useful for ordering the phenomenological data. It must be emphasized that, because of the complexity of the natural environment, the application of results from laboratory studies to natural systems must be tested. Usually there is a difference between chemical reactions studied in the laboratory and those observed in nature; this difference helps to identify those environmental effects that have been overlooked in the laboratory and indicate where additional investigations may be needed.

Interrelation of P, Fe and Mn in Aquatic Transformations

Fe and Mn are among the major components of the earth's crust, while phosphorus is a minor constituent of the lithosphere. Although these elements seldom occur as dissolved species at concentrations larger than 10^{-4} M, they play important roles in limnology and oceanography and in water supplies; they are discussed together in this chapter because their aquatic transformations (dissolution, transport, distribution, precipitation and accumulation) are interrelated and are interdependent with those of other significant components of natural waters.

10-2   Phosphorus

This section sets out the important known chemical properties of the forms of phosphorus in natural water systems. This discussion of phosphorus chemistry is not exhaustive. The goal is to provide a chemical orientation before considering the role of phosphorus in limnology and oceanography.

514

Phosphorus Species in Natural Water Systems

Phosphorus occurs in nature almost exclusively as phosphate, the fully oxidized state, with formal oxidation number V and coordination number 4. Phosphate occurs in all known minerals as orthophosphate, the ion form of which is represented as

$$\left[\begin{array}{c} O \\ | \\ O\!-\!P\!-\!O \\ | \\ O \end{array} \right]^{3-}$$

or $[PO_4]^{3-}$ (Table 10-1).

Phosphate is found in the dissolved form in natural waters as a result of the natural weathering and solution of the phosphate minerals, soil erosion and transport, soil fertilization and resultant phosphorus transport, biological transfer (assimilation and dissimilation processes involving phosphorus in agriculture, etc.), and use of soluble phosphate compounds in detergent manufacture, water treatment and industry (domestic and industrial wastewaters) (Table 10-1).

The condensed inorganic phosphates are not found occurring naturally in minerals. They are found in all plants and animals where they are synthesized

Table 10-1A   Dissolved Phosphorus Forms of Possible Significance in Natural Waters

Form	Representative Compounds or Species
Orthophosphate	$H_2PO_4^-$, HPO_4^{2-}, PO_4^{3-}, $FeHPO_4^+$, $CaH_2PO_4^+$
Inorganic condensed phosphates	
pyrophosphate	$H_2P_2O_7^{2-}$, $HP_2O_7^{3-}$, $P_2O_7^{4-}$, $CaP_2O_7^{2-}$, $MnP_2O_7^{2-}$
tripolyphosphate	$H_2P_3O_{10}^{3-}$, $HP_3O_{10}^{4-}$, $P_3O_{10}^{5-}$, $CaP_3O_{10}^{3-}$
trimetaphosphate	$HP_3O_9^{2-}$, $P_3O_9^{3-}$, $CaP_3O_9^-$
Organic orthophosphates	
sugar phosphates	Glucose-1-phosphate, adenosine monophosphate
inositol phosphates	Inositol monophosphate, inositol hexaphosphate
phospholipids	Glycerophosphate, phosphatidic acids, phosphatidyl choline
phosphoamides	Phosphocreatine, phosphoarginine
phosphoproteins	
Organic condensed phosphates	Adenosine-5'-triphosphate, coenzyme A
Phosphorus-containing pesticides	O_2N—⟨C₆H₄⟩—$OPS(OCH_3)_2$; etc.

Table 10-1B   Solid Phase Forms of Phosphorus of Possible Significance in Natural Water Systems

Form	Representative Compounds or Substances
Soil and rock mineral phases	
hydroxylapatite	$Ca_{10}(OH)_2(PO_4)_6$
brushite	$CaHPO_4 \cdot 2H_2O$
carbonate fluorapatite	$(Ca,H_2O)_{10}(F,OH)_2(PO_4,CO_3)_6$
variscite, strengite	$AlPO_4 \cdot 2H_2O$, $FePO_4 \cdot 2H_2O$
wavellite	$Al_3(OH)_3(PO_4)_2$
Mixed phases, solid solutions, sorbed species, etc.	
clay-phosphate (e.g., kaolinite)	$[Si_2O_5Al_2(OH)_4 \cdot (PO_4)]$
metal hydroxide-phosphate	$[Fe(OH)_x(PO_4)_{1-x/3}]$, $[Al(OH)_x(PO_4)_{1-x/3}]$
clay-organophosphate	$[Si_2O_5Al_2(OH)_4 \cdot ROP]$, clay-pesticide, etc.
metal hydroxide-inositol phosphate	$[Fe(OH)_3 \cdot$ inositol hexaphosphate]
Suspended or insoluble organic phosphorus	
bacterial cell material	Inositol hexaphosphate or phytin, phospho-
plankton material	lipid, phosphoprotein, nucleic acids,
plant debris	polysaccharide phosphate
proteins	

enzymatically and constitute a part of the polyphosphate pool. Man-made condensed inorganic phosphates, produced by dehydration and condensation of orthophosphates, are substantial components of synthetic detergents (pyrophosphate and tripolyphosphate builder compounds). Polyphosphates and metaphosphates (e.g., so-called "hexametaphosphate") are used in water conditioning operations for scale prevention and corrosion control.

Organic phosphorus compounds in solids and in natural waters are the products of biological growth. Almost no information is available to identify the specific compounds or groups of compounds which may make up a dissolved organic phosphorus fraction in waste effluents, agricultural soil drainage water or surface waters. Studies on phosphorus distributions in lakes indicate that the soluble organic phosphorus may amount to from 30 to 60% of the total phosphorus. Observations on phosphorus movement in soil have shown that substantial fractions of the mobile phosphorus in the soil solution consist of dissolved and colloidal organic phosphorus. A committee report [1] on the determination of various phosphate fractions in surface waters pointed out that organic phosphates frequently are a significant part of the total phosphate present. Inspection of the data in that report suggests the dissolved organic phosphorus amounted to as much as 25% of total phosphorus in

[1] Association of American Soap and Glycerine Producers, *J. Amer. Water Works Assoc.*, **50**, 1563 (1958).

some surface water samples. Rigler [2] studied the soluble organic or colloidal phosphorus fraction of waters from nine lakes and concluded that the fraction of the total phosphorus than can be described accurately as soluble organic is unknown (being perhaps between 15 and 30%) because it depends on the operational procedure used to separate soluble and insoluble forms. The Kaskaskia River in Illinois was found to contain apparent organic phosphorus amounting to from 15 to 30% of the total phosphorus [3]. Many river samples show significant concentrations of organic phosphorus in *both* dissolved and suspended form. The chemistry of organic phosphorus compounds in the soil is discussed in a recent volume on soil chemistry [4].

CATEGORIES OF PHOSPHORUS FRACTIONS. Hutchinson [5] has employed four categories in discussing the phosphorus cycle in lakes. These are (a) soluble phosphate phosphorus, (b) acid-soluble suspended (sestonic) phosphorus, (c) organic soluble (and colloidal) phosphorus and (d) organic suspended (sestonic) phosphorus. The categories for the phosphorus forms considered by some chemical oceanographers, for example, Strickland and Parsons [6], are based on two factors: their reactivity (in 5 min time) with acid molybdate and their particle size. Strickland and Parsons distinguish between the following four operational forms of phosphorus: (a) soluble reactive, (b) soluble unreactive, (c) particulate unreactive and (d) particulate reactive.

Our knowledge of the amounts of different phosphorus forms and species in natural waters is thus derived from various operational methods of the kinds outlined briefly above (Table 10-1). It needs to be kept in mind that there may not always be a simple correspondence between *chemical* species of phosphorus and these operationally defined forms.

Equilibria of Phosphates

The distribution of the several acid and base species of orthophosphates and condensed phosphates in solution is governed by pH. Information on the pH-dependent distribution of the several species is required in interpreting the solubility behavior, complex formation and sorption processes of phosphorus in water. Table 10-2 contains some pertinent equilibrium constants. The predominant dissolved orthophosphate species over the pH range 5 to 9 are $H_2PO_4^-$ and HPO_4^{2-}. Predominant pyrophosphate species at neutral pH are $H_2P_2O_7^{2-}$ and $HP_2O_7^{3-}$.

[2] F. H. Rigler, *Limnol. Oceanog.*, **9**, 511 (1964).

[3] R. S. Engelbrecht and J. J. Morgan, *Sewage Ind. Wastes*, **31**, 458 (1959).

[4] A. D. McLaren and G. H. Peterson, Eds., *Soil Biochemistry*, Dekker, New York, 1967.

[5] G. E. Hutchinson, *A Treatise of Limnology*, volume I, Wiley, New York, 1957.

[6] J. D. H. Strickland and T. R. Parsons, "A Manual of Sea Water Analysis," *Bull. Fisheries Res. Board Can.*, **125**, 1–185 (1960).

Table 10-2A   Acidity and Hydrolysis of Phosphates and Metal Ions†

No.	Equilibrium	Log Equilibrium Constant, 25°C‡
8	$H_3PO_4 = H_2PO_4^- + H^+$	-2.2
9	$H_2PO_4^- = HPO_4^{2-} + H^+$	-7.0
10	$HPO_4^{2-} = PO_4^{3-} + H^+$	-12.0
11	$H_3P_2O_7^- = H_2P_2O_7^{2-} + H^+$	-2.4
12	$H_2P_2O_7^{2-} = HP_2O_7^{3-} + H^+$	-6.6
13	$HP_2O_7^{3-} = P_2O_7^{4-} + H^+$	-9.3
14	$H_3P_3O_{10}^{2-} = H_2P_3O_{10}^{3-} + H^+$	-2.3
15	$H_2P_3O_{10}^{3-} = HP_3O_{10}^{4-} + H^+$	-6.5
16	$HP_3O_{10}^{4-} = P_3O_{10}^{5-} + H^+$	-9.2
17	$HP_3O_9^{2-} = P_3O_9^{3-} + H^+$	-2.1
18	$Fe^{3+} + H_2O = FeOH^{2+} + H^+$	-2.2
19	$FeOH^{2+} + H_2O = Fe(OH)_2^+ + H^+$	-4.7
20	$Fe^{3+} + 3H_2O = Fe(OH)_3(s) + 3H^+$	-6.0
21	$Al^{3+} + H_2O = AlOH^{2+} + H^+$	-5.0
22	$Al^{3+} + 3H_2O = Al(OH)_3(s) + 3H^+$	-9.5

† The equilibrium constants in this and related tables are from L. G. Sillén and A. E. Martell, *Stability Constants of Metal–Ion Complexes*, Special Publ., No. 17, The Chemical Society, London, 1964, which should be consulted for actual details.

‡ Acidity constants of most sugar phosphates, inositol phosphates and condensed organic phosphates are such that they exist as anionic species in neutral and alkaline solutions (not considering metal complexes).

Table 10-2*a* also contains equilibrium data on hydrolysis of iron(III) and aluminum ions. The equilibria are significant in considerations of solubility of iron or aluminum phosphate and of complex formation between phosphates and these metal ions.

COMPLEX FORMATION REACTIONS OF INORGANIC PHOSPHATES. Phosphate, pyrophosphate, triphosphate and higher polyphosphate anions are known to form complexes, chelates and insoluble salts with a number of metal ions. The extent of complexing and chelation between various phosphates and metal ions in natural waters will depend upon the relative concentrations of the phosphates and the metal ions, the pH and the presence of other ligands (sulfate, carbonate, fluoride, organic species) in the water. The reactions in Table 10-2*b* are representative of major inorganic phosphorus species. Phosphate concentrations are generally low. Complex formation involving these major cations and various phosphate anions will have small effect on the metal ion distribution, but may have significant effects on the phosphate distribution. Such metal ions as those of ferric iron, manganous manganese, zinc, copper, etc., are present in concentrations comparable to or lower than

Table 10-2B   Complex Formation by Phosphates

No.	Equilibrium	Log Equilibrium Constant†
23	$Ca^{2+} + HPO_4^{2-} = CaHPO_4(aq)$	2.7
24	$Mg^{2+} + HPO_4^{2-} = MgHPO_4(aq)$	2.5
25	$Fe^{3+} + HPO_4^{2-} = FeHPO_4^{+}$	8.3
26	$Ca^{2+} + P_2O_7^{4-} = CaP_2O_7^{2-}$	5.6
27	$Ca^{2+} + HP_2O_7^{3-} = CaHP_2O_7^{-}$	2.0
28	$Mg^{2+} + P_2O_7^{4-} = MgP_2O_7^{2-}$	5.7
29	$Na^{+} + P_2O_7^{4-} = NaP_2O_7^{3-}$	2.2
30	$Fe^{3+} + 2HP_2O_7^{3-} = Fe(HP_2O_7)_2^{3-}$	22.0
31	$Ca^{2+} + P_3O_{10}^{5-} = CaP_3O_{10}^{3-}$	8.1
32	$Mg^{2+} + P_3O_{10}^{5-} = MgP_3O_{10}^{3-}$	8.6

† Sugar phosphates and organic condensed phosphates form complexes with many cations. Stability constants for Ca^{2+} and Mg^{2+} are on the order of 10^3 to 10^4.

that of the phosphates. For these ions, complex formation might significantly affect distribution of the metal ion, the phosphates or both. Some representative computations on metal ion and phosphate complexes in simple model homogeneous systems are presented in Table 10-3. The calculations for a 1×10^{-5} M Fe(III)–phosphate system show that the degree of complexation is rather small at pH 5 (Example 1). For a 1×10^{-5} M Fe(III)–pyrophosphate system at pH 7, binding of pyrophosphate to soluble iron is extensive (Example 4).

A hypothetical solution containing 1×10^{-5} M total orthophosphate at pH 7 contains from 20 to 60% of the phosphate as a calcium complex for calcium levels of from 10^{-3} to 3×10^{-3} M (Example 2). A hypothetical pyrophosphate solution (no orthophosphate) of 1×10^{-5} M concentration would have approximately 90% of the phosphate bound at the same pH and at the higher calcium level (Example 3). These examples are of rather limited scope. However, they should serve to indicate that phosphate complex formation can occur under conditions encountered in natural waters. The chelates of the linear condensed phosphates may be of particular significance with respect to the rates of hydrolytic scission of P—O—P bonds in waters of different calcium and magnesium levels. Complexation and chelation of aqueous Fe(II), Fe(III) and Mn(II) may prove to be significant with respect to rates of iron and manganese oxygenation. Lastly, the degree of complexation between phosphates and metal ions must be considered in evaluating solubilities of phosphates under natural water conditions. This aspect is discussed below.

Table 10-3   Some Examples of Phosphato Metal Complexes

A. HOMOGENEOUS SOLUTION SYSTEMS

1. Ferric-orthophosphate

$Fe_T = [Fe^{3+}] + [FeOH^{2+}] + [Fe(OH)_2^+] + [FeHPO_4^+]$

$Fe' = Fe^{3+} + [FeOH^{2+}] + [Fe(OH)_2^+]$

$Fe_T = Fe' + [FeHPO_4^+]$

$\dfrac{Fe_T}{Fe'} = 1 + \dfrac{K[HPO_4^{2-}]}{\alpha}$ $[\alpha = f(\text{pH}), \text{cf. (viii) of Example 6-1 in Section 6-3}]$

If $Fe_T = 1 \times 10^{-5}\ M$ and $P_T = 1 \times 10^{-5}\ M$, less than 1% of iron and phosphate is complexed at pH 5.

2. Calcium-orthophosphate

$$Ca_T = [Ca^{2+}] + [CaHPO_4(aq)]$$
$$P_T = [CaHPO_4(aq)] + [HPO_4^{2-}] + [H_2PO_4^-] + \cdots$$

If $Ca_T = 1 \times 10^{-3}\ M$ and $P_T = 1 \times 10^{-5}\ M$ at pH 7, 20% of the phosphate is complexed with calcium; if $Ca_T = 5 \times 10^{-3}\ M$, about 60% is complexed.

3. Calcium-pyrophosphate

$Ca_T = [Ca^{2+}] + [CaP_2O_7^{2-}] + [CaHP_2O_7^-]$

$PP_T = [CaP_2O_7^{2-}] + [CaHP_2O_7^-] + [HP_2O_7^{3-}] + [P_2O_7^{4-}] + [H_2P_2O_7^{2-}] + \cdots$

$Ca_T = 1 \times 10^{-3}\ M$ and $PP_T = 1 \times 10^{-5}\ M$ at pH 7, 60% of the pyrophosphate is complexed by calcium. For $Ca_T = 5 \times 10^{-3}\ M$, approximately 90% of the pyrophosphate is complexed.

4. Ferric-pyrophosphate

$$Fe_T = Fe' + [Fe(HP_2O_7)_2^{3-}] = Fe' + \dfrac{K Fe'[HP_2O_7^{3-}]^2}{\alpha}$$

If $Fe_T = 1 \times 10^{-5}\ M$ and $PP_T = 1 \times 10^{-5}\ M$, more than 99% of the pyrophosphate is complexed with ferric iron at pH = 5.

B. HETEROGENEOUS PRECIPITATE-SOLUTION SYSTEMS

1. Ferric-orthophosphate

$$FePO_4(s) + H^+ = FeHPO_4^+$$

$$\dfrac{[FeHPO_4^+]}{[H^+]} = K = 10^{-2.7}$$

At pH 5, in the presence of solid ferric-phosphate, $[FeHPO_4^+] = 2 \times 10^{-8}\ M$. At pH 7, $[FeHPO_4^+] = 2 \times 10^{-10}\ M$.

2. Hydroxylapatite

$$Ca_5OH(PO_4)_3(s) + 4H^+ = 2Ca^{2+} + 3CaHPO_4(aq) + H_2O$$

$$\dfrac{[CaHPO_4(aq)]^3[Ca^{2+}]^2}{[H^+]^4} = 3 \times 10^2$$

At pH 7, in the presence of hydroxylapatite and with $[Ca^{2+}] = 1 \times 10^{-3}\ M$, $[CaHPO_4(aq)] = 3 \times 10^{-7}\ M$, or 0.009 mg liter^{-1}.

Solubility Relationships for Phosphates

Several insoluble forms of inorganic orthophosphate have been mentioned above in discussing the occurrence of phosphorus in nature. Equilibrium data required to describe the solubility relationships of four orthophosphate solid phases are given in Table 10-4. By employing the solubility equilibrium constants and the acidity constants it is possible to compute total phosphate solubility (P_T) under specified conditions (pH, calcium concentrations, etc.). For example, we can compute soluble P_T for pure $AlPO_4(s)$ in contact with pure water whose pH is adjusted by addition of acid or base. When $Al(OH)_3(s)$ or $Al_2O_3(s)$ forms, the soluble P_T is then governed by an additional equilibrium condition. Similar considerations apply to $FePO_4(s)$ solubility. Figure 10-1b shows solubility of total phosphate for $FePO_4(s)$ and $AlPO_4(s)$. (The possible influences of complex formation have been neglected in computing these solubility curves. This is elaborated upon below.) The minimum solubility of $AlPO_4(s)$ is ca. 10 μg liter^{-1} at pH 6. At pH 7 the solubility increases to 300 μg liter^{-1} and at pH 5 it increases to 30 μg liter^{-1}. $FePO_4(s)$ is more soluble than $AlPO_4(s)$ (100 μg liter^{-1} at pH 5 and ca. 1 mg liter^{-1} at pH 6).

However, these computations cannot always be used alone to account for total phosphate in solution at equilibrium. Phosphate anions enter the solid oxides or hydroxides of iron and aluminum (isomorphic replacement, solid solution, etc.).

Soluble phosphate appears not to be controlled by $CaHPO_4(s)$ in natural waters, at least where some calcium is present. Figure 10-1b shows a plot of P_T in equilibrium with solid $CaHPO_4$ in the presence of 40 mg liter^{-1} calcium (1×10^{-3} M). The solubility of hydroxylapatite, $Ca_5OH(PO_4)_3(s)$, is less than that of $CaHPO_4(s)$. Equilibria 1, 2 and 3 (Table 10-4) describe the

Table 10-4   Solubility Equilibria of Orthophosphates and Condensed Phosphates

No.	Equilibrium	Log Equilibrium Constant
1	$Ca_5OH(PO_4)_3(s) = 5Ca^{2+} + 3PO_4^{3-} + OH^-$	-55.6
2	$Ca_5OH(PO_4)_3(s) + 3H_2O$ $= 2[Ca_2HPO_4(OH)_2]_{surface} + Ca^{2+} + HPO_4^{2-}$	-8.5
3	$[Ca_2HPO_4(OH)_2]_{surface} = 2Ca^{2+} + HPO_4^{2-} + 2OH^-$	-27
4	$CaHPO_4(s) = Ca^{2+} + HPO_4^{2-}$	-7
5	$FePO_4(s) = Fe^{3+} + PO_4^{3-}$	-23
6	$AlPO_4(s) = Al^{3+} + PO_4^{3-}$	-21
7	$Ca_2P_2O_7(s) = Ca^{2+} + CaP_2O_7^{2-}$	-7.9

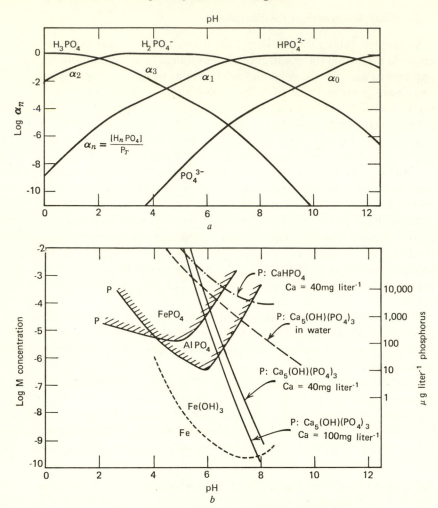

Figure 10-1 Effect of pH on various phosphorus forms.

solubility of hydroxylapatite. It is believed that the hydroxylapatite reacts with water to form a "surface complex" (reaction 2), and that this complex is in metastable equilibrium with the aqueous solution. Figure 10-1*b* shows three P_T curves under control of hydroxylapatite solubility. One curve describes the solubility of the apatite in water (acid or base for pH adjustment). The remaining two curves show the influence of calcium in solution upon P_T. In a hypothetical water with a calcium concentration of 40 mg liter^{-1} and pH of 7, total soluble phosphate is limited to ca. 10 μg liter^{-1}. A calcium level of 100 mg liter^{-1} lowers the maximum equilibrium P_T to 1 μg liter^{-1}. Hydroxylapatite — not $CaHPO_4(s)$ — should be considered as a likely stable solid phase in many natural waters, tending to buffer the total phosphate concentrations in these waters. Of course, these calculations should be extended to include such related solid phases as fluorapatite and carbonate fluorapatite.

Figure 10-1*b* indicates that elevation of the pH for a natural water containing typical levels of calcium should lead to apatite formation. In general, calcium carbonate formation can also be anticipated. The application of solubility equilibria to phosphate elimination has been discussed and demonstrated [7, 8]. Certain aspects of the kinetics of apatite formation will be discussed later (Section 10-6).

INFLUENCES OF COMPLEXES ON SOLUBILITY. The influences of phosphate complexes with metal ions can be illustrated by considering the cases of ferric phosphate and hydroxylapatite. In the presence of solid $FePO_4$, the concentration of the $[FeHPO_4^+]$ species is given by $10^{-2.7}[H^+]$ (Table 10-3*b*, Example 1). At pH 5 this amounts to $2 \times 10^{-8} M$, or less than about 1% of the total soluble phosphate. The influence of calcium phosphate complexing on hydroxylapatite solubility is also shown in Table 10-3*b*. $[CaHPO_4(aq)]$ is ca. $3 \times 10^{-7} M$ at pH 7 for a 40 mg liter^{-1} calcium level. This is of about the same order as the total phosphate found by ignoring complex formation. It appears that calcium complexation is significant in this instance. Further, a contribution to total solution phosphate is expected to come from the species $MgHPO_4(aq)$.

SOLUBILITIES OF CONDENSED PHOSPHATES. Pyrophosphate forms insoluble compounds with a number of bivalent metal ions (Mn^{2+}, Ca^{2+}, Mg^{2+}). The magnesium and manganese precipitates of pyrophosphate are employed for analytical purposes. For a water containing $10^{-3} M$ Ca^{2+} at pH 7, the maximum soluble concentration of total pyrophosphate is approximately $2 \times 10^{-5} M$ (1.2 mg liter^{-1} P), and over 50% is in the form of the calcium complex. However, precipitation of condensed phosphates as calcium or magnesium salts is probably not to be expected in lakes and rivers in view of the concentrations expected and found there.

Sorption of Phosphates on Surfaces

Examples of phosphate and polyphosphate sorption are well known. Threshold treatment with polyphosphate to prevent calcium carbonate deposition involves adsorption of polyphosphate anions on the $CaCO_3$

[7] W. Stumm, in *Advances in Water Pollution Research*, Volume 2, Macmillan, New York, 1964, p. 220.
[8] J. O. Leckie and W. Stumm, in *Advances in Water Quality Improvement*, E. Gloyna and W. W. Eckenfelder, Eds., Univ. of Texas Press, Austin, Tex., 1970.

crystals [9]; phosphate anions are taken up from water by alumina, by kaolinites and montmorillonites [10], and also by freshly precipitated ferric and aluminum hydroxides. A strong tendency toward chemical bonding between phosphate groups and metal ions in a solid lattice appears to be the underlying principle for these various "sorption" phenomena.

Van Olphen has described polyphosphate adsorption on clay plate edges as chemisorption involving specific interaction of Al^{3+} ions and

$$\begin{array}{c} O \\ | \\ -O-P-O- \\ | \\ O \end{array}$$

groups. Hsu [11] has presented arguments supporting similar chemical interactions as the basis for fixation of phosphate by aluminum and iron compounds in acidic soils. It appears reasonable to view precipitation of aluminum or iron phosphate and adsorption of phosphates or polyphosphates on ferric or aluminum oxides or hydroxides as resulting from the same kind of chemical forces: those involved in forming complex ions or insoluble salts. In this view adsorption is a special case of precipitation in which surface metal ions form metal–ion–phosphate bonds by reaction with solution species [11]. From the colloid-chemical standpoint, phosphate ions may serve as potential-determining ions. For $Al_2O_3 \cdot 3H_2O$ (gibbsite) and hydrous ferric oxide it has been suggested that the precipitation products should be represented as: $(Al,Fe) \cdot (H_2PO_4)_n(OH)_{3-n}$ [12]. It is important to recognize that the formation of a new solid phase is concordant with thermodynamic information on the oxides or hydroxides and phosphates. For example, the equilibrium

$$Al(OH)_3(s) + H_2PO_4^- = AlPO_4(s) + OH^- + 2H_2O$$

has an equilibrium constant at 25°C of ca. $10^{-2.5}$. Formation of the aluminum phosphate is possible at pH 6 if the total solution phosphate is greater than 3×10^{-6} M (ca. 0.09 mg liter^{-1}). When a bulk or surface phase of iron or aluminum orthophosphate is formed, the pH dependence of phosphate precipitation should tend to conform to the solubility diagrams for $FePO_4$ and $AlPO_4$ (Figure 10-1b). For phosphate sorption onto $Al(OH)_3(s)$ or $Fe(OH)_3$ there is generally an increase of uptake as the pH is decreased (increasing positive charge on oxide–hydroxide surface).

Sorption of phosphates and polyphosphates onto clay minerals appears to involve at least two mechanisms, chemical bonding of the anions to positively

[9] B. Raistrick, *Discussions Faraday Soc.*, **5**, 234 (1949).
[10] H. van Olphen, *An Introduction to Clay Colloid Chemistry*, Wiley-Interscience, New York, **1963**.
[11] P. H. Hsu, *Soil Sci.*, **99**, 398 (1965).
[12] B. W. Bache, *Soil Sci.*, **15**, 110 (1964).

charged edges of the clays and substitution of phosphates for silicate in the clay structures. In general, high phosphate adsorption by clays is favored by a lower pH. Maximum sorption of orthophosphate on montmorillonite occurs at pH 5–6. Sorption by kaolinite is maximum at a pH near 3. The sorption of phosphates and polyphosphates by clay particles is fundamental to the peptizing ability which these substances show in clay suspensions.

Hydrolysis of Condensed Phosphates

Hydrolytic scission of pyrophosphate and tripolyphosphate is thermodynamically favorable and yields orthophosphate as the eventual product. However, the rates of reversion can be slow. The physical-chemical and biochemical factors that influence the rate of hydrolytic breakdown of the condensed phosphates have been summarized by Van Wazer [13] and discussed by Clesceri and Lee [14]. Among the most significant factors are temperature, pH and enzymatic activity. In addition, metal–ion complexing of the condensed phosphates (Ca^{2+}, Mg^{2+}) appears to affect hydrolysis rates. The presence of bacteria, algae and enzymes causes a significant increase in the rate of condensed phosphate degradation. While half-lives of pyrophosphate or tripolyphosphate in distilled water are in the order of 4000–5000 days, the hydrolysis rates in natural lake and river waters are 100 to 1000 times faster.

10-3   Chemistry of Iron and Manganese

These elements belong to the Subgroup VIIA in the Periodic Table and, like other transition elements, they exist in a number of oxidation states. The chemistry of aqueous iron primarily involves the II and III oxidation state. In the case of manganese the II and IV states are perhaps of greatest importance in connection with manganese in natural waters. MnO_4^- need not be considered because it is thermodynamically unstable; it is reduced by water.

The Mn^{3+} ion is unstable in aqueous solution because it is subject to disproportionation as shown by the approximate equilibrium constant, at 25°C, $2Mn^{3+} + 2H_2O = Mn^{2+} + MnO_2(s) + 4H^+$; $\log K \simeq 9$. Strong complexing agents, however, might stabilize Mn in the III oxidation state. The mineralogy of the oxides, hydroxides, hydrous oxides and sulfides of Fe and Mn is varied

[13] J. R. Van Wazer, *Phosphorus and Its Compounds*, Volume I, Interscience, New York, 1958.
[14] N. L. Clesceri and G. F. Lee, *Intern. J. Air Water Pollution*, **9**, 723, 743 (1965).

Table 10-5   Iron and Manganese Species and Their Free Energy of Formation

	$\bar{G}_f^\circ$, 25°C, kcal mole^{-1}†
Fe(II)	
$\quad$ Fe^{2+}(aq)	-20.3 (NBS)
$\quad$ FeSiO$_3$(s)	-257 (L)
$\quad$ Fe$_2$SiO$_4$(s)	-319.8 (L)
$\quad$ α-FeS(s)	-23.32 (NBS)
$\quad$ FeS$_2$(s) (pyrite)	-39.9 (NBS)
$\quad$ Fe(OH)$_2$(s)	-115.57 (NBS)
$\quad$ FeCO$_3$(s) (siderite)	-160.5 (S)
Fe(III)	
$\quad$ Fe^{3+}(aq)	-2.53 (NBS)
$\quad$ Fe(OH)$_3$(s)	-177.1 (NBS)
$\quad$ (amorph)-FeOOH(s)	-111.1 ($I = 3\ M$ NaClO$_4$) (Sc)
$\quad$ α-FeOOH(s) (goethite)	-113.7 ($I = 3\ M$ NaClO$_4$) (Sc)
$\quad$ α-Fe$_2$O$_3$(s) (haematite)	-177.1 (NBS)
$\quad$ FePO$_4 \cdot 2$H$_2$O (strengite)	-279
Fe(II), Fe(III)	
$\quad$ Fe$_3$O$_4$ (magnetite)	-242.4 (NBS)
Mn(II)	
$\quad$ Mn^{2+}(aq)	-54.4 (L)
$\quad$ MnS(s)	-53 (L)
$\quad$ MnCO$_3$(s) (rhodochrosite)	-194.8 (M)
$\quad$ Mn(OH)$_2$(s) (pyrochrosite)	-147.3 (M)
Mn(IV): (MnO$_{1.7}$–MnO$_2$)	
$\quad$ MnO$_2$(s) (ramsdellite)	—
$\quad$ β-MnO$_2$(s) (pyrolusite)	—
$\quad$ γ-MnO$_2$(s) (nsutite) ‡	-109.1 (B)
$\quad$ "δ-MnO$_2$"(s) [manganate(IV)]	-108.3 (B)§
$\qquad$ (a) manganese(III) manganate(IV), Mn$_7$O$_{13} \cdot 5$H$_2$O‖	
$\qquad$ (b) sodium(II) manganese(III) manganate(IV), Na$_4$Mn$_{14}$O$_{27} \cdot 9$H$_2$O‖	
$\quad$ ZnMn$_3$O$_7 \cdot 3$H$_2$O (chalkophanite)	—
Mn(II) *and* (IV) *or* Mn(III)	
$\quad$ Mn$_3$O$_4$(s) (hausmannite)	-306.2 (B)
$\quad$ α-MnOOH(s) (grautite)	—
$\quad$ β-MnOOH(s) (feitknechtite)	—
$\quad$ γ-MnOOH(s) (manganite)	-133.3 (B)
$\quad$ γ-Mn$_2$O$_3$(s)	-132.2 (B)
Mn(VII)	
$\quad$ MnO$_4^-$	-111.3 (L)

† Literature data are from L, Latimer; NBS, National Bureau of Standards, Circular 500; S, Singer [31]; Sc, Schindler; B, Bricker [15]; M, Morgan [16].
‡ Composition: Mn$_{1-x}$Mn$_x$O$_{2-2x}$(OH)$_{2x}$.
§ Composition: (Na$_2$,Ca)Mn$_7$O$_{14} \cdot$H$_2$O, cf. Bricker [15].
‖ After R. Giovanoli, K. Bernhard and W. Feitknecht, *Helv. Chim. Acta*, **51**, 355 (1968).

and complex. X-ray identification has revealed a large group of mineral species. Table 10-5 lists the more pertinent compounds and species.

Among the oxides and hydroxides of iron we need to consider primarily $Fe(OH)_2$, Fe_3O_4(magnetite), amorphous $FeOOH$ and α-$FeOOH$(goethite). What is usually called "hydrous ferric oxide" or "ferric hydroxide" is more likely a poorly crystallized $FeOOH$. Data for the solubility and free energy of formation of $Fe(OH)_3(s)$, probably based on results obtained with active $FeOOH$ preparations, are nevertheless of operational value.†

The oxides and hydrous oxides of manganese that can be prepared from aqueous Mn^{2+} solution and their reaction paths are depicted in Figure 10-2 [15]. The preparative techniques that involve precipitation and oxidation have

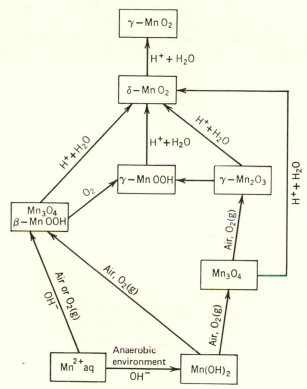

Figure 10-2   Paths of reaction in the system Mn–O_2–H_2O at 25°C. and 1 atm total pressure. From Bricker [15]. Reproduced with permission from American Mineralogical Society of America.

† Very recently it has been shown [V. Bvořák, W. Feitknecht and P. Georges, *Helv. Chim. Acta*, **52**, 501 (1969)] that basic carbonates of Fe(III) may be precipitated from HCO_3^--containing solutions.
[15] O. P. Bricker, *Amer. Mineral.*, **50**, 1296 (1965).

been described by Bricker [15] who obtained, in most cases, well-crystallized compounds that could be characterized by X-ray diffraction and electron microscopy. But the products obtained initially when Mn(II) is oxidized under various solution conditions (pH, alkalinity) in nature or in the laboratory are frequently not characterizable in terms of simple stoichiometry.

In Figure 10-3 oxygen uptake curves for 2.3×10^{-3} M Mn(ClO$_4$)$_2$ in the presence of 0.7, 1.1 and 18 equivalents of NaOH per Mn(II) equivalent are shown. Rapid consumption of O$_2$ is followed by an extended period of very slow oxygen uptake. After about 10 hr of oxygenation, the oxidation products were removed for X-ray analysis. For the systems with equivalent ratios of 0.7 and 1.1, each of the oxygenation products showed patterns strongly similar to the pattern exhibited by hausmannite, Mn$_3$O$_4$. Thus there is general agreement between the X-ray data and the O$_2$ consumption. The oxygenation product for the system with a base to Mn(II) equivalent ratio of 18 yielded an X-ray pattern which was closely similar to that which has been obtained for

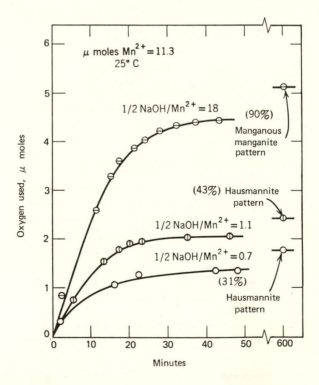

Figure 10-3   Manometric measurement of O$_2$ used in air oxidation of Mn^{2+} in presence of varying amounts of NaOH. From Morgan [16]. Reproduced with permission from J. Wiley and Sons.

manganous-manganite and "δ-MnO_2" (manganate IV) by several investigators. The maximum oxygen uptake per Mn indicates that the average degree of oxidation may be represented by $MnO_{1.90}$ This formulation is within the range of compounds for which the manganous-manganite pattern has been reported previously.

In addition to manometric observations of oxygen uptake as a measure of oxygenation stoichiometry, analytical methods have been employed to determine the principal composition of hydrous Mn oxides formed by air oxidation in dilute solutions (10^{-5} to 10^{-4} M). In the pH range 9–10, compositions of the oxidation products ranged from $MnO_{1.24}$ to $MnO_{1.43}$ [16].

Bricker [15] found the most highly oxidized compound that could be synthesized using air, oxygen or H_2O_2 to be γ-MnOOH. In order to prepare the nonstoichiometric forms of manganese dioxide it was necessary to acidify suspensions of Mn_3O_4 or one of the MnOOH polymorphs; then disproportionation of these compounds into Mn^{2+} and nonstoichiometric MnO_2 occurs. Fe_3O_4 (magnetite), similarly, does not usually occur with constant stoichiometry; lattice defects, isomorphic replacement and ion exchange reactions at its surface modify the Fe(II)/Fe(III) ratio.

Berner [17] has shown that tetragonal iron(II) sulfide is formed from aqueous sulfide solutions over a wide range of conditions. The substance described as hydrotroilite in many sedimentary occurrences is believed [17] to be — at least in part — poorly crystallized tetragonal FeS. Tetragonal FeS has been found [17] in sediments of natural waters. Furthermore, the disulfides (FeS_2) pyrite and marcasite occur in nature; in the laboratory they can be prepared by partial oxidation of tetragonal iron sulfide in acid solutions [17].

Fe(II) and Mn(II) as well as Fe(III) are known to be present in many silicate phases, but no reliable data on free energies of formation are available. As already illustrated for Fe(II) (Example 5-4 and Figure 5-8), $FeCO_3(s)$ (siderite) and $MnCO_3$ are more stable than $Fe(OH)_2$ or $Mn(OH)_2$ under the conditions of most natural waters.

Solubility and Complex Formation

Representative solubility diagrams of Fe(II) and Mn(II) in carbonate-bearing waters are given in Figures 5-8 and 10-4, respectively. The construction of these diagrams has been explained in Chapter 5, which also gives a solubility diagram for "ferric hydroxide" (or amorphous FeOOH) (Figure

[16] J. J. Morgan, in *Principles and Applications in Water Chemistry*, S. D. Faust and J. V. Hunter, Wiley, New York, 1967.
[17] R. A. Berner, *J. Geol.*, **72**, 293 (1964).

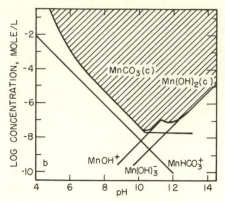

Figure 10-4   Solubility of Mn(II) in water calculated for 2×10^{-3} M total carbonic species.

5-3c). This figure illustrates that total soluble ferric iron, primarily present as $Fe(OH)_2^+$, cannot exceed concentrations of 1 μg Fe liter^{-1}. The solubility of higher-valent manganese oxides (MnO_x, where $1 < x \leq 2$) is so low that one cannot analytically detect soluble Mn, but some nonstoichiometric manganese oxides contain exchangeable Mn^{2+} and equilibrate in solution to yield Mn^{2+}. Ferric ions generally have a stronger tendency to form complexes than ferrous and manganous ions. Complex formation of Fe(III) with o-phosphate and many organic bases is well established. In order to evaluate the coordinating tendency of Fe(III), the relative affinities of Fe^{3+} for OH^- ions and other ligands need to be compared. Hydroxide ions often have a stronger affinity for Fe^{3+} than do organic or inorganic bases. As shown in Chapter 6, the extent of complex formation is thus pH dependent, and, within the pH range of natural waters, soluble or insoluble mixed Fe(III) complexes that may contain OH^- as well as other ligands can be formed. For example, Fe^{3+} interacts chemically with orthophosphate to form soluble phosphate–iron(III) complexes [e.g., $FeHPO_4^+$, $Fe_x(H_nPO_4)_y^{3x-(3-n)y}$]. Under slightly acid conditions pure $FePO_4(s)$ will be precipitated, whereas in the neutral and slightly alkaline pH range the precipitate is probably a metastable ferric compound containing both PO_4^{3-} and OH^- in variable proportions, depending upon the pH (Table 10-2b). The chemical interactions of Fe(III) with many organic bases are similar to that with orthophosphate.

The incorporation of coordinating anions into basic precipitates not only alters the solubility relations but also strongly affects the colloid chemical properties of the dispersed phase. An enhancement of the colloid stability of these suspensions frequently results. It is usually difficult to distinguish operationally by conventional means (membrane filtration, centrifugation, dialysis) between homogeneous phase soluble Fe(III) complexes and peptized Fe(III) dispersions (sols). In natural waters high concentrations of organic material are frequently associated with high concentrations of operationally "soluble" iron(III). Most of the so-called natural color of waters is probably ascribable to highly stabilized colloidal dispersions where the intensive yellow staining is caused partially by complex formation with hydrolyzed ferric iron.

The affinity of Mn(IV) for OH^- is so much larger than that of Fe(III) as evidenced, for example, by a comparison of the zero points of charge of the oxides (ferric oxide, $pH_{ZPC} = 6$–8, MnO_2, $pH_{ZPC} = 2$–3), that it is more difficult within the pH range of natural waters for other potential organic or inorganic ligands to compete successfully with OH^- ions for coordination to Mn(IV). Thus the chemical interaction of ligands with Mn(IV) is less prevalent than with Fe(III) in natural waters. Adsorption of tensioactive organic material on the surface of the manganese oxides, however, is to be anticipated, especially in view of the large specific surface area exhibited by these hydrous oxides.

Redox Equilibria

Thermodynamic considerations are useful for obtaining a general understanding of the potential reactions of iron and manganese. Activity ratio diagrams for iron and manganese (Figure 10-5) give a rapid survey on the stability relations at pH = 7 and for $[HCO_3^-] = 10^{-3} M$ and $[SO_4^{2-}] = 10^{-3} M$. Because of the uncertainty of free energy data, especially for the various iron oxides, the positions of the lines are not exact. There is considerable uncertainty about the pϵ values at which equilibrium between iron(III) oxide and Fe_3O_4 occurs (Example 10-1). In the pϵ range 8–10, MnO_2, Mn_3O_4 and $MnCO_3$ are of similar relative activity. It appears from the diagram that under the specified conditions Mn_3O_4 may not occur as a stable phase. The diagrams, furthermore, suggest that iron and manganese sulfides start to be formed as one passes to pϵ values lower than -2 to -3. Even if sulfides are being formed, the sulfate concentration in many natural waters does not vary appreciably; hence the assumption of a constant sulfate concentration is justified. But if pϵ values drop further (below p$\epsilon = -3$ at pH = 7), $[SO_4^{2-}]$ does not remain constant but decreases.

A more extensive summary of limiting stability relations for iron and manganese in the light of presently available thermodynamic information is given by the pϵ–pH diagrams shown in Figure 10-6 for a set of specified conditions. It is not necessary to show a large number of concentration contours on a pϵ–pH diagram; after a region of a given oxidation state is established, it is advantageous to use concentration pH diagrams. From the relative positions of the MnO_2–Mn^{2+} and O_2–H_2O lines it is apparent that MnO_2 becomes unstable with respect to the oxidation of water at pH values somewhat below 3.

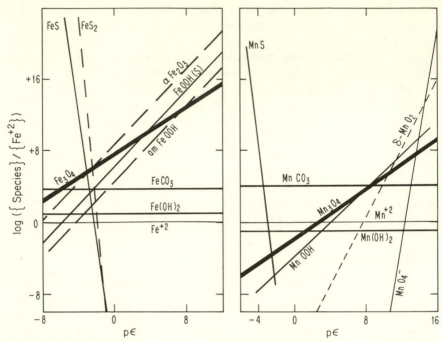

Figure 10-5 Activity ratio diagrams for pH = 7.0, $C_T = 10^{-3} M$ and $[SO_4^{2-}] = 10^{-3} M$.

Additional redox equilibria are mentioned briefly because of their significance with respect to the behavior of Fe and Mn in nature. Thermodynamically, all organic compounds are potentially suitable reductants for both Fe(III) and Mn(IV). Benzidine, *o*-tolidine and related redox indicators are useful in determining higher Mn oxides. For example, the metastable oxidized form of the *o*-tolidinium couple is colored intensely. At low pH values *o*-tolidine is not oxidized by Fe(III). Sulfide species are also significant reductants of Mn oxides. Thermodynamically, Cl_2 might be expected to serve as an oxidant for Fe(II) and Mn(II). Under alkaline conditions OCl^- and ozone might even oxidize Mn(IV) to MnO_4^-. In fact, MnO_4^- can be formed by ozonation of Mn-bearing waters.

Example 10-1 $Fe_3O_4(s)$–$FeOOH(s)$ *Equilibrium as a Possible Regulator of* p_{O_2} *in Atmosphere.*   L. G. Sillén [*Acta Chem. Scand.*, **18**, 1016 (1954)] has hypothesized that the reaction

$$12FeOOH(s) = 4Fe_3O_4(s) + 6H_2O(l) + O_2(g) \qquad (i)$$

which might also be written as

$$12Fe(OH)_3(s) = 4Fe_3O_4(s) + 18H_2O(l) + O_2(g) \qquad (ii)$$

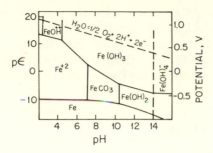

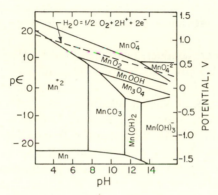

Figure 10-6 $p\epsilon$–pH Diagram for Fe and Mn. $C_T = 2 \times 10^{-3}\ M$, activity of soluble metal ion species, $10^{-5}\ M$.

might possibly participate in the control of the partial pressure of O_2 in the atmosphere. If O_2 is added to the system at a steady rate by photosynthesis, the withdrawal of C will at first increase but finally there will be a steady state where the backward reaction (i) consumes the O_2 added.

Does $p_{O_2} = 0.21$ atm correspond to this steady state and, if so, how far is this from the equilibrium value of (i)?

Using the Latimer values for $\Delta G°$ of (ii), we calculate an equilibrium constant, $K = p_{O_2} \simeq 10^{-1.5}$ atm. This value is very close to the present value of $p_{O_2} \simeq 10^{-0.7}$ atm.

As Sillén points out, the result depends very much on the free energy data used; we must remember that they occur with high stoichiometric coefficients [i.e., the standard free energy of formation of FeOOH(s) or $Fe(OH)_3$(s) is uncertain by a few kcal mole^{-1} and is multiplied by 12]. Hence any combination of other free energy data may give estimates for p_{O_2} different from that calculated above by tens of orders of magnitude. According to Sillén: "It is consequently hard to decide whether reaction (i) has fixed the p_{O_2} of our atmosphere in a steady state more or less close to equilibrium, or whether it acts at present only as a brake slowing down the steady increase of p_{O_2}. At any rate, this reaction seems so important for understanding the oxygen balance of our planet that it would be worthwhile to make a more precise study of its thermodynamics, under conditions comparable to those in the system (air + sea + sediments)."

Kinetics of Redox Reactions

The processes of Fe(II) and Mn(II) oxidation by oxygen and the removal of these ions from solution have been investigated [16, 18, 19]. Some of the findings of these studies are presented herein. An extensive description of the experimental procedures and a discussion of experimental results are available elsewhere [16, 18].

All experiments were conducted with dissolved Fe(II) or Mn(II) concentrations of less than $5 \times 10^{-4} M$. In each series of experiments the pH was controlled by continuously bubbling CO_2- and O_2-containing gas mixtures through test solutions of known alkalinity. The results of representative kinetic experiments are shown in Figure 10-7. Figures *a* and *b* show the course of Fe(II) and Mn(II) disappearance from solution at different pH values. It is evident that the reaction rates are strongly pH dependent. Oxidation of Fe(II) is very slow below pH 6. Measurable (within hours) oxygenation of Mn(II) is observed only at pH values above ca. 8.5.

The rate of oxygenation of Fe(II) in solutions of pH ≥ 5.5 was found to be first order with respect to the concentrations of both Fe(II) and O_2 and second order with respect to the OH ion. Thus a 100-fold increase in the rate of reaction occurs for a unit increase in pH. Catalysts (especially Cu^{2+}, Co^{2+}) in trace quantities, as well as anions which form complexes with Fe(III) (e.g., HPO_4^{2-}), increase the reaction rate significantly. The oxygenation kinetics follow the rate law

$$\frac{-d[\text{Fe(II)}]}{dt} = k[\text{Fe(II)}][\text{OH}^-]^2 p_{O_2} \tag{1}$$

where $k = 8.0\ (\pm 2.5) \times 10^{13}\ \text{min}^{-1}\ \text{atm}^{-1}\ \text{mole}^{-2}$ at 20°C. Frequently it is more convenient to use the rate law in the form

$$\frac{-d\ln[\text{Fe(II)}]}{dt} = \frac{k_\text{H}[O_2(\text{aq})]}{[\text{H}^+]^2} \tag{2}$$

where k_H at 20°C $= 3 \times 10^{-12}\ \text{min}^{-1}\ \text{mole}^1\ \text{liter}^{-1}$.† For a given pH, the

† The rate constants given in Table II of [18] are in error by a factor of $(\ln 10)^2$.
[18] W. Stumm and G. F. Lee, *Ind. Eng. Chem.*, **53**, 143 (1961).
[19] J. J. Morgan and W. Stumm, *Proceedings of the Second Conference on Water Pollution Research*, Pergamon, Tokyo, 1964.

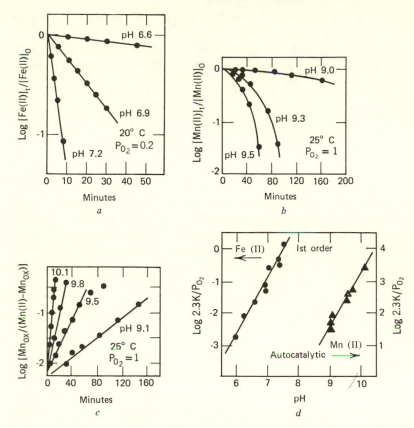

Figure 10-7   Oxidation of Fe(II) and Mn(II).

a: Oxygenation of Fe(II) in bicarbonate solutions.

b: Removal of Mn(II) by oxygenation in bicarbonate solutions.

c: Oxidation of Mn(II) in HCO_3^- solutions; autocatalytic plot.

d: Effect of pH on oxygenation rates.

rate increases about 10-fold for a 15°C temperature increase. For constant $[H^+]$, an activation energy of 23 kcal mole^{-1} can be calculated.

By comparing Figures 10-7*a* and 10-7*b* it is obvious that Mn(II) oxygenation does not follow the same rate law as Fe(II) oxygenation. The manner of the decrease of the Mn(II) concentration with time suggests an autocatalytic reaction. The integrated form of the rate expression

$$\frac{-d[Mn(II)]}{dt} = k_0[Mn(II)] + k[Mn(II)][MnO_2] \tag{3}$$

was found to fit the experimental data well (Figure 10-7c), thus lending support to an autocatalytic model.

The reaction might be visualized to proceed according to the following pattern (reactions are not balanced with respect to water and protons)

$$Mn(II) + \tfrac{1}{2}O_2 \xrightarrow{\text{slow}} MnO_2(s) \tag{4}$$

$$Mn(II) + MnO_2(s) \xrightarrow{\text{fast}} Mn(II) \cdot MnO_2(s) \tag{5}$$

$$Mn(II) \cdot MnO_2(s) + \tfrac{1}{2}O_2 \xrightarrow{\text{slow}} 2MnO_2(s) \tag{6}$$

Although many other interpretations of the autocatalytic nature of the reaction are possible, the following experimental findings are in accord with such a reaction scheme: (a) The extent of Mn(II) removal during the oxygenation reaction is not accounted for by the stoichiometry of the oxidation reaction alone, that is, not all the Mn(II) removed from the solution [as determined by specific analysis for Mn(II)] is oxidized (as determined by measurement of the total oxidizing equivalents of the suspension); (b) as pointed out before, the products of Mn(II) oxygenation are nonstoichiometric, showing various average degrees of oxidation ranging from ca. $MnO_{1.3}$ to $MnO_{1.9}$ (30–90% oxidation to MnO_2) under varying alkaline conditions [19]; and (c) the higher valent manganese oxide suspensions show large sorption capacities for Mn^{2+} in slightly alkaline solutions. The relative proportions of Mn(IV) and Mn(II) in the solid phase depend strongly on pH and other variables.

The X-ray patterns of the products in the presence of a large excess of base were identical with those of typical manganous manganite and δ-MnO_2 ($MnO_{1.8}$ to $MnO_{1.95}$). Oxidation products in solutions of pH of the order of 9.5 had a stoichiometric composition of ca. $MnO_{1.3}$ (Mn_3O_4); their X-ray patterns had some similarity to that of hausmannite, but were rather amorphous.

Both Mn(II) oxidation and removal rates follow the rate law of (3). The rate dependence on the O_2 concentration is the same as that of Fe(II). Thus k in (3) can be formulated as:

$$k = k'[OH^-]^2 p_{O_2} \tag{7}$$

The pH dependence of Fe(II) and Mn(II) oxidation are compared in Figure 10-7d. Metal ions (Cu^{2+}) and complex formers do not appear to have a marked effect upon the reaction rate, although catalytic effects of hydroxy carboxylic acids have been observed. It may be inferred from the catalytic influence of MnO_2, however, that surface catalysis by other active interfaces can influence the reaction rate.

Qualitative and quantitative experiments have confirmed that S(−II) compounds (HS^-, cysteine) as well as a variety of organic substances, especially those that contain hydroxy and/or carboxylic functional groups (e.g., phenols, polyphenols, gallic acid, tannic acid), can reduce both ferric iron and MnO_2 reasonably fast in synthetic solutions (minutes to hours). Such observations are especially interesting in the case of iron, where the same type of substances that reduce ferric iron can also catalyze the oxygenation rate. This apparent contradiction may be explained by the following kind of reaction sequence:

$$Fe(II) + \tfrac{1}{4}O_2 + org \longrightarrow Fe(III)\text{–org complex} \qquad (8)$$

$$Fe(III)\text{–org complex} \longrightarrow Fe(II) + oxidized\ org \qquad (9)$$

$$Fe(II) + \tfrac{1}{4}O_2 + org \longrightarrow Fe(III)\text{–org complex} \qquad (10)$$

Such a reaction pattern has been observed with phenols, "tannic acid" and cysteine. In these cases the ferrous–ferric system acts merely as a catalyst for the oxidation of organic material by oxygen. If conditions (pH, concentrations) are such that the rate of Fe(II) oxygenation is slow in comparison to the Fe(III) reduction by the organic material, a relatively high steady-state concentration of Fe(II) can be maintained in the system as long as the organic material is not fully oxidized. Under such circumstances organic material retards the over-all oxidation of Fe(II) while it hastens the specific Fe(II) oxygenation step. Light can modify such redox reactions; in the presence of certain organic complex formers, light will tend to enhance the reduction of ferric iron.

OXIDATION OF FERROUS IRON IN ACID SOLUTIONS. There are a number of instances in nature where iron-bearing waters of pH considerably below 5 are encountered. Of special concern are those waters in coal mining regions where pH values of 3 are not uncommon. Iron oxidation in such acidic systems is not characterized by the same kinetic relationships which describe the reaction in neutral waters. The insert of Figure 10-8 depicts the rate of oxidation at pH $= 2$. These results can be fitted by a first-order rate expression.† Only 5% of the reaction is complete in 150 days. Additional studies in the low pH range (pH < 4) show the rate of oxidation to be independent of pH. By combining the experimental results obtained by Singer [20] for acidic systems with those obtained by Lee [18] for neutral waters, we can plot the rate of oxygenation of ferrous iron over the entire pH range as in Figure 10-8. The oxidation reaction is catalyzed by interfaces and by light. In the presence of light the reaction is approximately 2–3 times as fast as in its absence. Substantial surface area concentrations ($S > 100$ m² liter^{-1}) are necessary to enhance the oxidation reaction markedly.

---

† The kinetic order of the reaction has not been established definitively.

[20] P. C. Singer and W. Stumm, *Proceedings of the Conference on Acid Mine Research*, 1968, p. 12.

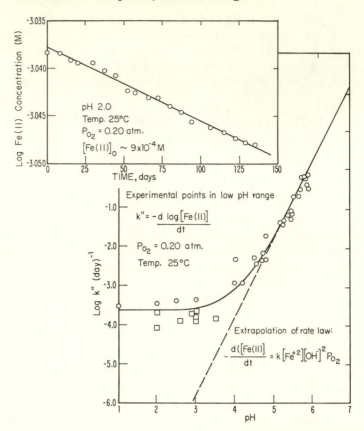

Figure 10-8 Oxygenation rate of ferrous iron as a function of pH. Insert: Experimental data for oxidation at pH = 2.0. At low pH the oxidation rate is independent of pH, while in the high pH range Equation 10-1 (second-order dependence on $[OH^-]$) is fulfilled. Experimental points are those obtained for the low pH region by Singer [20].

Hydrolysis of Ferric Iron

Oxidation of ferrous iron is followed by the hydrolysis to insoluble hydrous ferric oxide (see Section 6-2 and Example 6-1). The attachment of a hydroxo group to a ferric ion is presumably quite fast. Rate constants for the formation of simple Fe(III) complexes with Cl^-, SCN^- and SO_4^{2-} are in the order of $10^3 sec^{-1}$ [21]. The rate of dimerization, $2FeOH^{2+} \rightleftharpoons Fe_2(OH)_2^{4+}$, in solutions not supersaturated with respect to a solubility product of $Fe(OH)_3(s)$ ($pK_{s0} = -38$) is characterized by a rate constant of $10^2 sec^{-1}$ [21].

[21] H. Wendt, Z. Elektrochem., **66**, 235 (1935).

Some experimental results were obtained [22] by measuring residual $[Fe^{3+}]$ as a function of time with a ferro–ferri electrochemical cell. The measurements were carried out in solutions in which the ferric ion was generated in homogeneous solution by oxidizing a fraction of Fe^{2+} to Fe^{3+} in a ferrous iron solution using ozone as an oxidant. The experimental results displayed in Figure 10-9 have been fitted in terms of a second-order rate law. (Because other rate expressions could also fit the experimental data, no mechanistic interpretation must be derived from this form of representation.) In the low pH range, hydrolysis, as measured by the decrease in free $[Fe^{3+}]$, is

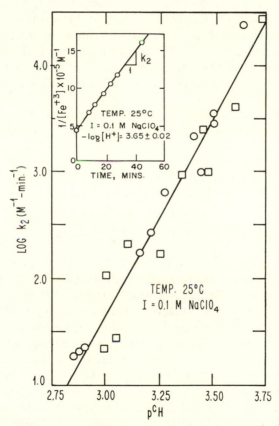

Figure 10-9   pH dependence of "second-order rate constant" for hydrolysis of Fe^{3+}. Insert: individual experimental run.

[22] P. C. Singer and W. Stumm, unpublished data.

much faster than the oxidation of Fe^{2+} to Fe^{3+}; as expected, the rate is strongly dependent upon $[OH^-]$. Most anions tested, for example, SO_4^{2-} and Cl^-, tend to increase the rate of hydrolysis.

The Oxidation of Pyrite; the Release of Acidity into Coal Mine Drainage Waters

The sulfur-bearing minerals which predominate in coal seams are the iron sulfide ores pyrite and marcasite. Both have the same ratio of sulfur to iron but their crystallographic properties are quite different. Marcasite has an orthorhombic structure while pyrite is isometric. Marcasite is less stable and more easily decomposed than pyrite. The latter is the most widespread of all sulfide minerals and, as a result of its greater abundance in the eastern United States, pyrite is recognized as the major source of acid mine drainage. $FeS_2(s)$ is used here as a symbolic representation for the crystalline pyritic agglomerates found in coal mines.

During coal mining operations pyrite is exposed to air and water; the following over-all stoichiometric reactions may characterize the oxidation of pyrite:

$$FeS_2(s) + \tfrac{7}{2}O_2 + H_2O = Fe^{2+} + 2SO_4^{2-} + 2H^+ \tag{11}$$

$$Fe^{2+} + \tfrac{1}{4}O_2 + H^+ = Fe^{3+} + \tfrac{1}{2}H_2O \tag{12}$$

$$Fe^{3+} + 3H_2O = Fe(OH)_3(s) + 3H^+ \tag{13}$$

$$FeS_2(s) + 14Fe^{3+} + 8H_2O = 15Fe^{2+} + 2SO_4^{2-} + 16H^+ \tag{14}$$

The oxidation of the sulfide of the pyrite to sulfate (11) releases dissolved ferrous iron and acidity into the water. Subsequently, the dissolved ferrous iron undergoes oxygenation to ferric iron (12) which then hydrolyzes to form insoluble "ferric hydroxide" (13), releasing more acidity to the stream and coating the stream bed. Ferric iron can also be reduced by pyrite itself, as in (14), where sulfide is again oxidized and acidity is released along with additional ferrous iron which may re-enter the reaction cycle via (12).

The concentration of sulfate or acidity in the water can be directly correlated with the amount of pyrite which has been dissolved. The introduction of acidity into the stream arises from the oxidation of $S_2(-II)$ and the ensuing hydrolysis of the resulting $Fe(III)$. The dissolution of 1 mole of iron pyrite leads ultimately to the release of 4 equivalents of acidity — 2 equivalents from the oxidation of $S_2(-II)$ and 2 from the oxidation of $Fe(II)$. The decomposition of iron pyrite is among the most acidic of all weathering reactions because of the great insolubility of $Fe(III)$.

The following model is proposed to describe the oxidation of iron pyrite in natural mine waters

$$
\begin{array}{c}
 \longrightarrow Fe(II) + S_2^{2-} \\[2pt]
a' \quad\quad\quad\quad\quad\quad\quad\quad + O_2 \\[6pt]
FeS_2(s) + O_2 \xrightarrow{\;a\;} SO_4^{2-} + Fe(II)
\end{array} \qquad (15)
$$

Fast

$+ O_2$ |b   c| $+ FeS_2(s)$

Slow

$Fe(III) \underset{d}{\rightleftarrows} Fe(OH)_3(s)$

The reactions shown are schematic and do not represent the exact mechanistic steps. The model is similar to and carries with it the same over-all consequences as that suggested by Temple and Delchamps [23]. The rate-determining step is a reactive step in the specific oxidation of ferrous iron, reaction b. As Figure 10-8 shows, the rate of oxidation of ferrous iron under chemical conditions analogous to those found in mine waters is very slow, indeed considerably slower than the oxidation of iron pyrite by ferric iron, reaction c. At pH 3, half-times for the oxidation of Fe(II) are on the order of 1000 days while in the case of oxidation of pyrite by Fe(III), half-times on the order of 20 to 1000 minutes were observed. Similar rates have been observed for this latter reaction by Garrels and Thompson [24].

To initiate the sequence, pyrite is oxidized directly by oxygen (a) or is dissolved and then oxidized (a'). The ferrous iron formed is oxygenated extremely slowly (b) and the resultant ferric iron is rapidly reduced by pyrite (c), releasing additional acidity and new Fe(II) to enter the cycle via (b). Once the sequence has been started, oxygen is involved only indirectly in the re-oxidation of Fe(II) (b), the oxygenation of FeS_2 (a) being no longer of significance.

Precipitated ferric hydroxide deposited in the mine and the streams serves as a reservoir for soluble Fe(III) (d). If the regeneration of Fe(III) by reaction b is halted so that the concentration of soluble Fe(III) decreases, it will be

[23] K. L. Temple and E. W. Delchamps, *Appl. Microbiol.*, **1**, 255 (1953).
[24] R. M. Garrels and M. W. Thompson, *Amer. J. Sci.*, **258**, A57 (1960).

replenished by dissolution of the solid $Fe(OH)_3$ and will be free to act again should it come in contact with additional FeS_2.

The following pertinent consequences of this model need to be emphasized.

1. Ferric iron cannot exist in contact with pyritic agglomerates. Fe(III) is rapidly reduced by iron pyrite. The exclusion of oxygen does inhibit regeneration of Fe(III) by the oxidation of Fe(II).

2. The over-all rate of dissolution of pyrite is independent of its surface structure.

3. Microorganisms influence the over-all rate of reaction by mediating the oxidation of ferrous iron since it, alone, is the rate-determining step [25]. Direct microbial oxidation of pyrite must be discounted. Microbial catalysis, as by the autotrophic iron bacteria *Thiobacillus* and *Ferrobacillus ferrooxidans*, seems to be ecolgically significant as evidenced by the few field investigations conducted. Numerous accounts of autotrophic iron oxidation prevail, but few quantitative reports of their actual activity in nature have appeared.

Perhaps the same cycle is responsible for the dissolution and leaching of other mineral sulfides found in copper and uranium mines. Microbial leaching of these other minerals has always been demonstrated in the presence of iron, pyrite being the most abundant and widespread of all mineral sulfides.

Colloid-Chemical Properties of Hydrous Oxides

Hydrous Fe and Mn oxides suspended in waters and present in sedimentary and soil concretions are frequently present in an X-ray amorphous or microcrystalline form, characterized by high specific surface areas; areas up to 300 $m^2 g^{-1}$ have been reported for "δ-MnO_2" [manganate(IV)] and for α- and γ-FeOOH. As already documented in Chapter 9 these hydrous oxides, especially at pH values higher than pH_{ZPC}, are capable of interacting with cations. Sorption of metal ions to these oxides may properly be interpreted as surface complex formation or as ion exchange; hydrogen ions or other cations are released as metal ions become adsorbed. The adsorption has been found to be strongly dependent on pH. While adsorption with Group I and II cations takes place predominantly in the diffuse part of the electric double layer, the transition and heavy metal ions become specifically attached to the surface. Murray et al. [26] estimate the specific sorption potential for Ni^{2+} and Cu^{2+} on Mn-manganite to be -1.9 kcal mole^{-1}.

[25] M. P. Silverman and D. G. Lundgren, *J. Bacteriol.*, **77**, 642 (1959); **78**, 325 (1959).
[26] P. J. Murray, T. W. Healy and D. W. Fuerstenau, in *Adsorption from Aqueous Solution*, Advances in Chemistry Series, No. 79, American Chemical Society, Washington, D.C., 1968.

Representative results of studies on the kinetics and equilibria of the pH-dependent sorption of Mn(II) on MnO_2 and $Fe(OH)_3$ suspensions are presented in Figure 10-10. Sorption capacities for Mn(II) at pH 8 are in the order of 1.0 and 0.3 moles of Mn(II) sorbed per mole of MnO_2 and $Fe(OH)_3$, respectively. Sorption capacities for Ni(II) and Zn(II) are slightly less than those for Mn(II), whereas significantly smaller tendencies for sorption of Mg^{2+} and Ca^{2+} ions prevail [27]. Cation sorption capacities of such magnitudes are comparable on a per weight basis with those of clays.

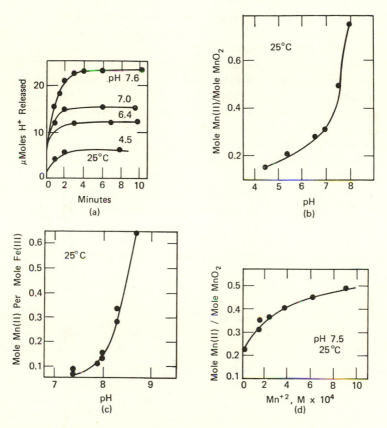

Figure 10-10   Sorption characteristics of Fe(III) and Mn(IV) oxides.
 a: Kinetics of Mn(II) sorption. Rates of H^+ release in Mn(II) sorption by MnO_2. (s)
 b: Mn(II) sorption by MnO_2 as a function of pH.
 c: Mn(II) sorption by $Fe(OH)_3$(s) as a function of pH.
 d: Sorption isotherm: Mn(II) on MnO_2.

[27] J. J. Morgan and W. Stumm, *J. Colloid Sci.*, **19**, 347 (1964).

Metal ions sorbed to hydrous oxides become released upon pH lowering or, obviously, upon reduction of the oxide. The desorption of metal ions bound to manganese oxides may also be facilitated by a partial reduction of the oxide; for example, the Mn(IV) in manganate(IV) can be reduced to Mn(II) without a change of phase [28], but the exchange capacity is likely to decrease.

10-4   Ground Water Circulation

Equilibrium or near-equilibrium conditions are more likely to prevail in ground waters than in surface waters because a relatively large surface area of solid minerals is exposed to slowly moving water and because transitory biological disturbances are rare.

Phosphorus occurs in ground waters almost exclusively in the form of inorganic orthophosphate ($H_2PO_4^-$ and HPO_4^{2-}). As Figure 10-1 suggests, in the near neutral pH range of typical ground waters, phosphate concentrations usually cannot exceed 10^{-6} M (30 μg liter^{-1} phosphorus) because of the solubility limitation by hydroxylapatite. Indeed phosphorus concentrations in ground waters seldom exceed a few μg liter^{-1}. Agricultural drainage waters, on the other hand, contain total phosphorus ranging from 0.05 to 1 mg liter^{-1} (2×10^{-6} to 3×10^{-5} M), but the phosphorus in these waters may be present in dissolved and suspended (precipitated, adsorbed living and dead bacteria, etc.) forms and as orthophosphate, polyphosphate and organic phosphate [29].

Similarly total iron and manganese concentrations encountered in ground waters as soluble species can be predicted from solubility equilibria. A ground water saturated with rhodochrosite ($MnCO_3$), siderite ($FeCO_3$) and calcite ($CaCO_3$) should contain soluble [Mn^{2+}], [Fe^{2+}] and [Ca^{2+}] in the molar ratio 1:1.5:120.

There is little discrepancy between field measurements of the solubility of Fe(II) and Mn(II) and the solubility of the respective carbonates predicted from calculations. Ghosh, O'Connor and Engelbrecht [30] sampled the influent ground water at eight water treatment plants in Illinois. After correcting their data for temperature and activity [31] we find a small but constant oversaturation (1.5 to 2.5 times) compared to the calculated solubility. Unless special precautions are taken, the measured pH value may not

[28] W. Feitknecht, H. R. Oswald and U. Feitknecht-Steinmann, *Helv. Chim. Acta*, **43**, 1947 (1960).

[29] R. J. Hannapel, W. H. Fuller and R. H. Fox, *Soil Sci.*, **97**, 421 (1964).

[30] M. M. Ghosh, J. T. O'Connor and R. S. Engelbrecht, J. Sanitary Engineering Division, American Society of Civil Engineers, Proc. Paper 4687, Feb. 1966.

[31] W. Stumm and P. C. Singer, J. Sanitary Engineering Division, American Society of Civil Engineers, Proc. Paper 4920, Oct. 1966.

represent the actual pH of the water in the aquifer; such discrepancy may result in an apparent supersaturation. Hem [32] in examining 20 ground waters for equilibrium with calcite and siderite ($FeCO_3$) presents a "dual saturation index" diagram for the two minerals. In general, we would expect a ground water in equilibrium with calcite to be in equilibrium with siderite, too, if both minerals were present in the same geological formation. Furthermore, we would expect similar conditions of real or apparent (Hem's data are uncorrected with respect to temperature) under- or oversaturation with respect to the two solid phases. The best straight line through Hem's data has a slope of roughly unity and passes through the origin if a value [22] of $pK_{s0} = 10.24$ ($I = 0$, 25°C) is adopted for the solubility product of siderite; this is indicative of equivalent saturation. The concentrations of dissolved oxygen and carbon dioxide in a ground water, near its source of recharge, reflect the partial pressures of O_2 and CO_2 in the soil gas which usually is enriched with CO_2 because of respiratory activity of microorganisms. Typically soil solutions contain CO_2 concentrations representative of partial pressures between 10^{-1} and 10^{-2} atm. Respiratory production of CO_2 is accompanied by a concordant consumption of O_2, but because O_2 is present at much higher concentrations than CO_2, the O_2 content changes relatively little. (A tenfold increase in the CO_2 content of air caused by oxidation of organic material is accompanied by a reduction in the O_2 content of only about 1%.) Hence, the pϵ level of a ground water in contact with a soil gas is not much different from that of a water in equilibrium with the atmosphere. Because of the higher CO_2 content of the soil gas, however, the total concentration of carbonic species becomes quite high. At the prevailing pϵ levels Fe and Mn are not sufficiently soluble ($< 10^{-8}$ M) to be analytically detectable (see Figure 7-9).

Fe^{2+} *and* Mn^{2+} *in Ground Water as a Result of Organic Pollution*

If, however, organic matter penetrates into the ground water, because of the much smaller solubility in water of O_2 than CO_2 the water will become depleted in O_2 and low pϵ values may be expected. Under these circumstances Fe and Mn become soluble as Fe^{2+} and Mn^{2+}, and Fe- and Mn-bearing minerals are attacked. Because Fe- and Mn-containing minerals are almost ubiquitous in usual ground water systems, the occurrence of soluble Fe and Mn in the water are primarily an indication of low pϵ (or O_2 depletion) in the aquifer. The absence of soluble Fe and Mn in a ground water cannot be used as an indication of the absence of Fe- and Mn-bearing minerals; more typically it is an indication that the water in the aquifer contains dissolved

[32] J. D. Hem, *U.S. Geol. Surv. Water Supply Papers*, **1459-C** (1960).

oxygen. Infiltration of organic wastes into an aquifer, man-made or natural contamination, impairs the quality of the ground water primarily because it imparts low $p\epsilon$ values to these waters and hence creates conditions where soluble Fe and Mn prevail. A well-documented case, where infiltration of wastes into soil from a beet sugar company has affected an aquifer over wide distances and has seriously impaired the water supply of communities, has been described by Minder [33].

Fe^{2+} is a reducing agent with respect to higher valent manganese oxides; that is, the latter are capable of oxidizing Fe^{2+}. Soluble Mn^{2+} will appear prior to Fe^{2+} upon progressive lowering of $p\epsilon$ in the water of an aquifer. Equilibrium calculations show that in the neutral pH range the presence of nitrate is incompatible with the presence of detectable concentrations of soluble Fe. On the other hand, not all the nitrate needs to be reduced in order to have Mn^{2+} in solution (see, for example, Figure 7-9).

Example 10-2   *Mixing of Ground Waters.*   Two ground waters and both exposed to and in equilibrium with an atmosphere exhibiting a p_{CO_2} of 10^{-2} atm. Ground water I enters a stratum containing siderite and no calcite. Ground water II enters a stratum containing only calcite. Solutions I and II, each having been in equilibrium with their respective solid phase, are intercepted by a well and thereby mixed in equal proportions. A problem of this type has been discussed by J. D. Hem (in *Principles and Applications of Water Chemistry*, S. D. Faust and J. V. Hunter, Eds., Wiley, New York, 1967).

(i) Compute $[Me^{2+}]$, $[HCO_3{}^-]$ and $[H^+]$ for each type of ground water.

(ii) Compute the composition of the mixture.

(iii) Is the mixture stable with respect to precipitation of $FeCO_3$? Assume 25°C.

The composition of solutions I and II may be computed with the help of (35) and (39) in Chapter 5 or with (ix) of Example 5-3. Using the latter equation as a first approximation

$$[Me^{2+}] \simeq 0.63 + K_{pso}^{1/3} p_{CO_2}^{1/3} \tag{i}$$

with ${}^+K_{pso} = K_{so}K_1K_HK_2{}^{-1}$. Using constants from Table 5-1 for the carbonate and $CaCO_3$ system and a value of 10.24 for pK_{so} of $FeCO_3$, the following values obtain (25°C, $I = 0$)

$$\log {}^+K_{pso}(CaCO_3) = -5.83$$
$$\log {}^+K_{pso}(FeCO_3) = -7.73$$

[33] L. Minder, *Schweiz. Z. Hydrol.*, **22**, 380 (1960).

considering furthermore that

$$[HCO_3^-] = 2[Me^{2+}] \qquad \text{(ii)}$$

$$[H^+] = K_1 K_H p_{CO_2}[HCO_3^-]^{-1} \qquad \text{(iii)}$$

the following results can be tabulated

$-\log$ Concentration (M) Calculated for 25°C

	HCO_3^-	Ca^{2+}	Fe^{2+}	H^+
Ground water I	(3.14) 3.10	—	(3.44) 3.40	6.7
Ground water II	(2.50) 2.44	(2.80) 2.74	—	7.3
Mixed water	2.66	3.04	3.70	7.1
After precipitation	(2.81)	(3.11)	(4.11)	(7.0)

Figures in parentheses give results obtained without correcting the constants for activities. With these results the ionic strength can be computed and equilibrium constants corrected for ionic strength. The following corrections have been applied:

$$p^{c+}K_{ps0} = p^+K_{ps0} - \frac{3\sqrt{I}}{1 + \sqrt{I}} \qquad \text{(iv)}$$

$$p^cK_1 = pK_1 - \frac{\sqrt{I}}{1 + \sqrt{I}} \qquad \text{(v)}$$

$$p^{c*}K_s = p^*K_s - \frac{2\sqrt{I}}{1 + \sqrt{I}} \qquad \text{(vi)}$$

where $^*K_s = K_{s0}K_2^{-1}$.

The mixed water is unstable because it is oversaturated with siderite. If it were in equilibrium with $FeCO_3(s)$, its equilibrium concentration would then be $-\log [Fe^{2+}] = 4.1$.

10-5 Transformations in Surface Waters

The transformations of Fe, Mn and P in lakes, reservoirs and streams are not characterized simply. Superimposed on the many physical-chemical variables are biological variables. Some of the principal effects of pH and oxidation state are summarized in Figure 10-11 while Figure 10-12, adapted from Phillips [34], depicts some of the phosphorus locations in a body of water in a schematic way and the exchange processes between the various locations.

[34] J. E. Phillips, in *Principles and Applications in Aquatic Microbiology*, H. Heukelekian and N. C. Dondero, Eds., Wiley, New York, 1964.

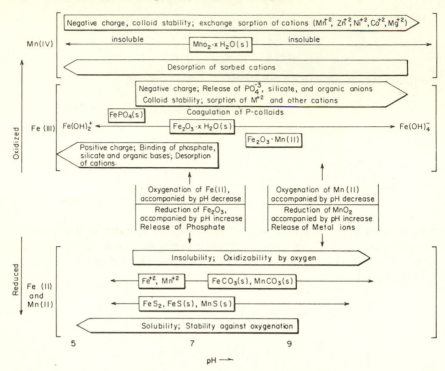

Figure 10-11 Major features of chemistry of iron, manganese and phosphorus summarized.

ABUNDANCE OF IRON AND MANGANESE. In surface waters a substantial fraction of iron and manganese is present in suspended form; in the surface layers of most natural waters, conditions of pH and O_2 concentration are such that any Fe^{2+} and Mn^{2+} introduced into these layers is readily oxidized to insoluble Fe(III) and to $MnO_x(s)$. These oxides settle into the deeper water layers. Complex-forming organic and inorganic bases can enhance the colloid stability. Usually surface waters that are high in concentrations of organic matter also contain substantial amounts of total ferric iron. For most natural waters, concentrations of "soluble" ferric iron far in excess of those predicted by solubility equilibria have been reported. Hence a substantial fraction of the iron in many waters is present either in suspended form or — although less likely (see Section 6-9) — as an organic complex. From what has been said about the size of $Fe(OH)_3$ particles (Section 6-9), it is clear that even membrane filtration cannot in general distinguish between dissolved and particulate Fe(III) species.

While concentrations of manganese in ground and surface waters are often very small, in certain situations, particularly in deep impoundments, these

Water

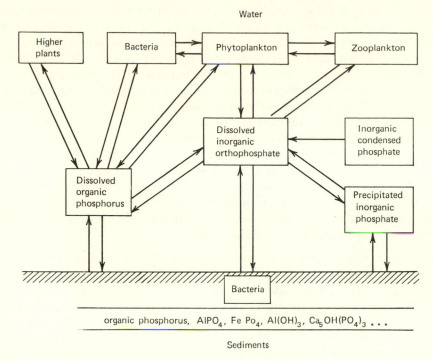

Figure 10-12 Phosphorus transformations, adapted from Phillips [34]. Schematic representation of some of the phosphorus locations in a body of water and of exchange processes between the various locations. Hayes and Phillips [51] have suggested that the turnover times (the time required for loss from a phase of as much phosphorus as is present in that phase) range from ca. 5 min for the exchange between dissolved inorganic phosphate and phytoplankton to many days for some of the other exchanges. The slowest processes appear to involve the water–sediment exchanges (~ 15 days for abiotic processes and ~ 3 days for bacterially-mediated processes). Exchange between inorganic phosphate and dissolved organic phosphorus appears to be fairly rapid (~ 8 hr). These figures were proposed for lake waters. They should be viewed as merely suggestive for a general orientation. Reproduced with permission from J. Wiley and Sons.

may show extreme temporal and spatial variability. Because of the limited solubility of $MnCO_3$ and MnS, soluble manganese concentrations seldom exceed 20 mg liter^{-1}.

ABUNDANCE OF PHOSPHORUS. Total phosphorus concentrations in U.S. rivers now range from roughly 0.01 to 1 mg liter^{-1}, although such local factors as waste disposal and industrial activity (e.g., mining) can cause higher values. Condensed inorganic phosphates in surface waters are highly variable, in part because they are subject to continual hydrolytic breakdown to orthophosphate. Rigler [2] reported total phosphorus in nine Canadian lakes ranging from 0.005 to 0.113 mg liter^{-1}. Mean values of total P in unpolluted lake surface waters as summarized by Hutchinson [5] are on the order of 0.01 to 0.04 mg liter^{-1} (3×10^{-7} to 13×10^{-7} M), but levels as high as 2 mg liter^{-1} are not uncommon in lake waters receiving heavy domestic or agricultural drainage. A weighted-average phosphorus content for the oceans of 2.3 μM (0.07 mg liter^{-1} P) has been reported [35], but concentrations more than ten times as high have been found in some deeper waters.

Table 10-6 gives a nutrient budget for the ocean (cf. Emery, Orr and Rittenberg [36]). According to these estimates, phosphorus and nitrogen are present in the ocean at 100 times their estimated annual use by phytoplankton; this annual use, however, far exceeds the annual contribution to the ocean. Accordingly, the major proportion of the nutrients must be regenerated from organic debris. Phosphorus is brought to the ocean and deposited in the sediments in approximate balance. The difference observed in the case of nitrogen may be attributed to denitrification.

RESIDENCE TIME OF DISSOLVED PHOSPHATE. A very large fraction of the phosphorus in a given system may be inside living organisms at any given time. Residence time of dissolved phosphate, that is, the average time phosphorus atoms remain in solution, varies from 0.05 to 200 hr [37]. A system having a short residence time may be low in dissolved phosphate, as in the sea, or it may be very active biologically, as in algal blooms. When both conditions occur together, as in small lakes, the residence time becomes very short. The concentration of dissolved phosphate in natural waters gives little indication of phosphate availability. Pomeroy [37] and others have suggested that the flux of phosphate is more important than the concentration of dissolved phosphate in maintaining high rates of production.

Table 10-6   Nutrient Budget in the Ocean [36]

	Millions of Metric Tons	
	Nitrogen	Phosphorus
Reserve in ocean	920,000	120,000
Annual use by phytoplankton	9,600	1,300
Annual contribution		
by river, dissolved	19	2
suspended	0	12
by rain	59	0
Annual loss to sediments	9	13

[35] A. C. Redfield, *Amer. Sci.*, **46**, 205 (1958).
[36] K. O. Emery, W. L. Orr and S. C. Rittenberg in *Essays in Natural Science in Honor of Captain Allan Hancock*, Univ. of So. California Press, Los Angeles, p. 299 (1955).
[37] L. R. Pomeroy, *Science*, **131**, 1173 (1960).

Temporal and Spatial Distribution

Obviously, the concentration of such elements as phosphorus, nitrogen, iron and manganese as a function of time and position cannot simply be computed from a knowledge of the initial concentration, the inflow concentrations and the hydrographic characteristics. The circulation of these elements is regulated not only by hydraulic and hydrographic conditions, but depends strongly on the interrelation of biota and chemical environment. The behavior of such elements exemplifies the additional significance of oxidation and reduction, of complex formation and of surface chemical reactions.

Although it is not possible to give a satisfactory account of all the variables which are responsible for the observed behavior, an attempt is made to sketch briefly the transformation of Fe, Mn and P in lakes.

LAKES. The rate of sedimentation of insoluble Fe and Mn oxides is influenced by many factors, among which the colloid chemical nature of the ferric precipitate is perhaps most significant. Organic bases can form surface complexes on the insoluble ferric oxide and enhance the colloid stability. If the hypolimnetic waters are devoid of dissolved oxygen (as in many eutrophic lakes during the periods of stagnation) conditions favorable for the reduction of ferric oxide to ferrous iron prevail. Fe(III) which has not been reduced in the waters overlaying the sediments may become reduced to Fe(II) at the mud interface. The iron content of the water overlaying the deposits will progressively increase during the stagnation periods. During the fall and spring circulation, when dissolved O_2 becomes more or less evenly distributed, most of the ferrous iron is reoxidized to insoluble Fe(III).

A considerable enrichment of Mn is found in the hypolimnia of productive lakes while very little soluble Mn is found in the surface layers. It might appear reasonable to construct a picture for the Mn(II)–MnO_2 cycle similar to that given for iron. However, oxygenation of Mn(II) at the low concentrations encountered does not occur as readily as Fe(II) oxygenation within the pH range of most natural waters. A certain amount of Mn(II) oxygenation can occur in the immediate vicinity of surfaces which have a higher pH than the bulk of the solution, for example, at the surfaces of positively charged colloids (aluminum oxide) or on surfaces of algae. According to Goldschmidt [38], the mean geochemical Fe:Mn ratio for the accessible lithosphere is 50:1. Although the Fe:Mn concentration ratio in lakes tends to be lower than the geochemical ratio, the quantity of Mn present is significantly less than that of Fe. In view of the demonstrated high sorption capacity of ferric oxide for Mn(II), it can be postulated that a considerable portion of the Mn can participate in the cycle without even undergoing redox reactions. Under

[38] V. M. Goldschmidt, *Geochemistry*, Clarendon Press, Oxford, 1954.

aerobic conditions Mn(II) is sorbed by the ferric oxide. The fraction of the Mn which has had a chance to be oxidized will adsorb additional Mn(II), which will then settle, together with the ferric oxide, into the hypolimnion. Enrichment of soluble Mn(II) in the hypolimnion may occur by the following processes: (a) A decrease in pH (heterotrophic activity) will shift the exchange-adsorption equilibrium in such a way as to desorb Mn(II) from the higher-valent manganese oxides and from ferric oxide; (b) reduction of ferric oxide releases sorbed Mn(II); (c) ferrous iron and other reductants [S(−II), organic material] can reduce manganese oxides and render them soluble.

Figure 10-13 gives an example of the seasonal variation of Mn in lake Mendota (from Delfino and Lee [39]). These data are in accord with the general discussion given above. Most of the total Mn recorded in this figure was present as Mn(II). While traces of Mn oxides were detected in all oxygenated waters, in anoxic waters Mn(II) was observed exclusively. The Mn oxides found in the epilimnetic waters were X-ray amorphous. The dynamics in the variation of soluble Mn(II) is well documented in Figure 10-14 (from

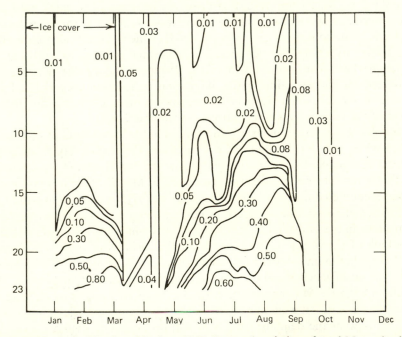

Figure 10-13 Lake Mendota Mn data, 1967. Seasonal variation of total Mn at the deep hole station, total depth 23 m. Data in milligrams liter^{-1}. From Delfino and Lee [39]. Reproduced with permission from American Chemical Society.

[39] J. J. Delfino and G. F. Lee, *Environ. Sci. Technol.*, **2**, 1101 (1968).

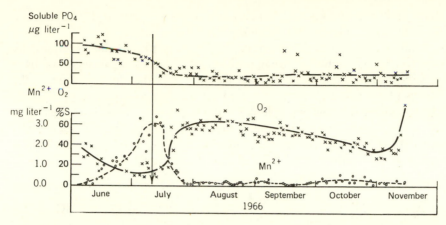

Figure 10-14   [O_2], [Mn^{2+}] and soluble P in a reservoir. From Bernhardt and Hötter [40]. Measurements made 5–10 cm above sediments. The arrow indicates time where artificial aeration was started. The decrease in soluble phosphorus accompanied by an increase in [O_2] is probably related to P precipitation by Fe(III). Reproduced with permission from Schweizerbart Verlay, Germany.

Bernhardt and Hötter [40]). Interestingly, the data in this figure indicates — as do observations by Delfino and Lee — that Mn^{2+} is present even if dissolved oxygen is not fully depleted. This illustrates that redox processes do not couple kinetically with one another very readily. The presence of O_2, even if available only in trace concentrations, indicates a redox intensity at which Mn oxides could not become reduced to Mn(II); however, organic substances that are present at these low oxygen levels may readily reduce Mn oxides.

During fall and spring recirculation when dissolved oxygen becomes more or less evenly distributed within the lake, most of the soluble and suspended bivalent Fe and Mn are reoxidized to insoluble oxides which through sedimentation are deposited in the sediments. The continuous sequence of circulation and stagnation accompanied by oxidation and reduction as well as precipitation and dissolution leads to a progressive accumulation of Fe and Mn in the lake sediments.

IRON AND PHOSPHATE. The interaction between physical processes (convection, sedimentation), the biological effects and chemical processes determines the spatial distribution of phosphate. Usually pronounced correlations of the seasonably variable concentrations of iron and phosphate in the hypolimnetic waters are observed. Two mechanisms may account primarily for the progressive accumulation of soluble phosphate in oxygen-deficient bottom layers

[40] H. Bernhardt and G. Hötter, *Arch. Hydrobiol.*, **63**, 404 (1967).

during the stagnation period: (a) decomposition of sedimented plankton and (b) reduction of phosphate-containing Fe(III) precipitates.

Soluble phosphorus is taken up by phytoplankton and is eventually returned to the water on decomposition of algae or organisms. We have already drawn attention to the fact (Section 8-7) that in lakes and in the oceans phosphate and nitrate are used up simultaneously in relatively constant proportion by the phytoplankton and that these elements are released in constant proportion upon decomposition of the plankton. Although variations in the P–N ratio occur, these stoichiometric relations are most readily explainable in terms of the simplified reaction given before

$$106CO_2 + 16NO_3^- + HPO_4^{2-} + 122H_2O + 18H^+ (+ \text{ trace elements, energy})$$

$$\underset{\text{respiration}}{\overset{\text{photosynthesis}}{\rightleftharpoons}} \underset{\text{algae}}{\{C_{106}H_{263}O_{110}N_{16}P_1\}} + 138O_2 \quad (16)$$

Accordingly, the plankton acts as a conveyor of P into the deeper water layers.

In many instances this accumulation may be more complex. Phosphate has a strong tendency to interact with ferric iron. The solubility diagram (Figure 10-1) shows that the solubility of $FePO_4$ is minimal at pH 5.5 but increases remarkably with increasing pH. Most natural lake waters contain larger concentrations of iron than phosphorus. Under these circumstances, a "mixed" ferric hydroxo-phosphate can be precipitated. The molar ratio of OH^- to PO_4^{3-} in the precipitate will increase with increasing pH and will be roughly proportional to the $[OH^-]^3/[PO_4^{3-}]$ ratio in solution, that is, the capacity of Fe(III) to bind phosphate will be much larger at low pH than at high pH (see (44)–(47) in Section 9-5). Most phosphate will be precipitated by excess Fe(III) immediately after circulation. It is reported [5] that an oxidized mud surface does not merely hold phosphate but prevents diffusion of phosphate and Fe(II) from deeper layers in the mud, because the ferrous iron is always in excess and, when oxidized, precipitates all the phosphate.

The simplified scheme given here is similar to the earlier hypothesis proposed by Einsele [41] and others. Many reported inconsistencies between such a hypothesis and individual observations can be understood by considering the interaction of phosphate with ferric iron to be a chemical one; its extent depends rather strongly on the prevailing concentrations of Fe(III), P and H^+.

H_4SiO_4 interacts with ferric iron in a way similar to phosphate. Soluble silicato–iron(III) complexes have a stability similar to that of soluble phosphato–iron(III) complexes. In laboratory experiments considerable portions of silica can be removed by ferric oxide precipitation in the pH range 5–7. It is thus plausible that the circulation of silica is partially determined by that of

[41] W. Einsele, *Arch. Hydrobiol.*, **33**, 361 (1938).

iron, although biological activity (diatoms) is usually more important in influencing the distribution of this element.

FACTORS INVOLVED IN TRANSPORTATION, ACCUMULATION AND LOSS OF CATIONS BY HYDROUS METAL OXIDES. The distribution of minor metallic elements, such as Cu^{2+}, Co^{2+}, Ni^{2+} and Zn^{2+}, and also radionuclides, such as Cs, Sr and Ru, frequently resembles that of nutrients; the cycle of such elements has been studied very little. Since most of these elements are not involved in redox cycles their nonconservative behavior and distribution have been accounted for partially in terms of their uptake by organisms and their sorption on abiotic seston. In view of the fact that the ion exchange capacities of hydrous manganese and ferric oxides are large, it is to be expected that these oxides, which are themselves participants in oxidation–reduction cycles, will profoundly modify the distribution of elements which can become incorporated into these oxides.

It is well known that the fate of radioactive waste components is determined to a large extent by the fate of nonradioactive waste components. The radioactive elements enter into chemical or surface-chemical combination with nonradioactive materials, which are generally present in concentrations that greatly exceed those of the radionuclides. While the role of clays and the biota in affecting the transport of radionuclides and other pollutants is commonly recognized, the significance of Fe and Mn among the factors that are involved is frequently overlooked. Cation exchange capacities of seawater and fresh water sediments are in the range of 30–60 meq/100 g. It has been shown by Goldberg [42] that the amounts of trace elements in seawater sediments are proportional to the iron or manganese content of the sediments. The uptake of cations by hydrous oxides is similar in concept to cation uptake by clays (ion exchange, increase in cation exchange capacity with increased pH and representation of the distribution equilibria by mass law type equations). A further similarity between clays and hydrous oxides is that both are capable of binding anions [ion exchange on edges of the clay lattice and complex formation between Fe(III) oxides and bases]. A significant difference between clays and hydrous oxides of Fe and Mn is that the latter materials participate in the redox cycle. A reduction of the hydrous Fe and Mn oxides leads to an instantaneous release of the sorbed elements. The observations by Mortimer [43], Hutchinson [5] and others that oxidized sediments have much greater adsorptive powers than reduced sediments lend support to the idea that Fe and Mn play a significant role in influencing the distribution and the transport of a variety of metal ions of both the naturally occurring and pollutant kinds. The influence of oxidizing and reducing conditions upon the distribution of

[42] E. D. Goldberg, *J. Geol.*, **62**, 249 (1954).
[43] C. H. Mortimer, *J. Ecol.*, **29**, 280 (1941); **30**, 147 (1942).

metal ions has been examined by Gorham and Swaine [44]. Jenne [45] proposed that the hydrous oxides of Mn and Fe in general furnish the principal control in the fixation of Co, Ni, Cu and Zn in soils and fresh-water sediments. The common occurrence of these oxides as coatings allows them to exert surface chemical activity far out of proportion to their total concentrations.

MANGANESE NODULES. Part of the Mn and Fe supplied to the oceans is deposited in the form of ferromanganese minerals, the so-called manganese nodules. They are composed primarily of ferrous and ferric oxide (goethite), manganite and "δ-MnO_2" [manganate(IV)]. According to Buser and Grütter [46], in these concretions layers of MnO_2 alternate with disordered layers of manganous or ferric hydroxides. Manganese nodules containing up to 22% manganese were recently also found at several localities in Lake Michigan [47]. Chemical concepts already presented in this chapter on the surface-catalyzed oxygenation of Mn(II) and the pH-dependent sorption of Mn^{2+} and Fe^{2+} on MnO_2 provide the basis for a plausible hypothesis to explain the formation and the physical-chemical characteristics of these materials. Material resembling ferromanganese aggregations can be formed in the laboratory by suspending a calcite crystal in an aerated HCO_3^- solution (pH 7–8) containing traces of Fe(II) and Mn(II). Ferric oxide will coat the crystal surface and become an adsorbent for Mn^{2+} ions, which are then very slowly oxidized to Mn oxides; these can in turn react with ferrous iron or additional Mn(II). Because of the large specific surface areas involved, appreciable amounts of organic material may become associated with these minerals and may affect the growth of the concretions. The cation exchange capacity of the hydrous oxides provides a ready explanation for the fact that the nodules contain appreciable proportions of Cu^{2+}, Ni^{2+}, Zn^{2+}, Pb^{2+} and rare earths. Detailed discussions on the formation and properties of ferromanganese minerals have been given by Goldberg [48], Arrhenius [49] and Wangersky [50].

BIOLOGICAL MEDIATION OF IRON AND MANGANESE TRANSFORMATIONS. The discussion thus far has emphasized certain chemical factors effective in the

[44] E. Gorham and D. J. Swaine, *Limnol. Oceanog.*, **10**, 166 (1965).

[45] E. A. Jenne, Advances in Chemistry Series, No. 73, *Trace Inorganics in Water*, American Chemical Society, Washington, D.C., 1968, p. 337.

[46] W. Buser and A. Grütter, *Schweiz. Mineral. Petrog. Mitt.*, **36**, 49 (1956).

[47] R. Rossman and E. Callender, *Science*, **162**, 1123 (1968).

[48] E. D. Goldberg, in *The Sea*, Volume II, M. N. Hill, Ed., Wiley-Interscience, New York, 1963.

[49] G. Arrhenius, in *The Sea*, Volume III, M. N. Hill, Ed., Wiley-Iinterscience, New York, 1963.

[50] P. J. Wangersky, in *Radioecology*, Reinhold, New York, 1963.

ecological relations of these elements. This should not be taken as minimizing the significance of the activity of the large biomass in profoundly modifying the distribution of the elements. It is well documented that there are bacteria and algae capable of depositing manganese and ferric oxides from ferrous and manganous solutions; their metabolism appears to be linked to oxygenation processes. It is important to keep in mind that organisms influence the kinetics and the thermodynamics of reactions. Microbial mediation has often been invoked for reactions which can be observed to occur as readily in sterile systems. Biological activity frequently influences the pertinent chemical reactions in an indirect way, for example, by yielding metabolites capable of reducing Mn oxides or by facilitating reactions which lower the pH, thus bringing about the release of sorbed Mn(II) and other cations.

The existence of autotrophic bacteria which mediate Fe(II) oxidation in acid solutions (e.g., *Ferrobacillus ferrooxidans*) is well established. As Example 10-3 illustrates, substantial amounts of Fe(II) must be oxidized in order to provide sufficient energy for growth. In the near neutral pH range of most natural waters, oxidation of Fe(II) is very fast in the absence of bacterial catalysis.

Example 10-3 *Autotrophic Iron Bacteria; Cell Yield.* Estimate the quantity of iron(II) that has to be oxidized through mediation by autotrophic iron bacteria in order to synthesize 1 g of bacterial mass.

The free energy released by the oxidation of ferrous iron, using the data available in Latimer, is

$$4Fe^{2+} \quad + O_2 + 4H^+ = 4Fe^{3+} \quad + 2H_2O \qquad (i)$$
$$\bar{G}_f^\circ \quad 4(-20.3) \quad (0) \quad 4(0) \quad 4(-2.53) \quad 2(-56.7)$$

$\Delta G^\circ = -42.4$ kcal or -10.6 kcal mole^{-1} of Fe(II) oxidized.

At pH 3, since

$$\Delta G = \Delta G^\circ + RT \ln Q \qquad (ii)$$

where Q is the reaction quotient, the free energy released per mole is

$$\Delta G = -10.6 + 1.364 \log \frac{1}{10^{-3}} = -6.5 \text{ kcal mole}^{-1}$$

For synthesis of cell material from CO_2 (assuming the assimilated end product to be glucose), the free energy required is

$$6CO_2 \quad + 6H_2O(l) \quad = C_6H_{12}O_6 + 6O_2$$
$$\bar{G}_f^\circ \quad 6(-92.3) \quad 6(-56.7) \quad (-217.0) \quad 6(0)$$

$\Delta G^\circ = +677$ kcal mole^{-1} of glucose or $+113$ kcal mole^{-1} of carbon synthesized.

Assuming a 36% efficiency for microbial conversion of energy, as is common in autotrophic processes, the stoichiometry of the autotrophic oxidation of ferrous iron is

$$\frac{6.5 \text{ kcal}/55.8 \text{ g Fe(II) oxidized}}{113 \text{ kcal}/12 \text{ g carbon assimilated}} \times 0.36 = \frac{1}{224} \qquad \text{(iii)}$$

Hence, 1 g of organic carbon is synthesized for every 224 g of Fe(II) oxidized. If we consider the thermodynamic relationships in another way,

$$1 \text{ mole of Fe(II) oxidized} \implies \frac{6.5 \text{ kcal}}{112 \text{ kcal mole}^{-1} \text{ of carbon}}$$

$$\times 0.36 \times \frac{12 \text{ g}}{\text{mole of carbon}} \qquad \text{(iv)}$$

$$\implies 0.25 \text{ g of carbon synthesized}$$

Lamanna and Mallette (*Basic Bacteriology*, Williams & Wilkins, Baltimore, 1965) approximate that 1 g of bacteria contains 10^{12} to 10^{13} bacterial cells. Therefore, 1 mole of Fe(II) yields approximately 10^{12} bacterial cells or a yield Y of

$$Y = \frac{-dB}{dS} = 10^{12} \text{ cells mole}^{-1} \text{ of Fe(II) oxidized}$$

10-6   Phosphate Exchange with Sediments

Sediments act as reservoirs of phosphorus in natural systems. According to Hayes and Phillips [51], the dynamic interaction between water and sediment may be represented as

$$\begin{matrix} \text{P in aqueous phase} & \rightleftharpoons & \text{P in solid phase} \\ \text{(a small fraction of the whole)} & & \text{(a large fraction of the whole)} \end{matrix} \qquad (17)$$

The phosphate concentration in waters overlying sediments is buffered by solubility and adsorption or ion exchange equilibria at the sediment–water interface. The nature of the phosphate control is not fully understood, although there is general agreement that chemical interactions of phosphate with Fe(III), Al(III) (clays) and Ca^{2+} are relevant. Microorganisms influence the extent of these reactions, perhaps to a large extent indirectly, (a) by maintaining a substantial fraction of phosphorus in the form of particulate or dissolved organic phosphorus, and (b) through control of environmental factors which determine solubility equilibria (pH, pϵ, type of residual organic matter).

[51] F. R. Hayes and J. E. Phillips, *Limnol. Oceanog.*, 3, 459 (1958).

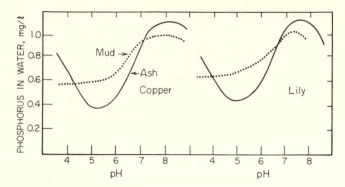

Figure 10-15 Effect of pH on ability of mud, or equivalent ash, to remove added inorganic phosphate from water (copper pond and lily pond). After MacPherson, Sinclair and Hayes [52]. Two grams of lake muds or their ashes were equilibrated with 1 liter of solution containing 1 mg liter^{-1} of P initially. The results given here were corrected for blanks (i.e., P leached from sediments with solutions containing no phosphorus). Reproduced with permission from the American Society of Limnology.

EQUILIBRIUM PARTITION OF INORGANIC PHOSPHATE. Figure 10-15 illustrates the results of one of the experiments of MacPherson, Sinclair and Hayes [52] on the equilibration of phosphate with oxidized lake sediment mud or its ash as a function of pH. This figure may be compared with the equilibrium solubility diagram (Figure 10-1). Qualitatively, the general shape of the curve describing the pH dependence of residual inorganic phosphate suggests that in the pH range 4.5 to 6.5 phosphate tends to be bound to the solid phase by Fe(III) and Al(III), either by precipitation or by adsorption (FePO$_4$, clays). Very little phosphate becomes sorbed between pH 7 and 9, but at higher pH values the tendency for precipitation of phosphate becomes enhanced and appears to be related to the progressively decreasing solubility of calcium phosphate (apatite) in accordance with the solubility predictions (Figure 10-1) and the calculations of Example 10-4.

Example 10-4 *Conversion of Calcite into Apatite.*   Under what condition (pH, concentration of inorganic P) can calcite be converted into hydroxylapatite? The simplifying assumptions $P_T \ll [Ca^{2+}]$ and $[Ca^{2+}] = C_T$ may be used.

The reaction may be written as

$$10CaCO_3(s) + 2H^+ + 6HPO_4^{2-} + 2H_2O$$
$$= Ca_{10}(PO_4)_6(OH)_2(s) + 10HCO_3^- \qquad (i)\dagger$$

---

† Apatite is represented here with a formula corresponding to its unit cell (cf. Table 10-4).
[52] L. B. MacPherson, N. R. Sinclair and F. R. Hayes, *Limnol. Oceanog.*, **3**, 318 (1958).

The equilibrium constant K for this reaction may be computed conveniently from

$$\log K = -\log K_{s0(\text{apatite})} + 10 \log K_{s0\,(\text{calcite})}$$
$$- 10 \log K_{\text{HCO}_3^-} + 6 \log K_{\text{HPO}_4^{2-}} + 2 \log K_{\text{W}} \quad \text{(ii)}$$

The equilibrium constant for (i) is, of course, at least as uncertain as the solubility product of apatite. With a value of K_{s0} for $\text{Ca}_{10}(\text{PO}_4)_6(\text{OH})_2(\text{s})$ of 10^{-12}, an equilibrium constant of $K = 10^{32}$ is obtained. Hence the free energy of this conversion, $\Delta G = RT \ln (Q/K)$, where Q is the quotient of the reactants, may be used to predict under what conditions (i) is possible. For example, at pH $= 8$ and $[\text{HCO}_3^-] = 10^{-3}\,M$, a $10^{-4}\,M$ solution of HPO_4^{2-} would, thermodynamically speaking, convert calcite into apatite because ΔG is approximately $-30\,\text{kcal mole}^{-1}$ of apatite formed. Figure 10-16 plots concentrations of soluble phosphate, P_T, necessary to convert CaCO_3 into apatite as a function of pH.

Although (i) represents an oversimplification (apatite is not necessarily formed as a pure solid phase), the tentative result obtained suggests that the phosphorus concentration at the sediment–water interface, especially at higher pH values, is buffered by the presence of hydroxyapatite.

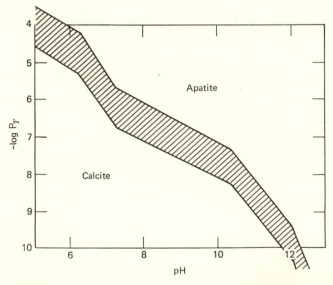

Figure 10-16 Phosphate necessary to convert $\text{CaCO}_3(\text{s})$ into $\text{Ca}_{10}(\text{PO}_4)_3(\text{OH})_2(\text{s})$. Calculated by assuming that the solution remains in saturation equilibrium with $\text{CaCO}_3(\text{s})$; furthermore, $P_T \ll C_T$ and, $[\text{Ca}^{2+}] = C_T$. The equilibrium constant for (i) has been taken as 10^{28} (upper line) and 10^{30} (lower line), respectively.

The mineral or chemical composition of phosphorus compounds tends to be different in different environments [53]. Variscite ($AlPO_4 \cdot 2H_2O$) and strengite ($FePO_4 \cdot 2H_2O$), appear to be more common in soils and fresh water sediments, whereas apatite prevails in certain marine sediments. Apparently the solubility product of apatite is not exceeded on the deep-ocean floor; it is established that skeletal apatite is slowly dissolving [49]. In areas of high organic productivity, in eutrophic lakes and in the shallow areas of the ocean, especially the tropical ocean, calcium phosphate minerals (substituted apatites, such as francolite $Ca_{10}[PO_4CO_3]_6F_2$) are deposited.

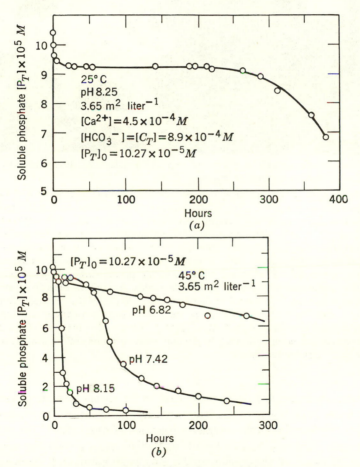

Figure 10-17   Heterogeneous nucleation of apatite on calcite.
a, b: Disappearance of phosphate from solution of calcite in equilibrium with p_{CO_2}.

[53] B. W. Nelson, *Science*, **158**, 917 (1967) [see also discussion by G. Müller, *Science*, **163**, 812 (1969)].

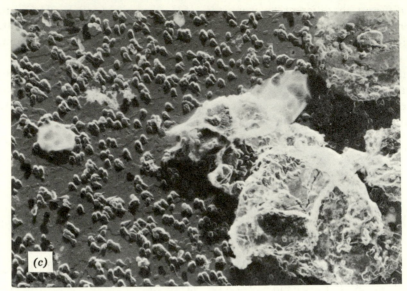

c: Electron micrograph of platinized replica of calcite single crystal covered with crystallites of hydroxapatite ($\times$ 20,000).

Representative curves on the kinetics of the formation of apatite on $CaCO_3$ (equilibrated with its solution) illustrate that the reaction involves three steps: (*1*) chemisorption of phosphate accompanied by heterogeneous formation of nuclei of amorphous calcium phosphate, (*2*) a slow transformation of these nuclei into crystalline apatite and (*3*) crystal growth of apatite. Surface diffusion on the surface of the crystal probably determines the rate of crystal growth. The nucleation reaction leads to a constant initial surface coverage of 1 P atom per 60 Å^2. The lag period during which a phase transformation occurs in the calcium phosphate nuclei, as well as the subsequent apatite crystal growth reaction, depend strongly on the $[CO_3^{-2}]/[PO_4^{-3}]$ ratio. Competition of CO_3^{2-} and PO_4^{3-} for adsorption at the growth sites most likely accounts for the reduced rates of reaction at high $[CO_3^{2-}]/[PO_4^{3-}]$ ratios. The presence of apatite on the calcite surface has been confirmed by grazing angle electron diffraction. (From Leckie and Stumm, unpublished data.)

In lakes, estuaries and the shallow areas of the oceans, sediments participate in the cycle of phosphorus. In water quality management the existence of a phosphorus reservoir in the sediments must not be overlooked. The phosphate in the sediments may be capable of supporting algal growth for some time, even if all inputs of phosphates to the lake were eliminated. Pomeroy et al. [54] have shown for the estuary of Dobay (Georgia) that 10 cm of sediment

[54] L. R. Pomeroy, E. E. Smith and C. M. Grant, *Limnol. Oceanog.*, **10**, 167 (1965).

contain enough exchangeable phosphate to replace that in the water at least 25 times. The algal blooms observed in tidal creeks may continue for days or weeks at the observed high rates of photosynthesis supported by phosphates from the sediments [54].

Kinetics of Phosphate Exchange with Sediments

There is little quantitative information available on the rates of release and fixation of phosphates by sediments. Pomeroy et al. [54] have estimated that the daily exchange across a submerged and undisturbed surface of sediments is of the order of 1 μmole P day^{-1} m^{-2}.

APATITE FORMATION. A few typical results on the fixation of phosphate by Ca^{2+} in the presence of calcite are given in Figure 10-17, where the disappearance of total soluble phosphorus is plotted as a function of time. These investigations have shown that calcite is not simply transformed into apatite but that calcite rather acts as a well-matched surface for the nucleation of apatite.

Author Index

Subject Index

Acetic acid, 99
  in methane fermentation, 333
Acid and Base, acidity constants, 78
  ampholyte, 92
  Brønsted concept, 70, 240–243
  buffer intensity, 108–114
  clays, 485
  conjugate acid base pairs, 72
  diprotic system, 92
  hard and soft, 259–262
  hydrolysis, 72
  ionization fraction, 102–108
  Lewis concept, 73, 240–242, 258–262
  metal ions as acids, 72, 240–246
  metal oxide, 481–484
  mixture of two acids, 100–101
  neutralizing capacity, 108–114
  polyelectrolytic, 72, 481–484
  polyprotic, 72
  strength, 73
  strong acid and bases, 76, 90, 91
  titration curve, 102–108
  volatile diprotic acid, 102
  weak, 90, 91
Acidity, CO_2-acidity, 129–130
  definition, 129–131
  metal hydroxides, 256
  metal ions, 72, 240–246
  mineral, 129–130
  pyrite oxidation, 540–542
Acidity constants, composite, 77
  definitions, 73–79
  "mixed," 82, 85
  numerical values, 78
  operational, 82
  salt effects, 84, 85
Active compounds, 166
Activity, and pH, 79–86, 373
  and pH measurement, 371–373
  chemical potential, 29–32
  coefficient (definition), 29–32

Activity (*continued*)
  coefficient, 7, 8, 31, 83–85, 372–373
  convention, 31, 32, 40, 79–86, 218–221
  Debye-Hückel law, 83, 84, 217–221
  definition, 29–32
  fugacity, 7
  ionic strength, 81, 83
  measurement, 371–373
  potentiometric measurements, 370–373
  scales, 79–86
  scales (conversion), 218–221, 373
  solid phase, 203–217
  surface, 451–454, 467–470
Activity product, (*see* Solubility)
Activity ratio diagram, 185–197, 311,
      312, 333
Adsorption, against electrostatic repulsion,
      474
  clays, 486–493
  coagulation by, 501
  colloid stability, 502–507
  electrostatic, 467–469, 472
  free energy of, 472–474
  from solution, 450–454
  Gibbs adsorption, 451–453
  interaction, 467–474
  ion exchange, 485–493
  isotherms, 451–454
  of hydrolyzed metalions, 474, 503–506
  of polymers, 473–474, 503, 504
  on hydrous oxides and hydroxides,
      474–485
  on silica surfaces, 481–485
  specific, 472–474
  Stern theory, 472–474
Adsorption coagulation, 501
Albite, 52, 118, 390
  solubility, 399
Algae, elemental composition, 432
  growth of, 428, 436
  nutrients, 433, 435, 550, 554